Surveying

Surveying

Fifth Edition

Jack McCormac
Clemson University

JOHN WILEY & SONS, INC.

PROJECT EDITOR Jennifer Welter

ASSOCIATE DIRECTOR OF MARKETING Ilse Wolfe

SENIOR PRODUCTION EDITOR Valerie A. Vargas

SENIOR DESIGNER Dawn Stanley

PHOTO CREDIT ON COVER: DIGITAL VISION

This book was set in AGaramond by Argosy and printed and bound by Malloy Inc.
The cover was printed by Phoenix Color Corp.

This book is printed on acid free paper. ∞

ISBN 0-471-23758-2 (paperback)
ISBN 0-471-45239-4 (WIE)

Printed in the Unites States of America

10 9 8 7 6 5 4 3 2 1

Preface

The purpose of this book is to provide an introduction to surveying in such a manner as to interest the student in the subject. The book was written for a one-semester course for two- and four-year colleges and universities.

Quite a few changes and additions have been made with this edition. More emphasis is given to the use of total stations and other modern equipment. Today's surveyor is becoming more involved each year with the global positioning system (GPS) as well as with geographic information systems (GIS). As a result, an additional chapter has been added concerning each of these topics.

The majority of the Problems have been revised and some new ones added. Instructors who have adopted this book for their courses may obtain from the publisher a solutions manual for all of the text Problems. Please visit the Instructor Companion Site for this book, located at www.wiley.com/college/mccormac to register for a password.

Enclosed with this text is a CD-ROM which contains a set of programs entitled SURVEY. These programs, which are written in Windows®95, enable the surveyor to quickly handle several of the otherwise tedious and time-consuming math calculations which he or she faces almost daily. Included are programs for precision, land area, omitted measurements, radiation surveys, and for horizontal and vertical curves.

The author is very grateful to the following individuals who reviewed the manuscript: Yassir A. AbdelRazig, Florida State University, and Joseph A. Caliendo, Utah State University.

I would also like to thank two faculty members in the civil engineering department at Clemson for their kind advice and assistance during the preparation of this book. They are David Clarke and Wayne Sarasua. I am also indebted to Ronald K. Williams of Minnesota State University Moorhead, who prepared the latest version of the SURVEY software.

Jack McCormac
Clemson University

Contents

CHAPTER 8 Leveling, Continued 145

CHAPTER 9 Angles and Directions 161

CHAPTER 10 Measuring Angle and Directions With Total Stations 181

Surveying

CHAPTER

Introduction

1-1 SURVEYING

Surveying has been with us for several thousand years. It is the science of determining the dimensions and contour (or three-dimensional characteristics) of the earth's surface by measurement of distances, directions, and elevations. Surveying also includes staking out the lines and grades needed for the construction of buildings, roads, dams, and other structures. In addition to these field measurements, surveying includes the computation of areas, volumes, and other quantities, as well as the preparation of necessary maps and diagrams.

In the past few decades there have been almost unbelievable advances in the technology used for measuring, collecting, recording, and displaying information concerning the earth. For instance, until recently surveyors made their measurements with steel tapes, angle measuring devices called transits and theodolites, and elevation equipment called levels. In addition, the measurements obtained were presented with laboriously prepared tables and maps.

Today the surveyor uses electronic instruments to automatically measure, display, and record distances and positions of points. Computers are used to process the measured data and produce needed maps and tables with tremendous speed.

These developments have contributed to great progress in many other areas including geographic information systems (GIS), land information systems (LIS), the global positioning system (GPS), remote sensing and others. As a result many persons have long had a feeling that the word surveying was not adequate to represent all these new activities along with the traditional work of the surveyor.

1-2 GEOMATICS

In 1988 the Canadian Association of Aerial Surveyors introduced the term *geomatics* to encompass the disciplines of surveying, mapping, remote sensing, and geographic information systems. Surveying is considered to be a part of this new discipline. Geomatics is a word quite new to the English language and one which is not yet found in either the Oxford or Webster dictionaries.

There are quite a few definitions of geomatics floating around as might be expected considering the term's relative infancy. With each of the definitions, however, there is a common theme and that is "working with spatial data". The word spatial refers to space and the words "spatial data" refer to data that can be linked to specific locations in geographic space. *In this text, geomatics is defined as being an integrated approach to the measurement, analysis, management, storage, and presentation of the descriptions and locations of spatial data.* The term geomatics is rapidly being accepted in the English speaking world particularly in the colleges and universities of the United States, Canada, Australia, and the United Kingdom.

Although this text is primarily concerned with surveying, the reader needs to understand that surveying is part of the very large, modern and growing field of geomatics. As a result much emphasis is given herein to modern surveying equipment and methods for gathering and processing data.

1-3 FAMOUS SURVEYORS

Many famous persons in our history were engaged in surveying at some period in their lives. Particularly notable among these are several presidents—Washington, Jefferson, and Lincoln. Although the practice of surveying will not provide a sure road to the White House, many members of the profession like to think that the characteristics of the surveyor (honesty, perseverance, self–reliance, etc.) contributed to the development of these leaders. Today, surveying is an honored and widely respected profession. A knowledge of its principles and ethics is useful to a person whatever his or her future endeavors will be.[1]

A large number of surveyors, other than presidents, have served our country well. Included are persons such as Andrew Ellicott (who surveyed several of our state boundaries and designed the beautiful streets of Washington, D.C.), David Rittenhouse (a farmer, surveyor, clockmaker, and one of our earliest surveying instrument makers), and General Rufus Putman (an aide to General Washington during the Revolutionary War and the first Surveyor General of the United States). Surveying provided Henry David Thoreau (of Walden Pond fame) with his living for over a decade. Many other famous surveyors are mentioned throughout this book.

1-4 EARLY HISTORY OF SURVEYING

It is impossible to determine when surveying was first used, but in its simplest form it is surely as old as recorded civilization. As long as there has been property ownership there have been means of measuring the property or distinguishing one person's land from another. Even the Old Testament contains frequent references to property ownership, property corners, and property transfer. For instance, Proverbs 22:28: "Remove not the ancient landmark, which thy fathers have set." (Similar statements are made in Deuteronomy 19:14 and 27:17.) The Babylonians surely practiced some type of surveying as early as 2500 B.C. because archaeologists have found Babylonian maps on tablets of that estimated age. Evidence has also been found in the historical records of India and China that show that surveying was practiced in those countries in the same time period.

[1]For purposes of convenience, the terms *rodman, instrumentman,* and *chainman* are used in the text as the terms *rodperson, instrumentperson,* and *chainperson* are not widely used. The author does not mean this to minimize the role of women in the surveying profession.

The early development of surveying cannot be separated from the development of astronomy, astrology, or mathematics because these disciplines were so closely interrelated. In fact, the term *geometry* is derived from Greek words meaning earth measurements. The Greek historian Herodotus ("the father of history") says that surveying was used in Egypt as early as 1400 B.C. when that country was divided into plots for taxation purposes. Apparently, geometry or surveying was particularly necessary in the Nile valley to establish and control landmarks. When the yearly floods of the Nile swept away many of the landmarks, surveyors were appointed to replace them. These surveyors were called *harpedonapata*, or "rope stretchers" because they used ropes (with markers or knots in them at certain designated intervals) for their measurements.[2]

During this same period surveyors were certainly needed for assistance in the design and construction of irrigation systems, huge pyramids, public buildings, and so on. Their work was apparently quite satisfactory. For instance, the dimensions of the Great Pyramid of Gizeh are in error only about 8 in. for a 750–ft. base. It is thought that the rope–stretchers laid off the sides of the pyramid bases with their ropes and checked squareness by measuring diagonals. In order to obtain the almost level foundations of these great structures, the Egyptians probably either poured water into long, narrow clay troughs (an excellent method) or used triangular frames with plumb bobs or other weights suspended from their apexes as shown in Figure 1-1.[3]

FIGURE 1-1 Ancient leveling frame.

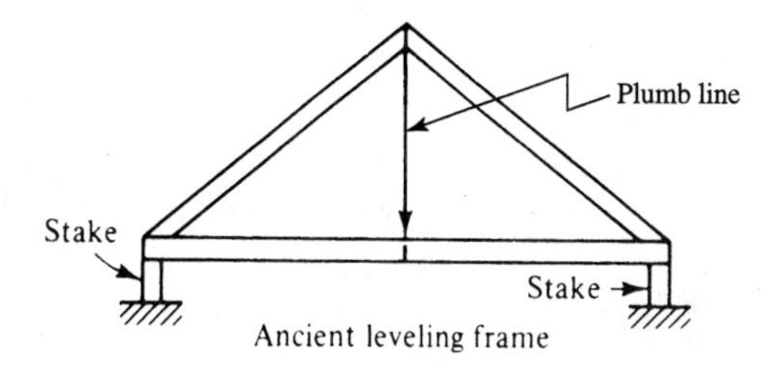

Each leveling frame apparently had a mark on its lower bar which showed where the plumb line should be when that bar was horizontal. These frames, which were probably used for leveling for many centuries, could easily be checked for proper adjustment by reversing them end for end. If the plumb lines returned to the same points, the instruments were in proper adjustment and the tops of the supporting stakes (see Figure 1-1) would be at the same elevation.

The practical-minded Romans introduced many advances in surveying by an amazing series of engineering projects constructed throughout their empire. They laid out projects such as cities, military camps, and roads by using a system of rectangular coordinates. They surveyed the principal routes used for military operations on the European continent, in the British Isles, in northern Africa, and even in parts of Asia.

Three instruments used by the Romans were the *odometer*, or measuring wheel, the *groma*, and the *chorobate*. The *groma*, from which the Roman surveyors received their name of gromatici, was used for laying off right angles. It consisted of two cross-arms fastened together at right angles in the shape of a horizontal cross, with

[2]C. M. Brown, W. G. Robillard, and D. A. Wilson, *Evidence and Procedures for Boundary Location*, 2nd ed. (New York: John Wiley & Sons, Inc., 1981), p. 145.
[3]A. R. Legault, H. M. McMaster, and R. R. Marlette, *Surveying* (Englewood Cliffs, N.J.: Prentice-Hall, Inc., 1956), p. 5.

plumb bobs hanging from each of the four ends (see Figure 1-2) The groma, which was pivoted eccentrically on a vertical staff, would be leveled and sights taken along its cross-arms in line with the plumb-bob strings.

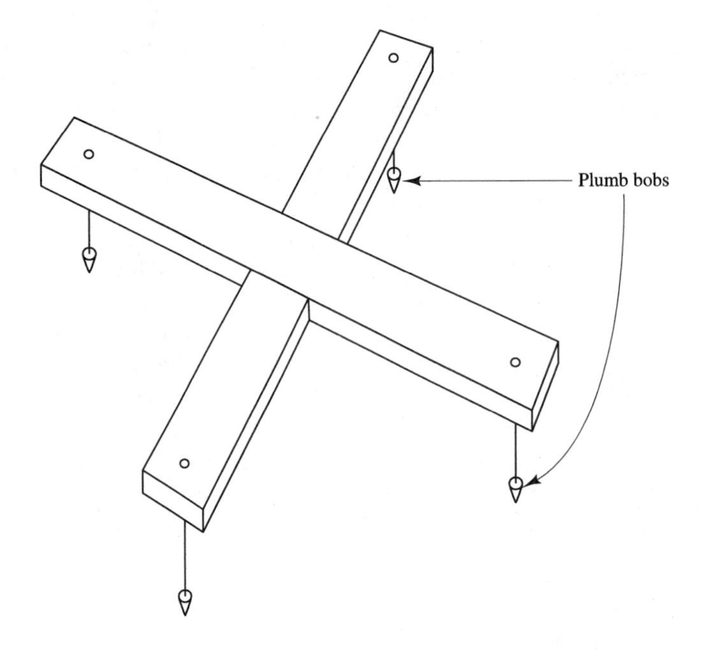

FIGURE 1-2 Groma, a Roman surveying device used for laying off right angles.

The *chorobate* (Figure 1-3) was an approximately 20-ft long wooden straight edge with supporting legs. It had a groove or trough cut into its top to hold water so that it could be used as a level.

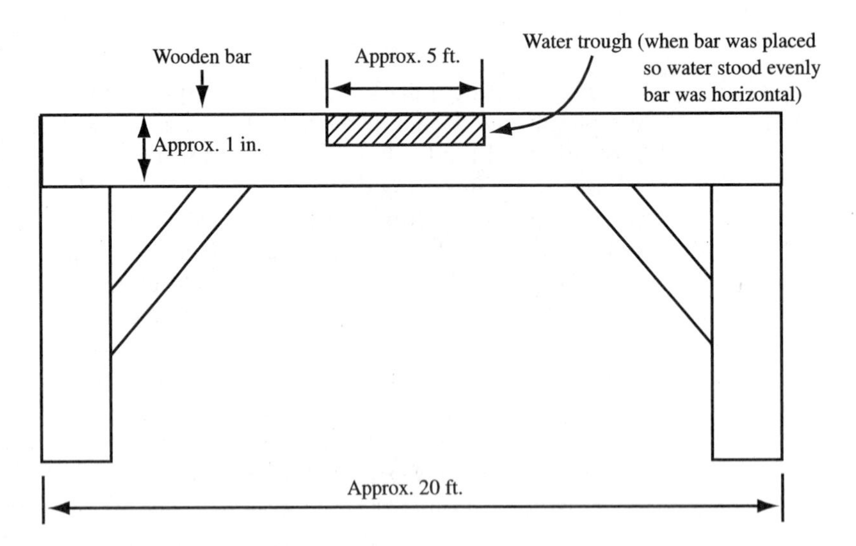

FIGURE 1-3 Chorobate, another Roman surveying device.

From Roman times until the modern era there were few advances in the art of surveying, but the last few centuries have seen the introduction of the telescope, vernier, theodolite, electronic distance-measuring equipment, computers, global positioning systems, and many other excellent devices. These developments will be mentioned in subsequent chapters. For a detailed historical list of early instrument development, the reader is referred to "Historical Notes upon Ancient and Modern Surveying and Surveying Instruments" by H. D. Hoskold.[4]

Many developments were made in the eighteenth and nineteenth centuries as various countries demanded better maps and information related to their national boundaries. In the U.S. a great deal of work was done in the surveying of the vast tracts of land held by the federal government. This work resulted in many advances in the areas of triangulation and surveying involving the three-dimensional characteristics of the earth's surface. One particular organization that had much to do with these advances was the U.S. Coast and Geodetic Survey, which was established in 1807. This group is now called the National Geodetic Survey (NGS) and is a part of the National Oceanic and Atmospheric Administration (NOAA) of the U.S. Department of Commerce.

During World War I and World War II, and the Korean and Vietnamese Wars of the 20th century, major advances were made in the development of surveying equipment needed for the preparation of maps. Similar advances have been made in recent decades in relation to space programs and missile development.

1-5 PLANE SURVEYS

In large-scale mapping projects, adjustments are made for the curvature of the earth and for the fact that north-south lines passing through different points on the earth's surface converge at the north and south poles. Thus, these lines are not parallel to each other except at the equator (see Figure 1-4). Plane surveys however, are made on such small areas that the effect of these factors may be neglected. The earth is considered to be a flat surface and north-south lines are assumed to be parallel. Calculations for a plane surface are relatively simple, since the surveyor is able to use plane geometry and plane trigonometry.

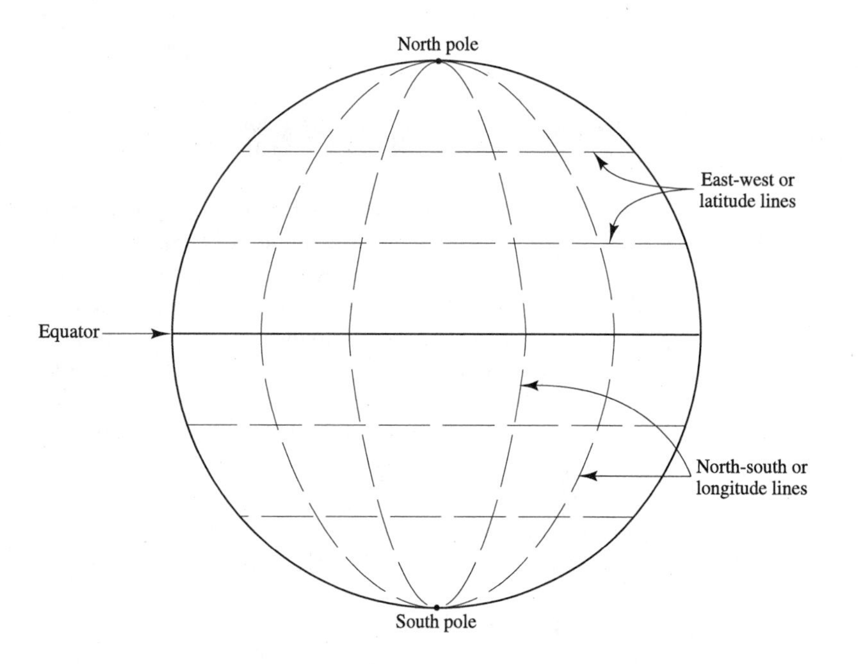

FIGURE 1-4 Drawing showing north–south and east–west lines for the earth's surface. Later these will be referred to as latitude and longitude lines.

[4] *Transactions of the ASCE,* 1893, vol. 30, pp. 135–154.

Surveys for farms, subdivisions, buildings, and in fact most constructed works are plane surveys. They should, however, be limited to areas of a few square miles at most. They are not considered to be sufficiently accurate for establishing state and national boundaries which involve very large areas.

It can be shown that an arc along the earth's curved surface of 11.5 miles in length is only approximately 0.05 ft longer than the plane or chord distance between its ends. As a result, it probably seems to the reader that such discrepancies are insignificant. Direction discrepancies due to the convergence of north-south lines are, however, much more significant than are distance discrepancies.

1-6 GEODETIC SURVEYS

Geodetic surveys are those that are adjusted for the curved shape of the earth's surface. (The earth is an oblate spheroid whose radius at the equator is about 13.5 miles greater than its polar radius.) Since they allow for earth's curvature, geodetic surveys can be applied to both small and large areas. The equipment used and the methods of measurement applied are about the same as they are for plane surveys. Elevations are handled in the same manner for both plane and geodetic surveys. They are expressed in terms of vertical distances above or below a reference curved surface, usually mean sea level (MSL).

Most geodetic surveys are made by government agencies, such as the former U.S. Coast and Geodetic Survey. *A list of mailing addresses of various organizations that are frequently contacted by surveyors is provided in Appendix A at the end of the book.* Although only a relatively small number of surveyors are employed by the National Geodetic Survey (NGS), their work is extremely important to all other surveyors. They have established a network of reference points throughout the United States that provides very precise information on horizontal and vertical locations. On this network all sorts of other surveys (plane and geodetic) of lesser precision are based.

1-7 TYPES OF SURVEYS

This section is devoted to a brief description of the various types of surveys. Most of these types of surveys employ plane rather than geodetic techniques.

Land surveys are the oldest type of surveys and have been performed since earliest recorded history. They are normally plane surveys made for locating property lines, subdividing land into smaller parts, determining land areas, and providing any other information involving the transfer of land from one owner to another. These surveys are also called *property surveys, boundary surveys,* or *cadastral surveys.* Today, the term *cadastral* is usually used with regard to surveys of public lands.

Topographic surveys are made for locating objects and measuring the relief, roughness, or three-dimensional variations of the earth's surface. Detailed information is obtained pertaining to elevations as well as to the locations of constructed and natural features (buildings, roads, streams, etc.) and the entire information is plotted on maps (called topographic maps).

Route surveys involve the determination of the relief and the location of natural and artificial objects along a proposed route for a highway, railroad, canal, pipeline, power line, or other utility. They may further involve the location or staking out of the facility and the calculation of earthwork quantities.

City or *municipal surveys* are made within a given municipality for the purpose of laying out streets, planning sewer systems, preparing maps, and so on. When the term is used, it usually brings to mind topographic surveys in or near a city for the purpose of planning urban expansions or improvements.

Construction surveys are made for purposes of locating structures and providing required elevation points during their construction (see Figure 1-5). They are needed to control every type of construction project. It has been estimated that 60% of the surveying done in the United States is construction surveying.[5]

FIGURE 1-5 Surveying on a construction site. (Courtesy of Carl Zeiss, Inc., Thornwood, N.Y.)

Hydrographic surveys pertain to lakes, streams, and other bodies of water. Shorelines are charted, shapes of areas beneath water surfaces are determined, water flow of streams is estimated, and other information needed relative to navigation, flood control, and development of water resources is obtained. These surveys are usually made by a governmental agency, for example, the National Geodetic Survey, the U.S. Geological Survey, or the U.S. Army Corps of Engineers.

Marine surveys are related to hydrographic surveys, but they are thought to cover a broader area. They include the surveying necessary for offshore platforms, the theory of tides, and the preparation of hydrographic maps and charts.

[5]R. C. Brinker and R. Minnick, eds., *The Surveying Handbook,* 2nd ed., New York: Van Nostrand Reinhold Company, Inc., 1995), p. 578.

Mine surveys are made to obtain the relative positions and elevations of underground shafts, geological formations, and so on, and to determine quantities and establish lines and grades for work to be done.

Forestry and geological surveys are probably much more common than the average layperson realizes. Foresters use surveying for boundary locations, timber cruising, topography, and so on. Similarly, surveying has applications in the preparation of geological maps.

Photogrammetric surveys are those in which photographs (generally aerial) are used in conjunction with limited ground surveys (the latter being used to establish or locate certain control points visible from the air). Photogrammetry is extremely valuable because of the speed with which it can be applied, the economy, the applications to areas difficult to access, the great detail provided, and so on. Its uses are becoming more extensive each year and is today used for a very large percentage of those surveys which involve significant acreage (roughly more than 20 to 40 acres depending on ground foliage and terrain).

Remote sensing is another type of aerial survey. It makes use of cameras or sensors that are transported either in aircraft or in artificial satellites.

As-built surveys are made after a construction project is complete to provide the positions and dimensions of the features of the project as they were actually constructed. Such surveys not only provide a record of what was constructed but also provide a check to see if the work proceeded according to the design plan.

The usual construction project is subject to numerous changes from the original plans due to design changes as well as to problems encountered in the field, such as underground pipes and conduits, unexpected foundation conditions, and other situations. As a result, the as-built survey becomes a very important document that must be preserved for future repairs, expansions, and modifications. For example, just imagine how important it is to know the precise location of water and sewer lines.

Control surveys are reference surveys. For a particular control survey a number of points are established and their horizontal and vertical positions are accurately determined. The points are established so that other work can conveniently be referenced or oriented to them.

The horizontal and vertical controls form a network over the area to be surveyed. For a particular project the horizontal control is probably tied to property lines, road center lines, and other prominent features. Vertical control consists of a set of relatively permanent points whose elevations above or below sea level have been carefully determined. (These points are called *bench marks.*)

In the decades to come, undoubtedly other special types of surveying will develop. Surveyors might very well have to establish boundaries under the ocean, in the Arctic and Antarctic, and even on the moon and other planets. Great skill and judgment by the surveying profession will undoubtedly be required to handle these tasks.

1-8 MODERN SURVEYING EQUIPMENT

During the last several decades there has been one revolution after another in the development of surveying equipment. With each revolution it seemed that an absolute all-time peak or zenith had been achieved with equipment—but each time something better came along. Several of these "zeniths" are listed below. They are each discussed in various sections throughout the text.

1. The first zenith was reached in the 1960s when electronic distance measuring (EDM) devices began to be commonly used. Surely this was the best equipment that would ever be available to the surveyor.
2. The next zenith was reached when angle measuring devices were combined with EDM instruments to form the so-called *total stations* (Figure 1-6). Was this not the ultimate development?

FIGURE 1-6 Surveying along city streets. (Courtesy of Nikon, Inc.)

3. Then automatic data collectors were developed for the total stations. They could be used to store measurements, to make computations, and to transfer or download measurement values to computers or plotters. Furthermore, information could be uploaded from computers to data collectors for use in the field. Was this the absolute summit in surveying equipment?
4. Another almost unbelievable development was the global positioning system (GPS). With the GPS system vertical and horizontal positions on earth could be obtained from radio signals broadcast from earth satellites. Surely this had to be the final step.
5. Recently robotic total stations have come on the market. There are instruments for determining positions which automatically take readings from bar coded rods. There are new laser alignment devices, instruments for automatically plotting maps, instruments that scan and take information from existing maps, and so on. What will the next zenith be?

It may seem to the reader that all the great developments possible in surveying equipment have been made. This cannot be the case, however, because after every great development in technology in the past there has been a greater one down the road. It is interesting to speculate what will happen next. Will it be complete robotic surveying? Will it be ----???

Today's surveying equipment is so superb that some students (particularly those who have previously worked with surveying crews) tend to think it is a waste of time to study "old-fashioned" fundamental surveying ideas such as errors, mistakes, taping, area calculations, and so on because with modern equipment they can take care of such items by merely pressing a few buttons. As we will see, however, a basic understanding of these topics is an essential background for the successful surveyor.

1-9 USE OF OLD SURVEYING EQUIPMENT

When properly maintained the old transits (used for measuring angles), levels (used for determining elevations), tapes, and other equipment will last a lifetime or longer. As a result much of the older or traditional equipment is still around and is still used for some surveying jobs (particularly construction work), despite the availability and common use of the modern equipment described in the preceding section. However, today's surveyor must face the reality that he or she must use up-to-date equipment to be economically competitive. As a result, little mention of older equipment is presented in the main body of this text but some information on the subject is given in the book's appendices.

1-10 MAINTENANCE OF EQUIPMENT

Although surveying equipment is manufactured with great care and precision (and can cost many thousands of dollars) it must be checked periodically and properly maintained to keep it in good adjustment. A very good rule to follow is that *the surveyor should not only keep his or her equipment in good adjustment but should also make measurements as though the equipment is not well adjusted.* (In this text methods are presented for measuring data in such a manner that instrument inadjustment errors will be substantially reduced.)

1-11 IMPORTANCE OF SURVEYING

As described in Section 1-4, it has been necessary since the earliest civilizations to determine property boundaries and divide sections of land into smaller pieces. Through the centuries the uses of surveying has expanded until today it is difficult to imagine any type of construction project that does not involve some type of surveying.

All types of engineers, as well as architects, foresters, and geologists, are concerned with surveying as a means of planning and laying out their projects. Surveying is needed for subdivisions, buildings, highways, railroads, canals, piers, wharves, dams, irrigation and drainage networks, and many other projects. In addition, surveying is required to lay out industrial equipment, set machinery, hold tolerances in ships and airplanes, prepare forestry and geological maps, and for numerous other applications.

The study of surveying is an important part of the training of a technical student even though that student may never actually practice surveying. It will appreciably help him or her to learn to think logically, to plan, to take pride in working carefully and accurately, and to record his or her work in a neat and orderly fashion. The student will learn a great deal about the relative importance of measurements, develop some sense of proportion as to what is important and what is not, and acquire essential habits of checking numerical calculations and measurements (a necessity for anyone in an engineering or scientific field). Moreover, an individual may be placed in a decision-making position regarding the retaining of surveying services. Without a basic understanding of the subject he or she may not be able to handle the situation.

1-12 SAFETY

A consideration of safety for surveying employees, clients, and outsiders is an extremely important topic not only from the point of view of physical suffering but also from that of economic loss. Personal injury and sickness cause a loss of efficiency for the company and greater expenses from the standpoint of less production and higher insurance costs. The reader understands that the more accidents that occur to a particular firm, the higher will be their insurance premiums. These premiums are today no small item in the cost of operating a business (even one with a splendid safety record).

Chapter 2 will cover the importance of planning in obtaining good surveys. Planning is also of extraordinary importance when we think of achieving safety. The surveyor needs to conduct periodic safety meetings and continuously discuss with his or her employees the dangers in the various types of surveys they conduct. Many hazards may be encountered when working near or on highways, on construction projects, near power lines, and on remote pieces of land with dangerous terrain features. It is absolutely essential for surveying parties to have first-aid kits and to have some personnel trained in first-aid procedures.

Surveyors should wear conspicuous clothing such as orange hunting vests so that they can easily be seen by motorists, fellow workers, and hunters. When in the area of construction projects, hard hats and safety shoes are a necessity. In snake-infested areas it is necessary to wear boots and/or leggings. Stings from wasps, bees, and yellow jackets, however, are a much more likely possibility. As a result, first-aid kits should include antidotes for persons who are allergic to such bites.

Ticks seem to reside everywhere trees are located, and their bites can cause life-threatening diseases. As a result, surveyors in tick-infested areas should use insect repellent, wear long-sleeved shirts, and tuck their pants into their boots. Furthermore, it is desirable for personnel to wear light-colored clothing so that ticks can easily be seen. Employees should also check each other for the presence of these pests.

One of the most dangerous situations for surveyors is work along highways. For such jobs it is absolutely necessary to make use of warning signs and flagmen. Of course, the best prevention of all is to stay off the roads and work with offsets if possible. When it is necessary to work along highways the work should be performed in accordance with the recommendations of the *Manual of Uniform Traffic Control Devices,* published by the U.S. Department of Transportation, Federal Highway Administration.[6] If these recommendations are followed, not only will the work be safer but the surveyor's liability will be reduced in case an accident does occur.

[6]Washington, D.C.: Superintendent of Documents, U.S. Government Printing Office, 1988, pp. 6F-1 to 6F-6.

The purpose of this section is not to list every possible safety precaution that should be taken, but rather to make surveyors conscious of safety and to try to make them think through the work to be done and the possible hazards that might be present. A few of many possible safety precautions follow:

1. Do not look at the sun through instrument telescopes unless special filters are used; serious and permanent eye injury may result.
2. Be aware of the danger of sunstrokes.
3. Wear gloves when working in briars or poison ivy or oak.
4. When cutting bushes, be extremely careful of the location of other persons and on the lookout for snakes.
5. Do not use steel tapes near electric lines.
6. Do not throw range poles.
7. Do not climb over fences while carrying equipment.
8. Obey safety laws of local, state, and federal governments.

The surveyor should also be familiar with the safety requirements of the Occupational Safety and Health Administration (OSHA) of the U.S. Department of Labor. This organization has developed safety standards and guidelines that apply to the various conditions and situations that can be encountered in all kinds of occupations including surveying.

1-13 LIABILITY INSURANCE

We are all familiar with stories of almost unbelievable court settlements which seem to occur so often in cases of personal injury. Similar settlements are frequently made in cases where surveying mistakes cause financial damages to clients. These cases are more common for construction projects than for rural land surveys.

For this discussion just imagine a surveyor who in preparing a topographic map is 2 ft off on elevations and that this mistake causes the contractor to have to fill in an acre or two of land with 2 ft of soil. What do you think the contractor is going to try to do with the bill for this extra expense? In a similar vein, imagine that a surveyor makes a mistake in locating a property line and as a result a building is erected which runs a few feet onto someone else's land. Who do you think is a prime candidate for paying damages?

From the preceding discussion the reader can see why it is so common today for industrial clients to require surveying companies to provide evidence of suitable liability insurance coverage. Unfortunately, this insurance is very expensive, deductibles on claims are quite large, and, as in other insurance situations, frequent claims lead to much higher premiums.

In some states if a surveyor causes financial injury to a client, that surveyor may lose his or her license unless satisfactory financial restitution is made to the client. A surveyor who carries a suitable liability policy will certainly be in a position to get more industrial jobs than a surveyor who does not have such insurance. (In other words, many persons and organizations may decide not to employ a surveyor who does not carry liability insurance.)

This discussion can be continued to cover personal injuries. The reader can understand that if a surveying employee (worker's compensation may be involved here) or an outsider is injured as a result of surveying activities, the surveyor or the surveyor's insurance company will be right in line for paying damages.

1-14 OPPORTUNITIES IN SURVEYING

There are few professions that need qualified people as much as does the surveying profession. In the United States, tremendous physical developments (subdivisions, factories, dams, power lines, cities, etc.) have created a need for surveyors at a faster rate than our schools are producing them. Construction, our largest industry, requires a constant supply of new surveyors.

For a person with a liking for a combination of outdoor and indoor work, surveying offers attractive opportunities (see Figures 1-5, 1-6, and 2-1). The jobs available in many disciplines in the United States are concentrated in large cities and in certain specific areas. This is not the situation in surveying because they are needed in all parts of the country—rural and metropolitan.

It now appears that beginning salaries for qualified surveyors, which have lagged behind civil engineering salaries in the past, are catching up and will perhaps pass them in the next few years. Despite this fact, most undergraduate surveying programs across the country are faced with decreasing enrollments. On the other hand, graduate school enrollments have been increasing. *A list of schools offering a bachelor's degree in surveying is provided in Appendix B at the end of the book.*

The U.S. Department of Labor claims that there is a demand for approximately 5000 new surveyors each year. Yet our colleges and universities are producing just over 400 graduates a year; this number includes both persons receiving two-year associate's degrees and those receiving bachelor's degrees. Furthermore, only about 100 of these people attend ABET (Accreditation Board for Engineering and Technology) accredited programs.[7]

You may wonder where the other surveyors are obtained. The answer is that they are people who may or may not have taken a course or two in surveying and who have worked their way up in the profession from rodman to instrumentman to draftsman and so on, eventually taking the registration exams. This route is becoming more difficult to follow and will probably be impossible in the near future in most states (if that is not already the case) because of increasing educational requirements for registration.

It is necessary for a person going into private practice to meet the licensing requirements of his or her state. These requirements, which are becoming stiffer each year, are discussed in detail in Chapter 24. Many people feel that by the year 2010, or soon thereafter, a bachelor's degree in surveying (and probably from an ABET-accredited school) will be required for licensing in almost all states.

Problems

1-1. What was the groma?

1-2. Define the term *surveying.*

1-3. What is geomatics?

1-4. Distinguish between plane and geodetic surveys.

1-5. List ten types of surveys.

[7]K. Thapa, "Why Should One Choose Surveying and Mapping as a Career," *P.O.B. Magazine,* Aug./Sept. 1993, vol. 18, no. 6, pp. 51-55.

1-6. What is a topographic survey?

1-7. Name and describe the two types of aerial surveys.

1-8. Distinguish between hydrographic and marine surveys.

1-9. Why is it absolutely necessary to make accurate surveys for underground mines?

1-10. Distinguish between horizontal and vertical control surveys.

1-11. Discuss the importance of liability insurance for the surveyor.

1-12. List five safety precautions that should regularly be taken by the surveyor.

CHAPTER

Introduction to Measurements

2-1 MEASUREMENT

In ordinary life, most of us are more accustomed to *counting* than to *measuring*. If the number of people in a room are counted, the result is an exact number without a decimal, say 9 people. It would be ridiculous to say that there are 9.23 people in a room. In a similar fashion a person might count the amount of money in his or her pocket. Although the result may contain a decimal such as $5.65, the result is still an exact value.

Surveying is concerned with measurements of quantities whose exact or true values may not be determined, such as distances, elevations, volumes, directions, and weights (Figure 2-1). If a person were to measure the width of his desk with a ruler divided into tenths of an inch, he could estimate the width to hundredths of an inch. If he were to use a ruler graduated in hundredths of an inch, he could estimate the width to thousandths of an inch, and so on. Obviously, with better equipment one can estimate an answer that is closer to the exact value but will never be able to determine the value absolutely. Thus a fundamental principle of surveying is that no measurement is exact and the true value of the quantity being measured is never known. (Exact or true values do exist, but they cannot be determined.) *Measurement is the principal concern of a surveyor.*

Although the exact value of a measured quantity is never known, we may know exactly what the sum of a group of measurements should be. For instance, the sum of the three interior angles of a triangle must be 180°, for a rectangle the sum of the four interior angles must be 360°, and so on. If the angles of a triangle are measured and total almost 180° we will learn to logically adjust or revise the individual angles slightly so their total is exactly 180°. In a similar manner we will learn to adjust series of vertical or horizontal measurements to some known totals.

2-2 NECESSITY FOR ACCURATE SURVEYS

The surveyor must have the skill and judgment necessary to make accurate measurements. This fact is obvious when one is thinking in terms of the construction of long bridges, tunnels, tall buildings, and missile sites or with the setting of delicate machinery, but it can be just as important in land surveying.

FIGURE 2-1 Surveying in the country. (Courtesy of Nikon, Inc.)

As late as a few decades ago land prices were not extremely high except in and around the largest cities. If the surveyor gained or lost a few feet in a lot or a few acres in a farm, it was usually not considered to be a matter of great importance. The instruments commonly used for surveying before this century were not very good compared to today's equipment, and it was probably impossible for the surveyor to do the quality of work expected of today's surveyor. (What will surveyors in future centuries think of twentieth-century surveying, and what wonderful equipment will be available to them?)

Today, land prices in most areas are very high, and evidently the climb has only begun. In many areas of high population and in many popular resort areas land is sold by so many dollars per square foot or so many hundreds or even thousands of dollars per front foot; therefore, the surveyor must be able to perform excellent work. Even in rural areas the cost of land is frequently "sky high."

2-3 ACCURACY AND PRECISION

The terms *accuracy* and *precision* are constantly used in surveying, yet their correct meanings are a little difficult to grasp.

Accuracy refers to the degree of perfection obtained in measurements. It denotes how close a given measurement is to the true value of the quantity.

Precision or *apparent accuracy* is the degree of refinement with which a given quantity is measured. In other words, it is the closeness of one measurement to another. If a quantity is measured several times and the values obtained are very close to each other, the precision is said to be high.

It does not necessarily follow that better precision means better accuracy. Consider the case in which a surveyor carefully measured a distance three times with a 100-ft steel tape and obtained the values 984.72 ft, 984.69 ft, and 984.73 ft. He did a very precise job and apparently a very accurate one. Should, however, the tape actually be found to be 100.30 ft long instead of 100.00 ft, the values obtained are not accurate, although they are precise. (The measurements could be made accurate by making a numerical correction of 0.30 ft for each tape length.) It is possible for the surveyor to obtain both accuracy and precision by exercising care, patience, and using good instruments and procedures.

In measuring distance, precision is defined as the ratio of the error of the measurement to the distance measured and it is reduced to a fraction having a numerator of unity or one. If a distance of 4200 ft is measured and the error is later estimated to equal 0.7 ft, the precision of the measurement is 0.7/4200 = 1/6000. This means that for every 6000 ft measured, the error would be 1 ft if the work were done with this same degree of precision.

A method frequently used by surveying professors to define and distinguish between precision and accuracy is illustrated in Figure 2-2. It is assumed that a person has been having target practice with his rifle. His results with target 1 were very precise because the bullet holes were quite close to each other. They were not accurate, however, as they were located some distance from the bull's-eye. The marksman fired accurately at target 2, as the holes were placed relatively close to the bull's-eye. The shots were not precise, however, as they were scattered with respect to each other. Finally, in target 3, the shots were both precise and accurate, as they were placed in the bull's-eye close to each other. *The objective of the surveyor is to make measurements that are both precise and accurate.*

FIGURE 2-2 Accuracy and precision.

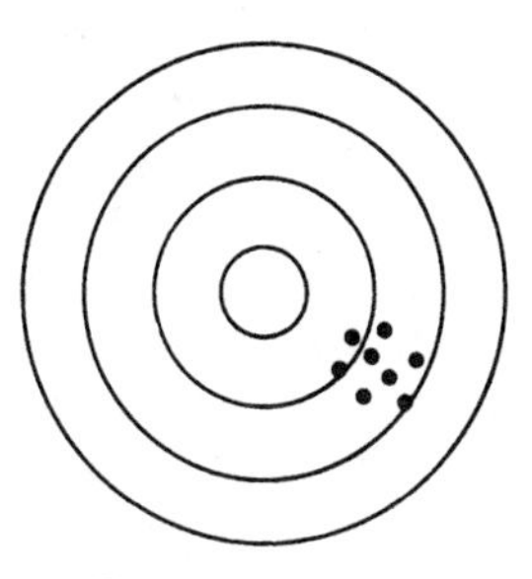

Target 1:
Good precision
Poor accuracy
(average away from bull's-eye)

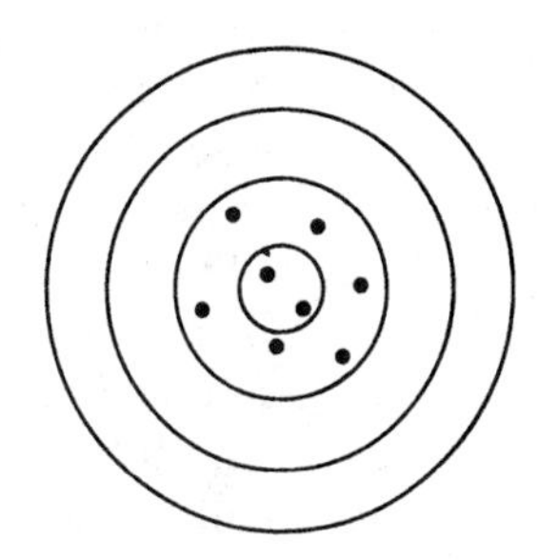

Target 2:
Poor precision
Good accuracy
(average in bull's-eye)

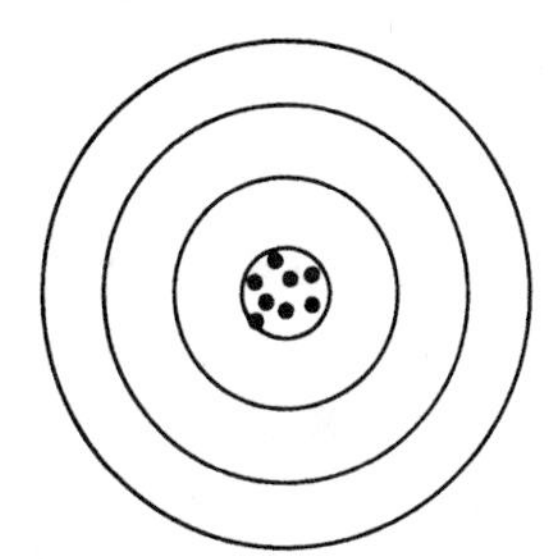

Target 3:
Good precision
Good accuracy
(average in bull's-eye)

2-4 ERRORS AND MISTAKES

There is no one whose senses are sufficiently perfect to measure any quantity exactly and there are no perfect instruments with which to measure. The result is that all measurements are imperfect. A major concern in surveying is the precision of the work. This subject is mentioned repeatedly as we discuss each phase of surveying.

The ever-present differences between measured quantities and the true magnitudes of those quantities are classified herein as either mistakes or errors. A *mistake* (or blunder) is a difference from a true value caused by the inattention of the surveyor. For instance, he may read a number as 6 when it is actually 9, may record the wrong quantities in the field notes, or may add a column of numbers incorrectly. The important point here is that mistakes are caused by the carelessness of the surveyor, and carelessness *can be eliminated* by careful checking. All surveyors will make occasional mistakes, but if they learn to apply carefully to the work the checks described in subsequent chapters, these mistakes will be eliminated. *Any true professional surveyor will not be satisfied with his or her work until he or she has made sure that any blunders are detected and eliminated.*

An *error* is a difference from a true value caused by the imperfection of a person's senses, by the imperfection of the equipment, or by weather effects. Errors *cannot be eliminated but they can be minimized* by careful work, combined with the application of certain numerical corrections.

2-5 SOURCES OF ERRORS

There are three sources of errors: people, instruments, and nature. Accordingly, errors in measurement are generally said to be personal, instrumental, and natural. Some errors, however, do not clearly fit into just one of these categories and may be due to a combination of factors.

Personal errors occur because no surveyor has perfect senses of sight and touch. For instance, in estimating the fractional part of a scale, the surveyor cannot read it perfectly and will always be either a little large or a little small.

Instrumental errors occur because instruments cannot be manufactured perfectly and the different parts of the instruments cannot be adjusted exactly with respect to each other. Moreover, with time the wear and tear of the instruments causes additional errors. *Although the past few decades have seen the development of more precise equipment, the goal of perfection remains elusive.* A reading from a scale will undoubtedly contain both a personal and an instrumental error. The observer cannot read the scale perfectly nor can the manufacturer make a perfect scale. Even though most readings for distances are shown digitally on our modern equipment they still contain instrumental errors.

Natural errors are caused by temperature, wind, moisture, magnetic variations, and so on. On a summer day a 100-ft steel tape may increase in length by a few hundredths of a foot. Each time this tape is used to measure 100 ft there will be a temperature error of those few hundredths of a foot. The surveyor cannot normally remove the cause of errors such as these, but can minimize their effects by using good judgment and making proper mathematical corrections of the results. Although this discussion has focused on tapes, natural errors affect measurements made with all types of surveying equipment.

2-6 SYSTEMATIC AND ACCIDENTAL OR RANDOM ERRORS

Errors are said to be systematic or accidental. A *systematic* or *cumulative error* is one that, for constant conditions, remains the same as to sign and magnitude. For instance, if a steel tape is 0.10 ft too short, each time the tape is used the same error (because of that factor) is made. If the full tape length is used 10 times, the error accumulates and totals 10 times the error for one measurement.

An *accidental, compensating,* or *random error* is one whose magnitude and direction is just an accident and beyond the control of the surveyor. For instance, when a person reads a tape, he or she cannot read it perfectly. One time he or she will read a value that is too large and the next time will read a value that is too small. Since these errors are just as likely to have one sign as the other, they tend to a certain degree to cancel each other or compensate for each other.

2-7 DISCUSSION OF ACCIDENTAL OR RANDOM ERRORS

Very often practicing surveyors and surveying students who have had some field experience say they don't need to study accidental errors and their accumulation or propagation. They seem to think that the subject is some college-professor-type-materials, which nobody in practice needs or uses. *However, as we shall see, they are wrong.*

The average student may have a little difficulty at first in understanding the elementary statistical material presented in this section and the next two. However, as this student moves into the later chapters of this textbook he or she will probably refer back on several occasions to this chapter. If this is done, a gradual understanding of this material and its importance in making measurements will be developed.

The quality of a measurement can be expressed by stating a relative error. For example, a distance may be expressed as 835.82 ± 0.06 ft. Such a statement indicates that the true distance measured probably falls between 835.76 and 835.88 ft, and its most probable value is 835.82 ft. The sign or direction of the probable error is not known and thus no correction can be made. The probable error can either be plus or minus, between the limits within which the error is likely to fall. Note that this form does not specify the magnitude of the actual error, nor does it indicate the error most likely to occur.

The term *probable* or *50% error* is sometimes used in discussing surveying measurements. If we say that a certain distance measurement has a 50% error of ±0.12 ft, we mean that there is a 50% chance that it contains an error of ±0.12 ft or less and a 50% chance that it contains a larger error.

How many clients would be happy if you told them that you had measured a distance for them and obtained a value of 632.34 ft and there was a 50% chance that the value was correct within ±0.12 ft? The answer is not very many. In fact, they would probably say "I want a 100% measurement, that is, one for which the given value has a 0% chance of being no further off than a certain value." (*There is no such thing as a 100% error, as there is always some chance that a measurement contains an error larger than any given value.*)

Continuing this discussion, we can refer to errors as being 60% errors, 68.3% errors, 90% errors, and so on. If we were to say that a given measurement is 632.34 ft and that there is a 90% error of ±0.09 ft, we are saying that there is a 90% chance that the error is ±0.09 ft or less and a 10% chance that it is larger. If we were to provide our client with a 90% error or 95% error or 99.7% error, he or she would probably be a little happier than with a 50% error.

2-8 OCCURRENCE OF ACCIDENTAL OR RANDOM ERRORS

The average student seems to have more trouble understanding the material in this section and the next two sections (2-9 and 2-10) than any other parts of the text. If this is the case, it is probably due to the absence of a course in statistics in their prior studies. Nevertheless, students will gradually begin to understand the material and its importance as they see its application to future situations in this text. When such situations arise, it might be well for any student to come back and reread these sections.

If a coin is flipped 100 times, the probability is that there will be 50 heads and 50 tails. Each time the coin is flipped there is an equal chance of it being heads or tails, and it is apparent that the more times the coin is flipped, the more likely it will be that the total number of heads will equal the total number of tails.

When a quantity such as distance is being measured, random errors occur due to the imperfections of the observer's senses and the equipment used. The observer is not able to make perfect readings. Each time a measurement is made it will be either too large or too small.

For this discussion it is assumed that a distance was measured 28 times, with the results shown in Table 2-1. In the first column of the table the values of the measurements are arranged in order of increasing value, while the number of times a particular measurement was obtained is given in the second column. The latter values are also referred to as the *frequencies* of the measurements.

The errors of the measurements are not known because the true value of the quantity is not known. The true value, however, is assumed to equal the arithmetic mean (or average) of the measurements. It is referred to as the *best value* or as the *most probable value.* The error of each measurement is then assumed to equal the difference between the measurement and the mean value. These are not really errors and they are referred to as *residuals* or *deviations.* For the measurements being considered here, the residuals are shown in the third column of Table 2-1.

TABLE 2-1

Measurement	Number or frequency of each measurement	Residual or deviation
96.90	1	–0.04
96.91	2	–0.03
96.92	3	–0.02
96.93	5	–0.01
96.94	6	0.00
96.95	5	+0.01
96.96	3	+0.02
96.97	2	+0.03
96.98	1	+0.04
Average = 96.94	28	

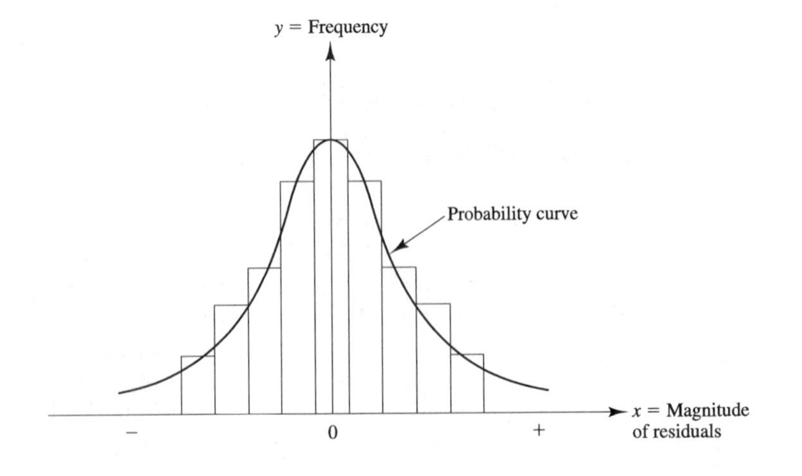

FIGURE 2-3 Histogram or frequency distribution diagram.

For a particular set of measurements of the same item, it is possible to take the residuals and plot them in the form of a bar graph, as shown in Figure 2-3. This figure, called a *histogram* or a *frequency distribution diagram*, has the magnitudes of the residuals plotted along the horizontal axis. Their signs, plus or minus, are shown by plotting them to the right or left of the origin (O). In addition, the number or frequency of the residuals of a particular size are shown vertically.

2-9 PROBABILITY CURVE

If an infinite number of measurements of an item could be taken and if the residuals of those values were plotted as a histogram and a curve drawn through the values, the results would theoretically fall along a smooth bell-shaped curve called a *probability curve* (also called the *Gauss curve* or the *normal error distribution curve*). Almost all surveying measurements conform to such a curve. This curve shows the relationship between the size of an error and the probability of its occurrence.

The probability curve provides the most accurate method for studying the precision of surveys and also provides us with the best available means for estimating the precision of future planned surveys. The curve can be represented with a mathematical equation (given at the end of this section), but the equation is seldom used in surveying measurements.

Theoretically, the probability curve would be perfectly bell-shaped if an infinite number of measurements were taken and plotted. Even where a fairly large number of measurements of a particular quantity is made, the histogram can be plotted and a curve drawn as shown in Figure 2-3. Such a curve will approximate the theoretical curve and can be used practically to estimate the most probable behavior of random errors. There is little approximation involved because a curve plotted for as few as five or six determinations of a quantity is practically the same as the theoretical curve based on an infinite number of measurements.

Fifty percent of the area underneath the probability curve of Figure 2-4 is shown shaded. There is a 50% chance that the error for a single measurement will fall within this area and a 50% chance that it will fall without. The value x_p shown in the figure is referred to as the *probable error* or 50% error. A particular measurement will have the same chance of having an error less than x_p as it does of being greater than x_p.

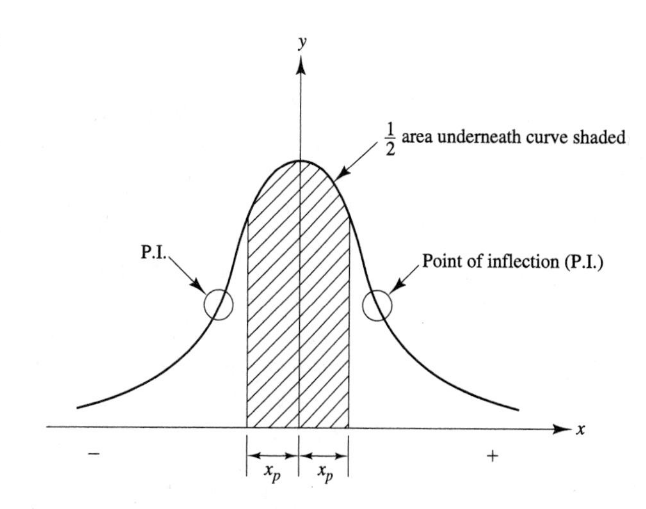

FIGURE 2-4 Histogram showing probable error and points of inflection, or standard deviation.

From statistics we can find that the average or 50% error can be determined by multiplying a constant 0.845 times the average numerical value of the residuals. The signs (±) of the residuals are not included in the average. Letting v equal a residual, the 50% error E_{50} is as follows:

$$E_{50} = \pm 0.845 v_{\text{average}}$$

The term *probable error* was commonly used in the past by surveyors and engineers but today is seldom used. Most of our references are now made to another value, called the standard error or standard deviation, a term that is defined in the next paragraph.

There are several ways in which errors may be denoted, but the most common one is to refer them to the *standard deviation* (σ), also called the *mean square error or standard error.* This value provides a practical means of indicating the reliability of a set of repeated measurements. If we examine the curves of Figures 2-4 and 2-5, we will see that there are points of inflection (P.I.) on each side of the curve, that is, points where the slope of the curve changes from concave to convex, or vice versa. The area underneath the probability curve between these points for a theoretical curve equals 68.3% of the total area. If a particular quantity is measured 10 times, it is anticipated that 68.3% or about 7 of the 10 measurements will fall between these values and 3 will not. The residuals at the P.I.s are called the *standard deviations* or *standard errors* and can be calculated from the following equation, which is derived by the method of least squares.

$$\sigma = \pm \sqrt{\frac{\Sigma v^2}{n - 1}}$$

where Σv^2 is the sum of the squares of the residuals and n is the number of observations.

If the student will examine the literature published by manufacturers of surveying equipment, he or she will note that they usually provide the standard error associated with the use of their equipment. Actually neither the standard (or 68.3%) value nor the probable (or 50%) value is often used as such by the surveyor. It is more likely that the 95.4% or 99.7% values will be used.

The probability of error at other positions on the curve can be determined from the following equation:

$$E_p = C_p \sigma$$

where E_p is the percentage error, C_p is a constant, and σ is the standard error. For instance, the 68.3% error occurs when $C_p = 1.00$ and the 95.4% error occurs when $C_p = 2.00$. It is impossible to establish an absolutely maximum error because this condition theoretically occurs at infinity, but many persons refer to the maximum error as being the 95.4% error (which occurs at 2.00σ), while others refer to the 99.9% error (which occurs at 3.29σ) as being the maximum. In the latter case, we see that 999 out of 1000 values would fall within this range. Several probabilities of error are summarized in Table 2-2 and represented graphically in Figure 2-5, Sometimes the values 2σ and 3σ (which correspond to the 95.4% and 99,7% errors, respectively) are referred to as being two standard deviations or three standard deviations, respectively.

TABLE 2-2 Probabilities for Certain Error Range

Error (=)	Probability (%)	Probability that error will be larger
0.50σ	38.3	2 in 3
0.6745σ	50.0	1 in 2
1.00σ	68.3	1 in 3
1.6449σ	90.0	1 in 10
1.9599σ	95.0	1 in 20
2.00σ	95.4	1 in 23
3.00σ	99.7	1 in 333
3.29σ	99.9	1 in 1000

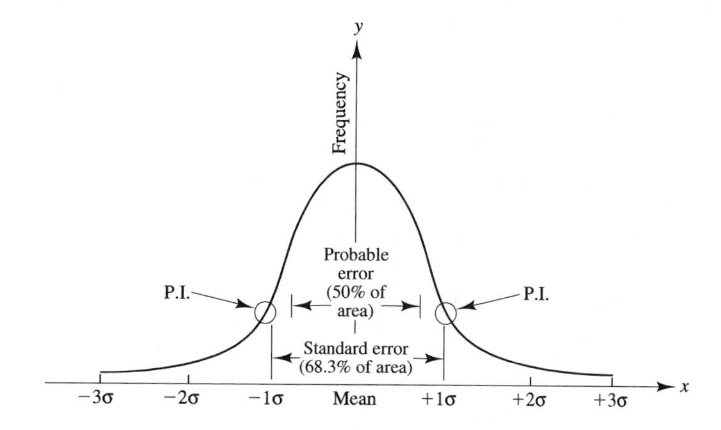

FIGURE 2-5 Probability curve.

The general equation of the probability curve for an infinite number of ordinates may be expressed as follows:

$$y = ke^{-h^2x^2}$$

where *y* is the probability of occurrence of a random error of magnitude *x*, *e* is the base of the natural logarithms 2.718, and *k* and *h* are constants that determine the shape of the curve.

2-10 PROPAGATION OF ACCIDENTAL OR RANDOM ERRORS

This section is devoted to the presentation of a set of example problems that apply the preceding discussion of random errors to practical surveying problems. Included are calculations for single quantities as well as values for series of quantities. The accumulation of random errors for various calculations is referred to as the *propagation of random errors.* To be able to develop equations for random error propagation for any but the simplest cases requires a substantial background in calculus. It is assumed in this section that the readers of this book do not necessarily have such a background, so the algebraic equations that result from such derivations are given in the paragraphs to follow and applied to simple numerical cases.

Measurements of a Single Quantity

When a single quantity is measured several times, the probable error or 50% error of any one of the measurements can be determined from the following equation:

$$E_p = C_p\sqrt{\frac{\Sigma v^2}{n - 1}}$$

where $C_p = C_{50} = 0.6745$. For other percent errors the values of C_p vary as shown in Table 2-2. Example 2-1 illustrates the application of the equation for different percent errors.

Example 2–1

Determine the 50%, 90%, and 95% errors for the distance measurements shown in the first column of the following table.

Solution

Measured value	Residual v	v^2
152.93	+0.02	0.0004
153.01	+0.10	0.0100
152.87	–0.04	0.0016
152.98	+0.07	0.0049
152.78	–0.13	0.0169
152.89	–0.02	0.0004
Avg = 152.91		$\Sigma v^2 = 0.0342$

$$E_{50} = \pm 0.6745\sqrt{\frac{0.0342}{6 - 1}} = \pm 0.056 \text{ ft}$$

$$E_{90} = \pm 1.6449\sqrt{\frac{0.0342}{6 - 1}} = \pm 0.136 \text{ ft}$$

$$E_{95} = \pm 1.9599\sqrt{\frac{0.0342}{6 - 1}} = \pm 0.162 \text{ ft}$$

Thus there is a 50% chance that the measurement will be within 0.056 ft of the actual value, a 90% chance that it will be within 0.136 ft and a 95% chance that it will be within 0.162 ft.

At this point we present a few remarks concerning measurements which seem to be completely out of line with the other values. When residuals are determined for a set of measurements, they should be compared with the average value for those residuals. Should any of them be very large (say, four or five or more times the average value) they probably are mistakes and thus should be discarded and the calculations continued with the remaining ones. A value should not be eliminated just to make things neat or symmetrical. For a value to be deleted it should be set apart clearly from the other values so that a definite mistake is indicated.

Repeated Measurements of a Single Quantity

We may decide that we can take a 100-ft steel tape and measure a distance with a standard error of ±0.02 ft. Then the thought might occur to us: What will be the total estimated standard error if we were to measure 1000 ft with this tape with the accidental errors sometimes plus and sometimes minus? A study of statistics (not included) will provide the answer given below.

When a single quantity is measured several times or when a series of quantities is measured, random errors tend to accumulate in proportion to the square root of the number of measurements. This is referred to as the *law of compensation.* The next few paragraphs present several applications of this law.

Should a single quantity be measured several times with an estimated random error E in each measurement, the following equation can be used to estimate the total random error occurring in all the measurements. The term n represents the number of measurements.

$$E_{\text{total}} = \pm E\sqrt{n}$$

Example 2–2

If a distance is measured nine times and the estimated standard error in each measurement is ±0.02 ft, what is the estimated total standard error in the nine measurements?

Solution

$$\begin{aligned} E_{\text{total}} &= \pm E\sqrt{n} \\ &= \pm 0.02\sqrt{9} = \mathbf{\pm 0.06\ ft} \end{aligned}$$

A Series of Similar Measurements

The surveyor is usually involved with the measurement of not just one quantity but with a series of quantities. For instance, he or she usually measures at least several different angles, distances and elevations while working on a particular project.

The expression used for estimating the total error for the repeated measurement of a single quantity applies equally well to the situation where a series of different quantities are being measured. It is, of course, assumed that all measurements are made with equal precision, so they are equally reliable.

Example 2–3

A series of 12 angles was measured, each with an estimated error of ±20 seconds of arc. What is the total estimated error in the 12 angles?

Solution

$$E_{\text{total}} = \pm 20''\sqrt{12} = \pm 69'' = \mathbf{\pm 1'09''}$$

As one moves further into the study of surveying, he or she may face the problem of needing to measure a distance, a set of angles, and so on, with a total error not exceeding a certain limit. The same expression used in the preceding paragraphs may be used to solve a problem such as the one illustrated in Example 2–4.

Example 2–4

It is desired to tape a distance of 2000 ft with a total error of no more than ±0.10 ft. How accurately must each 100 ft be taped so that the desired limit is not exceeded?

Solution

$$E_{\text{total}} = E\sqrt{n}$$

$$\pm 0.10 = E\sqrt{20}$$

$$E = \mathbf{\pm 0.022\ ft}$$

From this discussion we can see that the more times we measure a value, the smaller will become the estimated error in our mean value. We can see that the error of the mean varies inversely as the square root of the number of measurements. Thus, to double the precision of a particular measured quantity, four times as many measurements should be taken. To triple the precision, nine times as many measurements should be taken.

A Series of Unrepeated Measurements

When a series of independent measurements are made with probable errors of E_1, E_2, E_3, . . . , respectively, the total probable error can be computed from the following equation:

$$E_{\text{total}} = \pm\sqrt{E_1^2 + E_2^2 + \dots + E_n^2}$$

Example 2–5 illustrates the application of this equation.

Example 2–5

The four approximately equal sides of a tract of land were measured. These measurements included the following probable errors: ±0.09 ft, ±0.013 ft, ±0.18 ft, and ±0.40 ft, respectively. Determine the probable error for the total length or perimeter of the tract.

Solution

$$E_{\text{total}} = \pm\sqrt{(0.09)^2 + (0.013)^2 + (0.18)^2 + (0.40)^2}$$
$$= \pm\mathbf{0.45\ ft}$$

The reader should carefully note the results of the preceding calculations, where the uncertainty in the total distance measured (±0.45 ft) is not very different from the uncertainty given for the measurement of the fourth side alone (±0.40 ft). *It should be obvious that there is a little advantage in making very careful measurements for some of a group of quantities and not for the others.*

2-11 SIGNIFICANT FIGURES

This section might also be entitled "judgment" or "sense of proportion," and its comprehension is a very necessary part of the training of anyone who takes and/or uses measured quantities of any kind. When measurements are made, the results can be precise only to the degree that the measuring instrument is precise. This means that numbers which represent measurements are all approximate values. For instance, a distance may be measured with a steel tape as being 465 ft, or more precisely as 465.3 ft, or with even more care as 465.32 ft, but an exact answer can never be obtained. The value will always contain some error.

The number of significant figures that a measured quantity has is not (as is frequently thought) the number of decimal places. Instead, it is the number of certain digits plus one digit that is estimated. For instance, in reading a steel tape a point may be between 34.2 and 34.3 ft (the scale being marked at the 1/10-ft points) and the value is estimated as being 34.26 ft. The answer has four significant figures. Other examples of significant figures follow:

36.00620 has seven significant figures.

10.0 has three significant figures.

0.003042 has four significant figures.

The answer obtained by solving any problem can never be more accurate than the information used. If this principle is not completely understood, results will be slovenly. Notice that it is not reasonable to add 23.2 cu yd of concrete to 31 cu yd and get 54.2 cu yd. One cannot properly express the total to the nearest tenth of a yard as 54.2 because one of the quantities was not computed to the nearest tenth and the correct total should be 54 cu yd.

A few general rules regarding significant figures follow.

1. Zeros between other significant figures are significant, as, for example in the following numbers, each of which contains four significant figures: 23.07 and 3008.
2. For numbers less than unity, zeros immediately to the right of the decimal are not significant. They merely show the position of the decimal. The number 0.0034 has two significant figures.
3. Zeros placed at the end of decimal numbers, such as 24.3200, are significant.
4. When a number ends with one or more zeros to the left of the decimal it is necessary to indicate the exact number of significant figures. The number 352,000 could have three, four, five, or six significant figures. It could be written as $35\bar{2}{,}000$, which has three significant figures, or as $352{,}00\bar{0}$, which has six significant figures. It is also possible to handle the problem by using

scientific notation. The number 2.500×10^3 has four significant figures, and the number 2.50×10^3 has three.

5. When numbers are multiplied or divided or both, the answer should not have more significant figures than those in the factor, which has the least number of significant figures. As an illustration, the following calculations should result in an answer having three significant figures, which is the number of significant figures in the term 3.25.

$$\frac{3.25 \times 4.6962}{8.1002 \times 6.152} = 0.306$$

It is desirable to carry calculations to one or more extra places during the various steps; as the final step, the answer is rounded off to the correct number of significant figures.

6. For addition or subtraction, the last significant figure in the final answer should correspond to the last full column of significant figures among the numbers. An addition example follows:

$$\begin{array}{r} 33.842 \\ 361.3 \\ 81.24 \\ \hline 476.382 \end{array} = 476.4$$

7. When measurements are recorded in one set of units (such as 23,664 sq ft, which has five significant figures) it can be converted to another set of units with the same number of significant figures (such as 0.54325 acre).

2-12 FIELD NOTES

Perhaps no other phase of surveying is as important as the proper recording of field data. No matter how much care is used in making field measurements, the effort is wasted unless a clear and legible record is kept of the work. For this reason we shall go to some length to emphasize this aspect of surveying.

For most of surveying history field data were hand printed in special field books. In the past few decades, however, this procedure has dramatically changed as automatic data collectors interfaced with modern surveying instruments have become available. Nevertheless the use of hand printed notes is still an important topic and will probably remain so for a good many more years. The use of automatic data collectors is discussed in the next section of this chapter while hand printed notes are discussed in this section.

The cost of keeping a surveying crew or party in the field is appreciable. Including food and lodging it may run as high as $1000 per day. If the notes are wrong, incomplete, or confusing, much time and money have been wasted and the entire party may have to return to the job to repeat some or all of the work.

It is absolutely essential for good field notes to be prepared to document surveying work. The notes may be made manually and/or electronically. Both methods are discussed herein. *Note keeping is such an important part of surveying that the most competent person in the party (i.e., the party chief) usually handles the task.*

Field books may be bound or loose-leaf, but bound books are usually used. Although it seems that the loose-leaf types have all the advantages (such as the capability of having their pages rearranged, filed, moved in with other sets of notes, shifted back and forth between the field and the office.), the bound ones are more commonly used because of the possibility of losing some of the loose-leaf sheets. Long-life bound books capable of withstanding rough use and bad weather situations are the usual choice.

There are several kinds of field notebooks available, but the usual ones are 4 5/8 in. × 7 1/4 in., a size that can easily be carried in a pocket. This characteristic is quite important because the surveyor needs his hands for other work. Examples of field-book notes are shown throughout the text, the first one being Figure 3-1 in the next chapter. A general rule is that measured quantities are shown on the left-hand pages, and sketches and miscellaneous notes are shown on the right-hand pages.

In keeping notes the surveyor should bear in mind that on many occasions (particularly in large organizations) persons not familiar with the locality will make use of the notes. Some details may appear so obvious to the surveyor that he or she does not include them, but they may not be obvious at all to someone back in the office. Therefore, considerable effort should be made to record all the information necessary for others to understand the survey clearly. With practice the information needed in the field notes will be learned.

An additional consideration is that surveying notes are often used for purposes other than the one for which they were originally developed, and therefore they should be carefully preserved. To maintain good field notes is not an easy task. Students are often embarrassed in their first attempts at making accurate and neat notes, but with practice the ability can be developed. The following items are absolutely necessary for the successful recording of surveying information:

1. The name, address, and phone number of the surveyor should be printed in India ink on the inside and outside of the field-book cover.
2. The title of the job, date, weather, and location should be recorded. When surveying notes are being used in the office, it may be helpful to know something of the weather conditions at the time the measurements were taken. This information will often be useful in judging the accuracy of a particular survey. Was it 110°F or –10°F? Was it raining? Were strong winds blowing? Was it foggy or dusty or snowing?
3. The names of the party members, together with their assignments as instrument man ($\overline{\wedge}$), rodman (Ø), notekeeper (N), and so on should be recorded. As a minimum the first initial and full last name of each person should be provided. Sometimes court cases require that these persons be interviewed many years after a survey is done.
4. Field notes should be organized in a form appropriate to the type of survey. Because other people may very well use these notes, generally standard forms are used for each of the different types of surveys. If each surveyor used his or her own individual forms for all surveys, there would be much confusion back in the office. Of course, there are situations where the notekeeper will have to improvise with some style of nonstandard notes.
5. Measurements must be recorded in the field when taken and not trusted to memory or written on scraps of paper to be recorded at a later date. Sometimes it is necessary to copy information from other field notes. In such cases the word "COPY" should be clearly marked on each page with a note giving the source.
6. Frequent sketches are used where needed for clarity, preferably drawing lines with a straightedge. Because field books are relatively inexpensive compared to the other costs of surveying, crowding of sketches or other

data does not really save money. (The sketches need not always be drawn to scale, as distorted sketches may be better for clearing up questions.)

7. Field measurements must not be erased when incorrect entries are made. A line should be drawn through the incorrect number without destroying its legibility and the corrected value written above or below the old value. Erasures cause suspicion that there has been some dishonest alteration of values, but a crossed-out number is looked upon as an open admission of a blunder. (Imagine a property case coming up for court litigation and a surveying notebook containing frequent erasures being presented for evidence.) It is a good idea to use a red pencil for making additions to the notes back in the office to distinguish them clearly from values obtained in the field.
8. Notes are printed with a sharp medium-hard (3H or 4H) pencil so that the records will be relatively permanent and will not smear. Field books are generally used in damp and dirty situations and the use of hard pencils will preserve the notes. A clipboard and a clear plastic sheet of acetate can minimize field weather problems. The lettering used on sketches is to be arranged so that it can be read from the bottom of the page or the right-hand side.
9. The type of instrument and its number should be recorded with each day's work. It may later be discovered that the measurements taken with that instrument contained significant errors that could not be accounted for in any other way. With the instrument identified the surveyor may be able to go back to the instrument and make satisfactory corrections.
10. A few other requirements include numbering of pages, inclusion of a table of contents, drawing arrows on sketches indicating the general direction of north, and clear separation of each day's work by starting on a clean page each day. Should a particular survey extend over several days, cross-references may be necessary between the various pages of the project. The numbering system generally used for surveying notes is to record the page number in the upper right corner of each right-hand page. A single page number is used for both the right- and left-hand sides.

Finally, it is essential that notes be checked before leaving the site of the survey to make sure that all required information has been obtained and recorded. Many surveyors keep checklists in their field books for different types of surveys. Before they leave a particular job, they refer to the appropriate list for a quick check. Imagine the expense in time and money of having a survey crew make an extra trip to the site of a job some distance away in order to obtain one or two minor bits of information that had been overlooked.

2-13 ELECTRONICALLY RECORDED NOTES

We discuss at various places in this book the availability and uses of modern surveying instruments such as electronic distance-measuring devices, total station instruments, and similar equipment. Electronic data collectors are available which when used with these other modern instruments automatically display and record various distance and angle measurements at the touch of a button (Figure 2-6). Furthermore, it is often possible to transfer the data stored to field or office computers.

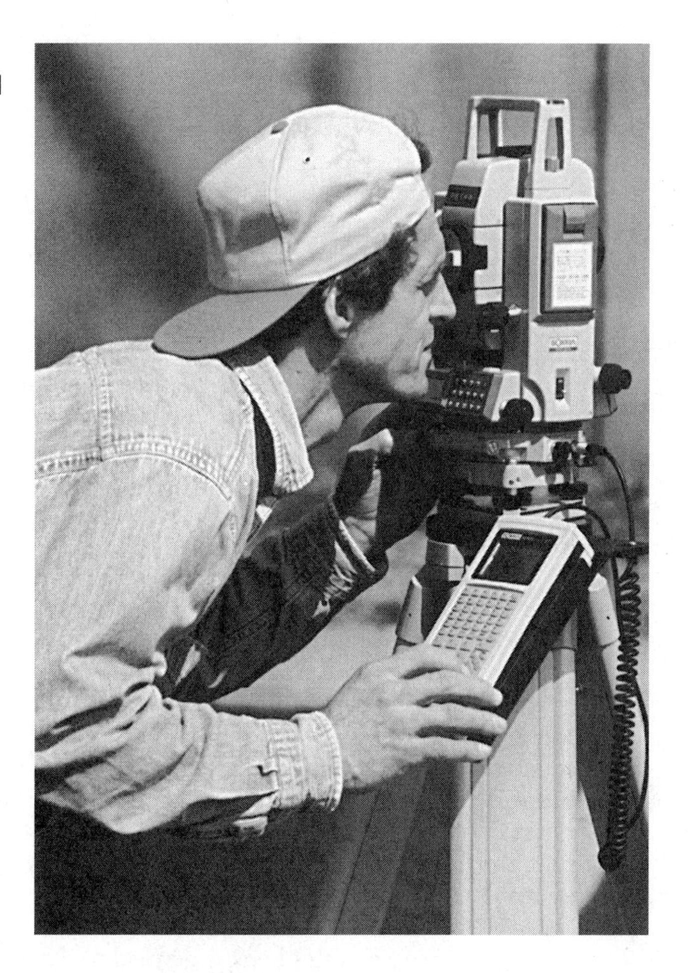

FIGURE 2-6 SDR 33 Electronic field book used with total station equipment which is discussed in Chapter 10. (Courtesy Sokkia Corporation.)

You can see that this capability enables us to reduce the blunders that might otherwise occur in recording information and in transferring it to the office.

It thus appears that data collectors make the notekeeper's work easier because he or she will be able to devote more time to preparing sketches and to writing descriptions and other nonnumerical information. There may not be as much extra time as you may think, however, because with modern equipment, measurements are made much more rapidly.

When automatic data collectors are used less time is required for processing and displaying field work. They are particularly useful when a large amount of data is being collected. A disadvantage, however, is that sketches cannot be placed in them and transferred to computers.

Data collectors are about the size of pocket calculators. The surveyor uses a keyboard to enter the usual information as to date, time, weather, instrument numbers, and so on. The usual data collector has been programmed so as to allow its user to follow a certain set of steps in inputting the needed information for a particular type of survey.

There are some potential problems with automatically recorded notes in relation to legal situations. For instance, it is not yet clear whether the courts, which readily accept hand-prepared field books, will accept magnetic tapes. Furthermore, there are problems with the potentials of deliberate altering of the tapes and their possible inadvertent destruction by human mistakes or by other electronic equipment.

2-14 OFFICE WORK AND DIGITAL COMPUTERS

Field surveying measurements provide the basis for large amounts of office work. Some of this paperwork includes precision computations, preparation of property drawings called plats, calculations for and drawings of topographic maps, and computation of earthwork volumes. A large percentage of these items are commonly handled today with digital computers. Once a surveyor has used a successful computer program with a personal computer or a programmed desk or pocket calculator for some part of office work (such as the calculation of land areas), he or she will be reluctant to ever again make those calculations with an ordinary hand-held calculator.

2-15 PLANNING

If we are to perform accurate surveys, we need to employ good equipment, good procedures, and good planning. Accurate surveys can be made with old and perhaps out-of-date equipment, but time and money can be saved with modern instruments. Good planning is the most important item necessary for achieving economy and is also very important for achieving accuracy.

Perhaps no topic in this book is of more importance than the few words presented in this chapter on the subject of planning. The underlying thought in the author's mind as he prepared this chapter was planning. Not only was he thinking of organizing the parties and equipment but also selecting the procedures to be used. In other words, if we want a survey to be done with a precision of 1/30,000, what procedures and equipment will we have to use?

A great deal of space in this book is devoted to the terms *accuracy* and *precision.* A thorough understanding of these subjects will enable the surveyor to improve his or her work and reduce the costs involved by careful planning.

The planning of a survey for a particular project includes the selection of equipment and methods to be used, reconnaissance of the area for existing surveying monuments, selection of possible locations for surveying stations, and so on.

The longer a construction project lasts, the more expensive it becomes. The owner's money is tied up longer, resulting in increases in interest costs and postponement of the time when business operations can begin and produce some return on investment.

Surveying, which is one of the many phases of a construction project, needs to be done as quickly as possible for the reasons mentioned above. The cost of accurate surveying is frequently 1% to 3% of the overall cost of the construction project—but if it is done poorly, the percentages may be many times these values. Thus carefulness in a survey must not be slighted, even though speed may be essential.

In general, surveyors would like to obtain precisions that are better than those required for the particular project at hand. However, economic limitations (i.e., what the employer is willing to pay) will frequently limit those desires.

Problems

2-1. The sides of a closed figure have been measured and found to have a total length of 5633.49 ft. If the total error in the measurements is estimated to be equal to 0.39 ft, what is the precision of the work? (Ans.: 1/14,445)

2-2. Repeat Problem 2-1 if the total distance is 4133.64 ft and the estimated total error is 0.15 ft.

2-3. What is a probable or 50% error?

2-4. What is the law of compensation relating to errors?

2-5. A quantity was measured 10 times with the following results: 3.751, 3.781, 3.755, 3.749, 3.750, 3.747, 3.748, 3.754, 3.746, and 3.745 ft. Determine the following:

a. Most probable value of the measured quantity.

b. Probable error of a single measurement.

c. 90% error.

d. 95% error. (Ans.: 3.753, ±0.007, ±0.017, and ±0.021)

2-6. The same questions apply as for Problem 2-5, but the following 12 quantities were measured: 157.20, 157.12, 157.16, 157.22, 157.28, 157.24, 157.22, 157.21, 157.15, 157.18, 157.31, and 157.25 ft.

2-7. A distance is measured to be 842.32 ft with an estimated standard deviation of ±0.16 ft. Determine the probable errors and the estimated probable precision at 2σ and at 3σ. (Ans.: ±0.11ft, ±0.32ft, 1/2632, 1/1755).

2-8. The following six independent length measurements were made (in feet) for a line: 642.349, 642.396, 642.381, 642.376, 642.368, and 642.343. Determine:

a. The most probable value.

b. The standard deviation of the measurements.

c. The error at 3.29σ.

For Problems 2-9 to 2-11, answer the same questions as for Problem 2-8.

Problem 2-9	Problem 2-10	Problem 2-11
196.349	202.62	754.36
196.506	202.28	754.35
196.339	202.78	754.39
196.362	202.61	754.37
196.347	202.63	754.40
196.341	202.72	754.29
196.352	202.56	754.33
196.501	202.65	754.27
		754.32
		754.42

(2-9 Ans.: 196.387, ±0.072, ±0.237) (2-11 Ans.: 754.35, ±0.05, ±0.16)

2-12. If the accidental error is estimated to be ±0.006 ft for each of 24 separate measurements, what is the total estimated error?

2-13. A random or accidental error of ±0.004 ft is estimated for each of 32 length measurements. These measurements are to be added together to obtain the total length.

a. What is the estimated total error? (Ans.: ± 0.023 ft)

b. What is the estimated total error if the random error is estimated to be ±0.007 ft per length? (Ans.: ±0.040ft)

2-14. It is assumed that the probable random error in taping 100 ft is ±0.020 ft Two distances are measured with this tape with the following results: 1416.33 and 1781.11 ft. Determine the probable total error in measuring each of the lines and then the total probable error for the two sides together.

2-15. A series of elevations was determined. The accidental error in taking each reading is estimated to be ± 0.005 ft. What is the total estimated error if 28 readings were taken? (Ans.: ± 0.026 ft)

2-16. A surveying crew or party is capable of taping distances with an estimated probable accidental error of ± 0.003 m for each 30-m distance or tape length. What total estimated probable accidental error should be expected if a total distance of 1900m is to be measured?

2-17. Two sides of a rectangle were measured as being 162.32 ft ± 0.03 ft and 207.46 ft ± 0.04 ft. Determine the area of the figure and the probable error of the area. Assume probable error = $\sqrt{A^2 E_b^2 + B^2 E_a^2}$ where A and B are lengths of sides. (Ans.: 33,674.9 sq ft ± 8.99 sq ft)

2-18. It is desired to tape a distance of 1100 ft with a total standard error of not more than ±0.10 ft.

a. How accurately should each 100-ft distance be measured so that the permissible value is not exceeded?

b .How accurately would each 100-ft distance have to be measured so that the 95% error would not exceed ±0.12 ft in a total distance of 2000 ft?

2-19. For a nine-sided closed figure the sum of the interior angles is exactly 1620°. It is specified for a survey of this figure that if these angles are measured in the field. their sum should not miss 1620° by more than ±1′. How accurately should each angle be measured? (Ans.: ± 20″)

CHAPTER 3

Distance Measurement

3-1 INTRODUCTION

One of the most basic operations of surveying is the measurement of distance. In surveying, the distance between two points is understood to be a horizontal distance. The reason for this is that most of the surveyor's work is plotted on a drawing as some type of map. A map, of course, is plotted on a flat plane and the distances shown thereon are horizontal projections. Land areas are computed on the basis of the same horizontal measurements. This means that if a person wants to obtain the largest amount of actual land surface area for each acre of land purchased, it should be purchased on the side of a very steep mountain.

Early measurements were made in terms of the dimensions of parts of the human body such as cubits, fathoms, and feet. The cubit (the unit Noah used in building his boat) was defined as the distance from the tip of a man's middle finger to the point of his elbow (about 18 in.); a fathom was the distance between the tip of a man's middle fingers when his arms were outstretched (approximately 6 ft). Other measurements were the foot (the distance from the tip of a man's big toe to the back of his heel) and the pole or rod or perch (the length of the pole used for driving oxen, later set at 16.5 ft). In England the "rood" (rod or perch) was once defined as being equal to the sum of the lengths of the left feet of 16 men, whether they were short or tall, as they came out of church one Sunday morning.[1]

Today all of the countries of the world except Burma, Liberia, and the United States use the metric system for their measurements. This system was developed in France in the 1790s. Shortly thereafter Thomas Jefferson, when he was the United States Secretary of State, recommended that the United States use the metric system but a bill to do that was defeated by one vote in Congress.[2]

The meter is supposedly equal to 1/10,000,000 of the distance from the equator to the north pole. Its application in the United States in the past has been limited almost entirely to geodetic surveys. In 1866 the U.S. Congress legalized the use of the metric system by which the meter was defined as being equal to 3.280833 ft (or

[1]C. M. Brown, W. G. Robillard, and D. A. Wilson, *Evidence and Procedures for Boundary Location,* 2nd ed. (New York; John Wiley & Sons, Inc., 1981), p. 268.
[2]P.R. Wolf and C.D. Ghilani, Elementary Surveying An Introduction to Geomatics, 10th ed. (Upper Saddle River, N.J.: Prentice Hall, 2002), p. 26.

39.37 in.) and 1 inch was equal to 2.540005 centimeters. These values were based on the length at 0°C of an International Prototype Meter bar consisting of 90% platinum and 10% iridium which is kept in Sèvres, France, near Paris. In accordance with the treaty of May 20, 1875, the National Prototype Meter 27, identical with the International Prototype Meter, was distributed to various countries. Two copies are kept by the U.S. Bureau of Standards at Gaithersburg, Maryland.

Reference to the prototype bar was ended in 1959 and the meter was redefined as being equal to 1,650,763.73 wavelengths of orange-red krypton gas, equal to 3.280840 ft. In 1983 the meter definition was changed again to its present value: the distance traveled by light in 1/229,792,458 sec. Supposedly, it is now possible to define the meter much more accurately. Furthermore, this enables us to use time (our most accurate basic measurement) to define length.[3]

Based on these values, before 1959 the foot was equal to 1/3.280833 = 0.3048006 m, whereas since that date it equals 1/3.280840 = 0.3048000 m. This very slight difference (about 0.0105 ft in 1 mile) is of no significance for plane surveys but does affect geodetic surveys, which often extend for many miles. As a large number of our geodetic surveys were performed before 1959, the earlier value (i.e., 1 m = 3.280833 ft) is used to define the *U.S. survey foot.*

Since September 30, 1992, all federal agencies of the United States have been required to use metric units for procurement, grants, and business-related activities. Since January 1994 the SI (System International) system must be used for the design of all new federal facilities. The projects most affected are control surveys for highways, bridges, dams, utilities, and government facilities.[4]

Although the various cities, counties, and states of the United States are not legally required at the present time to convert to the metric system, there are strong incentives provided by the federal government to push them in that direction. These incentives are in the form of certain matching funds provided by the federal government which will not be available unless the metric system is used on the projects involved. As a result it appears that our country will convert to SI units in the not-too-distant future. Some construction trades in the United States have already made the move.

To the surveyor the most commonly used metric units are the meter (m) for linear measure, the square meter (m^2) for areas, the cubic meter (m^3) for volumes, and the radian (rad) for plane angles. In many countries the comma is used to indicate a decimal; thus to avoid confusion in the metric system, spaces rather than commas are used. For a number having four or more digits, the digits are separated into groups of threes, counting both right and left from the decimal. For instance 4,642,261 is written as 4 642 261, and 340.32165 is written as 340.321 65.

To change surveying to metric units may seem at first glance to be quite simple. In a sense this is true in that if the surveyor can measure distance with a 100-ft tape, he or she can do just as well with a 30-m tape using the same procedures. Furthermore, almost all electronic distance-measuring devices provide distances in feet or meters as desired. However, we have several hundred years of land descriptions recorded in the English system of units and stored in our various courthouses and other archives. Future generations of surveyors will therefore never get away completely from the English system of units. As an illustration, today's surveyor still fre-

[3]R. C. Brinker and R. Minnick, eds., *The Surveying Handbook,* 2nd ed., (New York: Van Nostrand Reinhold Company, Inc., 1995), p. 43.

[4]A. Lane, "Metric Mandate," *Professional Surveyor,* Mar./Apr. 1992, vol. 12, no. 2, pp. 13–14.

quently encounters old land descriptions made in terms of so many chains (reference being made to the 66-ft chain).

There are several methods which can be used for measuring distance. These include pacing, odometer readings, stadia, taping, electronic distance measuring devices, and the global positioning system. The most common ones used today for surveying purposes are EDMs, GPS, and taping. The GPS method is being used more and more due to its accuracy and range. There is another method which should be included with this list and that is estimating. It is always wise for the surveyor to estimate distances in his or her head when possible. Such a procedure will often pick up large mistakes made in measuring or recording. All of the methods listed here are described in subsequent sections of this book. The method to be used for a particular survey is dependent on the purpose of the work. An experienced surveyor will be able to select the best method to be used considering the objectives of the survey and the cost involved.

Occasionally other units of length measurement are encountered in the United States. These include the following:

1. The *furlong* is defined as the length of one side of a square 10-acre field and equals 1/8th of a mile or 660 ft or 40 rods.
2. The *vara* is a unit of Spanish origin. In Spain it equals 32.8748 in., but in California it is 32.372 in., in Texas it is 33.3333 in., and in Florida it is 32.99312 in.
3. The *arpent* is a term which was used in land grants from the French kings. It was used to refer to either acreage or to length. The square arpent equals about 0.85 acres while the linear arpent equals one side of a square arpent. It is about 191.99 ft in some states (Alabama, Florida, Mississippi, and Louisiana), while in Arkansas and Missouri it is 192.50 ft.

3-2 PACING

The ability to pace distances with reasonable precision is useful to almost anyone. The surveyor in particular can use pacing to make approximate measurements quickly or to check measurements made by more precise means. By so doing he or she will often be able to detect large mistakes.

A person can determine the value of his or her average pace by counting the number of paces necessary to walk a distance that has been previously measured more accurately (e.g., with a steel tape). Here the author defines the pace as one step. Sometimes the term *stride* is used. A stride is considered to equal two steps or two paces. For most persons pacing is done most satisfactorily when taking natural steps. Others like to try to take paces of certain lengths (e.g., 3 ft), but this method is tiring for long distances and usually gives results of lower precision for short or long distances. As horizontal distances are needed, some adjustments should be made when pacing on sloping ground. Paces tend to be shorter on uphill slopes and longer on downhill ones. Thus the surveyor would do well to measure his or her pace on sloping ground as well as on level ground.

With a little practice a person can pace distances with a precision of roughly 1/50 to 1/200, depending on the ground conditions (slope, underbrush). For distances of more than a few hundred feet, a mechanical counter or *pedometer* can be used. Pedometers can be adjusted to the average pace of the user and automatically record the distance paced.

The notes shown in Figure 3-1 are presented as a pacing example for students. A five-sided figure was laid out and each of the corners marked with a stake driven into the ground. The average pace of the surveyor was determined by pacing a known distance of 400 ft, as shown at the bottom of the left page of the field notes. Then the sides of the figure were paced and their lengths calculated. This same figure or traverse is used in later chapters as an example for measuring distances with a steel tape, measuring angles with a total station, computing the precision of the latter measurements, and calculating the enclosed area of the figure.

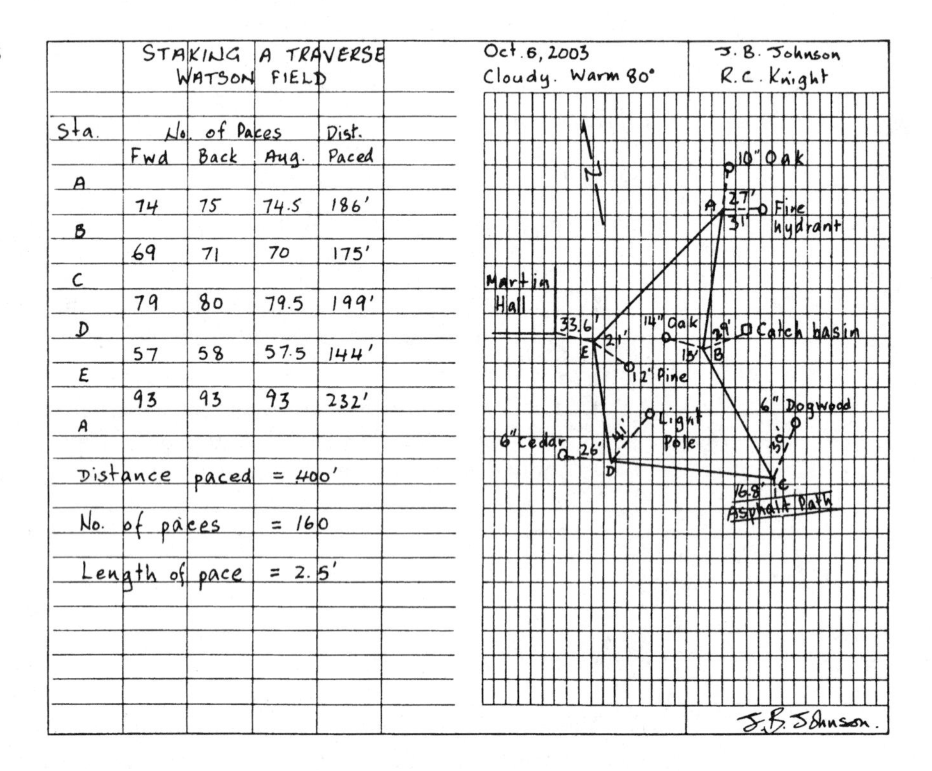

STAKING A TRAVERSE
WATSON FIELD

Sta.	No. of Paces Fwd	Back	Avg.	Dist. Paced
A				
	74	75	74.5	186′
B				
	69	71	70	175′
C				
	79	80	79.5	199′
D				
	57	58	57.5	144′
E				
	93	93	93	232′
A				

Distance paced = 400′
No. of paces = 160
Length of pace = 2.5′

FIGURE 3-1 Field notes for staking a traverse.

A note should be added here regarding the placement of the stakes. It is good practice for the student to record in his or her field book information on the location of the stakes. The position of each stake should be determined in relation to at least two prominent objects, such as trees, walls, sidewalks, so that there will be no difficulty in relocating them a week or a month later. This information can be shown on the sketch as illustrated in Figure 3-1.

You will note that a north arrow is shown in the figure. Every drawing of property should include such a direction arrow, as it will be of help (as to directions and positions) to anyone using the sketch.

3-3 ODOMETERS AND MEASURING WHEELS

Distances can be roughly measured by rolling a wheel along the line in question and counting the number of revolutions. An *odometer* is a device attached to the wheel (similar to the distance recorder used in a car) which does the counting and from the circumference of the wheel converts the number of revolutions to a distance. Such a device provides a precision of approximately 1/200 when the ground is smooth, along a highway, but results are much poorer when the surface is irregular.

The odometer may be useful for preliminary surveys, perhaps when pacing would take too long. It is occasionally used for initial route-location surveys and for quick checks on other measurements. A similar device is the *measuring wheel,* which is a wheel mounted on a rod. Its user can push the wheel along the line to be measured. It is frequently used for curved lines. Some odometers are available that can be attached to the rear end of a motor vehicle and used while the vehicle is moving at a speed of several miles per hour.

3-4 TACHYMETRY

The term *tachymetry* or *tacheometry,* which means "swift measurements," is derived from the Greek words *takus,* meaning "swift," and *metron,* meaning "measurement." Actually, any measurement made swiftly could be said to be tacheometric, but the generally accepted practice is to list under this category only measurements made with subtense bars or by stadia. Thus the extraordinarily fast electronic distance-measuring devices are not listed in this section.

Subtense Bar (Obsolete)

A tachymetric method that was occasionally used until a few decades ago for rural property surveys made use of the subtense bar (Figure 3-2). This obsolete device has been replaced by electronic distance-measuring devices. In Europe, where the method was most commonly used, a horizontal bar with sighting marks on it, usually located 2 m apart, was mounted on a tripod. The tripod was centered over one end of the line to be measured and the bar was leveled and turned so that it was made roughly perpendicular to the line.

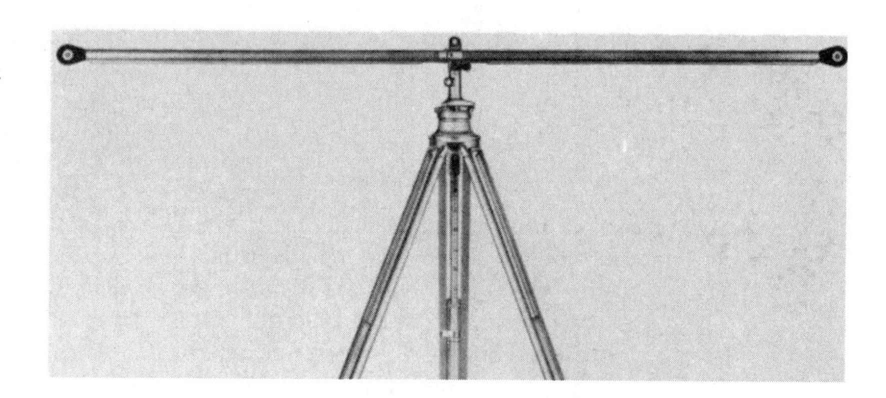

FIGURE 3-2 Subtense bar. (Courtesy of Leica, Inc., formerly Kern Instruments and Wild Heerbrugg Instruments.)

A theodolite (described in Appendix D) was set up at the other end of the line and sighted on the subtense bar. It was used to align the bar precisely in a position perpendicular to the line of sight by observing the sighting mark on the subtense bar. This mark could be seen clearly only when the bar was perpendicular to the line of sight. The angle subtended between the marks on the bar was carefully measured (preferably with a theodolite measuring to the nearest second of arc) and the distance between the ends of the line was computed. The distance D from the theodolite to the subtense bar was computed from the following expression:

$$D = \frac{1}{2} S \cot \frac{\alpha}{2} \quad \text{or} \cot \frac{\alpha}{2} \quad \text{since } S = 2\text{m}$$

in which S is the distance between the sighting marks on the bar and α is the subtended angle. These values are shown in Figure 3-3.

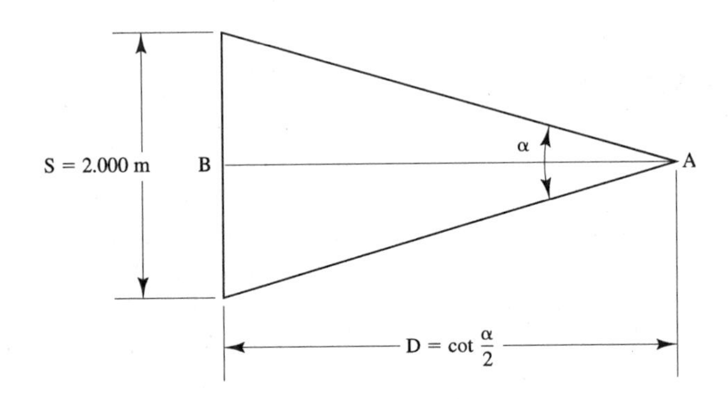

FIGURE 3-3 Distance measurement with a subtense bar.

For reasonably short distances, say less than 500 ft, errors were small and a precision of 1/1000 to 1/5000 was ordinarily obtained. The subtense bar was particularly useful for measuring distances across rivers, canyons, busy streets, and other difficult areas. It had an additional advantage in that the subtended angle was independent of the slope of the line of sight; thus the horizontal distance was obtained directly and no slope correction had to be made.

Stadia

Although the subtense bar was occasionally used in the United States, the stadia method was far more common. Its development is generally credited to the Scotsman James Watt in 1771.[5] The word *stadia* is the plural of the Greek word *stadium,* which was the name given to a foot race track approximately 600 ft in length.

Transit and theodolite telescopes (discussed in Appendix D) are equipped with three horizontal cross hairs which are mounted on the cross-hair ring. The top and bottom hairs are called stadia hairs. The surveyor sights through the telescope and takes readings where the stadia hairs intersect a scaled rod. The difference between the two readings is called the *rod intercept.* The hairs are so spaced that at a distance of 100 ft their intercept on a vertical rod is 1 ft; at 200 it is 2 ft; and so on, as illustrated in Figure 3-4. To determine a particular distance, the telescope is sighted on the rod and the difference between the top and bottom cross-hair readings is multiplied by 100. When one is working on sloping ground, a vertical angle is measured

[5] A. R. Legault, H. M. McMaster, and R. R. Marlette, *Surveying* (Englewood Cliffs, NJ: Prentice-Hall, Inc., 1956), p. 39.

and is used for computing the horizontal component of the slope distance. These measurements can also be used to determine the vertical component of the slope distance or the difference in elevation between the two points.

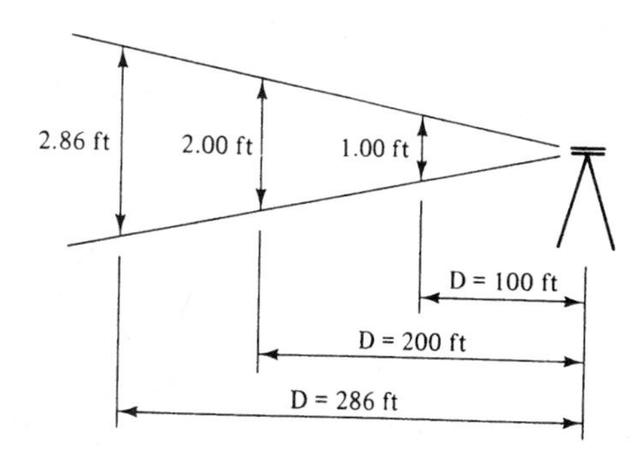

FIGURE 3-4 Stadia readings.

When the stadia method was used in surveying practice it had its greatest value in locating details for maps. Sometimes it was also used for making rough surveys and for checking more precise ones. Precisions of the order of 1/250 to 1/1000 were obtained. Such precisions are not satisfactory for land surveys. *Though the use of the stadia method is briefly described in Chapter 14 of this text it has been made obsolete by total station instruments as described in Chapters 10 and 14.*

The same principle can be used to estimate the heights of buildings, trees, or other objects, as illustrated in Figure 3-5. A ruler is held upright at a given distance such as arm's length (about 2 ft for many people) in front of the observer's eye so that its top falls in line with the top of the tree. Then the thumb is moved down so that it coincides with the point where the line of sight strikes the ruler when the observer looks at the base of the tree. Finally, the distance to the base of the tree is measured by pacing or taping. By assuming the dimensions shown in Figure 3-5, the height of the tree shown can be estimated as follows:

$$\text{height of tree} = \left(\frac{100}{2}\right)\left(6\,\text{in}\right) = 300\,\text{in} = 25\,\text{ft}$$

Artillery students were formerly trained in a similar fashion to estimate distances by estimating the size of objects when they were viewed between the knuckles of their hands held at arm's length.

FIGURE 3-5 Estimating the height of a tree.

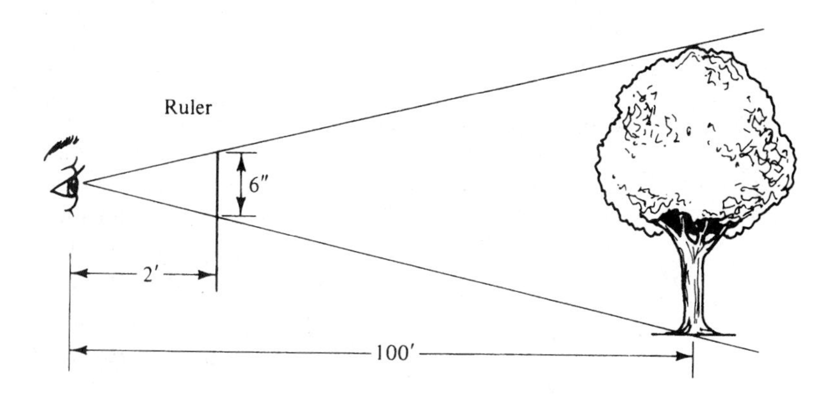

3-5 TAPING OR CHAINING

For many centuries, surveyors measured distances with ropes, lines, or cords that were treated with wax and calibrated in cubits or other ancient units. These devices are obsolete today, although precisely calibrated wires are sometimes used. For the first two-thirds of the 20th century the 100-ft steel ribbon tape was the common device used for measuring distances. Such measuring is often called *chaining*, a carryover name from the time when Gunter's chain was introduced. The English mathematician Edmund Gunter (1581–1626) invented the surveyor's chain (Figure 3-6) early in the seventeenth century. His chain, which was a great improvement over the ropes and rods used up until that time, was available in several lengths, including 33 ft, 66 ft, and 100 ft. The 66-ft length was the most common. (Gunter is also credited with the introduction of the words *cosine* and *cotangent* to trigonometry, the discovery of magnetic variations [discussed in Chapter 9], and other outstanding scientific accomplishments.[6])

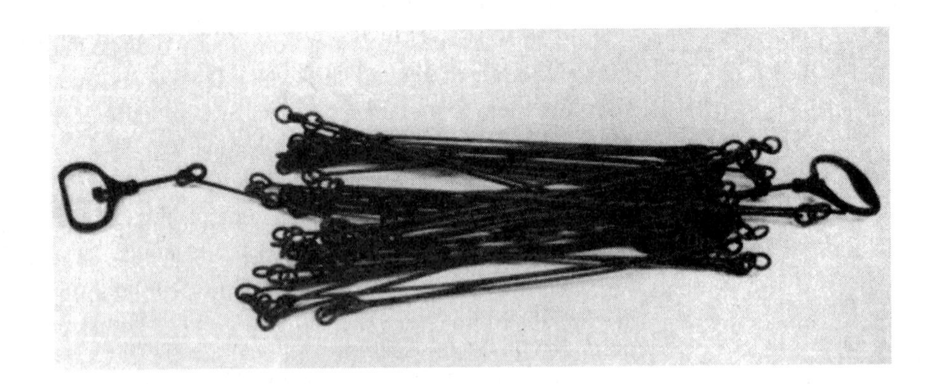

FIGURE 3-6
Old 33-ft chain.

The 66-ft chain, sometimes called the four-pole chain, consisted of 100 heavy wire links each 7.92 in. in length. In studying old deeds and plats the surveyor will often find distances measured with the 66-ft chain. He or she might very well see a distance given as 11 ch. 20.2lks. or 11.202 ch. The usual area measurement in the United States is the acre, which equals 10 square chains. This is equivalent to 66 ft by 660 ft, or 1/80 mi by 1/8 mi. or 43,560 sq ft.

Many of the early two-lane roads in the United States and Canada were laid out to a width of one chain, resulting in a 66-ft right-of-way. Steel tapes came into general use around the beginning of the twentieth century. They are available in lengths from a few feet to 1000 ft. For ordinary conditions, precisions of from 1/1000 to 1/5000 can be obtained, although much better work can be done by using procedures to be described later.

The general public, riding in their cars and seeing surveyors measuring distances with a steel tape, might think: "Anybody could do that. What could be simpler?" The truth is, however, that the measurement of distance with a steel tape, though simple in theory, is probably the most difficult part of good surveying. The efficient use of today's superbly manufactured surveying equipment for other surveying functions such as the measurement of angles is quickly learned. But correspondingly precise distance measurement with a steel tape requires thought, care, and experience.

[6]C. M. Brown, W. G. Robillard, and D. A. Wilson, *Evidence and Procedures for Boundary Location,* 2nd ed. (New York: John Wiley & Sons, Inc., 1981), p. 151.

In theory, it is simple, but in practice it is not so easy. Sections 3-9 to 3-11 are devoted to a detailed description of taping equipment and field measurements made with steel tapes.

3-6 ELECTRONIC DISTANCE MEASUREMENTS

Sound waves have long been used for estimating distances. Nearly all of us have counted the number of seconds elapsing between the flash of a bolt of lightning and the arrival of the sound of thunder and then multiplied the number of seconds by the speed of sound (about one-fifth of a mile per second). The speed of sound is 1129 ft/sec at 70°F and increases by a little more than 1 ft/sec for each degree Fahrenheit increase in temperature. For this reason alone, sound waves do not serve as a practical means of precise distance measurement because temperatures along a line being measured would have to be known to almost the nearest 0.01°F.

In the same way some distances for hydrographic surveying were formerly estimated by firing a gun and then measuring the time required for the sound to travel to another ship and be echoed back to the point of firing. Ocean depths are determined with depth finders which use echoes of sound from the ocean bottom. Sonar equipment makes use of supersonic signals echoed off the hulls of submarines to determine underwater distances.

It has been discovered during the past few decades that the use of either light waves, electromagnetic waves, infrared, or even lasers offers much more precise methods of measuring distance. Although it is true that some of these waves are affected by changes in temperature, pressure, and humidity, the effects are small and can be accurately corrected. Under normal conditions the corrections amount to no more than a few centimeters in several miles. Numerous portable electronic devices making use of these wave phenomena have been developed that permit the measurement of distance with tremendous precision.

These devices have not completely replaced chaining or taping, but they are commonly used by almost all surveyors and contractors. Their costs are at a point where the average surveyor must use them to remain economically competitive. For those who prefer to lease, arrangements may be made with various companies.

Electronic distance-measuring instruments (EDMs) have several important advantages over other methods of measurements (Figure 3-7). They are very useful in measuring distances that are difficult to access, for example, across lakes and rivers, busy highways, standing farm crops, canyons, and so on. For long distances (say, several thousand feet or more), the time required is in minutes, not hours as would be required for a typical taping party. Two people, easily trained, can do the work better and faster than the conventional four-person taping crew.

With EDMs the time required for a light wave to be sent along the required path (or a microwave to be sent along the path and reflected back to the starting position) is measured. From this information the distance may be determined as described in Chapter 5.

Desirably surveyors will rely on tapes for distances less than 100.00 ft and use EDMs for greater distances. Although the precisions obtained for small distances (100 ft or less) are probably better when steel tapes are used, the average surveyor today with EDMs rarely bothers to use tapes for any distances short or long. The supposed error occurring when an EDM is used is something in the order of ±(5 mm + 4 ppm) with the fixed instrument error ±5 mm being constant whether a small or long distance is involved. The 5 mm part is much more important in its

FIGURE 3-7 WILD D12002 electronic distance measuring instrument. (Courtesy of Leica, Inc., formerly, Kern Instruments and Wild Heerbrugg Instruments.)

effect on precisions for short distances than for long ones, as is the error occurring due to the fact that reflectors cannot be set exactly over points. Table 5-1 in Chapter 5 provides a typical set of precisions obtained with EDMs for both short and long distances.

The surveyor of today and tomorrow should become proficient with the steel tape, even though it is highly probable that he or she will eventually use the tape only a very small percentage of the time. Otherwise, the chances of making blunders and large errors will be magnified when the tape is used.

3-7 GLOBAL POSITIONING SYSTEM

With the global positioning system (discussed in detail in Chapters 15 and 16) positions on the earth's surface can be quickly and accurately located by measuring distances to orbiting satellites. This is done by determining the time required for radio signals broadcast by the satellites to travel to the points in question. Once the positions of the various points (in coordinates or latitudes and longitudes) are determined, the distances between them can easily be computed with accuracies which can approach or even exceed 1/1,000,000.

3-8 SUMMARY OF MEASUREMENT METHODS

Table 3-1 presents a brief summary of the various methods for measuring distances. There is a great variation in the precision obtainable with these different methods and the surveyor will select one that is appropriate for the purposes of the particular survey.

TABLE 3-1 Methods for Measuring Distances

Method	Precision	Uses
Pacing	1/50 to 1/200	Reconnaissance and rough planning
Odometer	1/200	Reconnaissance and rough planning
Subtense bar	1/1000 to 1/5000	Seldom used in United States and then only when taping is not feasible because of terrain and when electronic distance-measuring devices are not available
Stadia	1/250 to 1/1000	Formerly used for mapping, rough surveys, and for checking more precise work
Ordinary taping	1/1000 to 1/5000	Ordinary land surveys and building construction (still used today for short distances)
Precision taping	1/10,000 to 1/30,000	Excellent land surveys, precise construction work, and city surveys (rarely used today)
Base-line taping	1/100,000 to 1/1,000,000	Formerly used for precise geodetic work performed by the National Geodetic Survey
Electronic distance measurement	1/20,000 to ±1/300,000	In the past use was primarily for precise government geodetic work but is today commonly used for all types of surveys including land development, land surveys, and precise construction work
Global positioning system	up to and >1/1,000,000	Established to enable planes, ships, and other military groups to quickly determine their positions; increasingly being used for locating important control points and in many other phases of surveying including construction.

3-9 EQUIPMENT REQUIRED FOR TAPING

A brief discussion of the various types of equipment normally used for taping is presented in this section, and the next two sections are devoted to the actual use of the tapes. (Many surveyors still refer to the steel tape as a "chain" and the less precise rollup or woven tape as a tape.) A taping or chaining party should have at least one 100-ft steel tape, two range poles, a set of 11 chaining pins, a 50-ft woven tape, two plumb bobs, and a hand level. These items are discussed briefly in the following paragraphs.

Steel Tapes

Steel tapes are most commonly 100 ft long, approximately 5/16 in. wide, 0.025 in. thick, and they weigh 2 or 3 lb. They are either carried on a reel or done up in 5-ft loops to form a figure 8, from which they are thrown into a convenient circle approximately 9 in. in diameter. These tapes are quite strong as long as they are kept straight, but if they are tightened when they have loops or kinks in them, they will break very easily. If a tape gets wet, it should be wiped with a dry cloth and then again with an oily cloth.

The ends of steel tapes are made with heavy brass loops that provide a place to attach leather thongs or tension handles, enabling the user to tighten or tension them firmly. (If the leather thong is missing or breaks, and if a tension handle is not available, one of the chaining pins can be inserted through the loop and used as a handle for tightening the tape.)

Surveyor's tapes are marked at the 1-ft points from 0 to 100 ft. Older tapes have the last foot at each end divided into tenths of a foot, but the newer tapes have an extra foot beyond 0 which is subdivided. Tapes with the extra foot are called *add* tapes; those without the extra foot are called *cut* tapes (see Figure 3-10). Several variations are available: for example, tapes divided into feet, tenths, and hundredths for their entire lengths; tapes with the 0- and 100-ft points about 1/2 ft from the ends, and so on. Needless to say, the surveyor must be completely familiar with the divisions of the tape and its 0- and 100-ft marks before he or she does any measuring.

Cut tapes were used for many decades by a large number of surveyors but a large percentage of those surveyors did not like them because their use frequently led to arithmetic mistakes. For example, it was rather easy to add the end reading instead of subtracting it. Mistakes seem to be less frequent when add tapes are used, and as a consequence, add tapes have become the common type manufactured.

Tapes can be obtained in various lengths other than 100 ft. The 300- and 500-ft lengths are probably the most popular. The longer tapes, which usually consist of 1/8 in.-wide wire bands, are divided only at the 5-ft points, to reduce costs. They are quite useful for rapid, precise measurements of long distances on level ground. The use of long tapes permits a considerable reduction in the time required for marking at tape ends, and it also virtually eliminates the accidental errors that occur while marking.

Although the 100-ft tapes have been used by surveyors for quite a few decades, it seems likely that they will, in the next few years, begin to be replaced by metric steel tapes. Probably these tapes will be available in lengths of 20, 30, and 50 m, and others. It is highly possible that the 30m (98.4-ft) length will become the common metric tape because its length is so close to that of the present 100-ft tape. Perhaps the metric tapes will be divided into decimeters throughout their lengths, with the end decimeter divided into millimeters.

For very precise taping the Invar tape, made with 65% nickel and 35% steel, was formerly used. Although this type of tape had a very low coefficient of thermal expansion (perhaps 1/30th or less of the values for standard steel tapes), it was rather soft, easily broken, and cost about ten times as much as regular steel tapes. Invar tapes were at one time used for precise geodetic work and as a standard for checking the lengths of regular steel tapes. The Lovar tape had properties and costs somewhere in between those of the regular steel tapes and Invar tapes. Some tapes were made with a graduated thermometer scale that corresponded to temperature expansions and contractions. With different temperatures the surveyor could use a different terminal mark on the tape, thus automatically correcting for the temperature change.

Fiberglass Tapes

In recent years fiberglass tapes made of thousands of glass fibers and coated with polyvinyl chloride have been introduced on the market. These less expensive and more durable tapes are available in 50-ft, 100-ft, and other lengths. they are strong and flexible and will not change lengths appreciably with changes in temperature and moisture. They may also be used with little hazard in the vicinity of electrical equipment. When tension forces of 5 lb or less are applied, tension corrections are probably not necessary, but for values greater than 5 lb, length corrections are necessary.

Range Poles

Range poles are used for sighting points, for marking ground points, and for lining up tapemen in order to keep them going in the right direction. They are usually from 6 to 10 ft in length and are painted with alternate bands of red and white to make them more easily seen. The bands are each 1 ft in length and the rods can therefore be used for rough distance measurements. They are manufactured from wood, fiberglass, or metals. The sectional steel tubing type is perhaps the most convenient because it can easily be transported from one job to another.

Taping Pins (Also Referred to as Chaining Pins or Taping Arrows)

Taping pins are used for marking the ends of tapes or intermediate points while taping. They are easy to lose and are generally painted with alternating red and white bands. If the paint wears off, they can be repainted any bright color or they can have strips of cloth tied to them which can readily be seen. The pins are carried on a wire loop which can conveniently be carried by a tapeman, perhaps by placing the loop around his or her belt.

Plumb Bobs

A plumb bob is a pear-shaped or globular weight which is suspended on a string or wire and used to establish a vertical line. Plumb bobs for surveying were formerly made of brass to limit possible interference with the compasses that were used on old surveying equipment. (Iron or steel plumb bobs could cause errors in compass readings.) Plumb bobs usually weigh from 6 to 18 oz and have sharp replaceable points and a device at the top to which plumb-bob strings may be tied. Very commonly, plumb bobs are fastened to a *gammon reel.* This is a device that provides easy up-and-down adjustment of the plumb bob, instant rewinding of the plumb-bob string, and a sighting target.

Woven Tapes

Woven tapes (Figure 3-8) are most commonly 50 ft in length with graduation marks at 0.25-in. intervals. They can be either nonmetallic or metallic. Nonmetallic tapes are woven with very strong synthetic yarns and are covered with a specific plastic coating that is not affected by water. Metallic tapes are made with a water-repellent fabric into which fine brass, bronze, or copper wires are placed in the lengthwise direction. These wires strengthen the tapes and provide considerable resistance to stretching. (Because of the metallic wires, they should not be used near electrical units. For such situations nonmetallic woven or fiberglass tapes should be used.) Nevertheless, since all woven tapes are subject to some stretching and shrinkage, they are not suitable for precise measurement. Despite this disadvantage, woven tapes are often useful and should be a part of a surveying party's standard equipment. They are commonly used for finding existing points, locating details for maps, and measuring in situations where steel tapes might easily be broken (as along

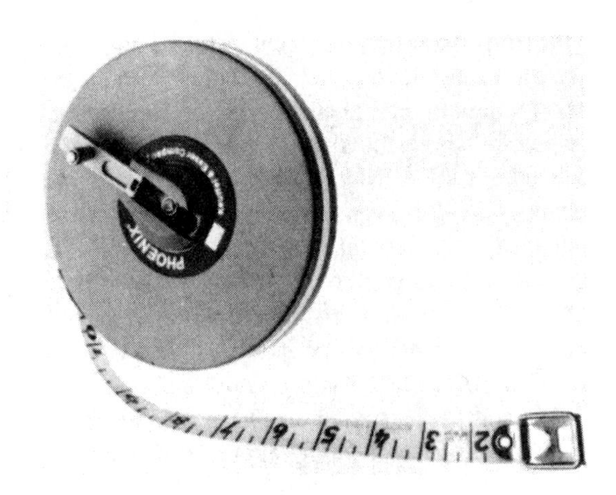

FIGURE 3-8 Woven tape. (Courtesy of Keuffel & Esser—a Kratos Company.)

highways) or when small errors in distance are not too important. Their lengths should be checked periodically or standardized with steel tapes.

Hand Levels

The hand level is a device which is very useful to the surveyer for helping him or her hold tapes horizontally while making measurements. It also may be used for the rough determination of elevation. It consists of a metal sighting tube on which is mounted a bubble tube (Figure 3-9). If the bubble is centered while sighting through the tube, the line of sight is horizontal. Actually, the bubble tube is located on top of the instrument and its image is reflected by means of a 45° mirror or prism inside the tube so that its user can see both the bubble and the terrain.

Spring Balances

When a steel tape is tightened, it will stretch. The resulting increase in length may be determined with the formula presented in Section 4-6. For average taping the tension applied can be estimated sufficiently to obtain desired precisions, but for very precise taping a *spring balance* or *tension handle* is necessary. The usual spring balance can be read up to 30 lb in 1/2-lb increments or up to 15 kg in 1/4-kg increments. If a tape is held above the ground, it will sag, with a resulting shortening of the distance between its ends. This effect may be counteracted by increasing the tension applied to the tape. This topic is also addressed in Section 4-6.

Clamping Handles

Leather thongs are usually placed through the loops provided at the ends of the tapes. With these thongs or with spring balances attached to the same loops, the tapes may be tensioned to desired values. When only partial lengths of tapes are used, it is somewhat difficult to pull the tape tightly. For such cases *clamping handles* are available. These have a scissors-type grip which enables one to hold the tape tightly without damaging it.

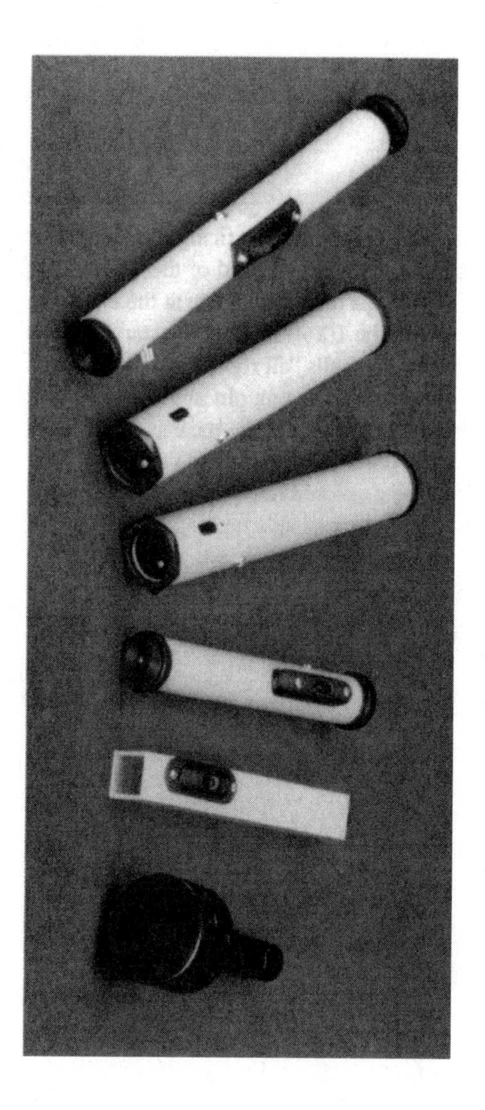

FIGURE 3-9 Hand levels. (Courtesy of Topcon America Corporation.)

3-10 TAPING OVER LEVEL GROUND

Ideally a steel tape should be supported for its full length on level ground or pavement. Unfortunately, such convenient conditions are usually not available because the terrain being surveyed may be sloping and/or covered with underbrush. If taping is done on fairly smooth and level ground where there is little underbrush, the tape can rest on the ground. The taping party consists of the head tapeman and the rear tapeman. The head tapeman leaves one taping pin with the rear tapeman for counting purposes and perhaps to mark the starting point. The head tapeman takes the zero end of the tape and walks down the line toward the other end.

When the 100-ft end of the tape reaches the rear tapeman, the rear tapeman calls "tape" or "chain" to stop the head tapeman. The rear tapeman holds the 100-ft mark at the starting point and aligns the head tapeman (using hand and perhaps voice signals) on the range pole which has been set behind the ending point. Ordinarily, this "eyeball" alignment of the tape is satisfactory, but use of a telescope is better and will result in improved precision. Sometimes there are places along a line where the tapeman cannot see the end point and there may be positions where they cannot see the signals of the instrumentman. For such cases it is necessary to set intermediate line points using the telescope before the taping can be started.

It is necessary to pull the tape firmly (see Sections 4-6 and 4-7). This can be done by wrapping the leather thong at the end of the tape around the hand, by holding a taping pin that has been slipped through the eye at the end of the tape, or by using a clamp. When the rear tapeman has the 100-ft mark at the starting point and has satisfactorily aligned the head tapeman, he or she calls "all right" or some other such signal. The head tapeman pulls the tape tightly and sticks a taping pin in the ground at right angles to the tape and sloping at 20° to 30° from the vertical. If the measurement is done on pavement, a scratch can be made at the proper point, a taping pin can be taped down to the pavement, or the point may be marked with a colored lumber crayon, called *keel.*

The rear tapeman picks up his taping pin and the head tapeman pulls the tape down the line, and the process is repeated for the next 100 ft. It will be noticed that the number of hundreds of feet which have been measured at any time equals the number of taping pins that the rear tapeman has in his or her possession. After 1000 ft has been measured, the head tapeman will have used his eleventh pin, and he calls "tally" or some equivalent word so that the rear tapeman will return the taping pins and they can start on the next 1000 ft. This discussion of counting taping pins is of little significance because almost any distance over 100 or 200 ft will today be measured with an EDM, as will a large percentage of distances of 100 or 200 ft or less. This topic is discussed in Chapter 5.

When the end of the line is reached, the distance from the last taping pin to the end point will normally be a fractional part of the tape. For older tapes, the first foot of the tape (from 0 to 1 ft) is usually divided into tenths, as shown in Figure 3-10. The head tapeman holds this part of the tape over the end point while the rear tapeman moves the tape backward or forward until he has a full foot mark at the taping pin.

FIGURE 3-10 Readings of cut and add tapes.

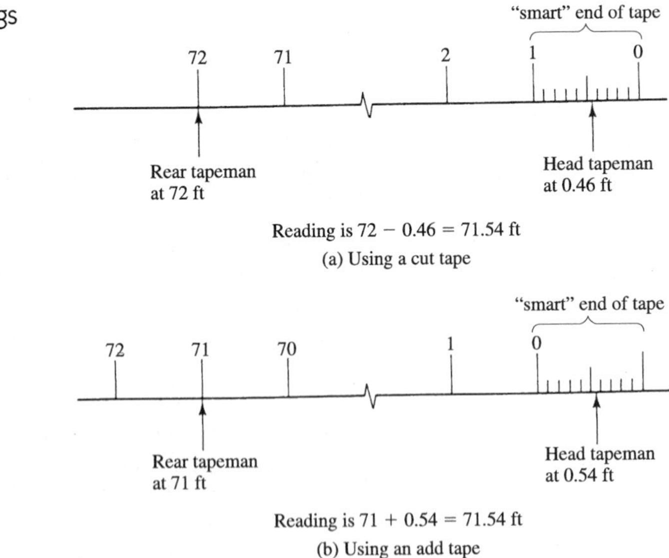

The rear tapeman reads and calls out the foot mark, say 72 feet, and the head tapeman reads from the tape end the number of tenths and perhaps estimates to the nearest hundredth, say 0.46, and calls this out. This value is subtracted from 72 ft to give 71.54 ft and the number of hundreds of feet measured before is added. These numbers and the subtraction should be called out so that the math can be checked by each tapeman.

For the steel tapes with the extra divided foot, the procedure is almost identical except that the rear tapeman would, for the example just described, hold the 71-ft mark at the taping pin in the ground. He or she would call out 71 and the head tapeman would read and call out plus 54 hundredths, giving the same total of 71.54 ft.

In review the rear tapeman reads the whole foot mark while the head tapeman reads the decimal value. Then the two readings are combined to establish the total distance.

A comment seems warranted here about practical significant figures as they apply to taping. If ordinary taping is being done and the total distance obtained for this line is 2771.54 ft, the 4 at the end is ridiculous and the distance should be recorded as 2771.5 or even 2772 ft because the work is just not done that precisely.

3-11 TAPING ALONG SLOPING GROUND OR OVER UNDERBRUSH

Should the ground be sloping there are three taping methods that can be used. The tape (1) may be held horizontally with one or both of the tapemen using plumb bobs as shown in Figures 3-11 and 3-12; or (2) it may be held along the slope, the slope determined, and a correction made to obtain the horizontal distance; or (3) the sloping distance may be taped, a vertical angle measured for each slope, and the horizontal distance later computed. This latter method is sometimes referred to as dynamic taping. Descriptions of each of these methods of measuring slope distances with a tape follow.

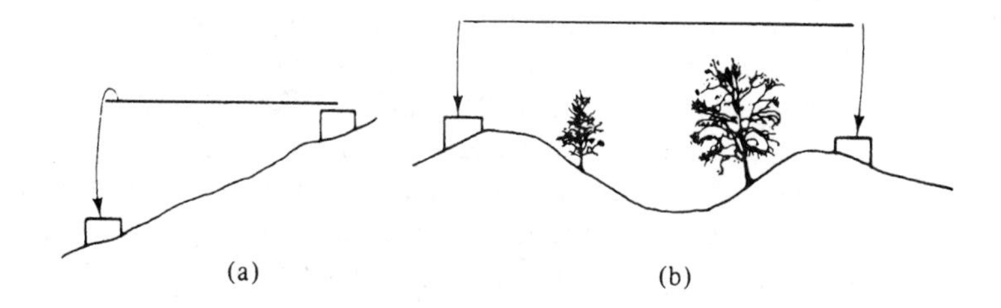

FIGURE 3-11 Holding the tape horizontally.

Holding the Tape Horizontally

Ideally, the tape should be supported for its full length on level ground or pavement. Unfortunately, such convenient conditions are often not available because the terrain being measured may be rough and covered with underbrush. For sloping, uneven ground or areas with much underbrush, taping is handled in a similar manner to taping over level ground. The tape is held horizontally, but one or both tapemen must use a plumb bob, as shown in Figure 3-12.

If taping is being done downhill, the rear tapeman will have to hold his or her plumb bob over the last point, while the head tapeman may be able to hold his or her end on the ground [Figure 3-12(a)]. If they are moving uphill, the head tapeman may be able to hold his or her end on the ground while the rear tapeman uses a

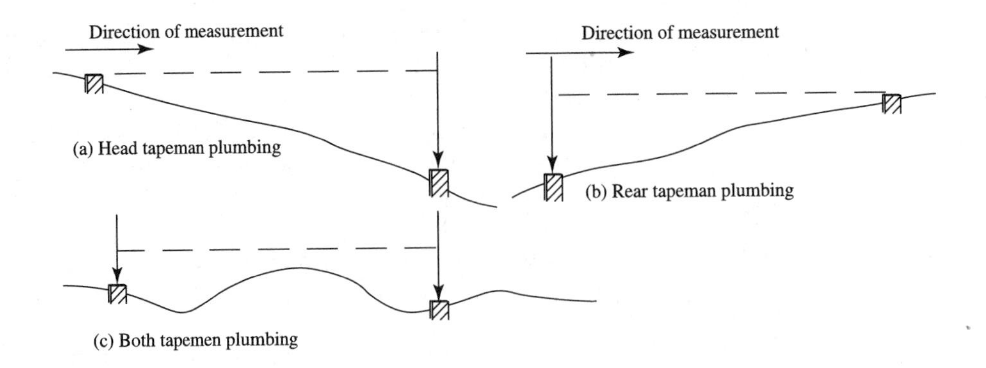

FIGURE 3-12 Holding the tape horizontally.

plumb bob [Figure 3-12(b)]. The head tapeman will always hold the so called "smart" end or divided end of the tape. Normally this is the zero end. Taping downhill is easier than taping uphill because the rear tapeman can hold the tape end on the ground at the last point instead of having to hold a plumb bob over the point while the head tapeman is pulling against him or her, as would be the case in taping uphill. If the measurement is over uneven ground or ground where there is considerable underbrush, both tapemen may have to use plumb bobs as they hold their respective ends of the tape above the ground [Figure 3-12(c)].

Considerable practice is required for a person to be able to do precise taping in rolling or hilly country. Although for many surveys the tapemen may estimate what is horizontal by eye, it pays to use a hand level for this purpose. Where there are steep slopes, it is difficult to estimate by eye when the tape is horizontal because the common tendency is for the downhill person to hold his or her end much too low, causing significant error. If a precision of better than approximately 1/2500 or 1/3000 is desired in rolling country, holding the tape horizontally by estimation will not be sufficient.

Another problem in holding the tape above the ground is the error caused by sagging of the tape (see Section 4-6). Note that both of these errors (tape not horizontal and sag) will cause the surveyor to get too much distance. In other words, either it takes more tape lengths to cover a certain distance or the surveyor does not move forward a full 100 ft horizontally each time that he or she uses the tape.

If the slope is less than approximately 5 ft per 100 ft (a height above the ground at which the average tapeman can comfortably hold the tape), the tapemen can measure a full 100-ft tape length at a time. If they are taping downhill, the head tapeman holds the plumb-bob string at the 0-end of the tape with the plumb bob a few inches above the ground. When the rear tapeman is ready at his or her end, the head tapeman is lined up on the distant point, and when the tape is horizontal and pulled to the desired tension, the head tapeman lets the plumb bob fall to the ground and sets a taping pin at that point.

In Figure 3-13 we see various methods of using a hand level. In parts (a) and (b) of the figure the tapeman on the left with the hand level bubble centered moves his head up or down until the line of sight through the hand level telescope hits the ground at the other tapeman's feet. This is the height at which he needs to hold his tape end (the other end being held on the ground) for the tape to be horizontal.

In part (c) the tapeman with the hand level stands with normal posture and sights (bubble centered) on the other tapeman. From the position where the line of sight strikes the other tapeman he can tell the height to which he needs to hold his tape end. For instance, if the line of sight hits the other tapeman's knee and they are both approximately the same height, the tapeman on the left is lower by the distance from his eye to his knee: that is, the difference in elevation shown in part (c) of the figure.

FIGURE 3-13 Using a hand level.

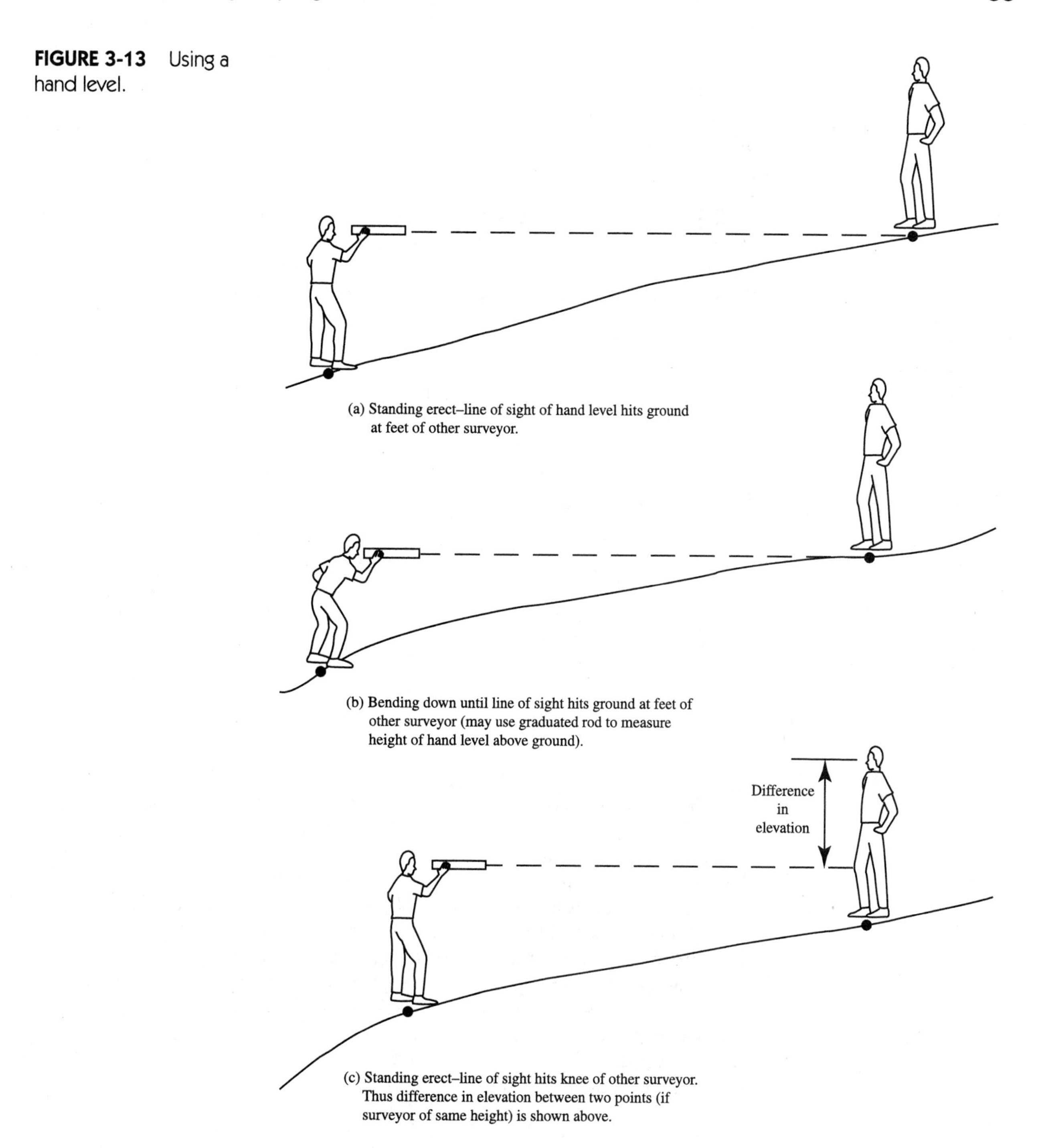

For slopes greater than approximately 5 ft per 100 ft, the tapeman will be able to hold horizontally only parts of the tape at a time. Holding the tape more than 5 ft above the ground is difficult, and wind can make it more so. If the tape is held at heights of 5 ft or less above the ground both forearms can be braced against the body and the tape can easily be pulled firmly without swaying and jerking.

Assuming they are proceeding downhill, the head tapeman pulls the tape along the line for its full length and then, leaving the tape on the ground, returns as far along the tape as necessary for them to hold horizontally the part of the tape from his or her point to the rear tapeman.

The head tapeman holds the plumb bob string over a whole foot mark, and when the tape is stretched, lined, and horizontal, lets the plumb bob fall and sets a taping pin. He holds the intermediate foot mark on the tape until the rear tapeman arrives, at which time he hands the tape to the rear tapeman with the foot mark that he has been holding. This careful procedure is followed because it is so easy for the head tapeman to forget which foot he was holding if he drops it and walks ahead. The tapemen repeat this process for as much more of the tape as they can hold horizontally until they reach the 0-end of the tape. This process is illustrated in Figure 3-14.

FIGURE 3-14 "Breaking tape" along a slope.

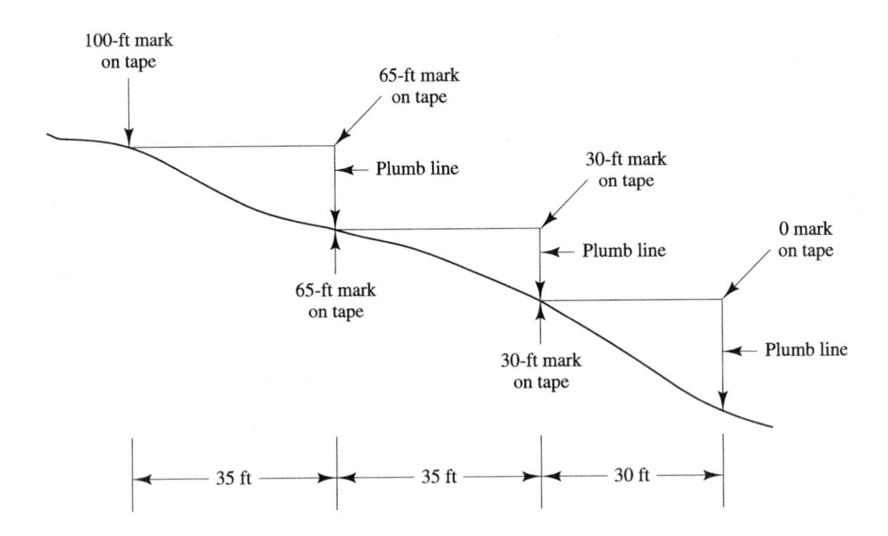

This process of measuring with sections of the tape is referred to as *breaking tape* or *breaking chain*. If the head tapeman follows the customary procedure of leaving a taping pin at each of the positions that he occupies when breaking tape, counting the number of hundreds of feet taped (as represented by the number of pins in the possession of the rear tapeman) would be confusing. Therefore, at each of the intermediate points the head tapeman sets a pin in the ground and then takes one pin from the rear tapeman. Instead of breaking tape, some surveyors find it convenient to measure the partial tape lengths and record those values in their field notes.

It is probably wise for a beginning surveyor to measure a few distances on slopes of different percentages holding the tape horizontal and then again with the tape along the slopes with no corrections made. These measurements should give him or her a feeling for the magnitude of slope errors.

Taping on Slopes and Making Slope Corrections

Occasionally, it may be more convenient or more efficient to tape along sloping ground with the tape held inclined along the slope. This procedure has long been common for underground mine surveys but to a much lesser extent for surface surveys. Slope taping is quicker than horizontal taping and is considerably more precise because it eliminates plumbing with its consequent accidental errors. Taping along slopes is sometimes useful when the surveyor is working along fairly constant, smooth slopes or when he wants to improve precision. Nevertheless, the method is generally not used because of the problem of correcting slope distances to horizontal values. This is particularly true in rough terrain where slopes are constantly varying and the problem of determining the magnitude of the slopes is difficult.

In some cases it may be impossible to hold the entire tape (or even a small part of it) horizontally. This may occur when taping is being done across a ravine (see Figure 3-15) or some other obstacle where one tapeman is much lower than the other one and where it is not feasible to "break tape." Here it may be practical to hold both ends of the tape on the ground.

FIGURE 3-15 Taping on slope.

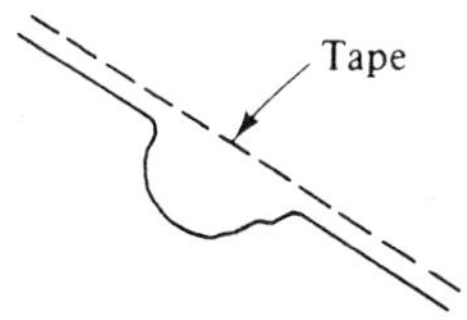

Tape

Slopes are often expressed in terms of *grade,* that is, the number of feet of vertical change in elevation per 100 feet of horizontal distance. Grade is expressed as a percentage and may be given a plus sign for uphill slopes and a negative sign for downhill slopes. Examples are shown in Figure 3-16.

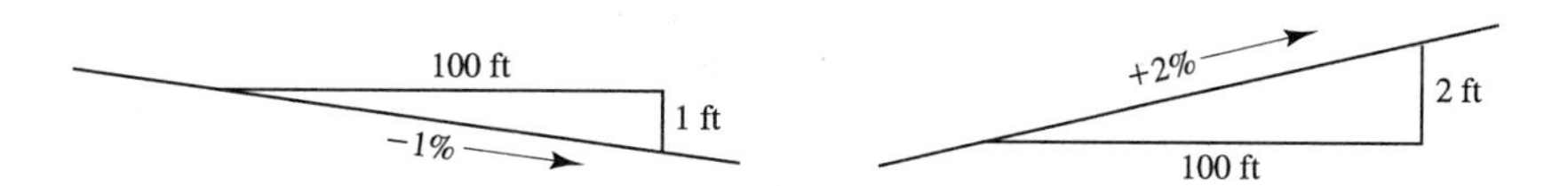

FIGURE 3-16 Grades.

When a person is measuring a distance with a tape held along a slope it will be necessary for him or her to determine the difference in elevation between the ends of the tape for each position of the tape or to measure the vertical angle involved. Once this is done, the horizontal distance for each measurement can be computed by trigonometry, the quadratic equation, or more easily by the approximate formula which is developed for that purpose in Section 4-5.

An almost obsolete handheld instrument called the clinometer was formerly used for measuring vertical angles and grades. With this instrument, which is illustrated in Figure 3-17, it was possible to measure vertical angles to approximately the nearest 10′ and to determine grades roughly. In effect the clinometer is a hand level to which a protractor is attached.

FIGURE 3-17 Hand level and clinometer (obsolete).

Dynamic Taping

With dynamic taping, which is very similar to the slope taping method, slope distances are measured. Then the vertical angle is measured and the horizontal distance computed.

3-12 REVIEW OF SOME TRIGONOMETRY

In surveying there are innumerable occasions where it is quite difficult and time consuming to measure certain distances and/or angles due to obstacles, weather conditions, time constraints, and so on. Frequently the use of simple trigonometric equations will enable us to quickly and easily compute those needed values without the necessity of having to measure them in the field. As an example it is assumed that a surveyor is working with a right triangle. If the lengths of one side of the triangle and one of the angles (not the 90° one) have been measured the two other side lengths and the missing angle can be quickly computed with trigonometric formulas.

The most common use of trigonometry in surveying involves the use of right triangle relations. Such a triangle is shown in Figure 3-18. Initially the angle α at corner A will be considered. In the figure the *hypotenuse* (L_{CA}) is the sloping side while the *adjacent side* (L_{AB}) is the side between the angle α and the right or 90° angle. The *opposite side* (L_{BC}) is the side which is opposite to or across from the angle.

FIGURE 3-18 A right triangle.

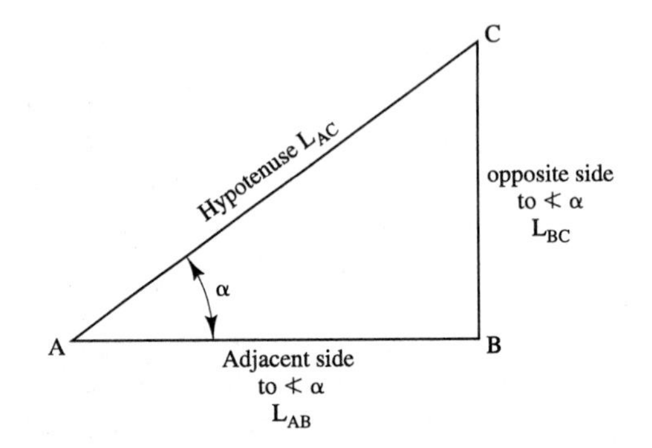

It can be proved that the relationships listed exist for right triangles with reference here being made to Figure 3-18 and angle α. These ratios are called the functions of the angles. The trigonometric terms sine, cosine, etc. are usually abbreviated as shown.

$$\text{sine } \alpha = \sin \alpha = \frac{\text{opposite side}}{\text{hypotenuse}} = \frac{L_{BC}}{L_{AC}}$$

$$\text{cosine } \alpha = \cos \alpha = \frac{\text{adjacent side}}{\text{hypotenuse}} = \frac{L_{AB}}{L_{AC}}$$

$$\text{tangent } \alpha = \tan \alpha = \frac{\text{opposite side}}{\text{adjacent side}} = \frac{L_{BC}}{L_{AB}}$$

$$\text{cotangent } \alpha = \cot \alpha = \frac{\text{adjacent side}}{\text{opposite side}} = \frac{L_{AB}}{L_{BC}}$$

$$\text{secant } \alpha = \sec \alpha = \frac{\text{hypotenuse}}{\text{adjacent side}} = \frac{L_{AC}}{L_{AB}}$$

$$\text{cosecant } \alpha = \csc \alpha = \frac{\text{hypotenuse}}{\text{opposite side}} = \frac{L_{AC}}{L_{BC}}$$

To commit these equations to memory is not difficult as one only has to memorize the first three and then realize that the cotangent is the invert of the tangent, the secant the invert of the cosine, and the cosecant the invert of the sine.

When the reader learns the names of the triangle sides as described here (hypotenuse, adjacent side, and opposite side), he or she will be able to easily apply the trigonometric relations to right triangles whatever their position (that is with the 90° angle on the left side or the right side or up on top of the triangle). If we are considering the angle at corner *C* in Figure 3-19, the opposite side is L_{AB} while the adjacent side is L_{BC}.

The values of the trigonometric functions for various angle sizes are commonly available with pocket calculators. As a result the author does not include tables for their values in this book. *When looking up the values of trigonometric functions with the usual calculator, the angles must first be converted to decimal form.* For example, the sine of 3′18″ becomes the sine of 3.3°.

In working with angles it is necessary for the student to realize that our angles are usually designated as being in degrees, minutes, and seconds, such as 54°24′38″. Then we must realize that to obtain the value of a trig function for a particular angle with the usual pocket calculator, it will be necessary to convert that angle to a decimal value. A sample calculation of this conversion follows:

$$54°24'38'' = 54°24' + \left(\frac{38}{60}\right)' = 54°24.6333'$$

$$= 54° + \left(\frac{24.6333}{60}\right)^{\circ} = 54.4106°$$

Several example problems (3–1 to 3–5) make use of the right triangle relations. It should be noted that some of the information obtained in these problems using trigonometric functions could have been obtained just as well with the pythagorean theorem.

Example 3–1

With reference to the triangle in Figure 3-19 the length of side *BC* and the size of the angle at corner *A* have been measured. Determine the length of side *AC* using trigonometric equations.

FIGURE 3-19

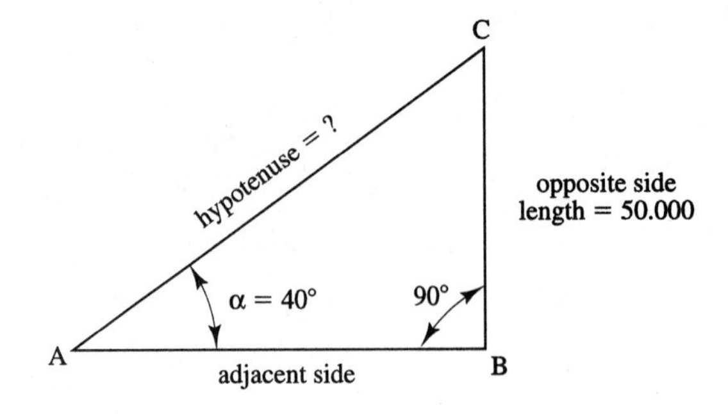

Solution

$$\sin\alpha = \sin 40^\circ = 0.64278761 \text{ with calculator}$$

$$\sin\alpha = \frac{\text{opposite side}}{\text{hypotenuse}}$$

$$0.64278761 = \frac{50.000}{L_{AC}}$$

$$L_{AC} = \mathbf{77.786\ ft}$$

Example 3–2

Determine the values of angles α and β in the right triangle of Figure 13-20.

FIGURE 3-20

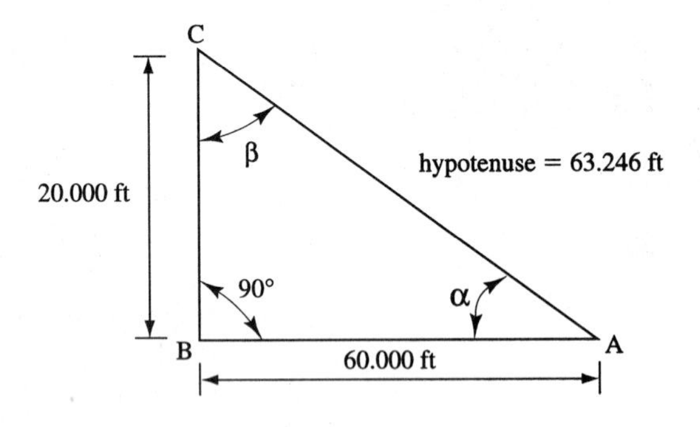

Solution

$$\sin\beta = \frac{L_{BA}}{L_{AC}} = \frac{60.000}{63.246} = 0.948676596$$

$$\beta = 71.5638^\circ = 71^\circ 33.8'$$

$$\alpha = 180^\circ - 90^\circ - 71^\circ 33.8' = \mathbf{18^\circ 26.2'}$$

Alternate Solution

$$\sin\alpha = \frac{L_{BA}}{L_{AC}} = \frac{20,000}{63.246} = 0.316225532$$

$$\alpha = 18.4348^\circ = 18^\circ 26.1'$$

Example 3–3

A 14.4 ft long ladder is leaned against a wall so that its base is 3.2 ft from the wall. (See Figure 3-21.) How far will the ladder reach vertically up the wall?

FIGURE 3-21

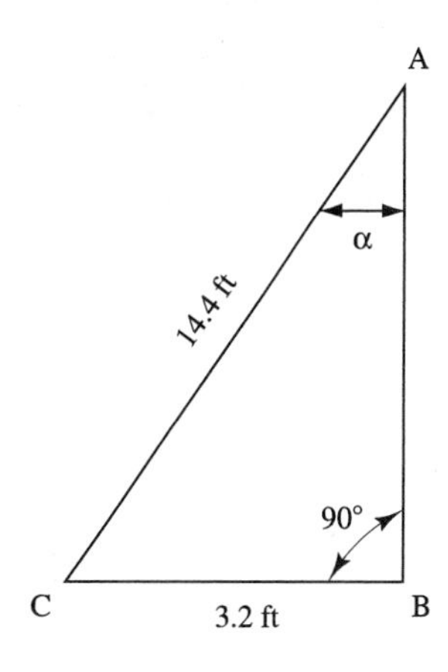

Solution

$$\sigma\iota\nu\,\alpha = \frac{L_{BC}}{L_{AC}} = \frac{3.2}{14.4} = 0.2222222$$

$$\alpha = 12.83958841^\circ$$

$$\cos\alpha = \frac{L_{AB}}{L_{AC}}$$

$$\cos 12.83958841^\circ = \frac{L_{AB}}{14.4}$$

$$L_{AB} = \mathbf{14.04\ ft}$$

Alternate Solution Using Pythagorean Theorem

$$(14.4)^2 = (3.2)^2 + (L_{AB})^2$$

$$L_{AB} = \sqrt{(14.4)^2 - (3.2)^2} = \mathbf{14.04\ ft}$$

Example 3–4

With modern electronic surveying equipment a distance has been measured from point A to point B and found to be 646.34 ft. The two points are not at the same elevation and the angle between a horizontal line and the sloping line has been measured and found to be 3°10′. Determine the horizontal distance between the two points and the difference in elevation between them (neglecting the effect of earth's curvature).

Solution

$$\sin 3°10' = \sin 3.16667° = \frac{\text{opposite side}}{\text{hypotenuse}}$$

$$0.055240626 = \frac{\text{opposite side}}{646.34} = \text{Diff. in elevation}$$

$$\text{Diff. in elevation}$$
$$= (646.34)(0.055240626) = 35.70 \text{ ft}$$

$$\cos 3°10' = \frac{\text{adjacent side}}{\text{hypotenuse}} = \frac{\text{Horiz. distance}}{\text{hypotenuse}}$$

$$0.998473071 = \frac{\text{Horiz. distance}}{646.34}$$

$$\text{Horiz. distance} = \mathbf{645.35 \text{ ft}}$$

Example 3–5

A surveyor needs to determine the distance between points A and B shown in Figure 3-22. Unfortunately this surveyor's electronic distance measuring equipment is being repaired. Therefore this surveyor lays off the 90° angle shown in the figure, sets point C 300 ft up the river as shown and measures the angle at C and finds it to be 54°18′. Calculate the distance AB.

FIGURE 3-22

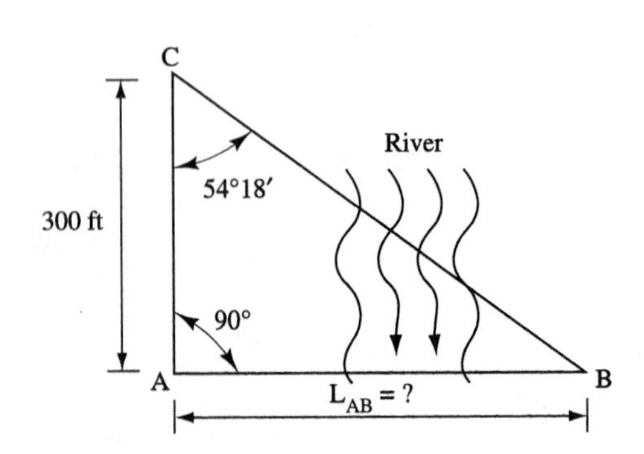

Solution

$$\tan 54°18' = \frac{\text{opposite side}}{\text{adjacent side}} = \frac{L_{AB}}{L_{AC}}$$

$$1.391647258 = \frac{L_{AB}}{300}$$

$$L_{AB} = \mathbf{417.494\ ft}$$

Surveying calculations pertain not only to right triangles but also to oblique triangles, other polygons, highway and railroad curves, and so on. Particularly common are oblique triangles (those which do not contain a 90° angle). Detailed trigonometric formulas for calculating the angles and side lengths for these triangles are provided in Table 2 of Appendix C of this book.

In Table 1 of Appendix C the trigonometric functions for right triangles for all sorts of situations are given. In this table two other trigonometric terms are encountered. These values, which are sometimes useful as described in Chapter 22 of this text (which deals with circular highway and railway curves), are the versed sign or versine and the external secant. In abbreviated form they are as follows:

$$\text{vers}\,\theta = 1 - \cos\theta$$

$$\text{ex sec}\,\theta = 1 - \sec\theta$$

Problems

3-1. Name six methods of measuring distances and list advantages and disadvantages of each.

3-2. List two situations where each of the following methods or instruments can be used advantageously for measuring distance: a) pacing; b) odometer; c) stadia; d) taping; e) EDM.

3-3. A surveyor counted the number of paces required to cover a 500-ft distance. The results were as follows: 188, 190, 187, and 191 paces. Then an unknown distance was stepped off four times, requiring 306, 308, 307, and 305 paces. Determine the average pace length and the length of the second line. (Ans.: 2.646 ft, 811 ft)

3-4. A surveyor paces a 200-ft length four times with the following results: 71, 72, 70, and 71.5 paces. How many paces will be necessary for this surveyor to lay out a distance of 340 ft?

3-5. Convert the following distances given in meters to feet using the latest meter definition which is based on the speed of light.

a. 236.40 m (Ans.: 775.59 ft)

b. 715.32 m (Ans.: 2346.85 ft)

c. 915.88 m (Ans.: 3004.86 ft)

3-6. Convert the following angles to decimal values:

a. 37°32′50″

b. 86°09′33″

c. 109°15′40″

3-7. Convert the following angles written in decimal form to degrees, minutes, and seconds:

a. 56.3498° (Ans.: 56°20′59″)

b. 32.0864° (Ans.: 32°05′11″)

c. 165.8424° (Ans.: 165°50′33″)

3-8. A 2-m subtense bar was set up at one end of a line, and a theodolite was set up at the other end. What was the horizontal length of the line if the following angle readings were taken on the bar: 0°23′16″, 0°23′15″, 0°23′17″, and 0°23′16″?

3-9. A 100-ft "cut" steel tape (zero end forward) was used to measure the distance between two stakes. If the rear tapeman is holding the 64-ft point and the head tapeman cuts 0.41 ft, what is the measured distance? (Ans.: 63.59 ft)

3-10. A 100-ft steel tape is 1/40 in. thick and 5/16 in. wide. If steel weighs 490 lb/ft^3, how much does this tape weigh?

3-11. Repeat Problem 3-10 if the tape is 0.030 in. thick and 3/8 in. wide. (Ans.: 3.83 lbs)

3-12. A surveyor has determined that side *AB* of the right triangle shown has a length of 317.64 ft and that the interior angle at corner *A* is 35°17′. Determine the lengths of the other sides of the triangle.

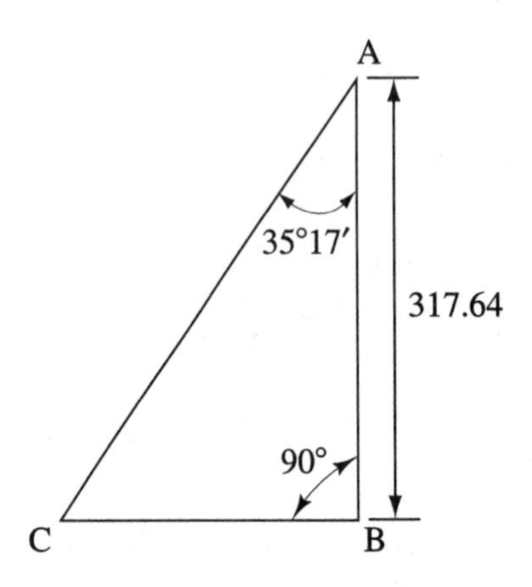

3-13. The hypotenuse of a right triangle is 81.11 ft long and one of the other sides has a length equal to 69.66 ft. Find the angle opposite to the 69.66 ft side. (Ans.: 59°11.2′).

3-14. The three sides of a triangle are 60, 80, and 100 ft. Determine the magnitudes of the interior angles.

3-15. A sloping earth dam rises 3.5 ft for every 10 ft of horizontal distance. What angle does the dam make with the horizontal? (Ans.: 19°17.4′)

3-16. A surveyor measures an inclined distance and finds it to be 1466.94 ft. In addition, the angle between the horizontal and the line is measured and found to be 2°56′30″. Determine the horizontal distance measured and the difference in elevation between the two ends of the line.

3-17. The angles at the corners of a triangular field are 32°, 58°, and 90° and the hypotenuse is 333.68 ft. How many feet of fencing will be needed to enclose this field? (Ans.: 793.48 ft)

3-18. It is desired to determine the height of a church steeple. Assuming that the ground is level, a 500.00 ft length is measured out from the base of the steeple and a 32°45′ vertical angle is determined from that point on the ground to the top of the steeple. How tall is the steeple?

3-19. Repeat Problem 3-18 if for an instrument set up 600.0 ft from a tower with its telescope center 5.000 ft above the ground. The telescope is sighted horizontally to a point 5.000 ft from the bottom of the steeple and then the angle to the top of the steeple is measured. It's 31°10′. How tall is the tower? (Ans.: 367.9 ft)

3-20. A section of a road with a constant 3% slope or grade (i.e., 3 ft vertically for each 100 ft horizontally) is to be paved. If the road is 24.000 ft wide and its total horizontal length is 760 ft, compute the road area to be paved.

3-21. Repeat Problem 3-20 if instead of having a 3% grade the road makes a 4° angle with the horizontal. (Ans.: 18, 285 ft^2)

3-22. We need to measure the height of the church steeple shown in the accompanying illustration. A horizontal distance has been measured out from the building as shown and the two vertical angles have been determined. How high is the church? Notice that it is unnecessary in this case to measure the height of the instrument telescope above the ground.

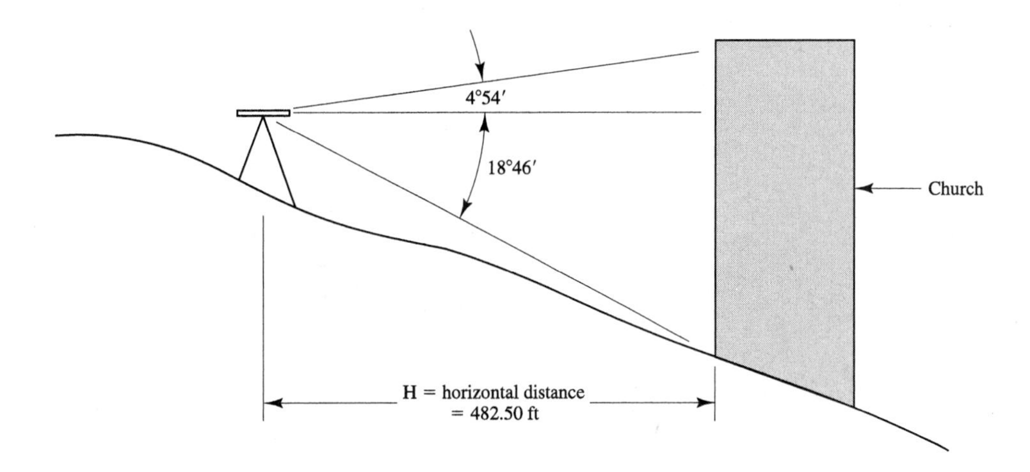

3-23. Repeat Problem 3-22 if the horizontal distance is 452.00 ft, the upper angle is 4°50′10″, and the lower angle is 15°32′40″. (Ans.: 163.97 ft)

3-24. Repeat Problem 3-22 if the height of the instrument is 5.00 ft above the ground and the slope distance from the center of the instrument to the bottom of the church is 512.16 ft. Assume the vertical angle from the telescope to the top of the steeple remains 4°54′. The bottom angle is not measured.

CHAPTER 4

Distance Corrections

4-1 INTRODUCTION

To the reader it may seem that the author has far too much to say about taping when almost all distance measurements are today made with electronic distance-measuring instruments alone or as part of total stations. This feeling may be even stronger after you study the discussions pertaining to the corrections of taping measurements for temperature or sag or other items and after you struggle through the detailed information presented concerning errors and mistakes. *The author, however, feels very strongly that if you learn this information concerning tapes, you will understand much better the entire measuring process, regardless of the surveying operation involved or the equipment used.*

The surveyor of today and tomorrow should become proficient with the steel tape, even though it is highly possible that he or she will eventually use the tape very little. Otherwise, the chance of making blunders and large errors will be magnified when the tape is used.

4-2 TYPES OF CORRECTIONS

The five major areas in which the surveyor may need to apply corrections either in measuring or in laying out lines with a tape are as follows:

1. Incorrect tape length or standardization error
2. Temperature variations
3. Slope
4. Sag
5. Incorrect tension

Once the appropriate errors are determined, the actual distance for a measured line can be determined by substituting into the following equation. The application of this equation will always be correct provided the correct algebraic sign is used for each correction.

$$\text{Actual distance} = \text{Measured distance} + \sum \text{corrections}$$

The next few sections of this chapter are devoted to a discussion of the first three of these corrections. To save space, only a few remarks are made concerning corrections for tape sag and tension.

4-3 INCORRECT TAPE LENGTH OR STANDARDIZATION ERROR

An important topic in surveying is the standardization of equipment, or the comparison of the equipment (whether it is a tape, an electronic distance-measuring instrument, or whatever) against a standard. In other words, has the equipment been damaged or shaken out of adjustment, have repairs or weather changes affected it, and so on? If so, the surveyor will need to adjust the equipment or make mathematical corrections to compensate for the resulting errors.

It is said that in ancient Egypt the workers on the pyramids were required to compare their cubit sticks against the standard or royal cubit stick each full moon. Those failing to do so were subject to death. Such a practice undoubtedly brought forth the best efforts from their personnel in the area of standardization.

Although steel tapes are manufactured to very precise lengths, with use they become kinked, worn, and imperfectly repaired after breaks. The net result is that tapes may vary by quite a few hundredths of a foot from their desired lengths. Therefore, it is wise to check them periodically against a standard. There are several ways in which this might be done. For instance, some surveying offices used to keep one standardized tape (perhaps an Invar type, previously described in Section 3-9) that was used only for checking the lengths of their other tapes. Some companies took a tape that had been standardized at 100.00 ft and used it to place marks 100.00 ft apart on a concrete curb, sidewalk, or pavement. The marks were frequently used for "standardizing" their tapes. They felt that the length between the marked points would not be changed appreciably as temperatures varied because of the mass of the concrete and the friction of the earth.

These practices were advisable for surveyors who had extensive practices and they yielded very satisfactory results for surveys where ordinary precision was desired, but they were probably not sufficient for extremely precise work. For such work, tapes can be mailed to the National Institute of Standards and Technology in Gaithersburg, Maryland. For a rather large fee, they will determine the length of a tape for specific tension and support conditions. They issue a certificate for each tape, giving its length to the nearest 0.001 ft at 68°F (20°C) for two conditions: one with the tape supported for its full length and subjected to a 12-lb pull, and the other with the tape supported at its ends only and subjected to a 20-lb pull.

Several municipal governments around the United States, various state agencies, and a good many universities will standardize tapes, very often free as a service to the public. In addition, the NGS has established approximately 200 base lines at various locations across the country where tapes and EDM equipment can be calibrated by the surveyor. A detailed description of one of these base lines is presented in Chapter 5. The distances between the monuments for the NGS base lines were measured with Invar tapes and/or electronic distance-measuring equipment with accuracies approaching 1/1,000,000. The locations of these base lines can be obtained by writing to the National Geodetic Survey, National Ocean Survey, Rockville, MD 20852 (http://www.ngs.noaa.gov/). This information is also available for each state from its surveying society.

If a tape proves to be in appreciable error from the standard, the surveyor must correct the measurements by the required amounts. He or she will note carefully whether a correction is positive or negative, as explained in the following paragraphs.

The important point to grasp in making corrections is that the tape "says zero ft at one end and 100 ft at the other end" even though its correct length is 99.94 ft, 100.10 ft, or some other value. If the surveyor (ignorant of the tape's true length) uses this tape 10 times, he believes that he has measured a distance of 1000 ft, but has really measured 10 times the actual tape length.

In measuring a given distance with a tape that is too long, the surveyor will not obtain a large enough value for the measurement and will have to make a positive correction. In other words, if the tape is too long, it will take fewer tape lengths to measure a distance than would be required for a shorter, correct-length tape. For a tape that is too short, the reverse is true and a negative correction is required. It should be simple enough to remember this rule: *Tape too long, add; tape too short, subtract.*

Examples 4-1, 4-2, and 4-3 illustrate the correction of distances which were measured with incorrect length tapes. The problem of Example 4-3 is stated backward from the ones of Examples 4-1 and 4-2, and the sign of the correction is therefore reversed.

Example 4-1

A distance is measured with a 100-ft steel tape and is found to be 896.24 ft. Later the tape is standardized and is found to have an actual length of 100.04 ft. What is the correct distance measured?

Solution

The tape is too long and a + correction of 0.04 must be made for each tape length as follows:

$$\text{Measured value} = 896.24 \text{ ft}$$

$$\text{Total correction} = +(0.04)(8.9624) = +\ 0.36 \text{ ft}$$

$$\text{Corrected distances} = \mathbf{896.60 \text{ ft}}$$

Alternative Solution

Obviously, the distance measured equals the number of tape lengths times the actual length of the tape. In this case, it took 8.9624 tape lengths to cover the distance and each tape length was 100.04 ft.

$$\text{Distance measured} = (8.9624)(100.04) = 896.60 \text{ ft}$$

Example 4-2

A distance is measured with a 100-ft steel tape and is found to be 2320.30 ft. Later the tape is standardized and is found to have an actual length of 99.97 ft. What is the actual distance?

Solution

The tape is too short. Therefore, the correction is minus.

$$\begin{aligned} \text{Measured value} &= 2320.30 \text{ ft} \\ \text{Total correction} = -(0.03)(23.2030) &= -\quad 0.70 \text{ ft} \\ \text{Corrected distance} &= \mathbf{2319.60 \text{ ft}} \end{aligned}$$

Example 4-3

It is desired to lay off a dimension of 1200.00 ft with a steel tape that has an actual length of 99.95 ft. What field measurement should be made with this tape so that the correct distance is obtained?

Solution

This problem is stated exactly opposite to the ones of Examples 4-1 and 4-2. It is obvious that if the tape is used 12 times, the distance measured (12×99.95) is less than the 1200 ft desired, and a correction of the number of tape lengths times the error per tape length must be *added.*

$$\begin{aligned} 12 \text{ tape lengths} = 12 \times 100.00 &= 1200.00 \text{ ft} \\ +12 \times 0.05 &= +\quad 0.60 \text{ ft} \\ \text{Field measurement} &= 1200.60 \text{ ft} \end{aligned}$$

Check

The answer can be checked by considering the problem in reverse. Here a distance has been measured as being 1200.60 ft with a tape 99.95 ft long. What actual distance was measured? The solution is as follows:

$$\begin{aligned} \text{Measured value} &= 1200.60 \text{ ft} \\ \text{Total correction} = -(0.05)(12.006) &= \quad -0.60 \text{ ft} \\ \text{Corrected distance} &= \mathbf{1200.00 \text{ ft}} \end{aligned}$$

One final example of this type of corrected measurement is shown with the field notes of Figure 4-1, where the sides of the traverse previously paced (shown in Figure 3-1) are measured with a tape of actual length 99.95 ft. In this case each of the sides was taped twice (forward and back), and an average value was obtained before the wrong-length tape correction was applied.

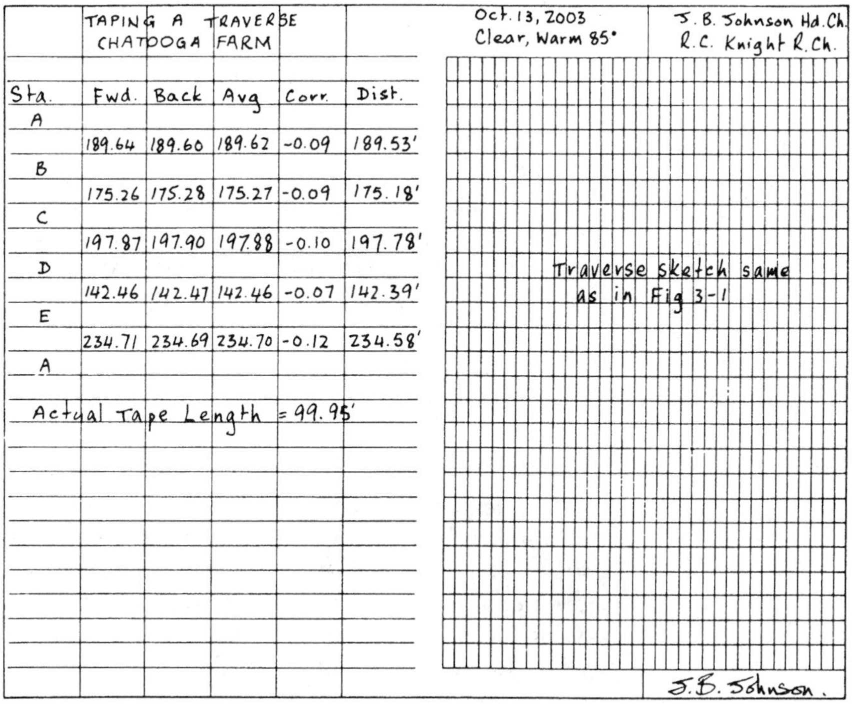

FIGURE 4-1
Taping field notes.

	TAPING A TRAVERSE CHATOOGA FARM					Oct. 13, 2003 Clear, Warm 85°	J. B. Johnson Hd. Ch. R. C. Knight R. Ch.
Sta.	Fwd.	Back	Avg	Corr.	Dist.		
A							
	189.64	189.60	189.62	-0.09	189.53′		
B							
	175.26	175.28	175.27	-0.09	175.18′		
C							
	197.87	197.90	197.88	-0.10	197.78′	Traverse sketch same as in Fig 3-1	
D							
	142.46	142.47	142.46	-0.07	142.39′		
E							
	234.71	234.69	234.70	-0.12	234.58′		
A							
Actual Tape Length = 99.95′							

J. B. Johnson.

4-4 TEMPERATURE VARIATIONS

Changes in tape lengths caused by temperature variations can be significant even for ordinary surveys. For precise work they are of critical importance. A temperature change of approximately 15°F will cause a change in length of approximately 0.01 ft in a 100-ft tape. If a tape is used at 20°F to lay off a distance of 1 mile and if the distance is checked the following summer with the same tape when the temperature is 100°F (no temperature correction being made), there will be a difference in length of 2.75 ft caused by the temperature variation. Such an error alone would be equivalent to a precision of 2.75/5280 = 1/1920 (not so good).

Steel tapes lengthen with rising temperatures and shorten with falling ones. The coefficient of linear expansion for steel tapes is 0.0000065 per degree Fahrenheit. This means that for a 1°F rise in temperature a tape will increase in length by 0.0000065 times its length.

As described in Section 4-3, the standardized length of a tape is determined at 68°F. A tape that is 100.000 ft long at the standard temperature will at 100° F have a length of 100.00 + (100 – 68) (0.0000065)(100) = 100.021 ft. The correction of a distance measured at 100°F with this tape can be made as described previously for wrong-length tapes. The correction of a tape for temperature changes can be expressed with this formula noting that it may be either plus or minus:

$$C_t = 0.0000065(T - T_s)(L)$$

In this formula, C_t is the change in length of the tape due to temperature change, T is the estimated temperature of the tape at the time of measurement, T_s is the standardized temperature, and L is the tape length.

If SI units are being used, the coefficient of linear expansion is 0.000 011 6 per degree Celsius (°C). The correction in length of a metric tape for temperature changes can be expressed by the formula

$$C_t = (0.000\ 011\ 6)(T - T_s)(L)$$

where T_s is the standardized temperature of the tape at manufacture (usually 20°C), T is the temperature of the tape at the time of measurement, and L is the tape length.

It will be remembered that the expressions for temperature conversion are as follows:

$$^\circ\mathrm{C} = \frac{5}{9}\left(^\circ\mathrm{F} - 32\right)$$

$$^\circ\mathrm{F} = \frac{9}{5}\left(^\circ\mathrm{C}\right) + 32$$

It is clear that a steel tape used on a hot summer day in the bright sunshine will have a much higher temperature than will the surrounding air. Actually, though, partly cloudy summer days will cause the most troublesome variations in length. For a few minutes the sun shines brightly and then it is covered for awhile by clouds, causing the tape to cool quickly, perhaps by as much as 20°F or 30°F. Accurate corrections for tape temperature variations are difficult to make because the tape temperature may vary along its length with sun, shade, dampness (in grass, on ground), and so on. It has been shown that the variation of a few degrees may make an appreciable variation in the measurement of distance.

For the best precision it is desirable to tape on cloudy days, early in the mornings, or late in the afternoons to minimize temperature variations. Furthermore, the very expensive Invar tapes, with their very small coefficients of expansion (0.0000001 to 0.0000002) are very helpful for precise work, but such measurement has been made obsolete by electronic distance-measuring equipment.

For very precise surveying, tape measurements are recorded and the proper corrections made. On cloudy, hazy days an ordinary thermometer may be used for measuring the air temperature, but on bright sunny days the temperature of the tape itself should be determined. For this purpose plastic thermometers attached to the tapes near the ends (so that their weights do not appreciably affect sag) should be used. (As mentioned previously some tapes have different end marks to be used, depending on the temperature.)

Regular steel tapes have not been used for geodetic work for quite a few decades because of their rather large coefficients of expansion and because of the impossibility of accurately determining their temperature during daytime operations. Until electronic distance-measuring devices were introduced, almost all of the base-line measurements of the National Geodetic Survey during the twentieth century were done with Invar tapes. Today, almost all of their length measurements are made with EDMs although GPS is being used more each year for this purpose.

4-5 SLOPE CORRECTIONS

Most tape measurements are made with tapes held horizontally, thus avoiding the necessity for making corrections for slope. In this section, however, measurements are assumed to be made on slopes. (You should realize that the correction formula developed in this section is applicable to plan or alignment measurements as well as to profile or slope measurements.)

In Figure 4-2 a tape of length s is stretched along a slope and it is desired to determine the horizontal distance h that has been measured. It is easy for tapemen to apply an approximate correction formula for most slopes. The expression, derived in this section, is satisfactory for most measurements, but for slopes of greater than approximately 10% to 15%, an exact trigonometric function or the Pythagorean theorem should be used. When a 100-ft slope distance is measured, the use of this approximate expression will cause an error of 0.0013 ft for a 10% slope and a 0.0064-ft error for a 15% slope.

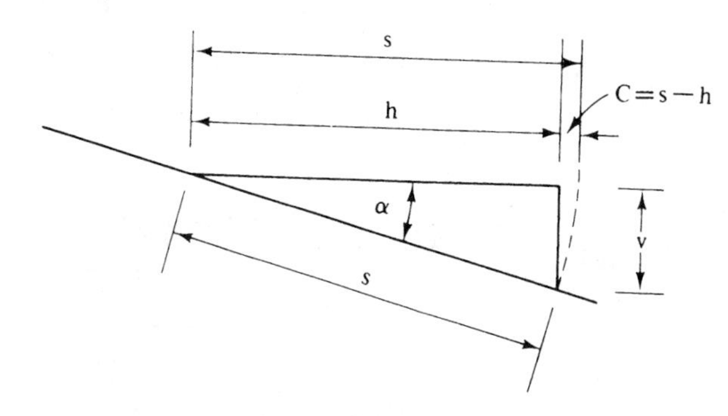

FIGURE 4-2
Taping along a slope.

It is very helpful to write an expression for the correction C shown in Figure 4-2. This value, which equals $s - h$ in the figure, is written in a more practical form by using the Pythagorean theorem as follows:

$$s^2 = h^2 + v^2$$

$$v^2 = s^2 - h^2$$

from which

$$v^2 = (s - h)(s + h)$$

$$s - h = \frac{v^2}{s + h}$$

and since $$C = s - h$$

thus $$C = \frac{v^2}{s + h}$$ *This correction always has a negative sign.*

For the typical 100-ft tape, s equals 100 ft and h varies from 100 ft by a very small value. For practical purposes, therefore, h can also be assumed to equal 100 ft when the slope correction expression is applied. It is written for 100-ft tapes in the form

$$C = -\frac{v^2}{200}$$

In taping it is normally convenient to measure a full tape length at a time. Therefore, in measuring along a slope it is often convenient for the head tapeman to calculate (probably in his or her head) the correction and set the taping pin that distance beyond the end of the tape so that they will have measured 100 ft horizontally. For a 6-ft vertical elevation difference

$$C = -\frac{6^2}{200} = -0.18 \text{ ft}$$

For tape lengths other than 100 ft, the correction equation can be written as

$$C = -\frac{v^2}{2s}$$

Sometimes for a long constant slope the tape is held on the ground and the correction is made for the entire length. Such a situation is illustrated in Example 4-4. *The reader should carefully note that the correction formula was derived for a single tape length.* For a distance of more than one tape length, the total correction will equal the number of tape lengths times the correction per tape length.

Example 4-4

A distance was measured on an 8% slope and found to be 2620.30 ft. What is the horizontal distance measured?

Solution

$$\text{Correction per tape length} = -\frac{(8)^2}{(2)(100)} = -0.32 \text{ ft}$$

$$\text{Total correction} = (26.2030)(-0.32) = -8.38 \text{ ft}$$

$$\text{Horizontal distance} = 2620.30 + (-8.38) = \mathbf{2611.92 \text{ ft}}$$

If the slope distance s is measured with a tape and an instrument is used to measure the vertical angle α from the horizontal to the slope, the horizontal distance can be obtained from the following equation:

$$H = s \cos \alpha$$

Example 4-5

A slope distance is measured with a steel tape and found to be 1240.32 ft. If the vertical angle is measured with a theodolite and found to be 3° 27′, what is the horizontal distance?

Solution

$$H = (1240.32)(\cos 3°27') = \mathbf{1238.07 \text{ ft}}$$

Example 4-6

It is desired to lay off a horizontal distance with a steel tape along a constant 4° 18′ slope. What should the slope distance be so that the resulting horizontal distance is 840.00 ft?

Solution

$$s = \frac{H}{\cos\alpha} = \frac{840.00}{\cos 4^\circ 18'} = \mathbf{842.37\ ft}$$

4-6 SAG AND TENSION CORRECTIONS

Sag

When a steel tape is supported only at its ends, it will sag into a curved shape known as the *catenary*. The obvious result is that the horizontal distance between its ends is less than when the tape is supported for its entire length.

To determine the difference in the length measured with a fully supported tape and one supported only at its ends or at certain intervals, the following approximate expression may be used:

$$C_s = -\frac{w^2 L^3}{24 P_1^2}$$

where C_s = correction in feet and is always negative
w = weight of tape in pounds per foot
L = unsupported length of tape in feet
P_1 =total tension in pounds applied to the tape

This expression, although approximate, is sufficiently accurate for many surveying purposes. It is applicable to horizontal taping or to tapes held along slopes of not more than approximately 10°.

To minimize sag errors, it is possible to use this formula and apply the appropriate corrections to the observed distance. Another and more practical procedure for ordinary surveying is to increase the pull or tension on the tape in order to compensate for the effect of sag. For very precise work, the tape is either supported at sufficient intervals to make sag effects negligible or it is standardized for the pull and manner of support to be used in the field.

Variations in Tension

A steel tape stretches when it is pulled, and if the pull is greater than that for which it was standardized, the tape will be too long. If less tension is applied, the tape will be too short. A 100-ft steel tape will change in length by approximately 0.01 ft for a 15-lb change in pull. Since variations in pull of this magnitude are improbable, errors cased by tension variations are negligible for all except the most precise chaining. Furthermore, these errors are accidental and tend to some degree to cancel. For precise taping, spring balances are used so that certain prescribed tensile forces can be applied to the tapes. With such balances it is not difficult to apply tensions within 1/2 lb or closer to desired values. As with all measuring devices, it is necessary periodically to check or standardize the tension apparatus against a known standard.

Despite the minor significance of tension errors, a general understanding of them is important to the surveyor and will serve to improve the quality of the work. The actual elongation of a tape in tension equals the tensile stress in psi over the modulus of elasticity of the steel (the modulus of elasticity of a material is the ratio of stress to strain and equals 29,000,000 psi or 2 050 000 kg/cm^2 for steel) times the length of the tape. In the following expression, the elongation of the tape in feet is represented by C_p, P_1 is the pull on the tape, A is the cross-sectional area in square inches, L is the length in feet, and E is the modulus of elasticity of the steel in psi.

$$C_p = \frac{P_1 / A}{E} L = \frac{P_1 L}{AE}$$

It will be noted that the tape has been standardized at a certain pull P, and therefore the change in length from the standardized situation is desired and the expression is written as

$$C_p = \frac{(P_1 - P)L}{AE} = \frac{(\Delta P)L}{AE}$$

After looking at this equation the reader can see a simple way to avoid its use. If the standard pull for a particular tape has been determined and the surveyor is using a scale to measure the pull applied to that tape it seems only logical that he or she will apply a pull equal to the standard value. If this is done there will be no need to make this correction as it should equal zero.

Normal Tension

If, when the tape is suspended, it is pulled very tightly, there will be an appreciable reduction in sag and some increase in the tape length because of tension. As a matter of fact, there is a theoretical pull for each tape at which the lengthening of the tape caused by tension equals its shortening caused by sag. This value is referred to as the *normal tension.* Its magnitude can be practically measured for a particular tape, or it can be computed theoretically as described in the following paragraphs.

A tape may be placed on a floor or pavement, tensioned at its standardized pull, and have its ends marked on the slab. The tape may then be held in the air above the slab supported at its ends only and pulled until its ends (as marked with plumb bobs) coincide with the marked points on the slab. The pull necessary to make the end points coincide is the normal tension. Its value may be measured with spring balances.

A theoretical method of determining the normal tension is to equate the expression for elongation of the tape caused by tension to the expression for shortening of the tape caused by sag. The resulting expression can be solved for P_1, the normal tension:

$$P_1 = \frac{0.204W\sqrt{AE}}{\sqrt{P_1 - P}}$$

P_1 occurs on both sides of the equation, but its value for a particular tape may be determined by a trial-and-error method. For a normal-weight 100-ft tape, this value will probably be in the range of 20 lb. As described in detail in Section 4-7, most distance measurements are too large because of the cumulative errors of sag, poor alignment, slope, and so on. As a result, overpulling the tape is a good idea for ordinary surveying because it tends to reduce some of these errors and improve the precision of the work. For such surveys, an estimated pull of approximately 30 lb is often recommended. For very precise surveying, the normal tension is applied to the tape by using accurate spring balances.

4-7 COMBINED TAPING CORRECTIONS

If corrections must be made for several factors at the same time (e.g., wrong-length tape, slope, temperature), the individual corrections per tape length may be computed separately and added together (taking into account their signs) in order to obtain a combined correction for all. Since each correction will be relatively small, it is assumed that they do not appreciably affect each other and each can be computed independently. Furthermore, the nominal tape length (100 ft) may be used for the calculation. This means that although the tape may be 99.92 ft long at 68°F and a temperature correction is to be made for a 40°F increase in temperature, the increase in tape length can be figured as (40)(0.0000065)(100) without having to use (40)(0.0000065)(99.92). Example 4-7 illustrates the application of several corrections to a single distance measurement.

Example 4-7

A distance was measured on a uniform slope of 8% and was found to be 1665.2 ft. No field slope corrections were made. The tape temperature at the time of measurement was 18°F. What is the correct horizontal distance measured if the tape is 100.06 ft long at 68°F?

Solution

The corrections per tape length are computed, added together, and then multiplied by the number of tape lengths.

$$\text{Slope correction / tape length} = -\frac{(8)^2}{(2)(100)} = -0.3200 \text{ ft}$$

Temp correction/tape length – (50)(0.0000065)(100) = –0.0325 ft

Standardization error/tape length = +0.0600 ft

Total correction/tape length = –0.2925 ft

Correction for entire distance = (16.652)(–0.2925) = –4.87 ft

Actual distance = Measured distance + Σ corrections
= 1665.2 + (–4.87) = **1660.3 ft**

4-8 COMMON MISTAKES MADE IN TAPING

Some of the most common mistakes made in taping are described in this section, and a method of eliminating each is suggested.

Reading Tape Wrong. A frequent mistake made by tapemen is reading the wrong number on the tape, for example, reading a 6 instead of a 9 or a 9 instead of a 6. As tapes become older these mistakes become more frequent because the numbers on the tape become worn. These blunders can be eliminated if tapemen develop the simple habit of looking at the adjacent numbers on the tape when readings are taken.

Recording Numbers. Occasionally, the recorder will misunderstand a measurement that is called out to him or her. To prevent this kind of mistake, the recorder can repeat the values aloud, including the decimals, as he or she records them.

Missing a Tape Length. It is not very difficult to lose or gain a tape length in measuring long distances. The careful use of taping pins, described in Section 3-10, should prevent this mistake. In addition, the surveyor can often eliminate such mistakes by cultivating the habit of estimating distances by eye, pacing, or better yet, by taking stadia readings whenever possible.

Mistaking End Point of Tape. Some tapes are manufactured with 0- and 100-ft points at the very ends of the tapes. Other tapes place them at a little distance from the ends. Clearly, tapemen should not make mistakes like these if they have taken the time to examine the tape before they begin to take measurements.

Making 1-Foot Mistakes. When a fractional part of the tape is being used at the end of a line, it is possible to make a 1-ft mistake. Mistakes like these can be prevented by carefully following the procedure described for such measurements in Section 3-10. Also helpful are the habits of calling out the numbers and checking the adjacent numbers on the tape.

4-9 ERRORS IN TAPING

In the following paragraphs we discuss briefly the common taping errors. As these errors are studied it is important to notice that the effect of most of them is to cause the surveyor to get too much distance. If the tape is not properly aligned, not horizontal, or sags too much, if a strong wind is blowing the tape to one side, or if the tape has shortened on a cold day, it takes more tape lengths to cover the distance.

Alignment of Tape. A good rear tapeman can align the head tapeman with sufficient accuracy for most surveys, although it is more accurate to use an instrument telescope to keep the tape on line. In some cases it is necessary to use one of these instruments when establishing new lines or when the tapemen are unable to see the ending point because of the roughness of the terrain. For the latter case it may be necessary to set up intermediate points on the line to guide the tapemen.

It is probable that most surveyors spend too much time improving their alignment, at least in proportion to the time they spend trying to reduce other more important errors. In taping a 100-ft distance the tape would have to be 1.414 ft out of line to cause an error of 0.01 ft. From this value it can be seen that for ordinary distances alignment errors should not be appreciable. As a matter of fact, experienced tapemen should have no difficult in keeping their alignment well within a foot of the correct line by eye, particularly when the lines are only a few hundred feet or less in length. They should be able to keep the tapes lined up with the range poles within at least 0.3 or 0.4 ft, which would cause an error of less than 0.001 ft for each tape length (a negligible error for most steel tape measurements).

Accidental Taping Errors. Because of human imperfections, tapemen cannot read tapes perfectly, cannot plumb perfectly, and cannot set taping pins perfectly. They will place the pins a little too far forward or a little too far back. These errors are accidental in nature and will tend to cancel each other somewhat. Generally, errors caused by setting pins and reading the tapes are minor, but errors caused by plumbing may be very important. Their magnitudes can be reduced by increasing the care with which the work is done or by taping along slopes and applying slope corrections to avoid plumbing.

Tape Not Horizontal. If tapes are not held in the horizontal position, an error results that causes the surveyor to obtain distances that are too large. These errors are cumulative and can be quite large when surveying is done in hilly country. Here the surveyor must be very careful.

If a surveyor deliberately holds the tape along a slope, he or she can correct the measurement with the slope correction formula

$$C = \frac{v^2}{2s}$$

which was presented in Section 4-5. It might be noticed that if one end of a 100-ft tape is 1.414 ft above or below the other end, an error of

$$\frac{(1.414)^2}{200} = 0.01\text{ ft}$$

is made. From this expression it can be seen that the error varies as the square of the elevation difference. If the elevation difference is doubled, the error quadruples. For a 2.828-ft elevation difference, the error made is

$$\frac{(2.828)^2}{200} = 0.04\text{ ft}$$

Incorrect Tape Length. These important errors were discussed in Section 4-3 and must be given careful attention if good work is to be done. For a given tape of incorrect length, the errors are cumulative and can add up to sizable values.

Temperature Variations. Corrections for variations in tape temperature were discussed in Section 4-4. Errors in taping caused by temperature changes are usually thought of as being cumulative for a single day. They may, however, be accidental under unusual circumstances with changing temperatures during the day and also with different temperatures at the same time in different parts of the tape. It is probably wiser to limit tape variations instead of trying to correct for them no matter how large they may be. Taping on cloudy days, early in the morning, or late in the afternoon or using Invar tapes are effective means of limiting length changes caused by temperature variations.

Sag. Sag effects (discussed in Section 4-6) cause the surveyor to obtain excessive distances. Most surveyors attempt to reduce these errors by overpulling their tapes with a force that will stretch them sufficiently to counterbalance the sag effects. A rule of thumb used by many for 100-ft tapes is to apply an estimated pull of approximately 30 lb. This practice is satisfactory for surveys of low precision, but it is not adequate for those of high precision because the amount of pull required varies for different tapes, different support conditions, and so on. It is also difficult to estimate by hand the force being applied. A better method is to use a spring balance for applying a definite tension to a tape, the tension required having been calculated or determined by a standardized test to equal the normal tension of the tape.

Miscellaneous Errors. Some of the miscellaneous errors that effect the precision of taping are (1) wind blowing plumb bobs; (2) wind blowing tape to one side, causing the same effect as sag; and (3) taping pins not set exactly where plumb bobs touch ground.

4-10 MAGNITUDE OF ERRORS

To get some feeling for the effects of common taping errors, consider Table 4-1. In this table various sources of errors are listed together with the variations that would be necessary for each error to have a magnitude of ±0.01 ft when a 100-ft distance is measured with a 100-ft tape.

TABLE 4-1 Errors of ±0.01 ft in 100-ft Measurements

Source of error	Magnitude of error
Incorrect tape length*	0.01 ft
Temperature variation*	15° F
Tension or pull variation*	15 lb
Sag*	7.5 in. sag at center line
Alignment*	1.4 ft at one end
Tape not level*	1.4 ft difference in elevation
Plumbing	0.01 ft
Marking	0.01 ft
Reading tape	0.01 ft

Source: J. F. Dracup and C. F. Kelly, *Horizontal Control As Applied to Local Surveying Needs* (Falls Church, Va.: American Congress on Surveying and Mapping, 1973), p.16.

*May be greatly minimized if sufficient field data are obtained and appropriate mathematical corrections are applied.

For the information shown in Table 4-2 it is assumed that a 100-ft steel tape is used to measure a distance of 100 ft. The ground is gently sloping, and the entire length of the tape can be held in a horizontal position at one time. It is further assumed that the measurement is subject to the set of accidental or random errors shown in the table. At the bottom of the table the magnitude of the most probable total random error is calculated as described in Chapter 2.

Should a distance be taped with the most probable error per tape length determined in Table 4-2 and found to be 2240.320 ft, the total probable error as described in Chapter 2 will equal

$$E_{\text{total}} = \pm E\sqrt{n} = (\pm 0.0119)\left(\sqrt{22.40320}\right) = \mathbf{\pm 0.056\ ft}$$

TABLE 4-2 Error Calculation Example

Source of error	Magnitude of error (ft)	Magnitude of error squared
Incorrect tape length	±0.005	0.000025
Temperature variation (15° F)	±0.009	0.000081
Tension or pull variation (5 lb)	±0.003	0.000009
Plumbing (0.005 ft)	±0.005	0.000025
Marking (0.001 ft)	±0.001	0.000001
Reading tape (0.001 ft)	±0.001	0.000001
		$\Sigma = 0.000142$

$$\text{Most probable error} = \sqrt{0.000142} = \pm 0.0119 \text{ ft}$$

$$\text{Corresponding precision} = \frac{0.0119}{100} = \frac{1}{8403}$$

Source: J. F. Dracup and C. F. Kelly, *Horizontal Control As Applied to Local Surveying Needs* (Falls Church, Va.: American Congress on Surveying and Mapping, 1973), p. 16.

4-11 SUGGESTIONS FOR GOOD TAPING

If the surveyor studies the errors and mistakes that are made in taping, he or she should be able to develop a few rules of thumb that will appreciably improve the precision of the work. Following is a set of rules that have proved helpful in the field:

1. Tapemen should develop the habit of estimating by eye the distances they are measuring because it will enable them to avoid most major blunders.
2. When reading a foot mark at an intermediate point on a tape, the tapeman should glance at the adjacent foot marks to be sure that he or she has read the mark correctly.
3. All points that the tapeman establishes should be checked. This is particularly true when the plumb bob has been used to set a point such as a tack in a stake.
4. It is easier to tape downhill whenever feasible.
5. If time permits (often it may not), distances should be taped twice, once forward and once back. Taping in the two different directions should prevent repeating the same mistakes.
6. Tapemen should assume stable positions when pulling the tape. This usually means feet widespread, leather thong wrapped around the hand (or use of taping clamp), standing to one side of the tape with arms in close to the body, and applying pull to the tape by leaning against it.
7. Since most taping gives distances that are too large, the surveyor can improve his or her work for ordinary surveys by pulling the tape very firmly, by estimating the smaller number when a reading seems to be halfway between two values, and perhaps even by setting taping pins slightly to the forward side.

4-12 TAPING PRECISION

Different kinds of surveys require different accuracies and many surveys must meet definite specifications. For instance, many states, cities, and towns have passed laws requiring certain minimum accuracy standards that must be met for work within their boundaries for various kinds of work. For instance, a common set of values might include the minimum values 1/5000 for rural land surveys and 1/10,000 for property within city boundaries.

The presentation of definite values that should be obtained for good, average, and poor taping is difficult because what is good under one set of conditions may be poor for another set of conditions. For example, a precision of 1/2500 is poor when taping is done along a level road, but it may very well be satisfactory when the work is done through heavy underbrush in mountainous terrain.

Below are presented some supposedly reasonable precision values that should be expected under ordinary taping conditions. The average value is probably sufficient for most preliminary surveys and the good value is desirable for most other surveys.

Poor 1/2500
Average 1/5000
Good 1/10,000

Taping can be done with precision much higher than 1/5000 if careful attention is paid to the reduction of the errors discussed previously in this chapter. Thus by carefully controlling tape tensions, precisely measuring tape temperatures, applying corrections, using hand levels to keep tapes horizontal or taping along slopes, and minimizing the use of plumb bobs but necessitating the measurement of the slopes and the application of the appropriate corrections, the surveyor will be able to tape with precisions of 1/10,000 and better.

Problems

For Problems 4-1 to 4-5, distances were measured with tapes assumed to be 100.00 ft long. Later the tapes were standardized and found to have different tape lengths. Determine the correct distances measured in each case.

	Measured distance (ft)	Correct tape length (ft)	
4-1.	1496.82	99.97	(Ans.: 1496.37 ft)
4-2.	843.56	100.06	
4-3.	1910.09	100.05	(Ans.: 1911.05 ft)
4-4.	1588.80	99.96	
4-5.	3942.18	99.98	(Ans.: 3941.39 ft)

For Problems 4-6 to 4-8, distances were measured with tapes assumed to be 30.0 m long. Later the tapes were standardized and found to have different lengths. Determine the correct distances measured in each case.

	Measured distance (m)	Correct tape length (m)	
4-6.	657.89	30.02	
4-7.	718.19	29.96	(Ans.: 717.23 m)
4-8.	1706.98	29.97	

4-9. The actual distance between two marks used at a university for standardizing tapes is 99.98 ft. When a certain tape was held along this line, the surveyor, thinking that the distance between the marks was 100.00 ft, observed that the tape was 100.03 ft long. What is the correct length of the tape?
(Ans.: 100.01 ft)

4-10. Repeat Problem 4-9 for values of 100.04, 100.00, and 99.96 ft, respectively.

4-11. Repeat Problem 4-9 for values of 30.02, 30.00, and 29.95 m, respectively.
(Ans.: 29.97 m)

For Problems 4-12 to 4-17, it is desired to lay off certain horizontal distances for building layouts. The length of the tapes used in Problems 4-12 to 4-15 are not 100.00 ft and not 30.00 m for Problems 4-16 to 4-17. Determine the field dimensions (or actual tape readings) that should be used with the incorrect tape lengths so that the correct dimensions are obtained.

	Desired dimension	Correct tape length	
4-12.	640.00 ft × 160.00 ft	100.06 ft	
4-13.	670.00 ft × 360.00 ft	100.04 ft	(Ans.: 669.73 ft × 359.86 ft)
4-14.	520.00 ft × 710.00 ft	99.96 ft	
4-15.	262.40 ft × 306.62 ft	99.93 ft	(Ans.: 262.58 ft × 306.83 ft)
4-16.	140.00 m × 220.00 m	30.06 m	
4-17.	220.00 m × 382.20 m	29.96 m	(Ans.: 220.29 m × 382.71 m)

4-18. It is desired to lay off a horizontal distance equal to 696.66 ft with a tape that is 50.10 ft long (not 50.00 as indicated on the scale). What should the recorded distance be?

4-19. A 50-ft woven tape is used to set the corners for a building. If the tape is actually 49.92 ft long, what should the recorded distances be if the building is to be 210.00 ft by 106.35 ft?
(Ans.: 210.34 ft × 106.52 ft)

4-20. A distance is measured through rough country and found to be 4620.80 ft. If on the average a plumb bob is used every 50 ft with a probable error of ±0.02 ft, what is the probable total error in the whole distance?

4-21. Rework Problem 4-20 if the plumb bob is used every 30 ft on the average and the distance measured is 2836.20 ft.
(Ans.: ±0.19 ft)

For Problems 4-22 to 4-25, distances were measured with 100-ft steel tapes and their average temperatures estimated. From these values and the standardized tape lengths (100.000 ft at 68° F), determine the correct distances measured.

	Recorded distance (ft)	Average tape temperature at time of measurement (°F)	
4-22.	2854.42	28	
4-23.	733.65	108	(Ans.: 733.84 ft)
4-24.	2166.90	98	
4-25.	2362.88	8	(Ans.: 2361.96 ft)

4-26. A steel tape that has a length of 100.000 ft at 68°F is to be used to lay off a building with the dimensions 330.00 ft by 682.00 ft.

a. What should the tape readings be if the tape temperature is 18°F at the time of measurement?

b. Repeat part (a) assuming the tape temperature is 108°F.

4-27. A 1-acre lot (43,560 ft^2) is to be staked out on level ground with the dimensions 204.00 ft by 213.53 ft. The standardized length of the tape at 68°F is 100.06 ft. If the tape temperature is 18°F, what field dimensions should the survey party use to lay out this lot? (Ans.: 203.94 ft × 213.47 ft)

4-28. Repeat Problem 4-27 assuming the tape temperature is 98°F.

For Problems 4-29 to 4-31, distances were measured with 30-m steel tapes and their average temperatures estimated. From these values and the standardized tape length (30.000 m at 20°C), determine the correct distances measured.

	Recorded distance (m)	Average tape temperature at time of measurement (°C)	
4-29.	520.80	30	(Ans.: 520.86 m)
4-30.	876.54	8	
4-31.	1322.28	36	(Ans.: 1322.53 m)

4-32. a. A distance was measured at a temperature of 28°F and was found to be 3110.20 ft. If the tape has a standardized length of 99.96 ft at 68°F, what is the correct distance measured?

b. If the same distance was measured at a temperature of 18°F, what would be the probable tape reading?

4-33. A distance was measured with a 100-ft steel tape at a temperature of 108°F and found to be 3122.66 ft. The next winter the distance was remeasured with the same tape at a temperature of 18°F and was found to be 3120.15 ft. What part of the discrepancy between the two measurements should be caused by the temperature difference?

(Ans.: 1.83 ft)

4-34. a. A distance is measured as 2076.34 ft when the tape temperature is 38°F. If the same distance is measured again with the same tape at a temperature of 98°F, what distance should be expected, neglecting other errors?

b. If the tape is 100.00 ft long at 68°F, what is the "true" distance measured?

4-35. A 100.00-ft tape is used to measure an inclined distance and the value determined is 1242.46 ft. If the slope is 5%, what is the correct horizontal distance obtained using the slope correction formula? (Ans.: 1240.91 ft)

4-36. A 200.00-ft tape is used to measure an inclined distance and the value determined is 2372.86 ft. If the slope is 8%, what is the correct horizontal distance? Use an appropriate slope correction formula.

4-37. Repeat Problem 4-36 assuming the measured value is 3346.20 ft and the slope is 4%. (Ans.: 3343.52 ft)

4-38. A 30.00-m tape is used to measure an inclined distance and the value determined is 1293.60 m. If the slope is 6%, what is the correct horizontal distance? Use an appropriate slope correction formula.

4-39. The slope distance between two points is measured and found to be 943.77 ft. If the elevation difference between the points is 13.70 ft, what is the horizontal distance between them? (Ans.: 943.67 ft)

4-40. A distance is measured along a 6°34′ slope and found to be 861.09 ft. What is the horizontal distance measured?

4-41. A surveyor wants to lay off a 825.00-ft horizontal distance using a steel tape measuring along a slope that is 6°10′ from the horizontal. What slope distance must be measured? (Ans.: 829.80 ft)

4-42. Repeat Problem 4-41 assuming the horizontal distance required is 784.36 ft and the vertical angle is 5°36′.

4-43. Two tapemen were attempting to lay off a 100-ft distance. Unfortunately, the tape became caught on a root at the 40-ft point on the tape, causing a 1.30-ft offset. What actual distance did they lay off? (Ans.: 99.96 ft)

4-44. A distance was measured along a constant slope between two points A and B and found to be 574.88 ft. If the elevation of point A is 742.62 ft and that of B is 866.96 ft, what is the horizontal distance between the two points?

CHAPTER 5

Electronic Distance Measuring Instruments (EDMs)

5-1 INTRODUCTION

Over the last 100 years there has been a constant but gradual improvement in the quality of surveying equipment. Not only have better tapes been manufactured but better instruments for measuring elevations and directions have been produced. Within the past few decades the pace of improvement has quickened tremendously as the electronic distance-measuring instruments (EDMs) and the total station instruments have come of age.

Now there are available to the surveyor devices with which he or she can measure short distances of a few feet or long distances of many miles with extraordinary speed and precision. These devices save time and money and reduce the size of conventional survey parties. Furthermore, they may be used with the same facility where there are obstacles intervening, such as lakes, canyons, standing crops, swampy and timbered terrain, hostile landowners, or heavy traffic. Electronic distance-measuring devices have revolutionized distance measurement not only for geodetic surveys but also for ordinary land surveys.

Another important point is that EDMs automatically display direct readout measurements, with the result that mistakes are greatly reduced. At the push of a button the operator may have the value shown in either feet or meters. The distances obtained are slope distances, but with most of the instruments vertical angles are measured and horizontal distances are computed and displayed.

The original EDMs were individual units used only for measuring distances. Today, however, the most modern instruments are a part of total station instruments (see Chapter 10). Whether they are a part of a total station or whether they are an individual unit the principle of operation is the same.

For the surveyor to fully understand the electronic measurement of distances, he or she would need to have a good background in physics and electronics. Fortunately, however, a person can readily use electronic distance-measuring equipment without really understanding very much about the physical phenomena involved. Although the equipment is complex, its actual operation is automatic and requires less skill than is needed with tapes.

Electronic distance-measuring devices were seldom used until the 1960s. The equipment was heavy, expensive, and required rather sophisticated maintenance. Nevertheless, their tremendous advantages—the instant and precise measurement of distances up to 10 miles or more over forbidding terrain—inspired much research, which has led to far better models. As a result, the equipment has become lighter, cheaper, and more maintenance-free. Generally, the manufacturers do not recommend that the owners of EDMs service the units. They feel that more often than not, worse problems will be created, resulting in even higher repair costs. It is practical, however, to call the company when an instrument problem occurs and describe the difficulty. They may be able to explain the necessary adjustment over the phone without the necessity of returning the equipment to the factory.

5-2 BASIC TERMS

A few brief definitions that the reader will encounter in the sections to follow are presented below.

An *electronic distance-measuring device* is an instrument that transmits a carrier signal of electromagnetic energy from its position to a receiver located at another position. The signal is returned from the receiver to the instrument such that two times the distance between the two positions is measured.

Visible light is generally defined as that part of the electromagnetic spectrum to which the eye is sensitive. It has a wavelength in the range of 0.4 to 0.7 μm (micrometers or microns).

Infrared light has frequencies below the visible portion of the spectrum: they lie between light and radio waves with wavelengths from 0.7 to 1.2 μm. Nevertheless, infrared light is generally put in the light wave category because the distance calculations are made by the same technique.

An *electro-optical* instrument is one that transmits modulated light, either visible or infrared. It consists of a measuring unit and a reflector (Figure 5-1).

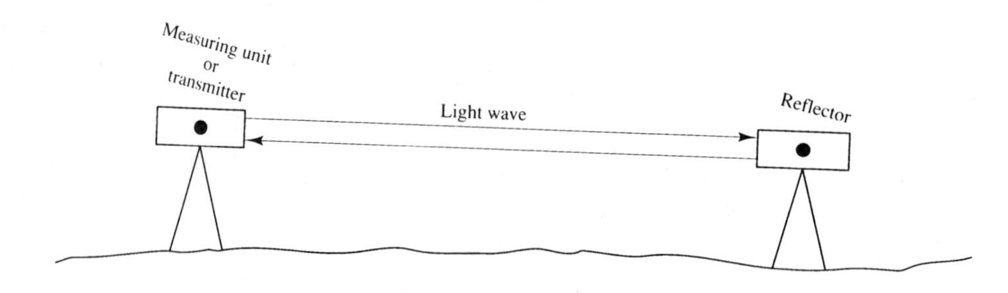

FIGURE 5-1 Electro-optical EDM.

A *reflector* (Figures 5-1 to 5-3) consists of several so-called retrodirective glass cube corner prisms mounted on a tripod. The sides of the prisms are perpendicular to each other within very close tolerances. Due to the perpendicular side, the prisms will reflect the light rays back in the same direction as the incoming light rays, hence the term *retrodirective.* Figure 5-2 illustrates this situation.

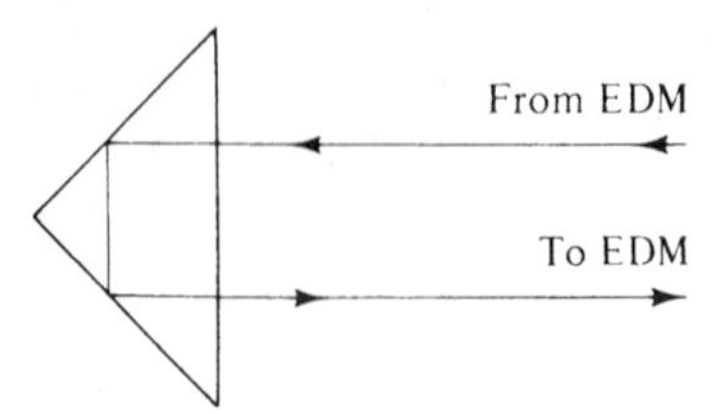

FIGURE 5-2 Reflection of visible or infrared light by a reflector.

The trihedral prisms mounted on a tripod will reflect light back to the transmitting unit even though the reflector is out of perpendicularity with the light wave by as much as 20°. The number of prisms used depends on the distance to be measured and on visibility conditions.

The distance-measuring capability of an electro-optical instrument can be increased by increasing the number of prisms used. In general, the distance that can be measured is doubled if the number of reflectors is squared. If a cluster of nine prisms is used instead of a cluster of three, the distance measured can probably be doubled. Using more than about 12 or 15 prisms does not help very much for most instruments. If more than 12 or 15 prisms are necessary another EDM with a greater range capability is probably needed. The distance-measuring capability of a particular EDM is affected not only by the number of prisms used but also by their quality and cleanliness.

FIGURE 5-3 Prism retroflectors. (Courtesy of Topcon America Corporation.)

A *laser* is one of several devices that produces a very powerful single-color beam of light. The word *laser* is an acronym for "light amplification by stimulated emission of radiation." Low-intensity light waves are generated by the device and amplified into a very intense beam that spreads only slightly even over long distances. The waves produced fall into the visible or into the infrared frequencies of the electromagnetic spectrum. *Personnel must be informed about the necessary precautions to the eyes when working with lasers and they must strictly adhere to these safety requirements.*

A *microwave* is an electromagnetic radiation that has a long wavelength and a low frequency and lies in the region between infrared and shortwave radio. The microwaves used in distance measurements have wavelengths from 10 to 100 μm.

5-3 TYPES OF EDMS

EDMs are classified as being electro-optical instruments or microwave instruments. The distinction between them is based upon the wavelengths of the electromagnetic energy which they transmit. The electro-optical instruments transmit light in short wavelengths of about 0.4 to 1.2 μm. This light is visible or just above the visible range of the spectrum. Microwave instruments transmit long wavelengths somewhere between 10 and 100 μm.

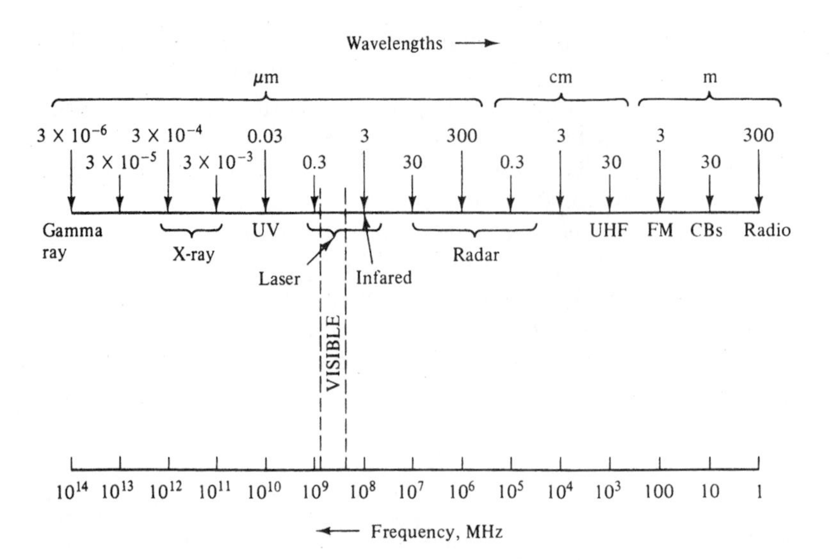

FIGURE 5-4 Electromagnetic radiation: frequencies and wavelengths.

Figure 5-4 shows in one sketch the relative frequencies and wavelengths for various forms of electromagnetic radiation, including gamma rays, lasers, radar, and others. In this figure Hz is an abbreviation for hertz, which is a unit of frequency equal to 1 cycle per second. The abbreviation MHz represents megahertz or 10^6 hertz.

The light-wave systems (including lasers and infrared) have a transmitter at one end of the line to be measured and a reflector at the other end. The reflector consists of one or more corner prisms, which were discussed in the last section. For short distances of a few hundred feet, bicycle reflectors or reflector tape are satisfactory.

Almost all short-range EDMs in use today for measurements up to a few miles are of the infrared type (Figures 5-5 and 5-6), although there are infrared instruments available that can be used for measurements up to and over 10 miles under normal conditions. There are laser EDMs available which are useful for short-range measurements as well as for much longer distances.

FIGURE 5-5 DM-80 infrared electronic distance module. (Courtesy of Cubic Precision.)

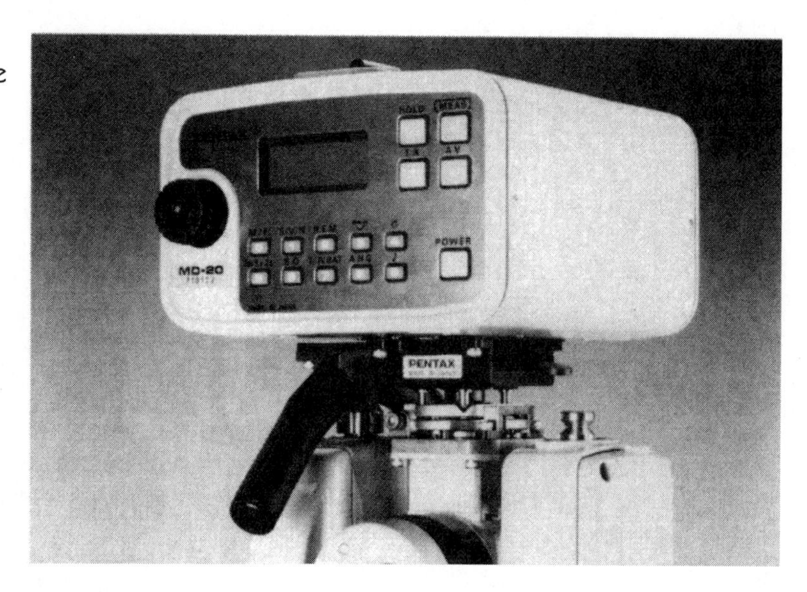

FIGURE 5-6 MD-20 infrared electronic distance meter. (Courtesy of Pentax Instruments.)

The reader should realize that almost all (probably over 90%) of the measurements made with EDMs are for distances of less than 1000 m. An even larger percentage of layout measurements are made for even shorter distances of perhaps 400 or 500 m or less.

One advantage that lasers have over the infrared type is that they are visible, and that makes them very useful for situations where sighting is difficult. With some laser equipment it is possible to sight the EDM to a point, place a red dot of light on the point with the laser, press a button, and measure the distance to the dot. This is very useful for measuring to points of difficult access, such as towers, church steeples, or bottoms of holes in the ground.

With microwave systems two instruments are necessary: a transmitting system and a receiving system. The beam is transmitted from one end of the line, received at the other end, and returned to the master instrument. Microwave instruments have the advantages that the waves penetrate through fog or rain and have a speech link between the two instruments. However, they are more affected by humidity than are the light-wave instruments. Another problem with microwave systems is the wider beam induced. This can cause some difficulties when surveys are being made inside buildings, in underground situations, or near water surfaces.

5-4 TIMED-PULSE INSTRUMENTS

Probably most of the readers of this book have on occasion wished that they had a device with which they could instantaneously measure the distance from their positions to some other point without the necessity of taping to that point or setting up an instrument or a reflector at the other point. Today several companies manufacture instruments with this capability.

These devices employ a timed-pulse infrared signal which is transmitted by a laser diode (Figure 5-7). To obtain a measurement, the time required for a signal to travel to and from an object is required. These amazing instruments can be used with or without prisms. If prisms are not used, the instruments can be used to measure distances up to 500 or 1000 ft depending on light conditions. With reflecting prisms their range is extended to several miles.

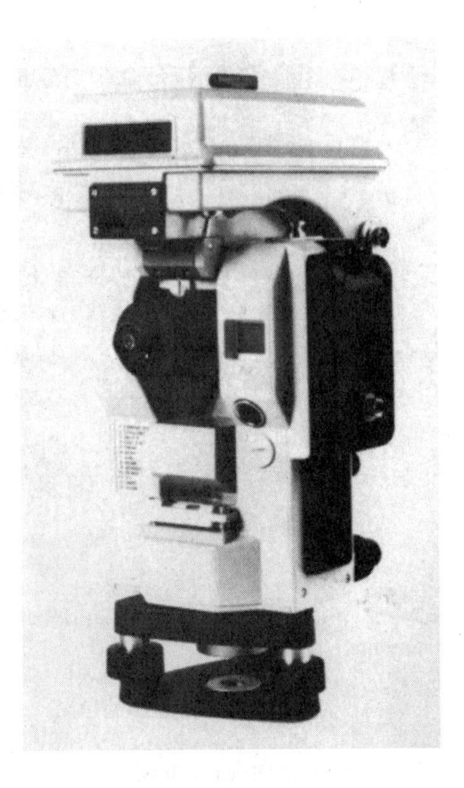

FIGURE 5-7 Eldi 10 infrared pulsed electronic distance meter connected to the ETh 3 electronic theodolite. (Courtesy of Carl Zeiss, Inc., Thornwood, N.Y.)

It is to be remembered that all objects are reflective, and if something moves onto the beam (such as a car or tree limb) the distance to that object will be determined and not the distance to the desired point. To help the user sight on the correct item, timed-pulse instruments make use of a visible laser beam which is used to identify the feature desired.

When prisms are not used, these EDMs can be used to obtain distances to topographic features that have vertical components, such as buildings, bridges, or stockpiles of materials. Best results are obtained if the item being observed has a smooth light-colored surface perpendicular to the beam. For such cases distances can be measured with a standard deviation of about ±(5mm + 3ppm, or parts per million) for distances up to 500 or 600 ft. For medium-dark to dark surfaces and to corners, edges, and inclined surfaces, the maximum distances are not as large and the accuracies are poorer.

Just think of all the measurements that can be taken with one of these instruments without having the rodmen climb all over tanks, buildings, stockpiles of materials, and other items to hold the reflectors. It can also be used to locate shorelines for hydrographic surveys, the inside walls of tunnels, and so on. Furthermore, if the equipment is interfaced with an electronic theodolite and data collector, areas and volumes can be computed, and cross sections and profiles plotted.

5-5 SETTING UP, LEVELING, AND CENTERING EDMS

Before setting up the EDM tripod some thought should be given to where you will need to stand in relation to the instrument to make the necessary observations. In other words, you need to determine how the tripod legs should be placed so that you can comfortably stand between them while sighting (see Figure 5-8).

FIGURE 5-8 Planning an instrument setup.

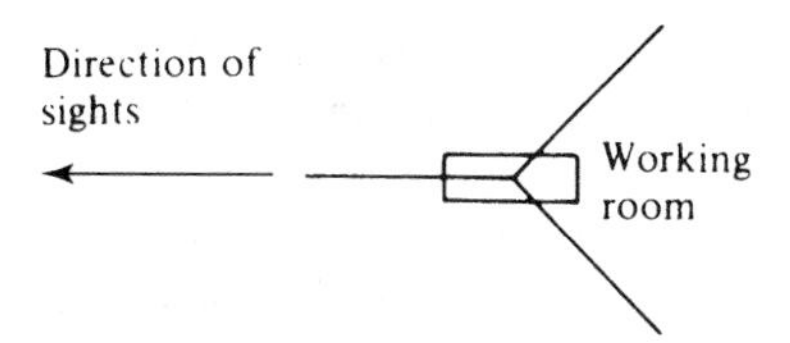

The tripod is ideally placed on solid ground, where the instrument will not settle as it most certainly will in muddy or swampy areas. For such locations it may be necessary to provide some special support for the instrument, such as stakes or a platform. The tripod legs should be well spread apart and adjusted so that the instrument is approximately level. The instrumentman walks around the instrument and pushes each leg firmly into the ground. On hillsides it is usually convenient to place one leg uphill and two downhill. Such a setup provides better stability.

There are two types of tripods available: extension-leg types and fixed-leg types. Fixed-leg tripods are more rigid and provide more stability during measurements. On the other hand, extension-leg tripods are more easily transported in vehicles and provide more flexibility in setting up.

It is necessary to set an EDM over a definite point, such as an iron pin or a tack in a stake, before a measurement is taken. To do this the tripod is first set up and roughly centered over the point in question with the tripod top or plate as level as possible. Then the *tribrach* (see Figure 5-9) is mounted on the tripod. The tribrach contains three leveling screws (which are enclosed and dust proof), an optical plummet, and a circular or bull's-eye level.

FIGURE 5-9 A tribrach. It contains level screws, level vial, and optical plummet. EDMs and other equipment can easily be interchanged on top of tribrach. (Courtesy Topcon America Corporation.)

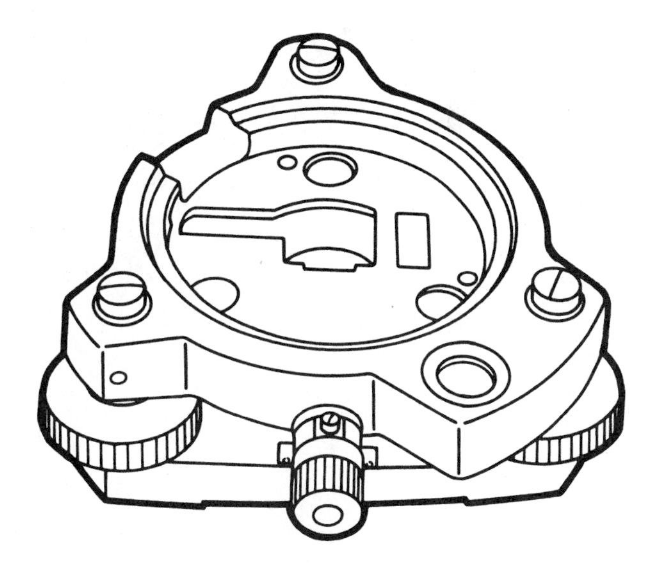

The optical plummet consists of a set of lenses and a mirror which enable the surveyor to look into a viewing port on the side of the instrument and, when the instrument is level, to see a point on the ground directly under the instrument center. In addition, a plumb bob suspended from a hook or chain beneath the instrument center may also be used to roughly center the instrument before the optical plummet is used. With the plumb bob and/or the optical plummet the instrument is centered over the point as closely as possible by adjusting the positions of the tripod legs. After the instrument is fairly close to being centered the plumb bob (if used) is removed.

Then the tribrach clamp is loosened and the tribrach slid on the flat tripod top until the optical cross hair is centered on the station point.

To level the tribrach the three leveling screws and the bull's-eye level are used. Although the bull's-eye levels are much less sensitive than the tube levels that will be discussed in Chapter 6, they are quite satisfactory for use with EDMs, total stations, and some other surveying equipment.

The bubble in the bull's-eye level is centered by adjusting one or more of the three leveling screws. For this discussion reference is made to Figure 5-10. If level screws #1 and #2 (shown in the figure) are rotated in opposite directions the bubble will move to the left or right along axis x-x. Rotation of level screw 3 will cause the bubble to move up or down along axis y-y. Once the bubble is roughly centered an automatic compensator takes over, levels the instrument, and keeps it level. The compensator is discussed in Section 6-7 of this text.

Finally the EDM is attached to the tripod with a threaded bolt that sticks up from the top of the tripod into the tribrach or leveling base of the instrument.

FIGURE 5-10 Centering the bubble.

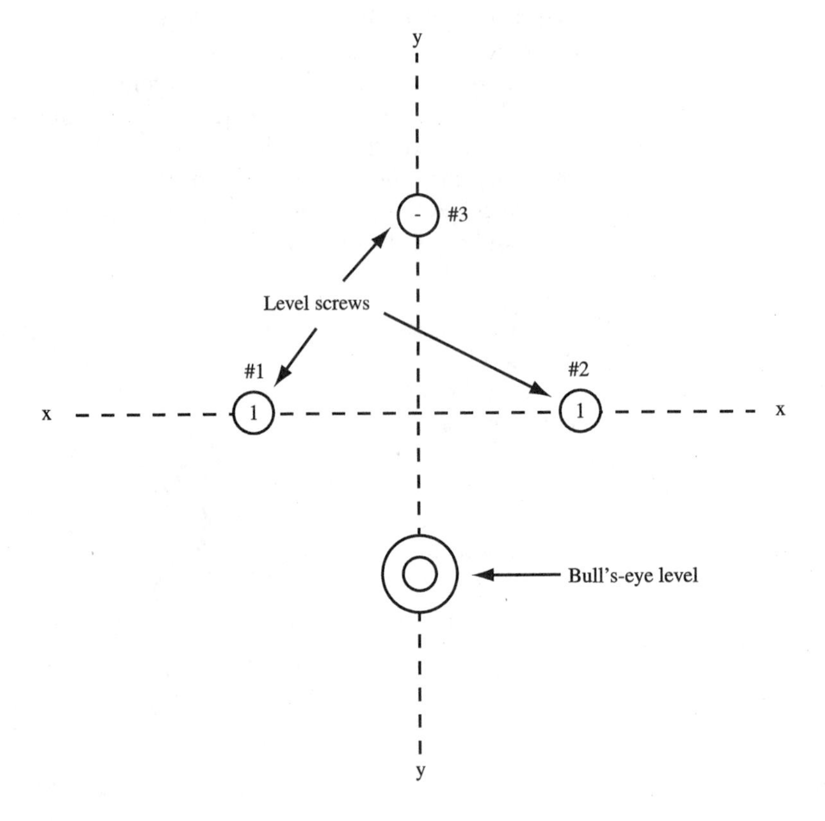

5-6 NECESSARY STEPS FOR MEASURING DISTANCES WITH EDMS

Battery packs are used to provide power for operating EDMs. The batteries must be fully charged before field work is undertaken. The wise surveyor will carry a spare battery pack on jobs to avoid the possibility of inconvenient delays. It should be noted that most EDMs can be operated on automobile batteries if desired.

EDMs are so completely automated that their use can be learned very quickly. To measure a distance with an EDM, it is necessary to set up the instrument and the reflectors, to sight on the reflectors, and finally to measure and record the value obtained. These steps are briefly described in this section.

1. The EDM is set up, centered, and leveled at one end of the line to be measured.
2. The prism assembly is placed at the other end of the line and is carefully centered over the end point. This is accomplished by holding the prism pole vertically over the point with the aid of a level which is attached to the rod, by fastening the prism pole in a tripod or bipod placed over the point, or by using a tripod with a tribrach to which the prism assembly is attached.
3. The height of the instrument to the axis of the telescope and the height of the prism center are measured and recorded. With the adjustable prism poles it is common practice to set the prism assembly to the same height as the EDM telescope.
4. The telescope is sighted toward the prism and the power is turned on.
5. The fine adjustment screws are used to point the instrument toward the reflector until the maximum strength returning signal is indicated on a signal scale.
6. The measurement is accomplished simply by pushing a button. The user may measure in feet or in meters, as desired. The display will be two places beyond the decimal if the measurement is in feet, or to three places beyond the decimal if in meters. If the measurements are recorded in a field book, it is not a bad idea to take an extra measurement using different units (feet or meters). Such a habit may enable one to pick up mistakes in recording, such as transposing numbers.
7. The values obtained are recorded in the field book or they may be recorded in an electronic data collector. Total station instruments (discussed in Chapter 10) automatically record the measurements taken.

5-7 ERRORS IN EDM MEASUREMENTS

Some people seem to have the impression that measurements made with electronic instruments such as EDMs and total stations are free of errors. The truth, of course, is that errors are present in any type of measurement, no matter how modern and up-to-date the equipment used may be. The sources of errors in EDM work are the same as for other types of surveying work: personal, natural, and instrumental.

Personal Errors

Personal errors are caused by such items as not setting the instruments or reflectors exactly over the points and not measuring instrument heights and weather conditions perfectly. For precise distance measurement with an EDM, it is necessary to center the instrument and reflector accurately over the end points of the line. Plumb bobs hanging on strings from the centers of instruments have long been used by surveyors for centering over points and are still in use. For very precise work, however, the *optical plummet* is preferred over the plumb bob. With the optical plummet, errors in centering can be greatly reduced (usually, to a fraction of a millimeter). Its advantage over the plumb bob is multiplied when there is appreciable wind. The axis of the optical plummet must be checked periodically under lab conditions.

Natural Errors

The natural errors present in EDM measurement are caused by variations in temperature, humidity, and pressure. Some EDMs automatically correct for atmospheric variables, for others it is necessary to "dial in" corrections into the instruments, and for still others it is necessary to make mathematical corrections. For instruments requiring adjustments, manufacturers provide tables, charts, and explanations in their manuals regarding how the corrections are to be made. For microwave instruments it is necessary to correct for temperature, humidity, and pressure, whereas for electro-optical instruments (Figure 5-11), humidity can be neglected. (The humidity effect on microwaves is more than 100 times its effect on light waves.) Meteorological data should be obtained at each end of a line and sometimes at intermediate points and averaged if a higher precision is desired. On hot sunny days it is desirable to protect both the EDM equipment and any meteorological equipment with umbrellas.

FIGURE 5-11 Geodimeter 220 electro-optical distance meter. (Courtesy of Geotronics of North America, Inc.)

Snow, fog, rain, and dust affect the visibility factor for EDMs and drastically reduce the distances that can be measured. The distance that can be measured with a particular instrument is sometimes affected by the shimmering phenomenon when sights are taken near the earth's surface. Obviously, keeping sights as high as possible above the ground will reduce this factor. It is also a good practice to keep electro-optical instruments pointed away from the sun, due to background radiation effects.

Whenever possible, surveyors using microwave equipment should try to avoid high-voltage power lines, microwave towers, and so on. Should they have to work in the vicinity of any of these structures, such information should be noted carefully in the field records since it may later be helpful should specifications not be met.

Instrumental Errors

Instrumental errors are usually quite small if the equipment has been carefully adjusted and calibrated. Each EDM has a built-in error, which varies from model to model. In general, the more expensive the equipment, the smaller the error.

When a measurement is made with an EDM, the beam goes from the electrical center of the instrument to the effective center of the reflector and then back to the electrical center of the instrument. However, the electrical center of an EDM does not coincide exactly with a plumb line through the center of the tribrach, nor does the effective center of the reflector coincide exactly with the point over which it is placed. The difference between the electrical center of the EDM and the plumb line is determined by the manufacturer and adjustments are made at manufacture to compensate for these errors.

The exact location of the effective center of the reflector is not easy to obtain because of the fact that light travels through the glass prisms more slowly than it does through air. The effective center is actually located behind the prisms and the distance needs to be subtracted from measured values. This error, which is a constant, may be compensated for at manufacture or it may be done in the field. It is to be realized that on some occasions reflectors made by different manufacturers (and thus having different constants) may be used with the EDM. If an EDM is supposed to be used with a prism with a 30-mm constant but is being used with a reflector with a 40-mm constant, it may be necessary to dial the 40 mm into the instrument.

5-8 CALIBRATION OF EDM EQUIPMENT

It is important to check EDM measurements periodically against the length of a National Geodetic Survey (NGS) base line or other accurate standard. From the differences in the values an *instrument constant* can be determined. This constant, which is a systematic error, enables the surveyor to make corrections to future measurements. Although a constant is furnished with the equipment, it is subject to change. It is as though we have an incorrect-length tape and have to make a numerical correction. With the value so determined, a correction is applied to each subsequent measurement. In addition, a record of the results and dates when the checks were made should be kept in the surveyor's files in case of future legal disputes involving equipment accuracy.

Unfortunately, there seems to be a rather large percentage of practicing surveyors who think that EDMs can continually be used accurately, without the necessity of calibration. However, just as other measuring devices, these instruments must be standardized periodically. Both electro-optical and microwave equipment should be checked against an accurate base line at frequent intervals. The electronically obtained distances should be determined while taking into consideration differences in elevations, meteorological data, and so on.

Most modern theodolites and automatic levels remain in good calibration for quite a few years when subjected to normal usage. Sadly, this is not the case for EDMs (aging of the electronic components is one reason) and they must be calibrated at least every few months, even if they are very carefully used. Recognizing the need for frequent calibration of these instruments, the NGS in 1974 began setting up calibration base lines around the United States. Today there are more than 300 such base lines.

The NGS compiles and publishes a description of base lines for each state showing location, elevations, horizontal distances, and other pertinent data. Copies may be obtained by writing to NGS, National Ocean Survey, Rockville, MD 20852. Copies are also on hand at each state's surveying society offices and on the Internet. A detailed description of one of these base lines and its location is presented in Figure 5-12.

FIGURE 5-12 Description of a sample base line.

US DEPARTMENT OF COMMERCE - NOAA
NOS - NATIONAL GEODETIC SURVEY
ROCKVILLE MD 20852 - MAY 27, 1981

CALIBRATION BASE LINE DATA
BASE LINE DESIGNATION: CLEMSON
PROJECT ACCESSION NUMBER: G16441

QUAD: N340824
SOUTH CAROLINA
PICKENS COUNTY

LIST OF ADJUSTED DISTANCES (APRIL 22, 1981)

FROM STATION N	ELEV.(M)	TO STATION N	ELEV.(M)	ADJ. DIST.(M) HORIZONTAL	ADJ. DIST.(M) MARK - MARK	STD. ERROR(MM)
0	228.600	150	229.071	149.9999	150.0006	.2
0	228.600	430	231.412	429.9949	430.0041	.4
0	228.600	1070	241.844	1069.9287	1070.0106	.6
150	229.071	430	231.412	279.9950	280.0048	.4
150	229.071	1070	241.844	919.9287	920.0174	.5
430	231.412	1070	241.844	639.9336	640.0186	.4

DESCRIPTION OF CLEMSON BASE LINE
YEAR MEASURED: 1981
CHIEF OF PARTY: WJR

THE BASE LINE IS LOCATED ABOUT 4.3 KM (2.7 MI) SOUTHEAST OF CLEMSON AND 2.9 KM (1.8 MI) WEST OF PENDLETON ALONG THE RIGHT-OF-WAY ON THE WEST SIDE OF UNITED STATES HIGHWAY 76 WHERE IT CROSSES THE ANDERSON-PICKENS COUNTY LINE. OWNERSHIP--MR. GEORGE WEATHERS, SOUTH CAROLINA HIGHWAY DEPARTMENT, PRE-CONSTRUCTION ENGINEER, POST OFFICE BOX 191, COLUMBIA, SOUTH CAROLINA 29202, TELEPHONE 803-758-3414.

TO REACH THE BASE LINE FROM THE SOUTH CAROLINA STATE HIGHWAY 93 OVERPASS ON UNITED STATES HIGHWAY 76 EAST OF CLEMSON, GO SOUTH ON HIGHWAY 76 FOR 2.6 KM (1.65 MI) TO NEW HOPE ROAD ON THE RIGHT AND THE 1070 METER POINT IN THE SOUTHWEST ANGLE OF THE INTERSECTION. TO REACH THE OTHER MARKS AND 0 METER POINT, CONTINUE SOUTH ON HIGHWAY 76 FOR 0.64 KM (0.4 MI) TO THE 430 METER POINT ON THE RIGHT, CONTINUE SOUTH FOR 0.32 KM (0.2 MI) TO A SIDE ROAD RIGHT AND THE 150 METER POINT IN THE SOUTHWEST ANGLE OF THE INTERSECTION, AND CONTINUE 0.16 KM (0.1 MI) SOUTH TO THE 0 METER POINT ON THE RIGHT ABOUT 0.9 M (3.0 FT) LOWER THAN THE HIGHWAY AND 21.5 M (70.5 FT) SOUTH OF THE ANDERSON COUNTY LINE.

THE 0 METER POINT IS A STANDARD NATIONAL GEODETIC SURVEY DISK STAMPED 0 1980, SET INTO THE TOP OF A ROUND CONCRETE MONUMENT 38 CM (15 IN) IN DIAMETER FLUSH WITH THE GROUND, LOCATED 40.9 M SOUTHEAST OF TELEPHONE JUNCTION BOX NUMBER 6, 21.5 M SOUTH OF ANDERSON COUNTY LINE SIGN, 21.5 M EAST OF WEST EDGE OF THE WOODS, 3.65 M WEST OF WEST EDGE OF HIGHWAY 76, AND 1.15 M SOUTHEAST OF A METAL WITNESS POST.

THE BASE LINE IS A NORTH-SOUTH LINE WITH THE 0 METER POINT ON THE SOUTH END. IT IS MADE UP OF THE 0, 150, 430, AND 1070 METER POINTS WITH A POINT FOR THE CALIBRATION OF 100 FOOT TAPES SET SOUTH OF THE 0 METER POINT. ALL OF THE MARKS ARE SET ON A LINE PARALLEL TO THE HIGHWAY AND IN THE DITCH ON THE WEST SIDE OF THE ROAD. THIS BASE LINE IS NOT CONNECTED TO THE NATIONAL NOR THE LOCAL CONTROL NETWORKS.

THIS BASE LINE WAS ESTABLISHED IN CONJUNCTION WITH THE STATE OF SOUTH CAROLINA. FOR FURTHER INFORMATION, CONTACT THE DIRECTOR, SOUTH CAROLINA GEODETIC SURVEY, SOUTH CAROLINA DIVISION OF RESEARCH AND STATISTICAL SERVICES, OFFICE OF GEOGRAPHIC STATISTICS, 915 MAIN STREET, SUITE 203, COLUMBIA, SOUTH CAROLINA 29201. TELEPHONE 803-758-3604.

If a base line is not available, two points can be set up (as much as 5 miles apart for microwave equipment) and the distances between them measured. A point can be set in between the other two and the two segmental distances measured (see Figure 5-13). The sum of those two values should be compared with the overall length. Should the three points not be in a line, angles will be needed to compute the components of the two segments to compare with the overall straight-line distance between the end points. The instrument constant can be calculated as follows—noting that the constant will be present in each of the three measurements:

$$\text{instrument constant} = AC - AB - BC$$

It is also desirable to check barometers, thermometers, and psychrometers approximately once a month or more often if they are subject to heavy use. These checks can usually be made with equipment that is available at most airports.

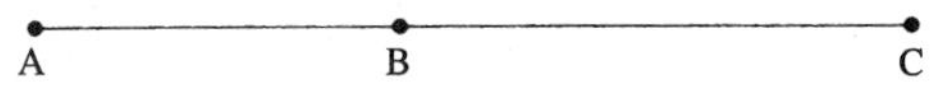

FIGURE 5-13 Calibrating EDM equipment.

A B C

5-9 ACCURACIES OF EDMS

The manufacturers of EDMs usually list their accuracies as a standard deviation. (It is anticipated that 68.3% of measurements of a quantity will have an error equal to or less than the standard deviation.) The manufacturers give values that consist of a fixed or constant *instrumental error* which is independent of distance, plus a *measuring error* in parts per million (ppm) that varies with the distance being measured.

EDMs are listed as having accuracies in the range from about ± (3 mm instrumental error + a proportional part error of 1 ppm) up to ±(10 mm + 10 ppm), where ppm is parts per million of the distance involved.

The first of these errors is of little significance for long distances but may be very significant for short distances of 100 or 200 ft or less. On the other hand, the proportional part error is of little significance for short or long distances. It can be seen that for short distances EDM equipment may on occasion not provide measurements as precise as those obtained by good taping.

It is assumed that a particular EDM manufacturer provides a standard deviation for one of their instruments as being equal to ±(5 mm + 5 ppm). With these data the estimated standard error and precision for the measurement of a 100-m distance with this instrument can be computed as follows, where the 100 m is converted to millimeters by multiplying it by 1000.

$$\text{Error} = \pm\left(5 + \frac{5 \times 100 \times 1000}{1{,}000{,}000}\right) = \pm 5.5\text{ mm}$$

$$\text{Precision} = \frac{5.5}{(100)(1000)} = \frac{1}{18{,}182}$$

In the same fashion the estimated values for standard error and precisions that would be obtained when measuring several other distances with this instrument are shown in Table 5-1.

TABLE 5-1 Typical Standard Error for Precision Values for an EDM

Distance measured [m(ft)]	Standard error (mm)	Precision
30 (98.4)	±5.15	1/5882
40 (131.2)	±5.20	1/7692
60 (196.8)	±5.30	1/11,321
100 (328.1)	±5.50	1/18,182
500 (1640.4)	±7.50	1/66,667

From the values shown you can see that depending on the precision desired, distances of 100 ft or less can theoretically be measured as precisely or more precisely with a steel tape than with this typical EDM.

Despite the fact that these short distances can on occasion be measured more precisely with tapes than with EDMs, tapes are not often used in practice. The average surveying crew is so accustomed to using EDMs that, if at all possible, they avoid the more difficult process of taping. In fact, they tape so rarely that they probably would find it quite difficult to achieve the high precisions desired and they would be quite prone to making mistakes.

5-10 COMPUTATION OF HORIZONTAL DISTANCES FROM SLOPE DISTANCES

All EDM equipment is used to measure slope distances. For most models the values obtained are corrected for the appropriate meteorological and instrumental corrections and then reduced to horizontal components. It is possible at the same time to determine the vertical components (or differences in elevations) for the slope distance. If the distance involved is rather short and/or the precision is not extremely high, the horizontal distance (h) equals the slope distance times the cosine of the vertical angle α:

$$h = s \cos \alpha$$

For greater distances and higher precision requirements, earth's curvature and atmospheric refraction will need to be considered. With many of the newer instruments, however, the computations are made automatically. As with taping along slopes, the horizontal values may be computed by making corrections with the slope correction formula (described in Section 4-5), by using the Pythagorean theorem, or by applying trigonometry. If the slope is quite steep, say greater than 10% to 15%, the slope correction formula (which is only approximate) should not be used.

To compute horizontal distances it is necessary either to determine the elevations at the ends of the line or to measure vertical angles at one or both ends. Example 5-1 illustrates the simple calculations involved when the elevations are known.

EXAMPLE 5-1

A slope distance of 1654.32 ft was measured between two points with an EDM. It is assumed that the atmospheric and instrumental corrections have been made. If the difference in elevation between the two points is 183.36 ft and if the heights of the EDM and its reflector above the ground are equal, determine the horizontal distance between the two points using:

a. The slope correction formula.
b. The Pythagorean theorem.

Solution

a. Using the slope correction formula:

$$C = \frac{v^2}{2s} = \frac{(183.86)^2}{(2)(1654.32)} = 10.22 \text{ ft}$$

$$h = 1654.32 - 10.22 = \mathbf{1644.10\ ft}$$

b. Using the Pythagorean theorem for right triangles:

$$h = \sqrt{(1654.32)^2 - (183.86)^2} = \mathbf{1644.07\ ft}$$

For Example 5-1 it was assumed that the distance measured was parallel to the ground; that is, the heights of the EDM and the reflector above the end points were equal. If these values are not equal, that fact must be accounted for in the calculations.

Some EDMs have the capability of vertical angle measurement, but with others it is necessary to set up a transit or theodolite to measure the angles. In the past the EDM was frequently mounted on a theodolite enabling the surveyor to measure the slope angle at the same time as the slope distance. As a result, the heights of the EDM, the reflector, the theodolite, and even the target would sometimes be different.

5-11 TRAINING OF PERSONNEL

Measurements obtained with EDMs may be extremely accurate or very inaccurate. The difference can usually be attributed to the amount of training (or lack of it) provided to personnel. The procedures used for field measurements vary somewhat with equipment from different manufacturers. As a result, such procedures are not described here. Each manufacturer provides an equipment manual in which operating instructions are given.

Despite the simplicity of EDM operation, however, it is advisable to have personnel trained as thoroughly as possible to get optimum results. No matter how fine and expensive the instruments used for a survey, they are of little value if they are not used knowledgeably. If a company spends $8,000 or $10,000 or more for EDM equipment but will not spend an extra few hundred dollars for detailed equipment training, the result will probably be poor economy.

A rather common practice among instrument companies is for one of their representatives to provide a few hours or even a day of training when an instrument is delivered. Such a short period is probably not sufficient to obtain the best results and to protect the large investment in equipment. Some manufacturers offer a week-long training course. Participation in such a course is normally a wise investment and will yield long-range dividends. In addition, surveyors should consider attending workshops or short courses on EDM theory and practice given by the National Geodetic Survey and by manufacturers and universities.

5-12 SUMMARY COMMENTS ON EDMS

EDMs enable us to measure short or long distances quickly and accurately over all types of terrain. Survey or traverse points can quickly be selected without having to choose those to which we can conveniently tape. If the instruments do not convert slope distances to horizontal components, we have to make the conversions. Furthermore, it becomes necessary to consider earth's curvature and atmospheric refraction in determining horizontal components if elevation differences at the ends of a line are more than a few hundred feet and/or if the required accuracy is >1/50,000. We particularly need to remember that all instruments get out of adjustment and thus need to be checked against a standard at frequent intervals.

An important and common use of EDMs and total stations is the measurement of lines without the necessity of setting up anywhere along the lines. For illustrative purposes it is desired to determine the distances *AB* and *BC* (see Figure 5-14) with the instrument set up at point *X*. The distances *XA*, *XB*, and *XC* are accurately determined and then the angles α and β shown in the figure are measured as described in subsequent chapters. Finally, the distances and directions of the lines *AB* and *BC* are computed as is also described in later chapters.

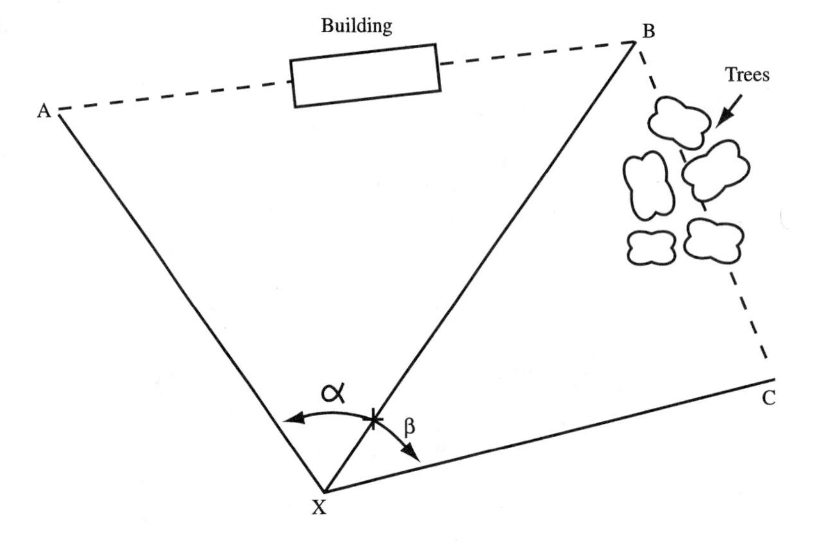

FIGURE 5-14 Determine the distances and directions of *AB* and *BC* without setting up along either line.

Problems

5-1. If the manufacturer of a particular EDM states that the purported accuracy with their instrument is ±(6 mm + 5 ppm), what error can be expected if a distance of 2400 m is measured? (Ans. = ±18mm)

5-2. Repeat Problem 5-1 assuming the distance measured is 2200 ft and the purported accuracy is ±(0.02 ft + 5 ppm).

5-3. What are the advantages of EDMs over steel tapes for distance measurement?

5-4. What atmospheric conditions need to be measured and factored in when using EDMs of the infrared type?

5-5. The slope distances shown were measured with an EDM. The vertical angles (measured from the horizontal) were also determined. Calculate the horizontal distance for each case.

	Slope distance (ft)	Vertical angle	
a.	626.56	+7°10′00″	(Ans.: 621.66 ft)
b.	1848.69	−3°20′20″	(Ans.: 1845.55 ft)

5-6. The slope distances shown were measured with an EDM. In addition, the slope percentages shown were determined. Compute the horizontal distances and differences in elevation.

	Slope distance (ft)	Slope percentage
a.	544.32	+4.8
b.	816.69	−5.2

5-7. The slope distances shown were measured with an EDM while the differences in elevation between the ends of each line were determined by leveling. What are the horizontal distances?

	Slope distance (ft)	Elevation difference (ft)	
a.	1296.42	−36.96	(Ans.: 1295.89 ft)
b.	816.67	−17.03	(Ans.: 816.49 ft)
c.	713.26	+13.44	(Ans.: 713.13 ft)

5-8. The horizontal length of a line is 1362.40 ft. If the vertical angle from the horizontal is −5°12′, what is the slope distance?

5-9. The slope distance between two points was measured with an EDM and found to be 1873.26 ft. If the zenith angle (the angle from the vertical to the line) is 97°10′22″, compute the horizontal distance. (Ans.: 1858.60 ft)

5-10. Repeat Problem 5-9 assuming the slope distance measured is 1502.66 ft and the zenith angle is 86°10′24″.

5-11. An EDM was set up at point A (elevation 770.22 ft) and used to measure the slope distance to a reflector at point B (elevation 886.42 ft). If the heights of the EDM and reflector above the points are 4.60 ft and 5.22 ft, respectively, compute the horizontal distance between the two points if the slope distance is 1429.86 ft. (Ans.: 1425.08 ft)

CHAPTER

Introduction to Leveling

6-1 IMPORTANCE OF LEVELING

The determination of elevations, known as *leveling,* is a comparatively simple but extremely important process. The significance of relative elevations cannot be exaggerated. They are so important that one cannot imagine a construction project in which they are not critical. From terracing on a farm or the building of a simple wall to the construction of drainage projects or the largest buildings and bridges, the control of elevations is of the greatest importance.

6-2 BASIC DEFINITIONS

Presented below are a few introductory definitions that are necessary for the understanding of the material to follow. In this chapter and the next, additional definitions are presented as needed for the discussion of leveling. Several of these terms are illustrated in Figure 6-1.

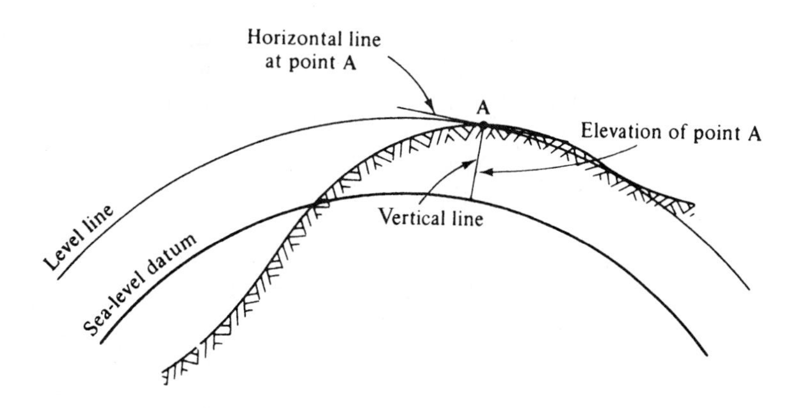

FIGURE 6-1 Some leveling terms.

A *vertical line* is a line parallel to the direction of gravity. At a particular point it is the direction assumed by a plumb-bob string if the plumb bob is allowed to swing freely. Because of the earth's curvature, plumb-bob lines at points some distance apart are not parallel, but in plane surveying they are assumed to be.

A *level surface* is a surface of constant elevation that is perpendicular to a plumb line at every point. It is best represented by the shape that a large body of still water would take if it were unaffected by tides.

The *elevation* of a particular point is the vertical distance above or below a reference level surface (normally, sea level).

A *level line* is a curved line in a level surface all points of which are of equal elevation.

A *horizontal line* is a straight line tangent to a level line at one point.

6-3 REFERENCE ELEVATIONS OR DATUMS

For a large percentage of surveys, it is reasonable to use some convenient point as a reference or datum with respect to which elevations of other points can be determined. For instance, the surface of a body of water in the vicinity can be assigned a convenient elevation. Any value can be assigned to the datum, for example, 100 ft or 1000 ft, but the assigned value is usually sufficiently large so that no nearby points will have negative elevations.

In the past many assumed reference elevations were used in the United States, even for surveys of major importance. An assumed value may have been given to the top of a hill, the surface of one of the Great Lakes, low-water mark of a river, or any other convenient point. These different datums created confusion and the current availability of the *sea-level datum* is a great improvement.

In the United States the sea-level datum is the value of mean sea level determined by averaging the hourly elevations of the sea over a long period of time, usually 19 years. This datum, which is mean sea level, is defined as the position the ocean would take if all tides and currents were eliminated. Its position is rising very slowly—probably because of the gradual melting of the polar ice caps and perhaps because of the erosion of ground surfaces. The change, however, is so slow that it does not prevent the surveying profession from using sea level as a datum. Over the past century the level of the world's oceans has risen by approximately 6 in. (15 cm).

In 1878, at Sandy Hook, New Jersey, the National Geodetic Survey (then the U.S. Coast and Geodetic Survey) began working on a transcontinental system of precise levels. In 1929, the agency made an adjustment of all first-order leveling in the United States and Canada, and established a datum for elevations throughout the country. This datum is referred to as the *National Geodetic Vertical Datum of 1929* (NGVD 29) or just the *sea-level datum of 1929*.

The use of the so-called "mean sea level" (NGVD 29) has several disadvantages. Our sea level is gradually rising around the world (perhaps up to 0.01 ft per year). Sea level is not the same thing in different places. For instance, on the west coast of the United States sea level is about 2 or 3 ft higher than it is on the east coast. In addition it varies up and down these coasts.

In Alaska the retreat of many of the great glaciers has removed tremendous loads from the ground in some locations. The ground has responded in those places by rising several feet. The removal of enormous quantities of water and oil from the ground in parts of certain areas of the country has caused subsidence of the land. Other items affecting sea levels are earthquakes, volcanoes, and the rise and fall of mountain chains.

Because of the above changes and as a result of additional leveling through the decades since NGVD 29 was established, the NGS has performed a new adjustment. It is referred to as the North American Vertical Datum of 1988 (NAVD 88) although it was not really completed until 1991. For this readjustment stations in Mexico were considered as well as those in the United States and Canada.

With NAVD 88 the NGS has fixed one mean sea level station along the St. Lawrence Seaway. It's known as Father Point Rimouski and to it all other elevations in the United States are referred. These later values vary slightly from the older "mean sea level values" of NGVD 29. Thus, it is absolutely necessary for surveyors to identify the datum to which their elevations are referred.

6-4 FIRST-, SECOND-, AND THIRD-ORDER SURVEYS

Surveys are frequently specified as being first-, second-, or third-order. Because these terms are occasionally used in various places in the text, a brief description of them is presented in this section so that the reader will not be puzzled when they are encountered.

The Federal Geodetic Control Committee (FGCC) has established a set of accuracy standards for horizontal and vertical control surveys. There are three major classifications and they are given in descending order as to accuracy requirements.

1. First-order surveys are made for the primary national control network, metropolitan area surveys, and scientific studies. (In detail they are very accurate surveys used for military defense, sophisticated engineering projects, dams, tunnels, and studies of regional earth crustal movements.)
2. Second-order surveys are done somewhat less accurately than are first-order surveys. They are used to densify the national network as well as for subsidiary metropolitan control. (In detail they are used for control along tidal boundaries, large construction projects, interstate highways, urban renewal, small reservoirs, and monitoring crustal movements.)
3. Third-order surveys are done somewhat less accurately than are second-order ones. They are general control surveys referring to the national network. (In detail they are used for local control surveys, small engineering projects, small-scale topographical maps, and boundary surveys.)

Let us assume that leveling is begun at one benchmark (BM), run to establish elevations at various points, and then run back to the original BM or to another one. We will find that there will be some discrepancy or *misclosure* between the given BM elevation and the measured value at the end of the route. The FGCC says that for first-order leveling misclosure should be no larger than 3mm $\sqrt{K}$ to 5mm $\sqrt{K}$ where K is the length of the leveling route in kilometers. For second-order surveys misclosures should be no greater than 6mm $\sqrt{K}$ to 8mm $\sqrt{K}$ and for third-order work no greater than 12mm $\sqrt{K}$. There are various classes within first- and second-order surveys thus explaining the ranges of values as 3mm $\sqrt{K}$ to 5mm $\sqrt{K}$.

The National Geodetic Survey establishes first- and second-order vertical control stations about 1 km apart in interrelated square grids that are 50 to 100 km on a side. Other governmental agencies (federal, state, county, and city) provide vertical control of lower orders. Information concerning first- and second-order control can be obtained by contacting the Director, NGS Information Center, National Ocean Survey, NOAA, Rockville, MD 20852, telephone (301) 443–8611.

One of the most important tasks of the National Geodetic Survey and the U.S. Geological Survey is the establishment of a network of known elevations throughout the country. As a result of their work, eventually there will be no area in the country that will be far from a point whose elevation in relation to sea level is known.

The National Geodetic and U.S. Geological Surveys have set up throughout the United States monuments whose elevations have been precisely determined. A monument, called a *bench mark,* is usually of concrete and has a brass disk embedded in it, as illustrated in Figure 6-2. For many years the practice was to record precise elevations on the monuments. Today, however, either the elevation is provided only to the nearest foot or no elevation at all is given. The reason is that the elevations of these monuments may be changed by frost action, earthquakes, vandalism, and so on. The elevations are carefully checked at intervals and the surveyor who needs the

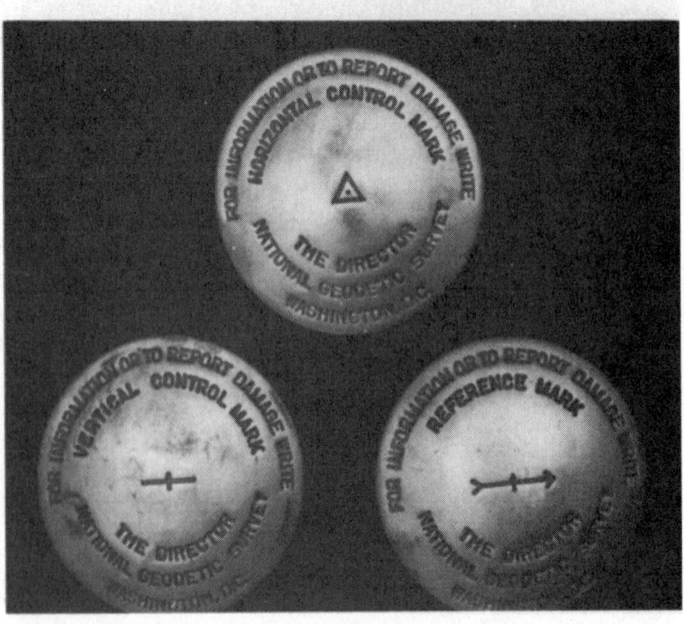

FIGURE 6-2 National Geodetic Survey brass disk marks are designed to be set in concrete or clamped to rods. (Courtesy of National Geodetic Survey.)

elevation of a particular monument should contact the agency that set it, even if a value is given on the monument. More than half a million of these monuments have been placed in the United States and its possessions.

6-5 METHODS OF LEVELING

There are three general methods of leveling: trigonometric, barometric, and spirit. Although the surveyor is concerned almost entirely with spirit leveling, a brief description of each type is given below.

Trigonometric leveling is leveling in which distances and angles are measured and elevation difference calculated by trigonometry. The method can be used to determine elevations of inaccessible points such as mountain peaks, church steeples, or offshore platforms. The procedure works very well for distances up to 800 or 1000 ft but for greater distances it may be necessary to consider the effect of earth curvature.

To determine the difference in elevation between two points it is necessary to measure either the horizontal distance (H) between the points or the slope distance (S) between them as well as the vertical angle (α) or the zenith angle (z). These values are shown in Figure 6-3 along with the necessary formulas for computing elevation differences.

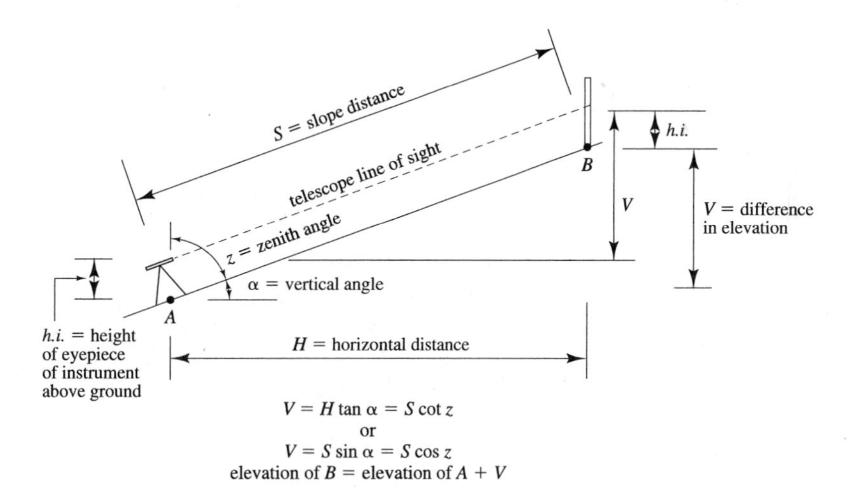

FIGURE 6-3 Measuring elevation difference V by trigonometry.

A surveying instrument is shown in the left side of the figure at point A. Notice that the height of the line of sight of the instrument's telescope is a distance *h.i.* above point *A*. You will then see that it is necessary to sight on a rod a distance *h.i.* above point *B* to obtain the correct angle measurement. If this is not done, the differences must be considered in the calculations.

If we sight to the top of a church steeple where we can't place a rod, we read the vertical or zenith angle to the top of the steeple. Then the elevation at the top of the steeple is equal to the ground elevation at the instrument plus the height of the instrument eyepiece above the ground plus the vertical distance V (H tan α) from the eyepiece to the top of the steeple.

Barometric leveling involves the determination of elevations by measuring changes in air pressure. Although air pressures can be measured with mercurial barometers, these instruments are cumbersome and fragile and are impractical for surveying purposes. Instead, the light and sturdy but less precise aneroid barometers commonly called *altimeters* are used.

Surveying altimeters have been manufactured with which elevations can be determined within about 2 ft. Such precision is sufficient only for preliminary or reconnaissance work. They do nevertheless offer the advantage that approximate elevations over a large area can quickly be determined. If careful procedures and larger aneroid barometers are used, much better results can be obtained.

Barometer readings at the same elevation vary with local air pressure conditions and are affected by temperature and humidity variations. If one barometer is used, it is adjusted at a known elevation and then readings are taken at other points. The barometer is then brought back to the original or starting point and read again. If its reading is different, it is necessary to distribute the difference around to the other points. A very similar procedure is illustrated in Section 8-1.

It is desirable to use more than one barometer. At least three are desirable for reasonably accurate elevation measurements. Ideally, one barometer is placed at a known elevation higher than that of the desired point, and one is placed at a lower point of known elevation. All of the barometers are read, and from the readings at the known elevations, corrections can be made at the point at which the elevation is being determined.

Spirit leveling, also called *direct leveling,* is the usual method of leveling. Vertical distances are measured in relation to a horizontal line, and these values are used to compute the differences in elevations between various points. A spirit level (discussed in Section 6-6) is used to fix the line of sight of the telescope. This line of sight is the assumed horizontal line with respect to which vertical distances are measured. The term *spirit leveling* is often used because the bubble tubes of many of the older levels were filled with alcohol.

If at this point you need a little more elaboration on this definition of spirit or direct leveling, please look ahead to Figure 7-1. There a level is set up and its bubble tube centered so that the instrument is level. A level rod (described in Section 6-8) graduated in foot or meter units increasing from its base is set up at a point of known elevation. It is then sighted through the leveled telescope and a reading taken. The elevation of the telescope line of sight equals the ground elevation plus the rod reading, which is the vertical distance from the ground up to the line of sight. Then the level rod is placed at some other point whose elevation is desired. A sight is taken on the rod at that new point and its elevation equals the level line-of-sight elevation minus the rod reading. This process can be repeated over and over. This is direct or spirit leveling.

6-6 THE LEVEL

A level consists of a high-powered telescope (20 to 45 diameters) with a spirit level attached to it in such a manner that when its bubble is centered, the line of sight is horizontal. The purposes of the telescope are to fix the direction of the line of sight and to magnify the apparent size of objects observed. The invention of the telescope has been credited to several people including the Dutch optician Hans Lippershey about 1607 and Galileo in about 1609. In colonial times telescopes were too large for practical surveying and they were not used on surveying instruments until the end of the nineteenth century. Their use tremendously increases the speed and precision with which measurements can be taken. These telescopes have a vertical cross hair for sighting on points and a horizontal cross hair with which readings are made on level rods. In addition, they may have top and bottom stadia hairs.

The telescope has three main parts: the objective lens, the eyepiece, and the reticle. The *objective lens* is the large lens located at the front or forward end of the telescope. The *eyepiece* is the small lens located at the viewer's end. It is actually a microscope that magnifies and enables the viewer to see clearly the image formed by the objective lens. The cross hairs and stadia hairs form a network of lines that is fastened to a metal ring called the *reticle, reticule,* or *cross-hair ring.* In older instruments the cross hairs were formed from spider webs or fine wires. In newer instruments, however, the cross hairs are formed by etching lines on a glass reticule. A line drawn from the point of intersection of the cross hairs and the optical center of the objective system is called the *line of sight* or *line of collimation* (the word *collimation* meaning lining up or adjusting the line of sight).

Some old telescopes are said to be *external focusing.* Their objective lens is mounted on a sleeve that moves back and forth as the focusing screw is turned. *Internal focusing* is used for modern telescopes. These instruments have a lens that moves back and forth internally between the objective lens and the reticule.

The level tube is an essential part of most surveying instruments. The closed glass tube is precisely ground on the inside surface to make the upper half curved in shape so that the bubble is stabilized. If the tube were not so curved, the bubble would be very erratic. The tube is filled with a sensitive liquid (usually a purified synthetic alcohol) and a small air bubble. The liquid used is stable and nonfreezing for ordinary temperature variations. The bubble rises to the top of the liquid against the curved surface of the tube. The tangent to the circle at that point is horizontal and perpendicular to gravity. The tube is marked with divisions symmetrical about its midpoint; thus when the bubble is centered, the tangent to the bubble tube is a horizontal line and is parallel to the axis of the telescope. For some very precise instruments the amount of fluid can be increased or decreased, enabling the surveyor to adjust and control the length of the bubble when temperature changes occur. Many instruments have circular bull's-eye levels as described in the next section.

6-7 TYPES OF LEVELS

Several types of levels are discussed in this section. These include the dumpy level, the automatic or self-leveling level, and the tilting level. In addition, a few comments are presented concerning construction site lasers used for elevation work.

Dumpy Level

The dumpy level was commonly used in surveying work until the past few decades. Although these excellent, sturdy, and long-lasting devices have very largely been replaced with more modern instruments, they are discussed here to help the student understand leveling.

Originally, the dumpy level had an inverting eyepiece and as a result was shorter (thus the name "dumpy") than its predecessors but had the same magnification power. A typical dumpy level with its various parts is shown in Figure 6-4. Its major components are its *telescope, level tube,* and *leveling head.* These and other parts are indicated in the figure. This level has a tube-type level vial and four leveling screws.

FIGURE 6-4 Dumpy level (an old instrument). (Courtesy of Berger Instruments.)

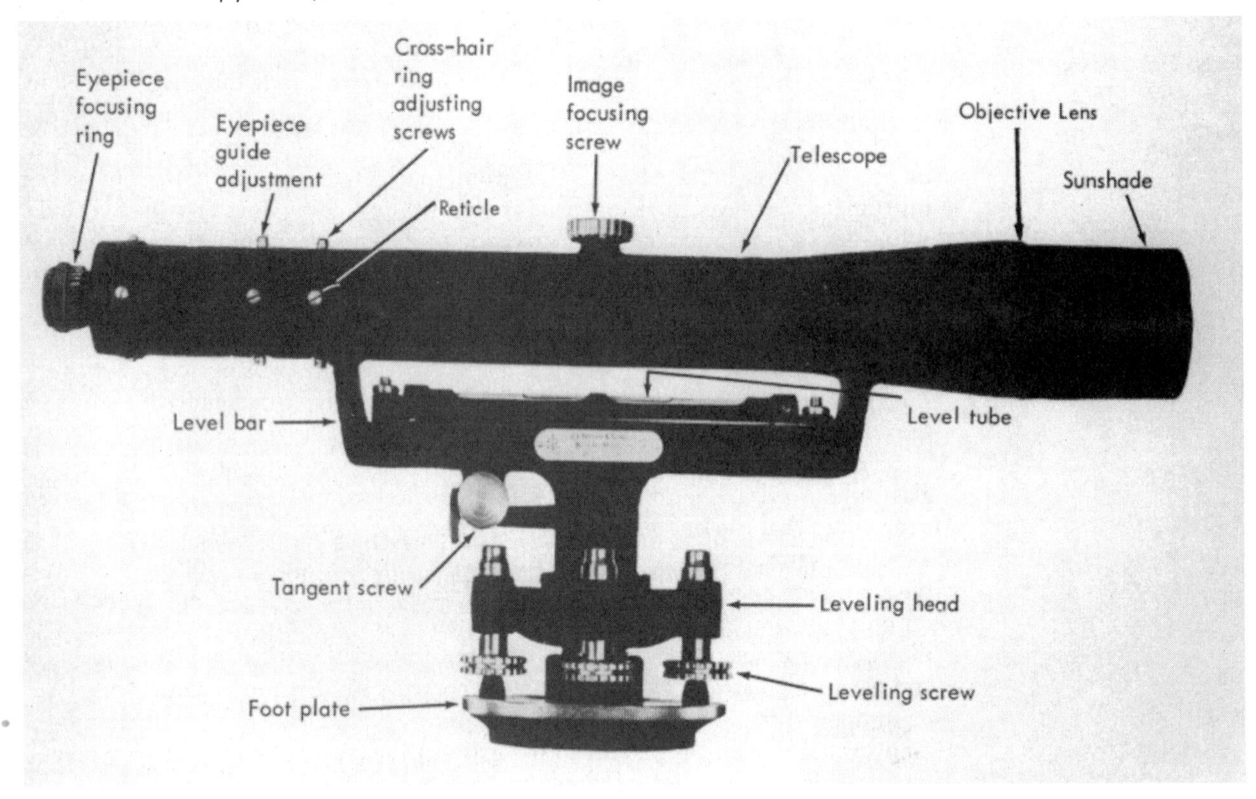

Wye Level

Another type of dumpy level that was formerly used was the wye level. This now obsolete level had its telescope supported in wye-shaped supports and held in place with curved clips. The telescope for this type of level could actually be removed and turned end for end for adjustment purposes.

Automatic or Self-Leveling Level

Automatic levels are the standard instruments used by today's surveyor. This type of level (an example of which is shown in Figure 6-5) is very easy to set up and use and is available with almost any desired range of precision. They are usually satisfactory for second-order leveling and may be satisfactory for first-order leveling if an optical micrometer (described in the next section of this chapter) is used. The automatic level has a small circular spirit or *bull's-eye* level and three leveling screws. The bubble is approximately centered in the bull's-eye level and the instrument itself automatically does the fine leveling.

The self-leveling level has a prismatic device called a *compensator* suspended on fine, nonmagnetic wires. When the instrument is approximately centered, the force of gravity on the compensator causes the optical system to swing almost instantaneously into a position such that its line of sight is horizontal.

The automatic level speeds up leveling operations and is particularly useful where the ground is soft and/or when strong winds are blowing because the instrument automatically relevels itself when it is thrown slightly out of level. When the

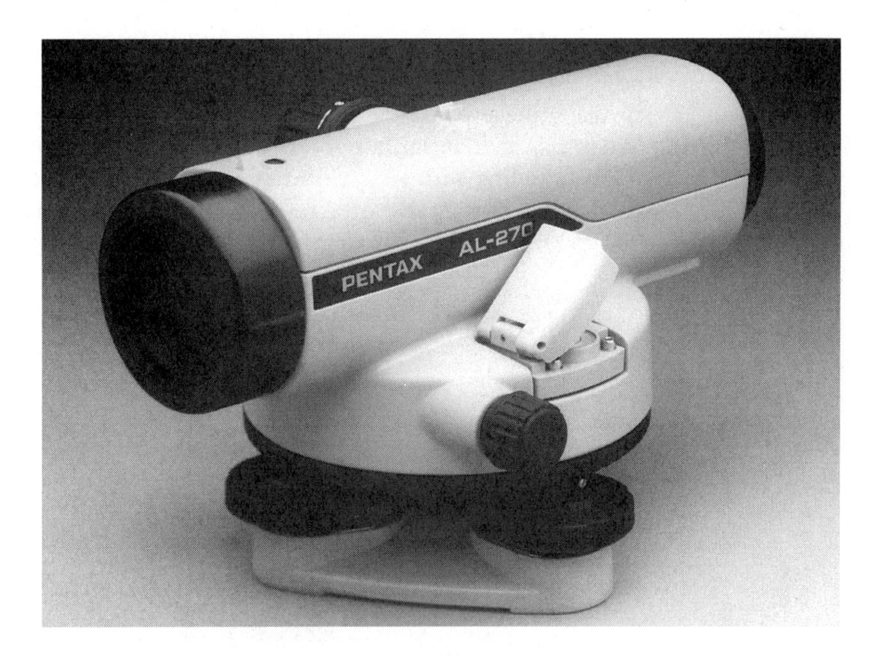

FIGURE 6-5 Pentax AL-270 automatic level. (Courtesy of Pentax Corporation.)

surveyor uses an ordinary level under these adverse conditions, he or she must constantly check the bubble to see that it remains centered.

If anything goes wrong with the compensator, the instrument will not level itself and future readings taken with it will be wrong. As a result the surveyor should periodically check to see if the compensator is working properly. This can be done in a few seconds by slightly turning one of the instrument's leveling screws, thus changing the rod reading. If the compensator is in proper working order the instrument will be automatically releveled and the rod reading will be returned to its original value.

Electronic Digital Level

The electronic digital level (Figure 6-6) is an automatic instrument because after its bull's-eye level is roughly centered the compensator will finish the leveling. The

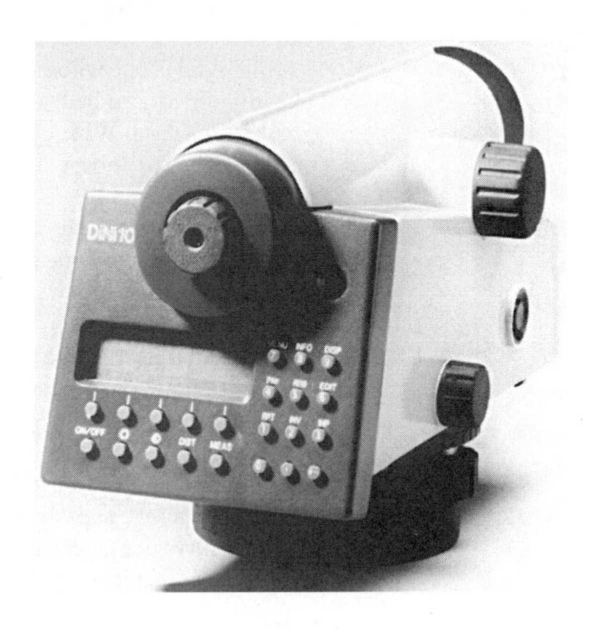

FIGURE 6-6 DiNi T total level station, which has electronic sensor for reading a bar code to determine elevations. It can also be used for distance and angle measurement. (Courtesy of Carl Zeiss, Inc.)

telescope and cross hairs of the instrument can be used to make readings as with other levels but it was primarily designed to make electronic readings. The surveyor sights on the level rod, one side of which is marked with a bar code (see Figure 6-7). When this is done and a button pressed the instrument will compare the image of the rod reading with a copy of the bar code that is kept in its memory. It will then display numerically the rod reading as well as the distance to the rod.

FIGURE 6-7 Portion of bar code used with electronic digital level.

Tilting Level

A tilting level is one whose telescope can be tilted or rotated about its horizontal axis. The instrument can be leveled quickly and approximately by means of a bull's eye or circular type level. With the telescope pointed at the level rod, the surveyor rotates a tilting knob that moves the telescope through a small vertical angle until the telescope is level.

The tilting level has a special arrangement of prisms that enables the user to center the bubble by means of a *split* or *coincident bubble*. The two halves of the bubble are actually the half ends of a single bubble and they coincide when the bubble is centered. A split image of the bubble is seen through a small microscope that is located next to the telescope eyepiece. As the leveling screws are adjusted and the bubble moves in its tube, the images of the two ends of the bubble move in opposite directions. When the instrument is correctly leveled, the two images will coincide in a continuous U-shaped curve. Manufacturers claim that split bubbles can be centered several times more precisely than the non-coincident type. Tilting levels are useful when a very high degree of precision is required. If they were used for ordinary, everyday surveying operations such as earthwork, the extra time required to bring the bubbles to coincidence probably could not be justified economically.

When the surveyor is using a dumpy level, he or she must check constantly to see if the bubble is centered, as it must be when a reading is taken. Using a tilting level while sighting through the telescope, the surveyor can at the same time look through a window on one side of the eyepiece and see if the bubble is coincident. The use of this type of level not only saves time, but also results in better accuracy as the surveyor needs only to use one screw to keep the line of sight horizontal as the telescope is turned about its vertical axis.

Laser Level

The laser is effectively used for several leveling operations. It is commonly used to create a known reference elevation or point from which construction measurements can be taken.

The lasers used for surveying and construction fall into general classes: single-beam lasers and rotating-beam lasers. A single-beam laser projects a string line that can be seen on a target regardless of lighting conditions. The line may be projected in a vertical, horizontal, or inclined direction. The vertical line provides a very long plumb line, which builders have needed throughout history. Horizontal and inclined lines are very useful for pipelines and tunnels.

A rotating-beam laser, which provides a plane of reference over open areas, can be rotated rapidly or slowly or can be stopped and used as a single beam. Today's rotating lasers are self-leveling and self-plumbing, thus providing both horizontal and vertical reference planes. The laser beam will not come on until the instrument is level. If the instrument is bumped out of position, the beam shuts off and will not

FIGURE 6-8 AL-50 electronic level, which provides an automatic level laser reference permitting instant grade check throughout a construction site for workers with receivers. (Courtesy of Nikon, Inc.)

FIGURE 6-9 Pentax PLP-42/45 self leveling rotating laser. (Courtesy of Pentax Corporation.)

come back on until it is level again. Thus it is particularly advantageous when severe winds are occurring. The rotating beam can be set to provide a horizontal plane for leveling or a sloped plane, as might be required for the setting of road or parking lot grades. It is precise for distances up to 1000 ft. This means that fewer instrument setups are required, and for construction jobs, it means that the laser can be set up some distance away and out of the way of construction equipment.

The laser is not usually visible to the human eye in bright sunlight, and thus some type of detector is needed. The detector can either be a small handheld or rod-mounted unit that may be moved up and down the level rod, or it may be an automatic detector. The latter device has an electronic carriage that moves up and down inside the rod and locates the beam.

Lasers can be very useful for staking pipelines and parking lots, for setting stakes to control excavation and fills, for topographic surveys, and so on (see Figures 6-8 and 6-9). A detailed discussion of lasers is provided in an article by Tom Liolios.[1]

Transits or Theodolites Used as Levels

Although they were primarily used for angle measurement, transits and theodolites could also be used for leveling. The results were fairly precise but not as good as those obtained with standard levels with their better telescopes and their more sensitive bubble tubes.

[1]Tom Liolios. "Lasers and Construction Surveying." *P.O.B. Magazine,* Aug./Sept. 1981, vol. 6, no. 6, pp. 38–41 and 63.

Total Stations Used as Levels

As described in Chapters 10 and 14 total stations are commonly used today for determining various elevations. The results are quite good but not as accurate as those obtained with levels.

6-8 LEVEL RODS

There are many kinds of level rods available. They may be one piece, two piece, three piece, etc., while others (for ease of transporting) are either telescoping or hinged. Level rods are usually made of wood, fiberglass, metal, or combinations thereof. They usually are graduated from zero at the bottom. Level rods are usually read directly through the telescope by the instrumentman and are commonly called *self-reading rods.* Sometimes a sliding target is placed on the rod as shown in Figure 6-10. The rodman sets the target at the desired position in accordance with signals from the instrumentman and then makes the reading directly on the rod. These rods are commonly called *target rods,* although such a designation is technically incorrect, as the target is only an accessory to the regular self-reading level rod.

Level rods are frequently named for cities or states. For instance, there are the *Philadelphia rod,* the *Chicago rod,* the *San Francisco rod,* the *Florida rod,* and so on. The Philadelphia rod has one major advantage over these other rods, as will be seen

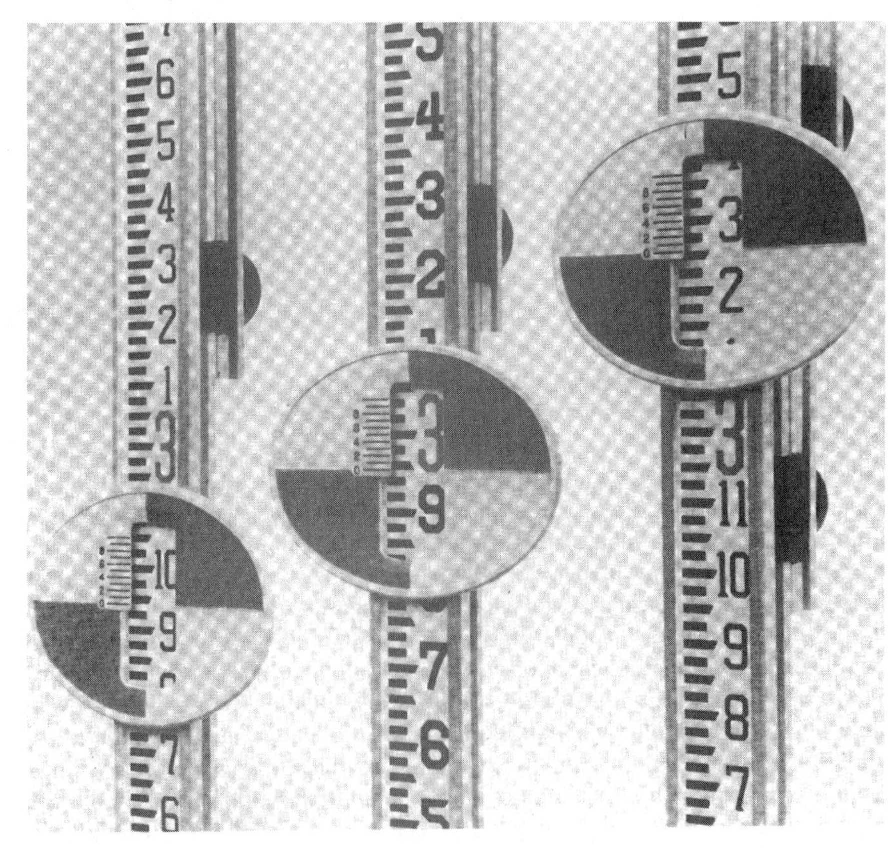

FIGURE 6-10 Level rods and targets. The rod in the center of the figure is divided into feet, tenths of a foot, and hundredths of a foot, while the ones to the left and right are divided into feet, inches, and eighths of an inch (convenient for construction work). The small scales (called verniers) seen on the targets enable the surveyor to make readings of fractional parts of the smallest divisions on the main scales. Verniers are discussed in Section 7-5. (Courtesy of Berger Instruments.)

in Section 7-7. This is the fact that the rodman can independently check the readings taken by the instrumentman. This rod, the most common one, is made in two sections. It has a rear section that slides on the front section. For readings between 0 and 7 ft, the rear section is not extended; for readings between 7 and 13 ft, it is necessary to extend the rod. When the rod is extended, it is called a *high rod.* The Philadelphia rod is distinctly divided into feet, tenths, and hundredths by means of alternating black and white spaces painted on the rod. The scales, or verniers, shown on the targets are discussed in Section 7-5. Figure 6-11 shows a Philadelphia rod and some sample readings on the rod taken through the telescope at locations where the telescope horizontal cross hair appears to intersect the rod.

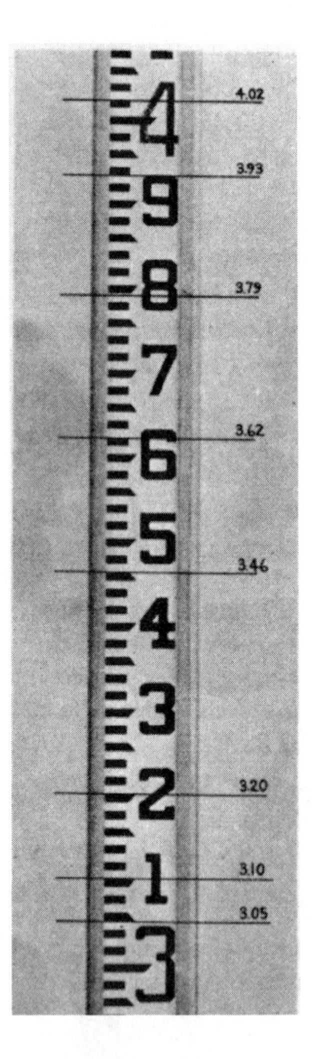

FIGURE 6-11 Various readings on a Philadelphia rod. Lines represent the horizontal cross hair of the instrument.
(Legault/McMaster/ Marlette, *Surveying*, © 1956, renewed 1983, p. 124. Reprinted by permission of Prentice Hall, Englewood Cliffs, New Jersey.)

The telescopes of many levels are equipped with stadia hairs, or three horizontal cross hairs. These hairs are used for three-wire leveling as described in Section 8-2 of this text. As a result the surveyor must be very careful to use the correct cross hair when readings are being taken.

The Chicago rod is 12 ft long and is graduated in the same way as the Philadelphia rod, but it consists of three sliding sections. The Florida rod is 10 ft long and is graduated in white and red stripes, each stripe being 0.10 ft wide. The newer fiberglass rods have oval or circular cross sections which fit together telescopically for heights up to 25 ft. When using these rods we must remember to be particularly careful while working in the vicinity of power lines. Also available for ease of transportation are tapes or ribbons of waterproofed fabric which are marked in the same way that a regular level rod is marked and which can be attached to ordinary wood strips. Once a job is completed, the ribbon can be removed and rolled up. The wood strip can be thrown away. The instrumentman can clearly read these various level rods through the telescope for distances up to 200 or 300 ft, but for greater distances he or she must use a target. A target is a small red and white piece of metal (see Figures 6-10 and 7-8) attached to the rod. The target has a vernier (see Section 7-5) that enables the rodman to take readings to thousandths of a foot.

When the Philadelphia rod is used as a target rod and the readings are 7 ft or less, the target is moved up and down until the horizontal cross hair bisects the target, or coincides with the line dividing the red and white colors on the target. At this time the target is clamped to the rod and the reading is taken by the rodman using the vernier on the target.

If the reading is higher than 7 ft, the target is clamped so that its center line is located at the 7-ft mark on the rod and the rear part of the rod is raised (and that moves the target with it) until the level horizontal cross hair bisects the target. Then the back and front strips of the level rod are clamped together and the reading taken on the rear face of the sliding part of the rod. The graduations on the rear face of the rod are numbered down from 7 to 13 ft. If the rod is extended 2 ft, the reading on the back will be $7 + 2 = 9$ ft. As the back section is pushed upward it runs under an index scale and vernier, which enables the rodman to estimate the reading on the rod's back side to thousandths of a foot.

Readings taken with a self-reading rod are almost as precise as those taken with a target rod, particularly if they are estimated to, say, the nearest 0.002 ft. The target rod, however, is quite useful in some situations: for long sights, in heavy woods, in rather dark places, during strong winds, and so on.

It used to be fairly common to use a target and vernier to make rod readings. This time-consuming procedure is today rather obsolete, however, and a reasonably economical device called an *optical micrometer* is available which enables the surveyor to make much more accurate readings.

The device is mounted in front of the telescope. It has a micrometer screw which can be used to turn a parallel-plate prism through a range equal to the least division of the rod. The surveyor sets up the instrument and turns the micrometer screw until the line of sight falls on a full rod division. This division is read and then is added to the micrometer reading (which is the proportional distance the horizontal cross hair was above the full rod division).

For very precise work almost all readings are taken in the SI system. The smallest divisions on the rods are about 5 mm. A tensioned Invar steel strip is attached to the rod face and a thermometer is used so that temperature corrections can be made to the readings. The optical micrometers on the levels enable us to make repeatable readings to about ±0.1 mm, which is about ±0.0003 ft.

6-9 SETTING UP THE LEVEL

Usually it is unnecessary to set up and center the level over a particular point. Convenient locations are selected so that sights to desired points can be easily made. The tripod is set up as was previously described for an EDM in Section 5-5 and the instrument is leveled as described in the next few paragraphs.

Automatic and Tilting Levels (Three-Screw Instruments)

For leveling the automatic and tilting three-screw levels the first step is to turn the telescope so that it is aligned over one of the screws and is thus perpendicular to a line through the other two. The bull's-eye bubble is centered by turning first one screw then the other two as was previously described for EDMs in Section 5-5. It is not necessary to rotate the telescope during leveling.

Four-Screw Instruments

An old four-screw level is attached to the tripod by means of a threaded base. After the instrument has been leveled as much as possible by adjusting the tripod legs, the telescope is turned over a pair of opposite leveling screws. In centering the bubble, opposite screws are used and the screws are turned in toward the center,)(, or in the opposite directions,)(. In each case, as the screws are turned, the bubble will follow the motion of the left thumb.

The bubble is roughly centered and then the telescope is turned over the other pair of leveling screws and the bubble is again roughly centered. The telescope is turned back over the first pair and the bubble is again roughly centered, and so on. This process is repeated a few more times with increasing care until the bubble is centered with the telescope turned over either pair of screws. If the level is properly adjusted, the bubble should remain centered when the telescope is turned in any direction. It is to be expected that there will be a slight maladjustment of the instrument that will result in a slight movement of the bubble; however, the precision of the work should not be adversely affected if the bubble is centered each time a rod reading is taken.

Two-Screw Instruments

Some newer levels have only two leveling screws. They are constructed similar to the three-screw levels, but one of the screws is replaced with a fixed point. This means that when the instruments are set up and leveled at a given point, they will be at a fixed elevation. If the instrument is moved slightly and releveled, it will return to the same elevation. To level the two-screw instrument, the telescope is turned until it is parallel to a line from the fixed point to one of the leveling screws. The instrument is leveled and then the telescope is turned to the other leveling screw; this instrument is again leveled and the process is repeated a few more times as needed.

6-10 SENSITIVITY OF BUBBLE TUBES

The divisions on bubble tubes were at one time commonly spaced at 1/10 in. intervals, but today they are usually spaced at 2-mm intervals. The student often wants to know how much the rod readings will be affected if the bubble is off center by one, two, or more divisions on the bubble tube when the readings are made. One way to find out is to place the level rod a certain distance from the level (say 100 ft) and take readings with the bubble centered; then with it one division off center; two divisions off center; and so on.

The *sensitivity* of the bubble tube may be expressed in terms of the radius of curvature of the bubble tube. Obviously, if the radius is large, a small movement of the telescope vertically will result in a large movement of the bubble. An instrument with a large radius of curvature is said to be sensitive. For very precise leveling very sensitive levels need to be used. The centering of the bubble for such instruments takes more time; therefore less sensitive instruments (which are more quickly centered) may be more practical for surveys of lesser precision.

Another way of expressing the sensitivity of the bubble tube is to give the angle through which the axis of the bubble tube must be rotated (usually given in seconds of arc) to cause the bubble to move by one division on the scale. If the movement of one division on the scale corresponds to 10″ of rotation, the bubble is said to be a 10″ bubble. For very precise levels, 0.25″ or 0.5″ bubbles or even more sensitive ones may be used. For ordinary construction leveling, 10″ or 20″ instruments may be satisfactory. A 20″ bubble with 2-mm divisions will have a 68-ft radius (see Figure 6-12). The average dumpy level had about a 20″ bubble and the radius of curvature was about 68 ft. For first-order leveling the levels used may have 2″ bubbles with a 680-ft radius of curvature.

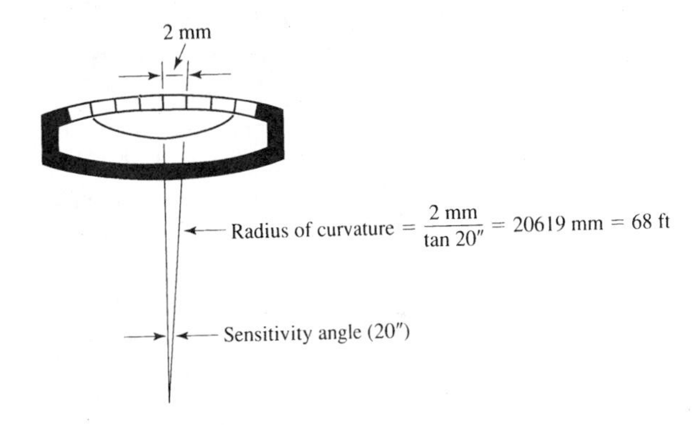

FIGURE 6-12 Sensitivity of bubble tubes. (A 20″ bubble is shown.)

6-11 CARE OF EQUIPMENT

Although surveying equipment is very precisely and delicately manufactured, it can be very durable when properly used and maintained. In fact, these instruments may last a lifetime in the hands of a careful surveyor, but a few seconds of careless treatment can result in severe or irreparable damage. Because of the long-lasting qualities of surveying equipment, the surveyor will occasionally have to use some rather antique levels and transits. This is not a disadvantage because a good old instrument will still be quite precise. Used transits and levels decline very little in value as the years go by. As a matter of fact, their selling prices in dollars often remain stable for several decades. However, finding parts for older instruments is sometimes a problem.

Several suggestions are made in the following paragraphs concerning the care of the instruments used for leveling.

The Level

Before the level is removed from its box, the tripod should be set up in a firm position. The user should observe exactly how the level is held in the box so that after use he or she can return it to exactly the same position. After the instrument is taken from its box, it should be handled by its base when it is not on the tripod. It should be carefully attached to the tripod. The surveyor must not permit anything to interrupt him or her until this task is completed. Many levels have been severely damaged when careless surveyors began to attach them to their tripods but then allowed their attention to be diverted before tightening was completed.

If possible, levels should not be set up on smooth hard surfaces, such as building floors, unless the tripod points can be either set in indentations in the floor or firmly held in place by other means, perhaps by triangular frames made for that purpose. Particular care should be exercised when the instruments are being carried inside buildings to avoid damage from possible collisions with doors, walls, or columns. In such locations, therefore, the level should be carried in the arms instead of on the shoulder. When working outside where it is normal to carry the instrument on the shoulder, the clamps should be left loose to allow the instrument to turn if it hits limbs or bushes. If the level is to be moved for long distances or over very rough terrain, it should be placed in its carrying case.

A level should never be left unattended unless it is in a protected location. If it is turned over by wind, cattle, children, or cars, the results will probably be disastrous. Some surveyors paint their tripod legs with very bright colors. Such a scheme is particularly wise when work is being done near heavy traffic. Another reason for not leaving instruments unprotected is thieves. Surveyors' levels, transits, EDMs, and total stations can be sold quickly and easily for good prices. In addition, the surveyor should protect instruments as much as possible from moisture. Waterproof hoods are desirable in case of sudden rain. If the level does get wet, all but the lens should be gently wiped dry. Moisture on the lens is usually permitted to evaporate because the lens can so easily be scratched while being dried. The lens should never be touched with anything other than a camel's-hair brush or, less desirably, a soft silk handkerchief. If the objective lens or the eyepiece lens becomes so dusty that it interferes with vision, it may be cleaned with a camel's-hair brush or with lens paper.

Leveling Screws

Perhaps the most common injury to levels by the novice surveyor is caused by applying too much pressure to the leveling screws. If the instrument is in proper condition, these screws should turn easily and it should never be necessary to use more than the fingertips for turning them.

To emphasize the "light touch" that should be used with leveling screws, many surveying instructors require that students carry out the following test: A leveling screw is loosened and a piece of paper is slipped underneath. Then the screw is tightened just enough to hold the paper in place against a slight tug of the hand. The instructor then explains that leveling screws should never be turned with any greater force than is required to hold the paper. This demonstration should make the student conscious of the need for care of these screws, which can become bound or even stripped if too much pressure is applied. If the threads are stripped, they usually have to be shipped back to the manufacturer for repair.

If the leveling screws do not turn easily, they may be cleaned with a solvent, such as gasoline, and the interior threads may be *very* lightly oiled with a light watch oil. When the level is taken indoors for storing or outdoors for use, its screws and clamps should be loosened because large temperature changes may cause severe damage.

Level Rod

The level rod should never be dragged on the ground or through water, grass, or mud, and its metal base should never be allowed to strike rocks, pavement, or other hard objects; such use will gradually wear away the metal base and will thereby cause leveling errors due to the change in length of the rod itself. Furthermore it must not be used to knock down weeds or brush. The Philadelphia rod must not be carried over the shoulder when it is fully extended, or leaned against trees or walls—it is just too flexible. It should be placed flat on the ground with the numbers facing upward.

Problems

6-1. Define: a vertical line, a level line, and a horizontal line.

6-2. Why is mean sea level gradually increasing?

6-3. Why are sea level values based on NAVD 88 different from those based on NGVD 29?

6-4. What is collimation?

6-5. What is spirit or direct leveling?

6-6. Distinguish between first-order, second-order, and third-order surveys.

6-7. How was the sea-level datum established in the United States?

6-8. What are the problems associated with barometric leveling?

6-9. What is a dumpy level?

6-10. What is one advantage of a two-screw level?

6-11. Describe a Philadelphia rod.

6-12. What is the sensitivity of a bubble tube? How is it determined?

6-13. What is a tilting level?

6-14. List several items which the surveyor must remember in caring for leveling equipment.

CHAPTER 7

Differential Leveling

7-1 THEORY OF SPIRIT LEVELING

For an introductory description of direct or spirit leveling it is assumed that the surveyor has set up the instrument and has leveled it carefully. He or she then sights on the level rod held by the rodman on some point of known elevation (this sight is called a *backsight,* or BS). If the backsight reading is added to the elevation of the known point, we will have the *height of the instrument* (HI), that is, the elevation of the line of sight of the telescope.

To illustrate this procedure, refer to Figure 7-1, in which the HI is seen to equal 100.00 + 6.32 = 106.32 ft. If the HI is known, the telescope may be used to determine the elevation of other points in the vicinity by placing the level rod on each point whose elevation is desired and by taking a reading on the rod for each point. Since the elevation of the point where the line of sight of the telescope intersects the rod is known (the HI), the rod reading called a *foresight* (FS) may be subtracted from the HI in order to obtain the elevation of the point in question. In Figure 7-1 a FS reading of 3.10 ft was taken on the rod. The elevation at the bottom of the rod at this new position is thus 106.32 – 3.10 = 103.22 ft.

The level may be moved to another area by using temporary points called *turning points* (TPs). The telescope is sighted on the rod held at a convenient turning point and a foresight is taken. This establishes the elevation of the point. Then the level can be moved beyond the TP and set up at a convenient location. A

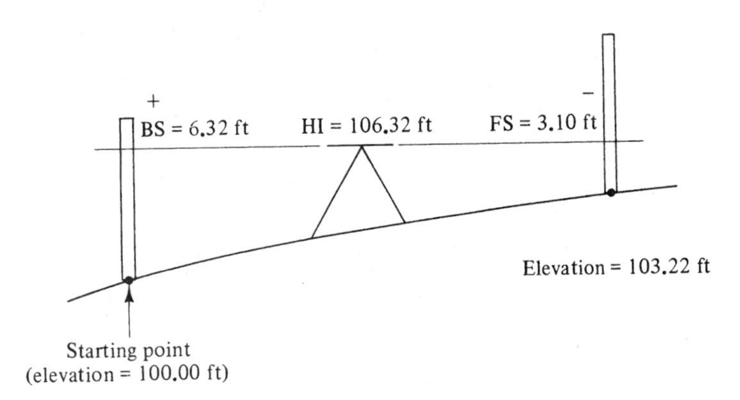

FIGURE 7-1 Direct or spirit leveling.

backsight is taken on the rod held at the turning point and the HI for the new instrument position is established. This process can be repeated over and over for long distances.

7-2 DEFINITIONS

A *bench mark* (BM) is a relatively permanent point of known elevation. It should easily be recognized and found and should be set fairly low in relation to the surrounding ground. It may be a concrete monument in the ground, a nail driven into a tree, an × mark in a concrete foundation, a bolt on a fire hydrant, or a similar object that is not likely to move. Bench marks that are to be permanent should be supported by structures that have completely settled and which extend below the frostline. The foundation of an old building usually meets these requirements very well. Parts of structures that must resist significant lateral forces, such as retaining walls, make rather poor bench marks. Such structures may move for years due to lateral earth pressure. Similar discussions can be made for the supporting foundations of poles and towers. Careful records should be made of bench marks because they may frequently be reused during the life of a job or for future work in the vicinity. They should be so completely and carefully described in the notes that another surveyor unfamiliar with the area can find and use them, perhaps years later.

A *turning point* (TP) is a temporary point whose elevation is determined during the process of leveling. It may be any convenient point on the ground, but it is usually wise to use a readily identifiable point such as a rock, a stake driven into the ground, or a mark on the pavement so that the level rod can be removed and put back in the same location as many times as required. It is essential that solid objects be used for turning points. Never should points be selected on soft or wet ground or grass, which will give or settle under the pressure of the rod. Should satisfactory natural objects not be available, a metal turning point pin, a wood stake, or the head of an ax or hatchet may be used if they are driven firmly into the ground. The usual procedure is to use the highest point on the object as the turning point.

A *backsight* (BS) is a sight taken to the level rod held on a point of known elevation (either a BM or a TP) to determine the height of the instrument (HI). Backsights are frequently referred to as *plus sights* because they are usually added to the elevations of points being sighted on to determine the height of the instrument.

A *foresight* (FS) is a sight taken to any point to determine its elevation. Foresights are often called *minus sights* because they are subtracted from HIs to obtain the elevations of points. *Notice that for any position of the instrument where the HI is known, any number of foresights may be taken to obtain the elevations of other points in the area.* The only limitations on the number of sights are the length of the level rod and the power of the telescope on the level. Normally, one cannot sight on the level rod held at a point whose elevation is greater than that of the HI. There are a few exceptions, such as when it is desired to determine the elevation of the underside of a bridge or the top of a mine shaft or tunnel. For such cases the level rod is inverted and held up against the point whose elevation is sought. Then the backsights are negative and the foresights are positive.

7-3 DIFFERENTIAL LEVELING DESCRIBED

Differential leveling, which is the process of determining the difference in elevation between two points, is illustrated in Figure 7-2, in which a line of levels is run from BM_1 to BM_2. The instrumentman sets up the level at a convenient point and backsights on the level rod held on BM_1. This gives the HI. The rodman moves to a convenient point (TP_1 in the figure) in the direction of BM_2. The instrumentman takes a foresight on the rod, thus enabling him or her to compute the elevation of TP_1. The level is then moved to a convenient location beyond TP_1 and a backsight is taken on TP_1. This gives the new HI. The rodman moves forward to a new location (TP_2), and so on. This procedure is continued until the elevation of BM_2 is determined.

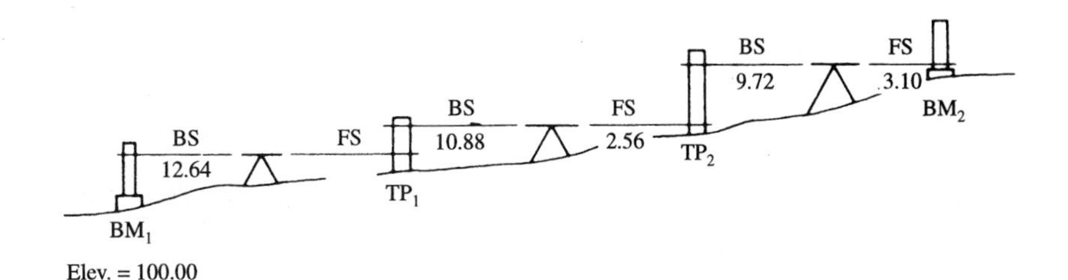

FIGURE 7-2 Differential leveling.

The initial backsight can be taken to a point of known elevation (a bench mark) or to a point of assumed elevation. If leveling is conducted starting with a bench mark, the elevations determined for subsequent points are actual or true elevations. If, on the other hand, the elevation of the starting point is assumed, the elevations determined for subsequent points will be in relation to that assumed elevation. However, the differential between the points will be valid even if the true elevations of all the points are unknown. This is "true" differential leveling. There are many applications in surveying where the important result of leveling is the point-to-point differential in elevations and not the true elevations of the points. This is particularly true if there are not available bench marks in the vicinity of the work.

It is very important in differential leveling to keep the lengths of the backsights and foresights approximately equal for each setup of the instrument. Such a practice will result in greatly reduced errors for cases where the instruments are out of adjustment and also for errors due to atmospheric refraction and earth's curvature (to be discussed in the next section). The easiest way to obtain roughly equal distances is by pacing, but stadia, EDMs, and taping are better. These topics are discussed further in Sections 7-6 and 7-8.

The usual form for recording differential leveling notes is presented in Figure 7-3 for the readings that were shown in Figure 7-2. *The student would be wise to study these notes very carefully before attempting leveling; otherwise, he or she may become confused in recording the readings.* In studying these notes, the student should particularly notice the math check. Since the backsights are positive and the foresights are negative, the surveyor should total them separately. The difference between these two totals must equal the difference between the initial and final elevations or else a math blunder has been made in the field book. *Since the math check is easy to make, there is no excuse for omitting it. For this reason, level notes are considered incomplete unless the check is made and shown in the notes.*

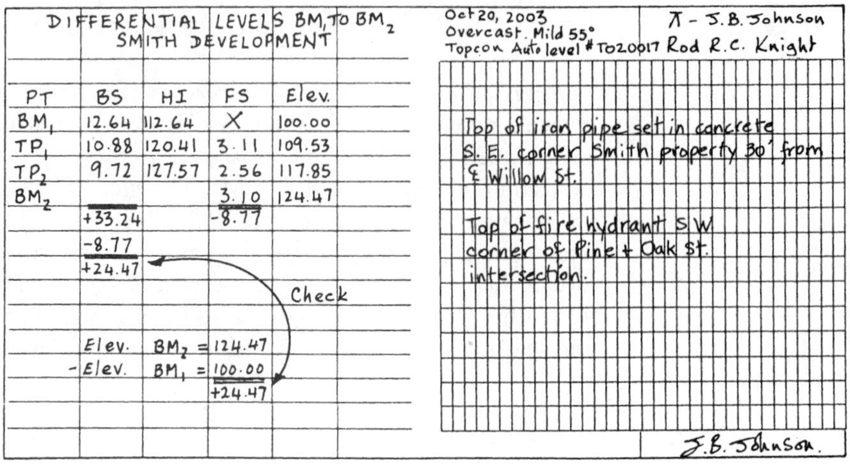

FIGURE 7-3 Differential level notes.

DIFFERENTIAL LEVELS BM₁ TO BM₂
SMITH DEVELOPMENT

PT	BS	HI	FS	Elev.
BM_1	12.64	112.64	X	100.00
TP_1	10.88	120.41	3.11	109.53
TP_2	9.72	127.57	2.56	117.85
BM_2			3.10	124.47
	+33.24		-8.77	
	-8.77			
	+24.47			Check
	Elev.	BM_2 =	124.47	
	- Elev.	BM_1 =	100.00	
			+24.47	

Oct 20, 2003 ⊼ - J.B. Johnson
Overcast. Mild 55°
Topcon Auto level #T020017 Rod R.C. Knight

Top of iron pipe set in concrete S.E. corner Smith property 30' from ℄ Willow St.

Top of fire hydrant S.W. corner of Pine + Oak St. intersection.

J.B. Johnson

7-4 EARTH'S CURVATURE AND ATMOSPHERIC REFRACTION

Up to this point we have in effect been assuming that when the instrument is leveled, its line of sight will represent a level line. In this section you will learn that this is not the case, due to the *earth's curvature.* The line of sight of the telescope will actually be perpendicular to a plumb line at one point only—that will be at the instrument location. You may then decide that the telescope line of sight is a horizontal line, but you will learn that this is not correct either, because of *atmospheric refraction.*

An expression for the amount of earth's curvature can easily be derived. Reference is made to Figure 7-4, where E is the average radius of the earth = 3959 miles and C is the departure of a horizontal line from the curved surface of the earth in a sight distance D. For the right triangle with sides E, D and hypotenuse $E + C$, we can write

$$D^2 + E^2 = \left(E + C\right)^2$$
$$= E^2 + 2EC + C^2$$

When leveling, the maximum sight distance is probably only a few hundred feet but certainly not more than a few miles. For such distances the value of C will at most be a few feet. Thus C^2 in the preceding expression is a negligible quantity in comparison with 3959 miles squared. Therefore C^2 is neglected and we have

$$D^2 + E^2 = E^2 + 2EC$$

$$C = \frac{D^2}{2E} = \frac{D^2}{\left(2\right)\left(3959 \text{ miles}\right)} = 0.0001263D^2 \text{ miles}$$

When we change to feet using D in miles we obtain

$$C = \left(0.0001263\right)\left(5280\right)\left(D^2\right) = 0.667D^2 \text{ ft}$$

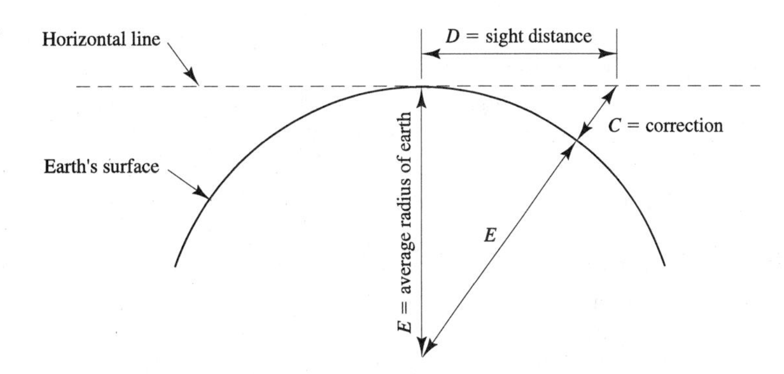

FIGURE 7-4 Earth's curvature.

When rays of light pass through air strata of different densities, they are refracted, or bent downward. This means that to see an object on the ground some distance away, a person actually has to look above it. The amount of refraction is dependent on temperatures, pressures, and relative humidities. It is greatest when the line of sight is near the ground or near bodies of water where temperature differences are large and therefore where large variations in air densities occur.

As the amount of refraction is difficult to determine, an average value is usually used. This value is approximately 0.093 ft in 1 mile (about one-seventh of the effect of the earth's curvature) and varies directly as the square of the horizontal distance.

Variations from the 0.093-ft value are not usually significant for the relatively short sight distances used in differential leveling—but they may have to be considered for very precise leveling and for extreme conditions for ordinary leveling. Refraction has been known to be as large as 0.10 ft in a 200-ft sight distance.

Because of earth's curvature, a horizontal line departs from a level line by 0.667 ft in 1 mile, also varying as the square of the horizontal distance. The effects of the earth's curvature and atmospheric refraction are represented in Figure 7-5.

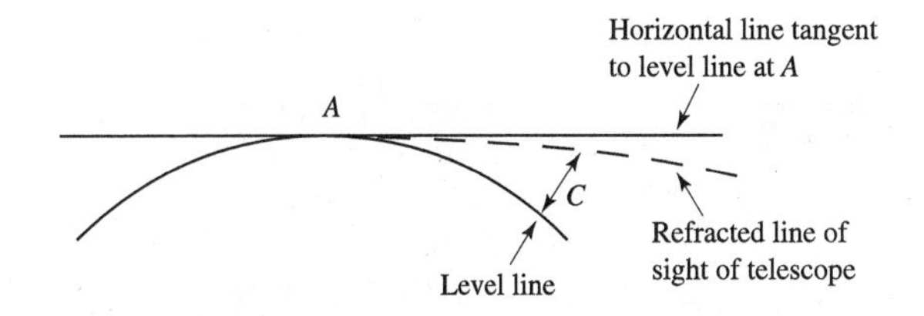

FIGURE 7-5 Earth's curvature and atmospheric refraction.

The combination of the earth's curvature and atmospheric refraction causes the telescope's line of sight to vary from a level line by approximately 0.667 minus 0.093, or 0.574 ft in 1 mile, varying as the square of the horizontal distance in miles. This may be represented in formula form as follows, where C is the departure of a telescope line of sight from a level line and M is the horizontal distance in miles:

$$C = 0.574M^2$$

For a telescope reading on a rod 100 ft away, the reading would theoretically be in error by

$$(0.574)\left(\frac{100}{5280}\right)^2 = 0.000206\,\text{ft}$$

Similarly, a reading on a rod at a 300-ft distance would be in error by 0.00185 ft. At a 1000-ft distance the error would be 0.0206 ft.

Leveling sight distances are rarely larger than a few hundred feet. In most modern instruments the optics limit sight distances to 300 ft or less. Beyond these distances the graduations on the rod are difficult to distinguish. For such distances errors due to earth's curvature and atmospheric refraction, calculated as shown above, are negligible except for the most precise surveys. However, errors due to instrument inadjustment (causing inclination of the telescope line of sight) may be very significant for such distances.

If the backsight distance were exactly equal to the foresight distance for each setup of the instrument, it could be seen that errors caused by atmospheric refraction and earth's curvature would theoretically cancel each other. Each of the readings would be too large by the same amount, and since the same error would be added with the backsight and subtracted with the foresight, the net result would be to cancel them. *A similar discussion can be made for minimizing errors caused by levels being out of adjustment. Such errors may be far larger than those caused by atmospheric refraction and earth's curvature.* Particularly significant is the error produced if the axis of the telescope is not parallel to the axis of the bubble tube. However, if BS and FS distances are kept approximately equal for each instrument setup such errors will be greatly reduced.

For surveys of ordinary precision, it is reasonable to neglect the effect of the earth's curvature and atmospheric refraction. The instrumentman, however, might like to, by eye, make BS and FS distances approximately equal. For precise leveling, it is necessary to use more care in equalizing the distances. Pacing or even stadia or EDMs or taping may be used.

For SI units the correction for earth's curvature and atmospheric refraction is given by the following equation, in which C is in meters and k is the distance in kilometers:

$$C = 0.0675k^2$$

7-5 VERNIERS

A *vernier* is a device used for making readings on a divided scale closer than the smallest divisions on the scale. The vernier, which was invented by the Frenchman Pierre Vernier in 1620, is a short auxiliary scale attached to or moved along the divided scale.

Most of the targets used on level rods have verniers on them with which rod readings can be estimated to the nearest 0.001 ft. Figure 7-6 illustrates the construction and the reading of level rod verniers. The numbers on this particular rod are for 3.1 ft and 3.2 ft and thus the divisions in between are the 0.01-ft divisions. The vernier is shown to the right of the rod and is so constructed that 10 divisions on the vernier cover 9 divisions on the rod. Therefore, each division on the vernier is 0.009 ft.

In this figure the 0 or bottom mark on the vernier coincides with the 3.10-ft mark on the rod. It will be noted that the next division of the vernier falls 1/10 of a division short of the next mark on the rod (or its 3.11-ft mark). The second division on the vernier falls 2/10 of a division short of the 3.12-ft mark on the rod.

If the vernier is moved up until its first division coincides with the first division on the rod or the 3.11 ft mark, the bottom of the vernier will be located at 3.101 ft on the rod. In the same manner, if the vernier is moved up until the second division on the

FIGURE 7-6 Level rod vernier.

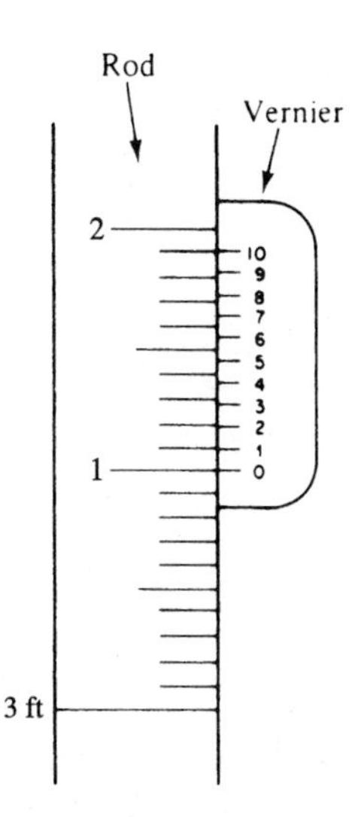

vernier coincides with the second mark on the main scale (3.120 ft), the bottom of the vernier will be located at 3.102 ft on the rod.

To read a level rod vernier, the bottom of the vernier is lined up with the horizontal cross hair and the reading is determined by counting the number of vernier divisions until a division on the vernier coincides with a division on the rod. This reading is added to the last division on the rod below the bottom of the vernier. In Figure 7-7 the rod reading at the bottom of the vernier is between 3.120 and 3.132 ft. The sixth division up on the vernier coincides with a division on the rod, so the rod reading is 3.120 plus the vernier reading 6 equals 3.126 ft. This also could be read as $3.180 - (6)(0.009) = 3.126$ ft.

FIGURE 7-7 Sample level rod reading using a vernier.

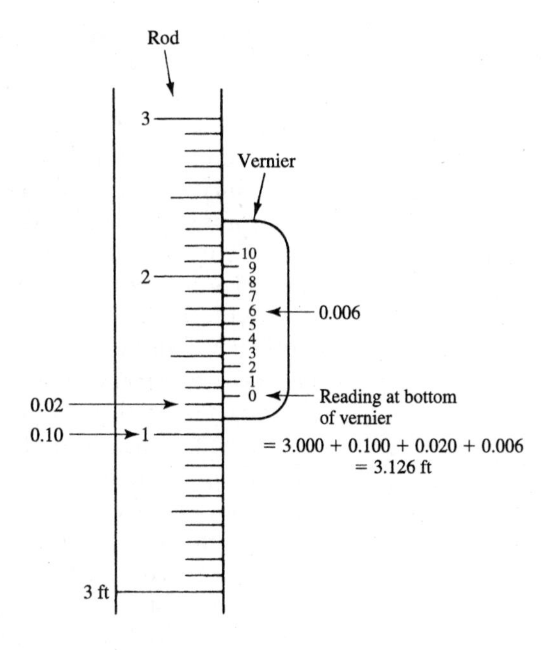

For the level rod and vernier described here, the smallest subdivision that can be read equals one-tenth of the scale division on the rod. Ten divisions on the vernier cover nine divisions on the rod, and the smallest subdivision that can be read with the vernier is

$$\frac{0.010}{10} = 0.001\,\text{ft}$$

If another vernier is used for which 20 divisions of the vernier cover 19 divisions of the rod, the smallest subdivision that can be read is 1/20 of the scale division, or

$$\frac{0.010}{20} = 0.0005\,\text{ft}$$

From this information an expression can be written for the smallest subdivision that can be read with a particular vernier. Letting n be the number of vernier divisions, s the smallest division on the main scale, and D the smallest subdivision that can be read, the following equation can be written:

$$D = \frac{s}{n}$$

Historically in surveying, one of the principal uses of verniers was for angle measurements, and for this reason the subject of verniers is continued in Appendix D of this book where transits and theodolites are discussed. Verniers are not used on our total stations.

7-6 LEVEL ROD TARGETS

For long sights or for situations in which readings to the nearest 0.001 ft are desired, a level rod target may be used. Targets are small circular or elliptical pieces of metal approximately 5 in. in diameter painted red and white in alternate quadrants. They are clamped to the rods. As shown in Figure 7-8, a vernier is part of the target.

The target is moved up or down as directed by the instrumentman until it appears to be bisected by the cross hair. At this point the horizontal cross hair coincides with the bottom of the vernier. The rodman takes the reading to the nearest 0.001 ft and this value is approximately checked by the instrumentman. An important item to remember about target readings is that although they are read to the nearest 0.001 ft, they can be no more accurate than the accuracy obtained in setting the target. In other words, for ordinary leveling "do not attach too much significance to readings taken to the nearest 0.001 ft." Although more accurate than those taken to the nearest 0.01 ft, the readings are probably not accurate to the third place. The reason is that the instrumentman just cannot signal the rodman to set the target on the rod as precisely as he or she can set the telescope cross hairs on the rod.

The precision of ordinary leveling can be increased somewhat by the laborious use of targets as described in the preceding paragraph. *A very simple, quick, and practical way in which precision can be appreciably improved is by merely limiting the lengths of sights, perhaps to 100 ft or less, and by having the instrumentman estimate the rod readings through the telescope to the nearest 0.002 ft.*

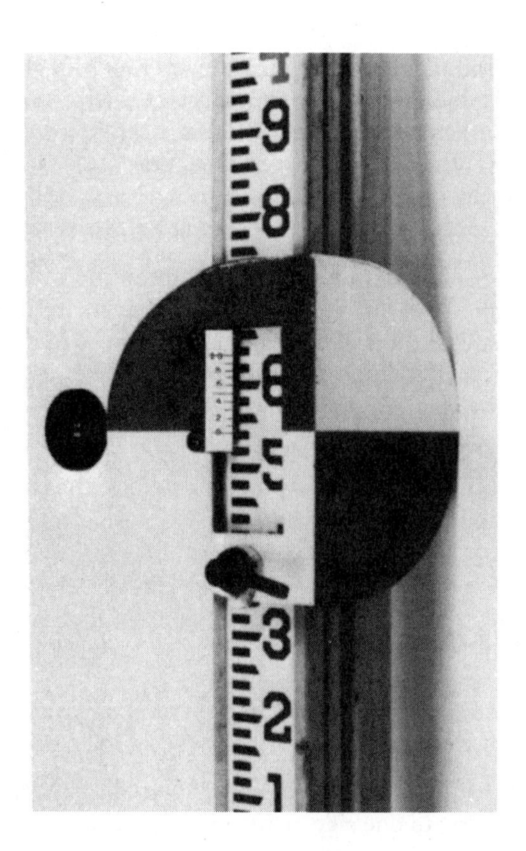

FIGURE 7-8 Level rod target.

7-7 COMMON LEVELING MISTAKES

The most common mistakes made in leveling are described in the following paragraphs.

Misreading the Rod. Unless the instrumentman is very careful, he or she may occasionally read the rod incorrectly; as, for instance, 3.72 ft instead of 4.72 ft. This mistake most frequently occurs when the line of sight to the rod is partially obstructed by leaves, limbs, grass, rises in the ground, and so on. There are several ways to prevent such mistakes. The instrumentman should always carefully note the foot marks above or below the point where the horizontal cross hair intersects the rod. If those red foot marks are not visible, he or she may ask the rodman to raise the rod slowly until a foot mark can be seen. To do this he or she may either call "raise for red" or give an appropriate hand signal, as described in Section 7-12.

An excellent procedure for the instrumentman is to call out readings as he or she takes them. The rodman, while still holding the rod properly, can point to the reading with a pencil. Obviously, the pencil should coincide with the horizontal cross hair if the reading was taken correctly. Another procedure is to use a target and have both persons take readings.

Another situation which can result in rod reading mistakes is the presence of stadia hairs in the telescope. The surveyor must be very careful to use the middle horizontal cross hair when leveling is being performed.

Moving Turning Points. A careless rodman causes serious leveling mistakes if he or she moves the turning points. The rodman holds the rod at one point while the instrumentman takes the foresight reading, and then while the level is being moved to a new position, the rodman puts the level rod down while he or she does something else. If, when the instrumentman is ready for the BS, the rodman holds the rod at some other point, a serious mistake can be made because the new location may have an entirely different elevation. Obviously, a good rodman prevents mistakes like this by using well-defined turning points or by clearly marking them with crayon (keel) if on pavement or by driving a stake or ax head into the earth.

Field Note Mistakes. To prevent the recording of incorrect values, the instrumentman should call out the readings as he or she reads and records them. This is particularly effective if the rodman is checking the readings with a pencil or with a target. To prevent addition or subtraction mistakes in level notes, the math check described in Section 7-3 should be carefully followed.

Mistakes with Extended Rod. When readings are taken on the extended portion of the level rod, it is absolutely necessary to have the two parts adjusted properly. If they are not, mistakes will be made.

7-8 LEVELING ERRORS

A brief description of the most common leveling errors and suggested methods for minimizing them are presented in the next several paragraphs.

Level Rod Not Vertical. When sighting on the rod, the instrumentman can see if the rod is leaning to one side or the other by means of the vertical cross hair in the telescope and, if necessary, can signal the rodman to straighten up. The instrumentman cannot, however, usually tell if the rod is leaning a little toward or away from the instrument. If the student thinks about this for a while, he or she will see that the smallest possible rod reading will occur when the rod is vertical. Many surveyors, therefore, have their rodmen slowly "wave" the rod toward and away from the instrument and then record the smallest reading observed through the telescope (Figure 7-9). A method used by some surveyors for ordinary leveling is to require each rodman to hold the level rod so that it touches his or her nose and belt buckle, but this practice is not as satisfactory as waving the rod. Some level rods, particularly those used for precise work, are equipped with individual circular levels (Figure 7-10) that allow the rodman to plumb the rod by merely centering the bubble. Other rods are equipped with conventional bubble tubes. The use of rod levels or other methods of plumbing is preferable to waving the rod. Waving the ordinary flat-bottomed rod can cause small errors due to the rotation of the rod about its edges instead of about the center of its front face. Waving will work best when the rod is placed on a rounded point.

Settling of Level Rod. It is essential to hold the level rod on firm definite points that will not settle and that are readily identifiable, so that the rodman, if called away

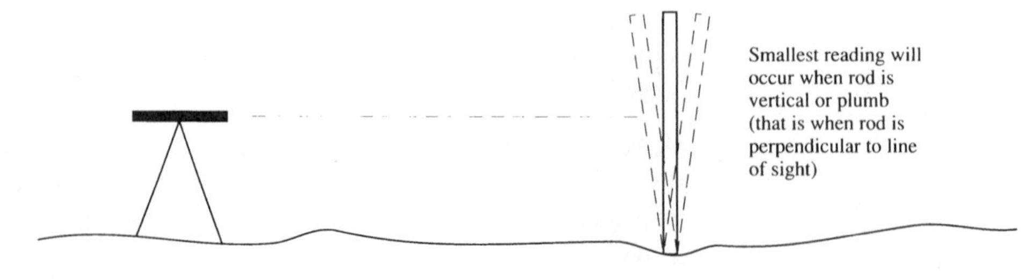

FIGURE 7-9 Waving the rod.

FIGURE 7-10 Rod level.
(Courtesy Chicago Steel Tape.)

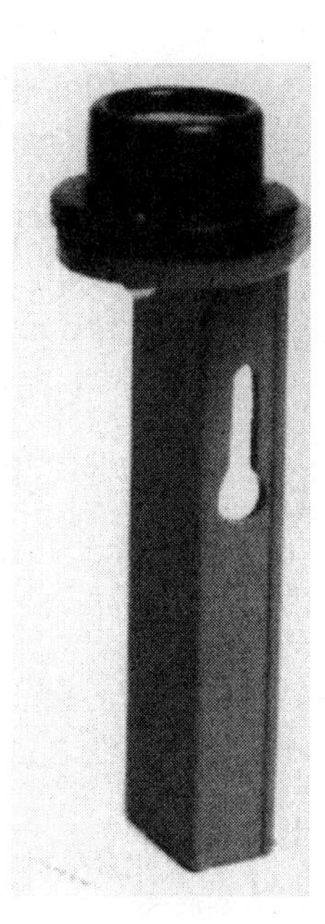

for some other work, may return to exactly the same spot. If such convenient points are unavailable, it may be necessary to take turning points on the ground, with the resulting possibility of settlement. To minimize this possibility, the rodman should hold the rod on a stake driven in the ground, on the head of an ax stuck in the ground, on a metal turning point pin, or on some other similar base.

Mud, Snow, or Ice Accumulation on Base of Rod. If the rodman is not careful, mud, snow, or ice may stick to the bottom of the rod. This can cause severe errors in leveling. The rod must not be dragged on the ground at any time and the bottom of the rod must be carefully cleaned when there is snow, ice, or mud.

Rod Not Fully Extended on High Rod. When the rear part of a Philadelphia rod is extended, it is called a high rod. Such an extension is necessary for readings from 7 to 13 ft. Frequently, level rods have been damaged by letting the upper part of the rod slide down so rapidly that the blocks on the two sections are damaged. The result is that the high rod readings may be in error and the rodman must carefully check the rod extension. If the extension is carelessly handled this should be classified as a mistake rather than as an error.

Incorrect Rod Length. If a level rod is of incorrect length (and no rod is of perfect length), the rod readings will be in error. If the length errors occur at the bottom of the rod, they will theoretically be canceled in the differential leveling process. Errors due to a misfit in extending the rod, however, will not cancel if some of the readings are taken above the joint and some of them below. This generally occurs when the surveyor is leveling up or down a slope. For such cases one tends to read on the high part of the rod for downhill shots and on the low part for uphill shots. Rod length should be checked periodically with a steel tape.

BS and FS Distances Not Equal. In Section 7-4 it was shown that if the lengths of backsights and foresights were kept equal for a particular setup, there would theoretically be no error caused by Earth's curvature and atmospheric refraction. For ordinary work it is sufficient to neglect or merely approximate by eye equal distances. For more precise work, it is necessary to pace distances or even use stadia or an EDM to keep BS and FS distances equal.

Errors due to instrument maladjustment are usually much more important than those due to atmospheric refraction and Earth's curvature. Particularly significant is the error produced if the axis of the bubble tube is not parallel to the line of sight of the telescope. However, if BS and FS distances are kept equal, such errors will be greatly minimized.

Bubble Not Centered on Level. If the bubble is not centered in the level tube when a reading is taken, the readings will be in error. It is surprising how easily this can happen. The instrumentman may sometimes brush against the instrument, the tripod legs may settle in soft ground, or the instrument may not be properly leveled or adjusted, with the result that when the telescope is turned the bubble does not stay centered. All of these factors mean that the instrumentman must be particularly careful. If the bubble is checked before and after readings are taken to be sure that it is centered, these errors will be reduced substantially. (It is to be remembered that automatic levels such as the one shown in Figure 7-11 will relevel themselves when they are thrown slightly out of level.)

FIGURE 7-11 AT24 automatic level. (Courtesy Topcon Positioning Systems, Inc.)

Settling of Level. In soft or swampy ground or asphalt there definitely will be some settling of the tripod. Between the time of the backsight and the foresight readings there will be settlement, with the result that the foresight reading will be too small. Special care should be taken in selecting the firmest possible places to set up the instrument. In addition, as little time as possible should be taken between the readings (use two rodmen, if possible). A further precaution in minimizing settlement errors is to take the foresight reading first on alternate setups.

Instrument Out of Adjustment. The surveyor will learn with experience to make simple checks constantly to see that the instruments are adjusted properly.

Two fairly common problems are mentioned here.

1. Should the line of sight of the telescope not be parallel to the axis of the bubble, tube errors will result in the rod readings. If, however, approximately equal BS and FS distances are used for each instrument setup, the approximately equal resulting errors will substantially cancel each other as the signs will be plus for one type of reading and negative for the other. Potential errors due to poor adjustment of this type may be far greater than those caused by combined earth's curvature and atmospheric refraction.
2. Should the horizontal cross hair not be truly horizontal, or in other words not be perpendicular to the vertical axis of the instrument when the instrument is leveled, rod reading errors will result. Such errors may be greatly reduced if readings are taken at the center of the cross hairs (where the vertical and horizontal hairs intersect).

Improper Focusing of Telescope (Parallax). If we look at the speedometer of a car from different angles, we will read different values. This is due to *parallax.* If the indicator and the speedometer scale were located exactly in the same plane, parallax would be eliminated.

Sometimes when we sight through a telescope, we find that if we move an eye a slight distance from one side to the other, there is an apparent movement of the cross hairs on the image or the object seems to move. Again, this is parallax and it can cause appreciable errors unless it is corrected. The surveyor should carefully focus the objective lens until the image and the cross hairs appear to be exactly in the same place, that is, until they do not move in relation to each other when the eye is moved back and forth. The distorting effect of parallax will then be prevented.

Heat Waves. On hot sunny days, heat waves from the ground, pavement, buildings, pipes, and other objects can seriously reduce the accuracy of work. Sometimes these waves are so intense that they cause large errors in rod readings. They may be so bad in the middle part of the day that work must be stopped until the waves subside. Heat wave errors can be minimized by reducing the lengths of sights. In addition, since the waves are worse near the ground, points should be selected so that the lines of sight are 3 or 4 ft or more above the ground.

Wind. Occasionally, high winds cause accidental errors because the winds actually shake the instrument so much that it is difficult to keep the bubble centered. These errors can be reduced by using shorter sight distances and by pushing the tripod shoes deeper in the ground and placing the legs farther apart.

7-9 SUGGESTIONS FOR GOOD LEVELING

After reading the lengthy list of mistakes and errors described in the preceding two sections, the novice surveyor may be as confused as the beginning golfer who is trying to remember 15 different things about his or her swing when trying to hit the ball. To perform good leveling, however, remember the following general rules:

1. Anchor tripod legs firmly.
2. Check to be sure that the bubble tube is centered before and after rod readings.
3. Take as little time as possible between BS and FS readings.

4. For each setup of the level, use BS and FS distances that are approximately equal.
5. Either provide rodmen with level rods that have level tubes (circular, conventional, etc.) with which the rods can be plumbed or have them wave the rods slowly toward and away from the instrument.
6. Use straight-leg (nonadjustable) tripods.
7. On sloping ground two of the tripod legs should be placed on the downhill side.

7-10 COMMENTS ON TELESCOPE READINGS

The instrumentman should learn to keep both eyes open when looking through the telescope. First, it is quite tiring to keep closing one eye all day to take readings. Second, it is convenient to keep one eye on the cross hair and the other eye open to locate the target.

If a person wears ordinary glasses for magnification purposes with no other corrections, it will not be necessary to wear glasses while looking through the telescope. The adjustment of the lens will compensate for the eye trouble.

7-11 PRECISION OF DIFFERENTIAL LEVELING

In this section we present as a guide the approximate errors that should result in differential leveling of different degrees of precision. It is assumed that levels of average condition and in good adjustment are used. *Rough leveling* pertains here to preliminary surveys in which readings are taken only to the nearest 0.1 ft and in which sights of up to 1000 ft may be used. In *average leveling,* rod readings are taken to the nearest 0.01 ft, and BS and FS distances may be approximately balanced by eye, particularly when leveling on long declines or upgrades, and sights up to 500 ft may be used. It is probable that 90% of all leveling falls into this category. In *excellent leveling,* readings are made to the nearest 0.001 ft, BS and FS distances are approximately equalized by pacing, and readings are taken for distances no greater than 300 ft.

The average errors will probably be less than the values given here. In these expressions M is the number of miles leveled and the values resulting from the expressions are in feet.

Rough leveling	$\pm 0.4\sqrt{M}$
Average leveling	$\pm 0.1\sqrt{M}$
Excellent leveling	$\pm 0.05\sqrt{M}$

For instance, if differential leveling is done over a route of 6 miles, the maximum error resulting from surveying of average precision should not exceed $\pm 0.1\sqrt{6} = \pm 0.24$ ft. These values are given for leveling done under ordinary conditions. Surveyors who frequently work in very hilly parts of the country might have some difficulty in maintaining these degrees of precision. If very rough leveling is done using the hand level, the maximum error can be limited to approximately $3.0\sqrt{M}$.

The error values listed here are approximate values, which in the opinion of the author represent the leveling that can be performed by the typical surveyor using his ordinary equipment. Section 8-2 of the next chapter describes methods of performing leveling even better than the excellent leveling mentioned here.

The reader will recall that in Section 6-4 of the last chapter that much higher standards were set for first-and second-order surveys by the Federal Geodetic Control Committee. Actually the value listed in this section for excellent leveling (± 0.05 ft $\sqrt{M}$) corresponds to the error value ($\pm$ 12mm $\sqrt{K}$) for third-order control surveys.

7-12 HAND SIGNALS

For all types of surveying it is essential for the various personnel to keep close communication with each other. Very often, calling back and forth is completely impractical because of the distances involved or because of noisy traffic or earthmoving machinery in the area. In the absence of walkie-talkies, therefore, a set of hand signals that is clearly understood by everyone involved is a necessity.

The instrumentman should remember that he or she has a telescope with which the rodman can be observed; the rodman, however, cannot see the instrumentman so clearly. As a result, the instrumentman must be very careful to give clear signals to the rodman. Following are some commonly used hand signals:

Plumb the Rod. One arm is raised above the head and moved slowly in the direction that the rod should be leaned [Figure 7-12 (a)].

Wave the Rod. The instrumentman holds one arm above his or her head and moves it from side to side. [Figure 7-12 (b)].

High Rod. To give the signal for extending the rod, hold arms out horizontally and bring them together over the head [Figure 7-12 (c)].

Raise for Red. Sometimes for very short sights the red foot marks will not fall within the telescope's field of view, and with the "raise for red" signal the instrumentman asks that the rod be raised a little so he or she can determine the correct foot mark. One arm is held straight forward, with palm up, and raised a short distance [Figure 7-12 (d)].

All Right. The arms are extended horizontally and waved up and down [Figure 7-12 (e)].

Pick Up the Instrument. The party chief may give this signal when a new setup of the instrument is desired. The hands are raised quickly from a downward position as though an object is being lifted [Figure 7-12 (f)].

Raise the Target. If one hand is raised above the shoulder with the palm visible, it means to raise the target [Figure 7-12 (g)]. If a large movement is needed, the hand is moved abruptly, but if only a small movement is needed, the hand is moved slowly.

Lower the Target. Lowering the hand below the waist means to lower the target [Figure 7-12 (h)].

Clamp the Target. The instrumentman, keeping one arm horizontal, moves his or her hand in vertical circles. This means to clamp the target. [Figure 7-12 (i)].

FIGURE 7-12 Hand signals.

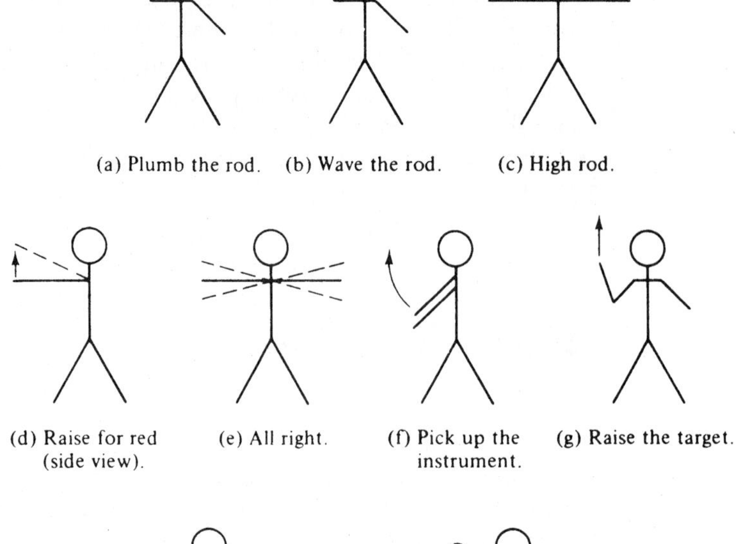

(a) Plumb the rod. (b) Wave the rod. (c) High rod.

(d) Raise for red (side view). (e) All right. (f) Pick up the instrument. (g) Raise the target.

(h) Lower the target.

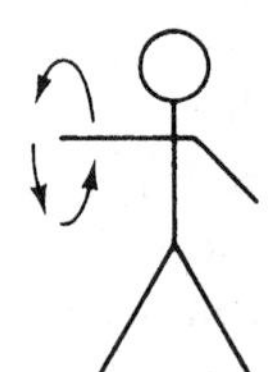

(i) Clamp the target.

Problems

7-1. Explain the difference between turning points and bench marks.

In Problems 7-2 to 7-4, complete and check the level notes shown.

7-2.

Station	BS	HI	FS	Elevation
BM_1	3.16			100.00
TP_1	6.12		5.98	
TP_2	2.47		6.48	
TP_3	4.06		8.36	
BM_2			7.42	

7-3.

Station	BS	HI	FS	Elevation
BM_1	4.654			519.850
TP_1	7.121		1.862	
TP_2	6.946		6.733	
TP_3	5.397		6.491	
TP_4	4.312		4.682	
BM_2			5.111	(Ans.: 523.401)

7-4.

Station	BS	HI	FS	Elevation
BM_1	5.111			642.230
TP_1	4.442		5.682	
TP_2	2.448		8.664	
BM_2	6.217		1.964	
TP_3	7.593		3.326	
BM_3			5.288	

In Problems 7-5 to 7-7, set up and complete differential level notes for the information shown in the accompanying illustrations. Include the customary math checks.

7-5.

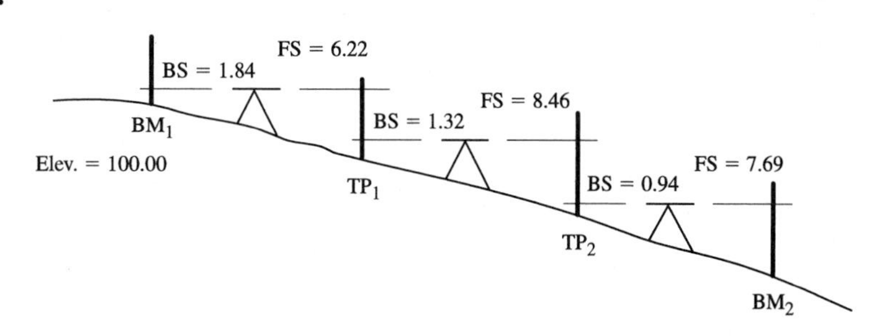

(Ans.: $BM_2 = 81.73$)

7-6.

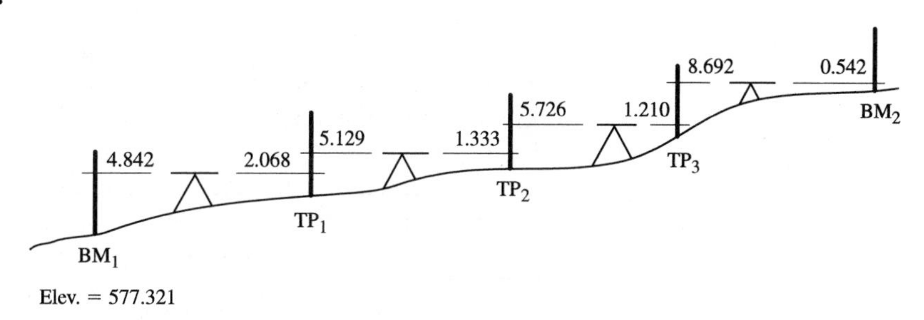

7-7.

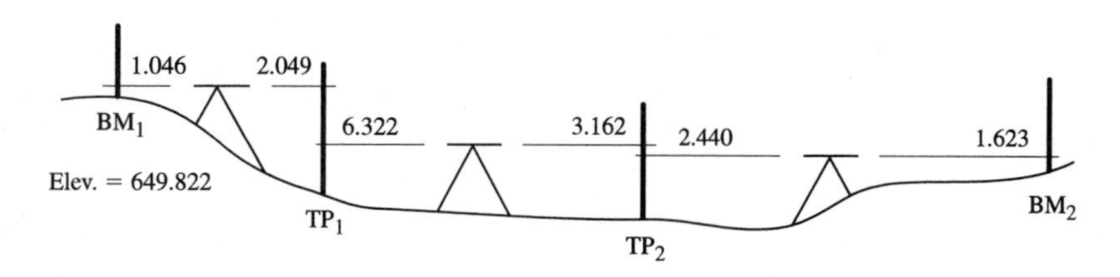

(Ans.: $BM_2 = 652.796$)

In Problems 7-8 to 7-10, the accompanying illustrations represent plans for differential leveling. The values shown on each line represent the sights taken along those lines. Prepare and complete the necessary field notes for this work (Y = instrument setup).

7-8.

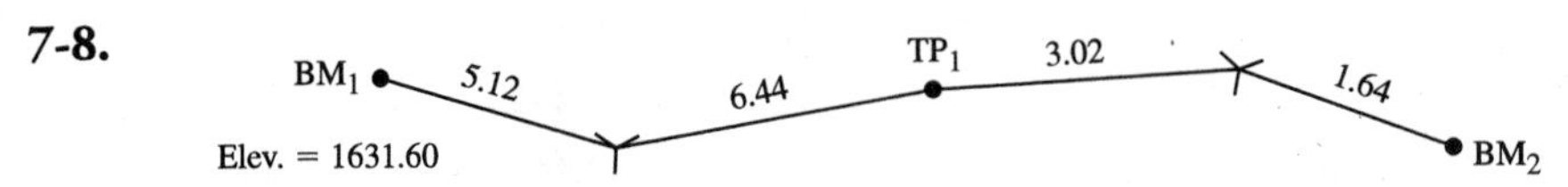

7-9.

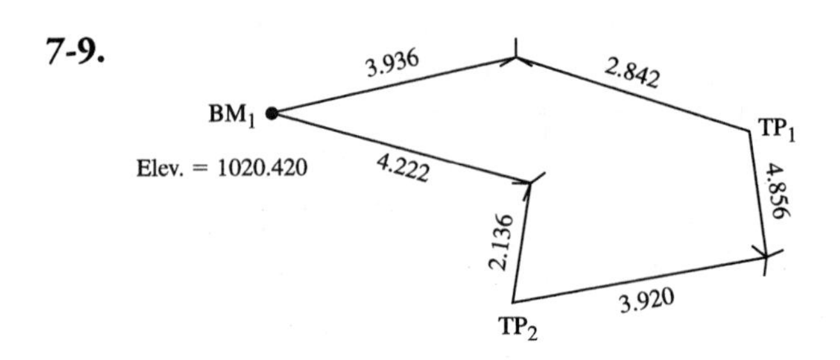

(Ans.: $BM_2 = 1020.364$)

7-10.

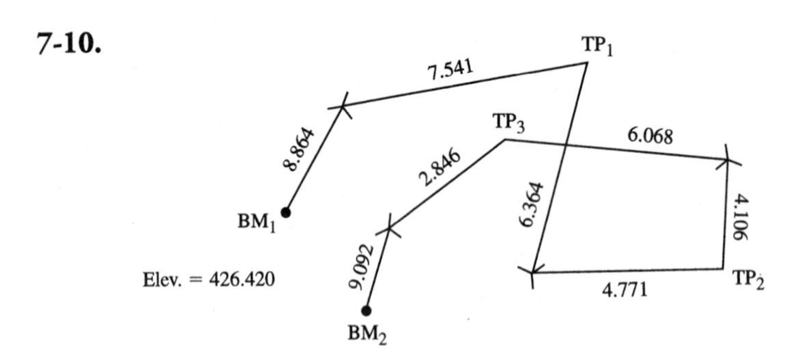

In Problems 7-11 to 7-15, the level rod readings are given in the order in which they were taken. In each case the first reading is taken on BM_1 and the last reading is taken on BM_2, the point whose elevation is desired. Set up the differential level notes, including the customary math check. The elevation of BM_1 is given under each problem number.

Problem 7-11	Problem 7-12	Problem 7-13	Problem 7-14	Problem 7-15
400.00	516.32	1453.960	766.228	542.397
5.44	6.32	8.620	4.116	6.733
7.32	9.16	5.684	7.322	3.641
5.12	4.10	9.484	2.460	8.364
8.48	8.72	2.462	8.121	4.462
5.62	6.23	12.020	5.482	9.108
6.44	7.77	5.394	9.648	2.126
7.42	5.09	7.442	5.424	5.326
5.33	10.32	4.888	6.656	3.344
↓	4.29	↓	3.360	↓
	6.65		1.026	

(Ans.: $BM_2 = 396.03$) (Ans.: $BM_2 = 1473.098$) (Ans.: $BM_2 = 558.355$)

7-16. In running a line of levels from BM_1 (elevation 907.32) to BM_2, the following readings were taken in the order given: 6.12, 7.09, 2.44, 5.98, 2.66, 6.87, 8.42, and 4.49. Set up and complete the level notes, including the math check.

7-17. In running a line of levels from BM_1 (elevation 526.32) to BM_2, the following readings were taken in the order given: 1.12, 3.68, 5.94, 4.68, 6.12, 2.86, 3.74, 8.49, 9.69, and 1.11. Set up and complete the level notes including the math check. (Ans.: $BM_2 = 532.11$)

7-18. A line of levels was run into a mine shaft. All of the points (BMs and TPs) were located in the shaft ceiling and readings were taken by inverting the level rod. Complete the resulting level notes shown, including the math check.

Station	BS	HI	FS	Elevation
BM_1	6.94			432.00
TP_1	8.29		3.84	
TP_2	7.73		5.76	
TP_3	7.02		2.64	
BM_2			5.74	

7-19. Repeat Problem 7-17 assuming that all points (BMs and TPs) were located in the top of a tunnel and were taken by inverting the level rod. (Ans.: $BM_2 = 520.53$)

For problems 7-20 to 7-22, complete and check the level notes.

7-20.

Station	BS	HI	FS	Elevation
BM_1	5.46			519.19
TP_1	8.11		3.11	
			9.55	
			6.48	
TP_2	7.02		3.06	
BM_2			8.44	

7-21.

Station	BS	HI	FS	Elevation
BM_{11}	6.68			1584.61
TP_1	4.41		1.96	
			4.24	
			3.18	
TP_2	6.36		6.96	
			6.12	
			5.11	
TP_3	7.42		9.64	
BM_{12}			8.64	(Ans.: $BM_{12} = 1582.28$)

7-22.

Station	BS	HI	FS	Elevation
BM_1	5.36			913.26
			7.42	
			5.66	
TP_1	7.40		4.21	
			2.63	
TP_2	3.68		7.32	
BM_2			2.12	

7-23. Compute the combined effect of the earth's curvature and atmospheric refraction for distances of 100 ft, 200 ft, 500 ft, 2000 ft, and 10 miles.
(Ans.: 0.0002; 0.0008; 0.0051; 0.0824; 57.40 ft)

7-24. A BS of 4.20 ft is taken on a level rod at a 100-ft distance, and an FS of 10.69 ft is taken on the rod held 1000 ft away. (a) What is the error caused by the earth's curvature and atmospheric refraction? (b) What is the correct difference in elevation between the two points?

7-25. In differential leveling from BM_1 to BM_2, the BS and FS distances for readings were as follows: BS 300 ft, FS 100 ft, BS 600 ft, FS 100 ft, BS 450 ft, FS 300 ft, BS 400 ft, and FS 200 ft. What is the error in the elevation of BM_2 caused by atmospheric refraction and the earth's curvature?
(Ans.: +0.014 ft)

7-26. What BS or FS distances for an instrument setup will cause an error due to the earth's curvature and atmospheric refraction equal to 0.005 ft? 0.02 ft? 0.10 ft?

7-27. Two towers *A* and *B* are located on flat ground and their bases have equal elevations above sea level. A person on tower *A* whose eye level is 18.0 ft above the ground can just see the top of tower *B*, which is 120 ft above the ground. How far apart are the towers? (Ans.: 20.06 miles)

7-28. A man whose eye level is 5.4 ft above the ground is standing by the ocean. He can just see the top of a lighthouse across the water. Neglecting tidal and wave effects, how high is the lighthouse if it's 18 miles away?

7-29. A surveyor is going to take a 6-mile sight across a lake from the top of one tower to a target on the top of another tower. It is desired to keep the line of sight 10 ft above the lake surface. At what equal heights above the shoreline should the instrument and the target be located? (Ans.: 15.17 ft)

7-30. A rotating light is located on the top of a lighthouse 110 ft above water level. How far is the light visible to a sailor on a ship assuming that his eye level is 16 ft above the water? Assume the water is calm (i.e., no appreciable waves).

7-31. Calculate the error involved in the following level rod readings if a 13.0-ft rod is assumed to be 6 in out of plumb at its top.

a. A BS of 11.800 ft (Ans.: 0.009 ft)

b. An FS of 4.640 ft (Ans.: 0.003 ft)

7-32. A line of levels is run from point *A* (elevation 619.32 ft) to point *B* to determine its elevation. The value so obtained is 696.20 ft. If a check of the rod after the work is done reveals that the bottom 0.03 ft has been worn away, what is the correct elevation of point *B* if there were 15 instrument setups?

7-33. The smallest divisions on a level rod are 1/10 of a foot. If 20 divisions on the vernier cover 19 divisions on the level rod, what is the least value readable? (Ans.: 0.005 ft)

7-34. The least divisions on a level rod are tenths of a foot. Describe a vernier that will enable the surveyor to read the rod to the nearest 0.01 ft.

7-35. For a level rod that is graduated to hundredths of a foot, design a target vernier so that it can be read to the nearest 0.002 ft.
(Ans.: 5 divisions on vernier covering, 4 divisions on level rod)

7-36. Repeat Problem 7-34 except that the rod to be used will be read to the nearest 0.005 ft.

CHAPTER

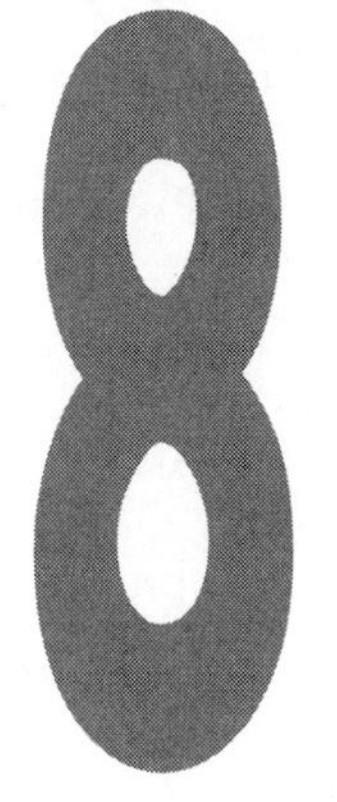

Leveling, Continued

8-1 ADJUSTMENTS OF LEVEL CIRCUITS

Levels over One Route

If a line of levels is run from one bench mark to set and establish the elevations of several bench marks some distance away, it will be necessary once the values are determined to level back to the starting bench mark (or to some other bench mark) to check the work. If this is not done, there may very well be serious undiscovered discrepancies in the work.

If the surveyor levels back to the starting bench mark or to another bench mark in the vicinity, he or she will in all probability (no matter how careful) obtain a different value. If the difference is reasonable (see Section 7-11), the surveyor will adjust proportionately the elevations that were set along the route. Such a procedure is described in the next few paragraphs. Should, however, the checked value contain a substantial and unacceptable discrepancy, it will be necessary to repeat the work.

Every time someone takes a reading with a level (Figure 8-1), errors will be involved. It thus seems logical to assume that corrections in a line of levels should be made in proportion to the number of instrument setups. In other words, the weight or relative importance given to a measured elevation should be inversely proportional to the number of instrument setups (the fewer setups, the smaller will be the total error and thus the greater weight given to the value determined).

Unless the terrain is rather unusual, there will be a fairly consistent relationship between the number of instrument setups and the distance leveled. As a result, corrections can ordinarily be made in proportion to the distances involved.

From the preceding discussion it follows that *the logical correction to the measured elevation of a particular point in a level circuit should be to the total correction as the distance to that point from the beginning point is to the total circuit distance.* For unusual terrain (as where slopes are very steep) corrections can be made in proportion to the number of instrument setups.

For this discussion, the line of levels shown in Figure 8-2 is considered. The surveyor starts from BM_1 with its known elevation and establishes BM_2, BM_3 and BM_4 and levels back to BM_1. The total distance leveled from BM_1 around the circuit and back to BM_1 was 15 miles. It is assumed that the total error in that distance was

+0.30 ft. If the error is assumed to be the same in any 1 mile of leveling as in any other mile, the following calculations can be made:

$$\text{Total correction to be made in 15 miles} = -0.30 \text{ ft}$$

$$\text{Correction per mile} = \frac{-0.30}{15} = -0.02 \text{ ft / mile}$$

Then the correction in the elevation of any one of the bench marks equals the number of miles from the beginning point (BM_1) to the bench mark in question times the correction per mile. For instance,

$$\text{Correction for } BM_2 \text{ elevation} = (3)(-0.02) = -0.06 \text{ ft}$$

$$\text{Correction for } BM_3 \text{ elevation} = (7)(-0.02) = -0.14 \text{ ft}$$

FIGURE 8-1 Nikon automatic level with built-in compensator which automatically levels the line of sight after instrument is roughly leveled. (Courtesy Nikon, Inc.)

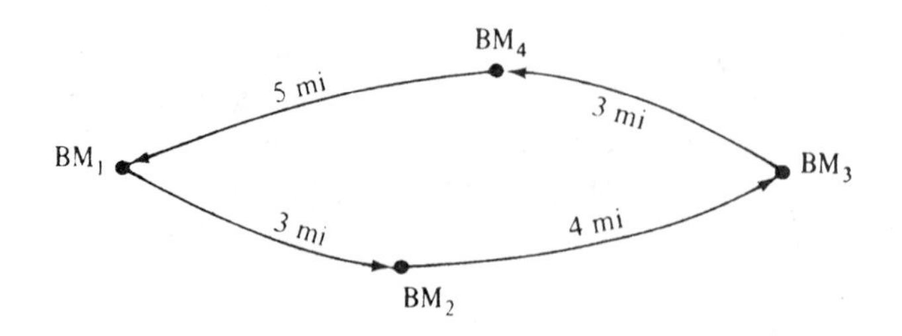

FIGURE 8-2 Line of levels.

The observed elevation for each of the bench marks in Figure 8-2 is shown in Table 8-1 together with the calculations of their most probable elevations.

TABLE 8-1 Elevations for Bench Marks in Figure 8-2

Point	Distance from BM_1 (mi)	Observed elevation (ft)	Correction (ft)	Most probable elevation (ft)
BM_1	0	200.00	0.00	200.00
BM_2	3	209.20	-0.06	209.14
BM_3	7	216.44	-0.14	216.30
BM_4	10	211.86	-0.20	211.66
BM_1	15	200.30	-0.30	200.00

Levels over Different Routes

If several lines of levels are run over different routes from a common beginning point to a common ending point where it is desirable to establish a bench mark, it is obvious that different results will be obtained. It is desirable to obtain the most probable elevation of this new point.

In comparing different lines of levels to the same point, the shorter a particular route, the greater will be the importance or the weight given to its results. In other words, the shorter a line, the more accurate its results should be. *Thus the weight of an observed elevation varies inversely as the length of its line.*

For the example shown in Figure 8-3, route *a* is run for 4 miles from BM_1 in order to establish BM_2, where an elevation of 340.21 ft is obtained. A second route, labeled *b*, is also run from BM_1 to BM_2 over a distance of 2 miles and a measured elevation of 340.27 ft is obtained.

FIGURE 8-3
Leveling over two routes.

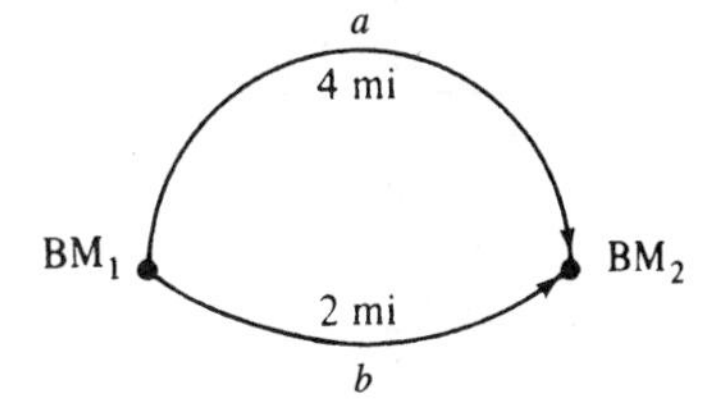

Because the second route is half as long as the first one, twice as much weight should be given to its observed value. From this information the most probable elevation of BM_2 can be obtained with the following expression:

$$\frac{(1)(340.21)+(2)(340.27)}{3} = 340.25 \text{ ft}$$

The preceding calculation is perfectly acceptable, but when several routes are involved and the numbers are not as simple, the calculations might be more convenient to handle in a table such as Table 8-2. Here the weight given to each route equals the reciprocal of its length in miles.

TABLE 8-2 Calculations for Bench Marks in Figure 8-3

Route	Length (mi)	Measured elevation of BM_2 (ft)	Measured elevation -340.00 (ft)	Weight of route	Weighted difference (ft)
a	4	340.21	0.21	$\frac{1}{4} = 0.25$	$\frac{0.25}{0.75} \times 0.21 = 0.07$
b	2	340.27	0.27	$\frac{1}{2} = 0.50$	$\frac{0.50}{0.75} \times 0.27 = 0.18$
				$\Sigma = 0.75$	$\Sigma = 0.25$

Most probable elevation of BM_2= 340.25 ft

Another example, somewhat more complicated, is presented in Table 8-3 and is shown in Figure 8-4. Again, BM_1 has a known elevation and it is desired to set and establish the elevation of BM_2. To accomplish this objective, four different routes are run from BM_1 to BM_2. The lengths of each of these routes and the values of the elevations obtained for BM_2 are given in the table and the most probable elevation of BM_2 is determined.

TABLE 8-3 Calculations for Bench Marks in Figure 8-4

Route	Length (mi)	Measured elevation of BM_2 (ft)	Measured elevation −106.00 (ft)	Weight of route	Weighted difference (ft)
a	1	106.50	0.50	$\frac{1}{1} = 1.00$	$\frac{1.00}{1.93} \times 0.50 = 0.26$
b	3	106.44	0.44	$\frac{1}{3} = 0.33$	$\frac{0.33}{1.93} \times 0.44 = 0.08$
c	2	106.52	0.52	$\frac{1}{2} = 0.50$	$\frac{0.50}{1.93} \times 0.52 = 0.13$
d	10	106.36	0.36	$\frac{1}{10} = 0.10$	$\frac{0.10}{1.93} \times 0.36 = 0.02$
				$\Sigma = 1.93$	$\Sigma = 0.49$

Most probable elevation of BM_2= 106.49 ft

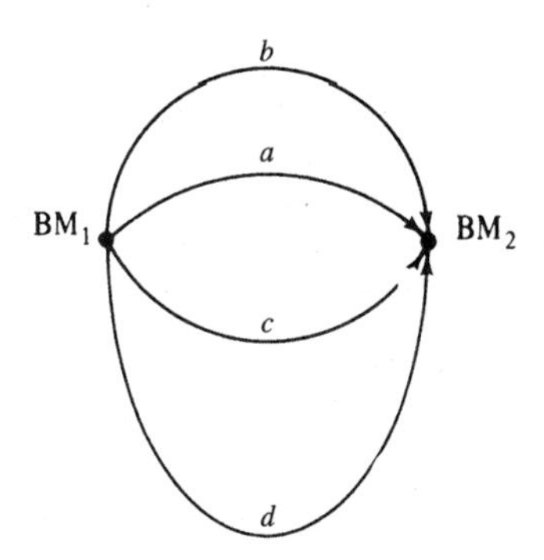

FIGURE 8-4 Leveling over several routes.

For a detailed discussion of level net adjustments the reader is referred to pages 157–167 of the book *Surveying for Civil Engineers,* 2nd ed., by Phillip Kissam.[1] Adjustments for complicated nets are usually performed with computers using the method of least squares.

8-2 PRECISE LEVELING

The term *precise leveling* is usually applied to the leveling practiced by the National Geodetic Survey. The purpose of this section is not to describe in detail how precise leveling is performed but rather to indicate methods by which the average surveyor working with ordinary equipment can substantially raise the precision of his or her work. The surveyor can apply one or more of the procedures presented in the following paragraphs with improvement in his or her leveling for each one used.

Miscellaneous

Several practices that improve the precision of leveling were mentioned in Chapter 7. These include the following: setting tripod legs firmly in solid ground, allowing the least time possible between BS and FS readings for each setup, using clearly marked solid TPs, limiting sight distances to 300 ft or less, careful plumbing of level rods, and avoiding leveling during strong winds and during severe heat waves. Another useful practice is to keep lines of sight at least 2 ft above the ground because of atmospheric refraction effects.

Shading the Level

If a level is used on warm days in the direct sunlight, the result may be unequal expansion of different parts of the level with consequent errors. For instance, if one end of a bubble tube becomes warmer than the other end, the bubble will move toward the warmer end. This problem may be greatly minimized by shading the instrument from the direct rays of the sun with an umbrella while observations are being made and while the instrument is being moved from one point to another (see Figure 8-5).

[1]New York: McGraw-Hill Book Company, 1981.

FIGURE 8-5 Precise leveling observations. (Courtesy of National Geodetic Survey.)

Three-Wire Leveling

A method called three-wire leveling is occasionally used when levels are equipped with stadia hairs in addition to the regular cross hairs. All three horizontal hairs are read and the average of the three readings is taken as the correct reading. It is helpful to balance fairly well the BS and FS distances, and this is easily done by reading the interval between the stadia hairs.

For many years three-wire leveling was used only for very precise leveling and not for ordinary work because it is such an agonizingly slow process when a target is used. Today, however, it is much more frequently used even for projects that require ordinary precision. A target is probably not used. The three readings are taken with the telescope and are either estimated by eye to about the nearest 0.002 ft, or better, they are taken with an optical micrometer (discussed in Section 6-8). To prevent mistakes in taking each set of three readings it is quite important to check the difference in readings between the top and middle hair and the middle and bottom hair. The differences should be almost identical.

Precise Leveling Rods

For ordinary leveling, the usual level rod is satisfactory, but precision can be improved by using a rod treated in some manner against contraction and expansion in order to minimize the effects of changes of temperature and humidity. The preferable solution involves the use of a graduated Invar tape that is independent of the main part of the rod so that it is free to slide in grooves on each side of the rod if the rod changes in length. Precise leveling rods are equipped with a thermometer and a bull's-eye level or a pair of ordinary levels placed at 90° to each other for plumbing the rod.

The National Geodetic Survey uses 3-m-long one-piece rods. They are marked in black-and-white checkerboard fashion so that the cross hairs will always fall on a white space, where its position can be precisely estimated. Furthermore, these rods have the characteristic of being clearly visible in both light and dark conditions. The smallest divisions on the rod are usually 5 mm. With optical micrometers and very precise levels, readings can be taken at least to ±0.1 mm.

Double-Rodded Lines

The use of two level rods and two sets of TP, BS, and FS readings may improve leveling precision a little. Two sets of notes are kept, and the observed elevations of the points in question are averaged. When this procedure is used, it is desirable to use TPs for the two lines which have elevation differences of at least 1 ft or more so that the possibility of making the same 1-ft mistakes in both lines is reduced. To minimize systematic errors, rods may be swapped between the lines at convenient intervals, for example, on alternate instrument setups. Double-rodded lines are particularly useful when it is necessary to run a quick set of differential levels in order to establish an elevation where there is not sufficient time to check the work by returning to the initial point or to some other point whose elevation has previously been established. An example set of notes for a double-rodded line is shown in Figure 8-6. In these notes the turning points are listed as H or L (high route or low route) to identify the BS and FS values on the two different lines.

FIGURE 8-6 Leveling notes for a double-rodded line.

DOUBLE RODDED LINE
ALONG GREEN STREET

Oct. 27, 2003
Clear, mild 60°
Nikon AL - 12 level # 7.

⊼ - J.B. Johnson
Rod - R.C. Knight
Notes - N.T. Hanson

Sta.	BS	HI	FS	Elev.	Mean	
BM₁	6.442	680.806	X	674.364		Conc. monument N.E. corner of intersection
	6.442	680.806				of Green and Oak Streets
TP₁ H	7.174	684.359	3.621	677.185		
TP₁ L	8.626	684.349	5.083	675.723		
TP₂ H	4.570	683.042	5.887	678.472		
TP₂ L	5.904	683.028	7.225	677.124		
TP₃ H	1.041	674.269	9.814	673.228		
TP₃ L	2.432	674.255	11.205	671.823		
BM₂			8.642	665.627	665.620	Spike in oak tree trunk 80'N of
BM₂			8.642	665.613		McDonalds on Green St.
	+42.631		-60.119			

CHECK

(Σ BS + Σ FS) ÷ 2 = (+42.631 − 60.119) ÷ 2 = − 8.744

674.364 − 665.620 = − 8.744

✓

8-3 PROFILE LEVELING

For purposes of location, design, and construction it is necessary to determine elevations along proposed routes for highways, canals, railroads, water lines, and similar projects. The process of determining a series of elevations along a fixed line is referred to as *profile leveling.*

Profile leveling consists of a line of differential levels with a series of intermediate shots taken during the process. The instrument is set up at a convenient point and a BS taken to a point of known elevation in order to determine the HI. If a bench mark of known elevation is not available, it is necessary either to establish one by differential levels from an existing BM or to set one up and give it an assumed elevation. (The latter procedure might not be satisfactory for some projects, for example, those involving water.)

After the HI is established, a series of foresights are taken along the center line of the project. These readings are referred to herein as *intermediate foresight readings* (IFS). Other surveyors may call them either *ground rod readings* or just rod readings. These readings are taken at regular intervals, say 50 or 100 ft, and at points where sudden changes in elevations occur, such as at the tops and bottoms of river banks, edges and center lines of roads and ditches, and so on. In other words, shots are taken where necessary to give a true picture of the ground surface along the route.

When it is no longer possible to continue with the IFS readings from the instrument position, it is necessary to take a FS on a TP and move the instrument to another position from which another series of readings can be taken. A portion of a typical set of profile level notes is shown in Figure 8-7. The author has included a partial math check for the differential leveling part of these notes. The check includes the BS and FS values from the elevation of BM_{78} to the HI elevation just beyond TP_2.

FIGURE 8-7 Profile leveling notes.

PROFILE LEVELING FOR PROPOSED CONGAREE HIGHWAY

Nov. 3, 2003 Overcast, Mild 55° Topcon Auto Level #R 20017 — ⊼ - J.B. Johnson, Rod - R.C. Knight, Notes - N.T. Hanson

Sta.	BS	HI	FS	IFS	Elev.	
BM_{78}	3.11	103.11	×	×	100.00	Brass monument 208' N of sta 0+00
0+00				8.6	94.5	
+50				7.3	95.8	
1+00				5.9	97.2	
+50				5.8	97.3	
2+00				6.1	97.0	
+50				7.3	95.8	
TP_1	4.73	105.67	2.17		100.94	Stone
3+00				3.7	102.0	
+22.7				3.2	102.5	Top of ditch
+24				8.7	97.0	Bottom of ditch
+35.6				8.6	97.1	Bottom of ditch
+38				5.5	100.2	Top of ditch
+50				5.3	100.4	
4+00				7.0	98.7	
+50				8.0	97.7	
TP_2	6.07	110.68	1.06		104.61	Stump
5+00				11.9	98.8	
+50				10.0	100.7	
	+13.91		−3.23			
	−3.23	Elev.	BM_{78}	= 100.00		
	+10.68	Elev.	HI	= 110.68		
	←	Checks	→	+10.68		N.T. Hanson

It will be noted that when surveys are made along fixed routes such as these, the distances from the starting points are indicated by stationing. The usual practice is to set stakes along the center line of the project at regular intervals, say 50 to 100 ft. Points along the route with even multiples of 100 ft are referred to as *full stations* as 0 + 00, 1 + 00, 2 + 00, and so on. Intermediate stations are referred to as *plus stations*. For instance, a point 234.65 ft from the beginning station would be designated as 2 + 34.65. Some surveyors like to assign a station of, say, 15 + 00 or 20 + 00 to the starting point of a project so that if the project is expanded in the other direction there probably will not be any negative stations.

When the SI system is used in the United States, some municipalities use 100 unit stations (1 station = 100 meters) while highway departments generally use 1000 unit stations (that is 1 station = 1000 m). In this text the author uses 1000 m for stations in SI problems.

It is usually unnecessary to take the intermediate foresight readings with as high a degree of precision as that needed for the regular FS values. If the surface is irregular, as in a field or a forest, readings to the nearest hundredth of a foot are unnecessary and perhaps a little misleading. They are usually taken to the nearest tenth of a foot, as in the example presented in Figure 8-7. When surfaces are smooth and regular, as along a paved highway, it is not unreasonable to take the IFS readings to the nearest 0.01 ft.

During profile leveling it is usually very wise to set a series of BMs because they can be very useful at a later date, for example, when grades are being established for construction. These control points should be set a sufficient distance from the proposed project center line so that they hopefully will not be obliterated during construction operations. When the profile is completed, it is necessary to check the work by tying into another BM or by running a line of differential levels back to the beginning point.

8-4 PROFILES

The purpose of profile leveling is to provide the information necessary to plot the elevation of the ground along the proposed route. A *profile* is the graphical intersection of a vertical plane, along the route in question, with the earth's surface. It is absolutely necessary for the planning of grades for roads, canals, railroads, sewers, and so on.

Figure 8-8 shows a typical profile of the center line of a proposed highway. It is plotted on profile paper especially made for this purpose, having horizontal and vertical lines printed on it to represent distances both horizontally and vertically. It is common to use a vertical scale much larger than the horizontal one (usually 10:1) in order to make the elevation differences very clear. In this figure a horizontal scale of 1 in. = 200 ft and a vertical scale of 1 in. = 20 ft are used. The plotted elevations for the profile are connected by freehand because it is felt that the result is a better representation of the actual ground shape than would be the case if the points were connected with straight lines.

The author has sketched a trial grade line in Figure 8-8 for the proposed highway. The major purpose of plotting the profile is to allow the establishment of grade lines. On an actual job, different grade lines are tried until the cuts and fills balance sufficiently within reasonable haul distances with grades that are not too excessive.

For highway projects, it is normally convenient to plot the plan and profile on the same sheet. The top half of the paper is left plain so that the plan can be plotted above the profile. The plan is not shown in Figure 8-8.

FIGURE 8-8 Profile and trial grade lines.

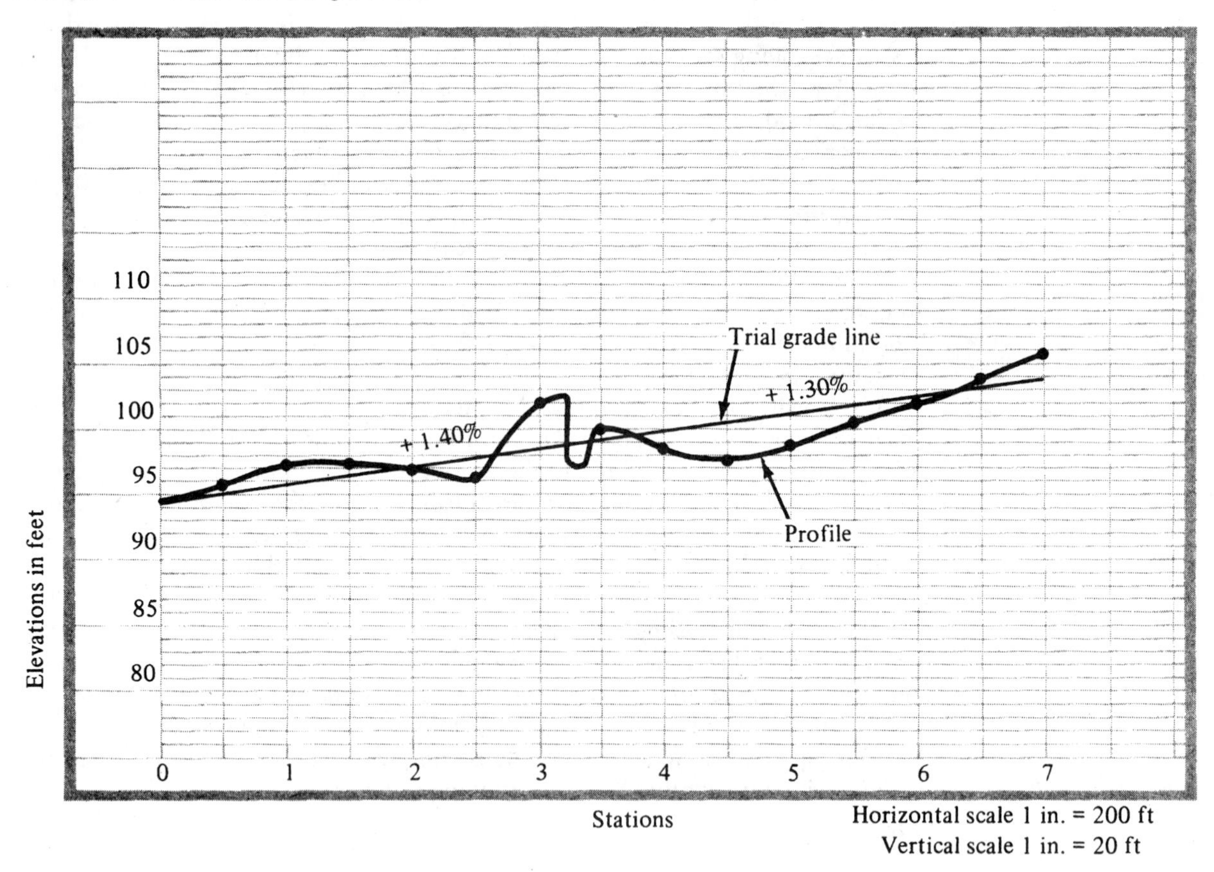

8-5 CROSS SECTIONS

Cross sections are lines of levels or short profiles made perpendicular to the center line of a project. They provide the information necessary for estimating quantities of earthwork. There are two general types of cross sections: the ones required for route projects such as roads and the ones required for borrow pits.

For route surveys, cross sections are taken at regular intervals such as the 50- or 100-ft stations and at sudden changes in the center-line profile. There is a tendency among surveyors to take too few sections, particularly in rough country. To serve their purpose, the sections must extend a sufficient distance on each side of the center line so that the complete area to be affected by the project is included. Where large cuts or fills seem probable, greater distances from the center line should be sectioned.

When cross sections are plotted in the office from field notes, the ground surface is plotted by connecting the elevations with straight lines. Thus if there are any changes or breaks in the slope of the ground it is important for the surveyor to take foresights to each of those points. To attempt to give the reader a feeling as to where sights should be taken, Figure 8-9 is presented. In this figure, suggested locations for rod readings are shown for three different locations along a proposed highway.

The necessary elevations can be determined with a regular level, with a hand level, or with a combination of both. Sometimes a hand level is held or fastened on top of a stick or board about 5 ft tall (called a Jacob staff) and is used to determine the needed elevations. Elevations are usually taken at regular intervals as, say, 25, 50, 75 ft, and so on, on each side of the route center line as well as at points where significant changes in

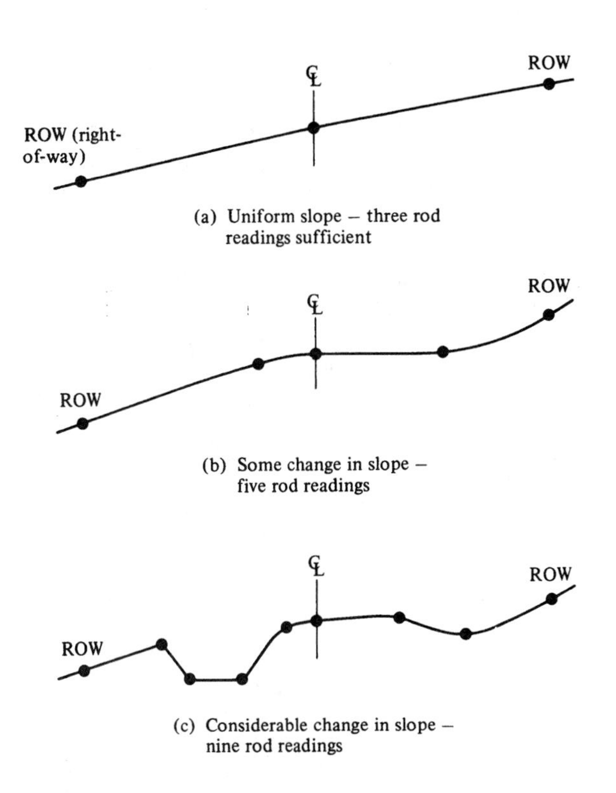

FIGURE 8-9 Suggested locations for rod readings for cross-section levels.

slopes or features (as streams, rocks, etc.) occur. *The height of the telescope line of sight above the ground is referred to as the h.i. throughout this book (the term HI is used to give the elevation of the line of sight with respect to a reference datum, normally sea level).*

In Figure 8-10 the level or Jacob staff is set up at the center line of the project. The height of the telescope of the instrument above the ground (h.i. in the figure) is measured and found to be 5.1 ft. At station 11+00 the rodman is sent out 15 ft to the left and an FS reading of 11.2 is taken. The difference in elevation from the center line to the point in question is thus -6.1 ft. Then the rodman moves out 18 more feet to a point 33 ft from the center line. The FS reading is 11.5 ft and the difference in elevation is -6.4 ft. If the difference in elevation becomes too large for the level rod height or if in going uphill the elevation differences are larger than the h.i. of the instrument, it will then be necessary to take one or more turning points to obtain all the necessary readings.

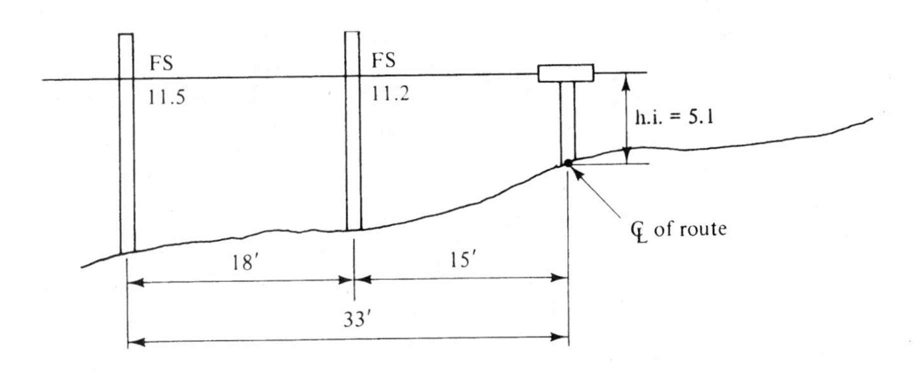

FIGURE 8-10 Taking cross sections.

A convenient form of recording cross-section notes is shown in Figure 8-11. This figure includes profile notes on the left-hand page and cross-section notes on the right-hand page. On the cross-section page, the numerators are the differences in elevations from the center-line stations to the points in question; the denominators are the distances from the center-line. The surveyor must record carefully the signs for these numbers. A plus sign is given to the points that are higher than the center line and a minus sign to those below.

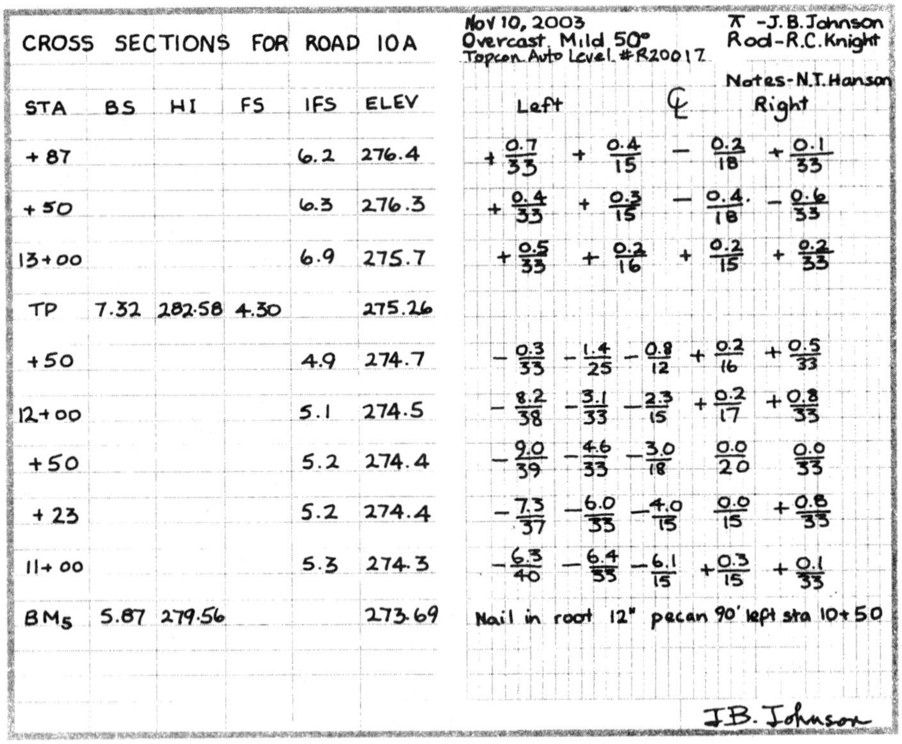

FIGURE 8-11 Profile and cross-section notes.

CROSS SECTIONS FOR ROAD 10A

Nov 10, 2003 Overcast, Mild 50° Topcon Auto Level #R20017

⊼ -J.B.Johnson Rod-R.C.Knight Notes-N.T.Hanson

STA	BS	HI	FS	IFS	ELEV	Left ℄ Right
+87				6.2	276.4	$+\frac{0.7}{33}$ $+\frac{0.4}{15}$ $-\frac{0.2}{18}$ $+\frac{0.1}{33}$
+50				6.3	276.3	$+\frac{0.4}{33}$ $+\frac{0.3}{15}$ $-\frac{0.4}{18}$ $-\frac{0.6}{33}$
13+00				6.9	275.7	$+\frac{0.5}{33}$ $+\frac{0.2}{16}$ $+\frac{0.2}{15}$ $+\frac{0.2}{33}$
TP	7.32	282.58	4.30		275.26	
+50				4.9	274.7	$-\frac{0.3}{33}$ $-\frac{1.4}{25}$ $-\frac{0.8}{12}$ $+\frac{0.2}{16}$ $+\frac{0.5}{33}$
12+00				5.1	274.5	$-\frac{8.2}{38}$ $-\frac{3.1}{33}$ $-\frac{2.3}{15}$ $+\frac{0.2}{17}$ $+\frac{0.8}{33}$
+50				5.2	274.4	$-\frac{9.0}{39}$ $-\frac{4.6}{33}$ $-\frac{3.0}{18}$ $\frac{0.0}{20}$ $\frac{0.0}{33}$
+23				5.2	274.4	$-\frac{7.3}{37}$ $-\frac{6.0}{33}$ $-\frac{4.0}{15}$ $\frac{0.0}{15}$ $+\frac{0.8}{33}$
11+00				5.3	274.3	$-\frac{6.3}{40}$ $-\frac{6.4}{33}$ $-\frac{6.1}{15}$ $+\frac{0.3}{15}$ $+\frac{0.1}{33}$
BM_5	5.87	279.56			273.69	Nail in root 12" pecan 90' left sta 10+50

J.B. Johnson

Many surveyors plot their profile and cross-section notes running up the page from the bottom. Such a form is quite logical in that, as the surveyor faces forward along the center line, the area to the right of the center line is shown on the right-hand side of the page and the area to the left is shown on the left-hand side. Thus the notes show the configuration of the ground as the surveyor looks at it. Close stations are at the bottom of the page and far stations are at the top of the page. The notes in Figure 8-11 are shown in this fashion.

8-6 MISTAKES IN NONCLOSED LEVELING ROUTES

Mistakes in leveling for closed level loops are an annoying problem, but they can be discovered and eliminated. Mistakes in leveling that is not part of a closed loop are a more serious problem. Examples of such cases include profile leveling, cross sections, construction grades, and so on. Just imagine the time and cost involved if we have to tear out a reinforced concrete footing for a building and rebuild it at a different elevation because of a mistaken FS reading. Similar remarks can be made about other construction work, such as bridge abutments, culverts, and other structures. The reader is reminded to be exceptionally careful in this type of work and to repeat many measurements.

Problems

8-1. Adjust the following unbalanced level circuit.

Point	Distance from BM_1 (miles)	Observed elevation (ft)	
BM_1	0	636.86	
BM_2	3	641.12	(Ans.: BM_2 = 640.94 ft)
BM_3	9	634.16	(Ans.: BM_3 = 633.62 ft)
BM_4	15	639.44	(Ans.: BM_4 = 638.54 ft)
BM_1	18	637.94	

8-2. Adjust the unbalanced level circuit shown in the accompanying table.

Point	Distance from BM_1 (miles)	Observed elevation (ft)
BM_1	0	734.94
BM_2	3	727.82
BM_3	7	721.28
BM_4	9	726.44
BM_5	15	722.29
BM_1	21	734.31

8-3. A line of levels was run to set the elevation of several bench marks. The following values were obtained:

Point	Distance from BM_1 (miles)	Observed elevation (ft)
BM_1	0	439.62
BM_2	3	438.77
BM_3	7	433.02
BM_4	10	437.64
BM_1	14	440.04

Is the closure satisfactory for a leveling requirement = $\pm 0.12\sqrt{M}$ where M is in miles and the expression provides permissible closure in feet? Adjust the elevations for all of the bench marks.

(Ans.: Closure is satisfactory)
(Ans.: BM_2 = 438.68 ft)
(Ans.: BM_3 = 432.81 ft)
(Ans.: BM_4 = 437.34 ft)

8-4. A closed loop of differential levels was run to establish the elevations of several bench marks with the following results. Determine the most probable elevation.

Distance (ft)		Observed elevation (ft)	
A to B	7000	BM_A	515.80
B to C	4000	BM_B	527.08
C to D	4500	BM_C	521.92
D to E	5000	BM_D	506.34
E to A	3000	BM_E	518.09
		BM_A	516.27

8-5. BM_1 has a known elevation and it is desired to establish the elevation of BM_2 by running differential levels over three different routes from BM_1, as shown in the accompanying table. What is the most probable elevation of BM_2?

Route	Length (miles)	Measured elevation of BM_2 (ft)	
a	2	1262.11	
b	4	1262.15	
c	8	1262.03	(Ans.: 1262.11 ft)

8-6. Several lines of levels are run over different routes from BM_1 in order to set BM_2 and establish its elevation. The lengths of these routes and the value of the elevations determined are shown in the accompanying table. Determine the most probable elevation of BM_2.

Route	Length (miles)	Measured elevation of BM_2 (ft)
a	3	922.84
b	5	922.92
c	11	922.81
d	17	922.90

8-7. Set up and complete the level notes for a double rodded line from BM_{11} (elevation 1642.324) to BM_{12}. In the following rod readings H refers to high route and L to low route: BS on BM_{11} = 7.342; FS on TP_1H = 2.306; FS on TP_1L = 4.107; BS on TP_1H = 9.368; BS on TP_1L = 11.162; FS on TP_2H = 3.847; FS on TP_2L = 5.111; BS on TP_2H = 8.339; BS on TP_2L = 9.619; FS on TP_3H = 4.396; FS on TP_3L = 5.448; BS on TP_3H = 7.841; BS on TP_3L = 8.896; FS on BM_{12} = 3.329. (Ans.: 1661.342)

8-8. The center line for a proposed highway has been staked. With an HI of 854.32 ft, the following intermediate foresights are taken at full stations beginning at Sta. 0 + 00: 3.1, 3.6, 4.1, 6.0, 6.9, 7.7, 8.6, 9.3, 9.4, 9.8, 9.2, 8.7, 8.2, 7.9, 8.0, 8.3, 8.8, and 9.4. Plot the profile for these stations.

8-9. Complete the following set of profile level notes.

Station	BS	HI	FS	IFS	Elevation
BM_1	3.10				1085.13
TP_1	6.14		7.66		
0 + 00				6.8	
1 + 00				6.6	
2 + 00				7.7	
2 + 70.5				8.3	
2 + 76.2				8.5	
3 + 00				7.1	
TP_2	11.86		5.85		
4 + 00				6.2	
5 + 00				5.8	
5 + 55.0				5.2	
5 + 65.2				4.9	
6 + 00				4.3	
BM_2			9.07		

(Ans.: $BM_2 = 1083.65$)

CHAPTER

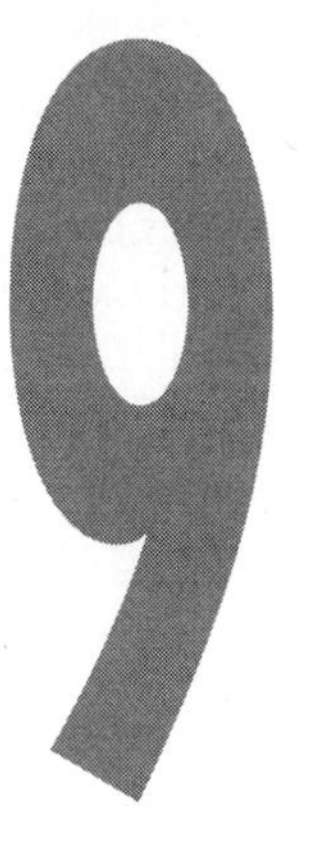

Angles and Directions

9-1 MERIDIANS

In surveying, the direction of a line is described by the horizontal angle that it makes with a reference line or direction. Usually, this is done by referring to a fixed line of reference called a *meridian.* There are three types of meridians: astronomic, magnetic, and assumed. An *astronomic meridian* is the direction of a line passing through the astronomic north and south poles and the observer's position as shown in Figure 9-1.

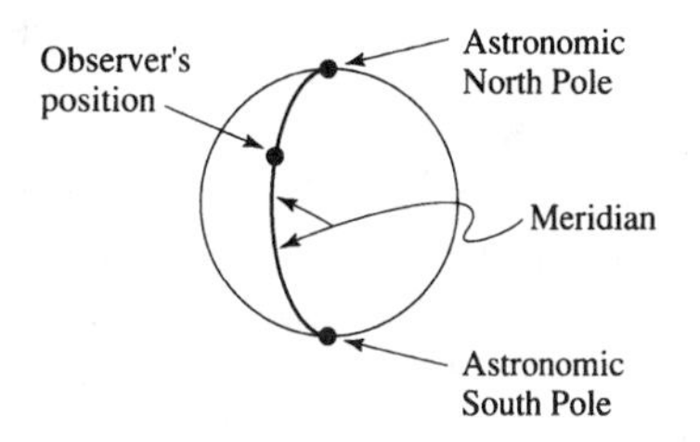

FIGURE 9-1 Astronomic meridian.

Astronomic north is based on the direction of gravity and the axis of rotation of the earth. It is determined from observations of the sun or other stars whose astronomical positions are known (the sun and the north star, Polaris, being the most common). Sometimes the term *geodetic north* is used. It is a direction determined from a mathematical approximation of the earth's shape. It is slightly different from astronomic north and that difference can be as much as 20 arc-seconds in some parts of the western United States. A *magnetic meridian* is the direction taken by the magnetized needle of a compass at the observer's position: an *assumed meridian* is an arbitrary direction taken for convenience.

Astronomic meridians should be used for all surveys of large extent and, in fact, are desirable for all surveys of land boundaries. They do not change with time and can be reestablished decades later. Magnetic meridians have the disadvantage of being affected by many factors, some of which vary with time. Furthermore, there is no precise method available for establishing what magnetic north was years ago in a given locality. An assumed meridian may be used satisfactorily for many surveys that are of limited extent. The direction of an assumed meridian is usually taken roughly in the direction of an astronomic meridian. Assumed meridians have a severe disadvantage—

the problem of reestablishing their direction if the points of the survey on the ground are lost.

For surveys of limited extent another type of meridian is sometimes used. A line through one point of a particular area is selected as a reference meridian (usually astronomic) and all other meridians in the area are assumed to be parallel to this, the so-called *grid meridian.* The use of a grid meridian eliminates the need for considering the convergence of meridians at different points in the area.

9-2 UNITS FOR MEASURING ANGLES

Among the methods used for expressing the magnitude of plane angles are the sexagesimal system, the centesimal system, and the methods using radians and mils. These systems are described briefly in the following paragraphs.

Sexagesimal System. In the United States, as in many other countries, the sexagesimal system in which the circumference of circles is divided into 360 parts or degrees is used. The degrees are further divided into minutes and seconds (1° = 60 minutes and 1 minute = 60 seconds). Thus an angle may be written as 36°27′32″. The National Geodetic Survey uses the sexagesimal system for angles and directions.

Centesimal System. In some countries, particularly in Europe, the centesimal system is used in which the circle is divided into 400 parts called *gon.* (These were until recently referred to as *grads.*) It will be noted that 100 gon = 90°. An angle may be expressed as 122.3968 gon (which when multiplied by 0.9 will give us 110.15712° or 110°09′25.6″).

The Radian. Another measure of angles frequently used for calculation purposes is the radian. One radian is defined as the angle inscribed at the center of a circle by an arc length exactly equal to the radius of the circle. This definition is illustrated in Figure 9-2. The circumference of a circle equals 2π times its radius r and thus there are 2π radians in a circle. Therefore, 1 radian equals

$$\frac{360^\circ}{2\pi} = 57.30^\circ$$

(One gon is equal to 0.01571 radian.)

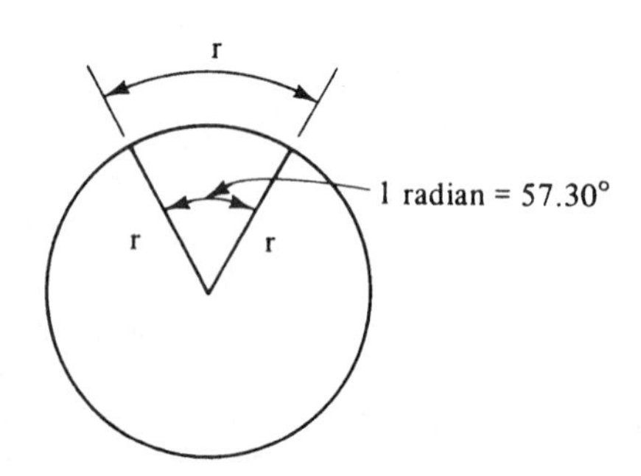

FIGURE 9-2 Radian.

The Mil. Another system of angular units divides the circumference of the circle into 6400 parts or mils. This particular system of angle measurement is used primarily in military science.

9-3 AZIMUTHS

A common term used for designating the direction of a line is the *azimuth*. The azimuth of a line is defined as the clockwise angle from the north or south end of the reference meridian to the line in question. For ordinary plane surveys, azimuths are generally measured from the north end of the meridian. This will be the case for the problems presented in this book. Azimuths are occasionally measured clockwise from the south end of the meridian for some geodetic and astronomic projects. When azimuths are measured from the south end of the meridian, that fact must be clearly indicated.

The magnitude of an azimuth can fall anywhere from 0 to 360°. The north azimuths of lines *AB, AC,* and *AD* are shown in Figure 9-3. The values are 60°, 172°, and 284°, respectively.

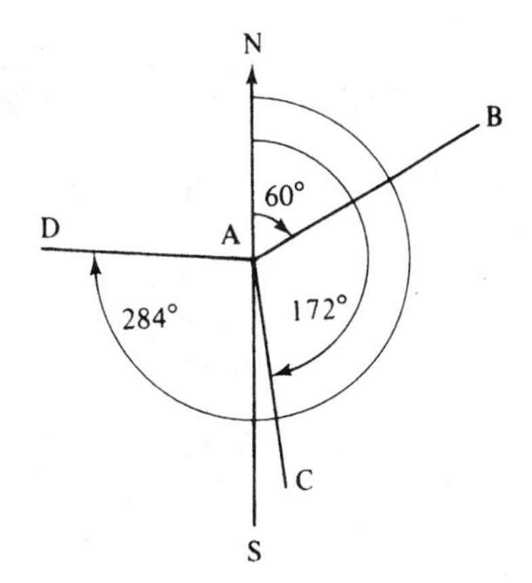

FIGURE 9-3 Azimuths.

Every line has two azimuths (forward and back). Their values differ by 180° from each other, depending on which end of the line is being considered. For instance, the *forward azimuth* of line *AB* is 60° and its *back azimuth* or the azimuth of line *BA,* which may be obtained by adding or subtracting 180°, equals 240°. Similarly, the back azimuths of lines *AC* and *AD* are 352° and 104°, respectively.

Azimuths are referred to as astronomic, magnetic, or assumed, depending on the meridian used. The type of meridian being used should be clearly indicated.

9-4 BEARINGS

Another method of describing the direction of a line is to give its *bearing*. The bearing of a line is defined as the smallest angle which that line makes with the reference meridian. It cannot be greater than 90°. In this manner, bearings are measured in relation to the north or south ends of the meridian and are placed in one of the quadrants so that they have values of NE, NW, SE, or SW.

In Figure 9-3 the bearing of line *AB* is N60°E, that of *AC* is S8°E, and that of *AD* is N76°W. As with azimuths, it will be noticed that every line has two bearings, depending on which end of the line is being considered. For instance, the bearing of line *BA* in Figure 9-3 is S60°W.

Depending on the reference meridians being used, bearings may be astronomic, magnetic, or assumed. Thus it is important, as it is for azimuths, to indicate clearly the type of reference meridian being used. It is correct to say that the bearing of a line is N90°E, but it is more common to say that it is due east. Similarly, the other three cardinal directions are usually referred to as due south, due west, and due north.

We have seen that directions may be given by either bearings or azimuths and we have learned that they are easily interchangeable. Until the last few decades American surveyors generally favored the use of bearings over azimuths and most legal documents reflected this favoritism. Today, however, programmable pocket calculators and electronic computers are used almost every day by surveyors. To make calculations with these devices it is usually simpler to use the straight numerical values of azimuths rather than getting involved with bearings, which require knowledge of quadrants and the signs of the trigonometric functions. *As a result, surveyors today generally use azimuths instead of bearings.*

9-5 THE COMPASS

Human beings have been blessed on earth with a wonderful direction finder, the magnetic poles. The earth's magnetic field and the use of the compass have been known to navigators and surveyors for many centuries. In fact, before the sextant and transit were developed, the compass was the only means by which the surveyor could measure angles and directions.

For many centuries there has been a legend that says that the compass was originally developed by a Chinese emperor who had to fight a battle in a heavy fog. The story goes that he invented a chariot that always pointed toward the south and was thereby able to locate his enemy. Actually, no one knows who first developed the compass, and the honor is claimed by the Greeks, the Italians, the Finns, the Arabs, and a few others. Regardless of its inventor, it is known that the compass was available to sailors during the Middle Ages.

The magnetic poles are not points but oval areas that are located not at the geographic poles but some distance away. Today the magnetic north pole is located approximately 1000 miles south of the astronomic north pole in the Canadian Arctic near Ellef Rinanes Island. It is moving in a northerly direction at about 9 miles per year. Perhaps the earth's magnetic field is produced by electrical currents originating in the hot liquid outer core of the earth. The flow of these currents seems to be continually changing as does the magnetic field.[1]

The compass needle lines up with magnetic north; in most places this means that the needle points slightly east or west of astronomic north, depending on the locality. The angle between astronomic north and magnetic north is referred to as the *magnetic declination.* (Navigators call it the *variation* of the compass, while others use the term *deviation.*) The magnetic lines of force in the northern hemisphere are also inclined downward from the horizontal toward the magnetic north pole. The magnetized needles of compasses are counterbalanced with a little coil of brass wire on their south ends to keep their north ends from dipping downwards toward the magnetic pole thus hitting the compass face. The so-called *angle of dip* of the needle varies from 0° to 90° at the magnetic poles. As a result the position of the coil of wire may have to be adjusted to balance the dip effect at different latitudes to

[1]L. Newitt, "Tracking the North Magnetic Pole," *Professional Surveyor,* July/Aug. 1997, vol. 17, no. 5, pp. 7–8.

keep the needle horizontal. Not only are the magnetic fields not located at the geographic poles of the earth, but magnetic directions are also subject to several variations: long-term variations, annual variations, daily variations, as well as variations caused by magnetic storms, local attractions, and so on.

Modern total stations are manufactured without compasses because of all the inaccuracies involved in using them. (Some of these instruments do have a *trough compass,* which consists of a magnetic needle mounted in a narrow box such that it can swing through a very short arc.) Despite the fact that compasses are not commonly available on our instruments today, a brief discussion of them is included in this chapter because a knowledge of their use can be very helpful to the modern surveyor. Such knowledge is particularly important for land surveys, where it is so often necessary to try to relocate old property lines whose directions were originally established with magnetic compasses.

9-6 VARIATIONS IN MAGNETIC DECLINATION

The angle of declination at a particular location is not constant but varies with time. For periods of approximately 150 years there is a gradual unexplainable shift in the earth's magnetic fields in one direction after which a gradual drift occurs in the other direction to complete the cycle in the next 150-year period. This variation, called the *secular variation,* can be very large, and it is quite important in checking old surveys whose directions were established with a compass. There is no known method of accurately predicting the secular change and all that can be done is to make observations of its magnitude at various places around the world. Records kept in London for several centuries show a range of magnetic declination from 11°E in 1580 to 24°W in 1820. The time period between extreme eastern and western declinations varies with the locality. It can be as short as 50 years or even less and as long as 180 years or more.

Maps are available that provide the values of magnetic declination throughout the world. The chart shown in Figure 9-4 was prepared from observations at thousands of stations around the globe (8500 in the United States alone). On this chart points of equal declination are connected with lines called *isogonic lines.* For some parts of the country magnetic declinations are zero and the lines connecting them are called *agonic lines.* It is thought that this chart provides magnetic declinations within about 15′ of their correct values for more than half of the country. For some extreme cases, however, they may be in error by as much as 1°. The chart shows that the declination varies from over 20°W in parts of Maine to over 20°E in parts of Washington State—a range of over 40°. Updated versions of this map, which are published approximately every five years, may be obtained from the U.S. Department of the Interior, Geological Survey, Box 25046, M.S. 968, Denver Federal Center, Denver, CO 80225-0046.

In addition to secular variations in the magnetic declinations, there are also annual and daily variations of lesser importance. The *annual variations* usually amount to less than a 1′ variation in the earth's magnetic field. Daily there is a swing of the compass needle through a cycle, causing a variation of as much as approximately one-tenth of 1°. The magnitude of this *daily variation* is still so small in comparison to the inaccuracies with which the magnetic compass can be read that it can be ignored.

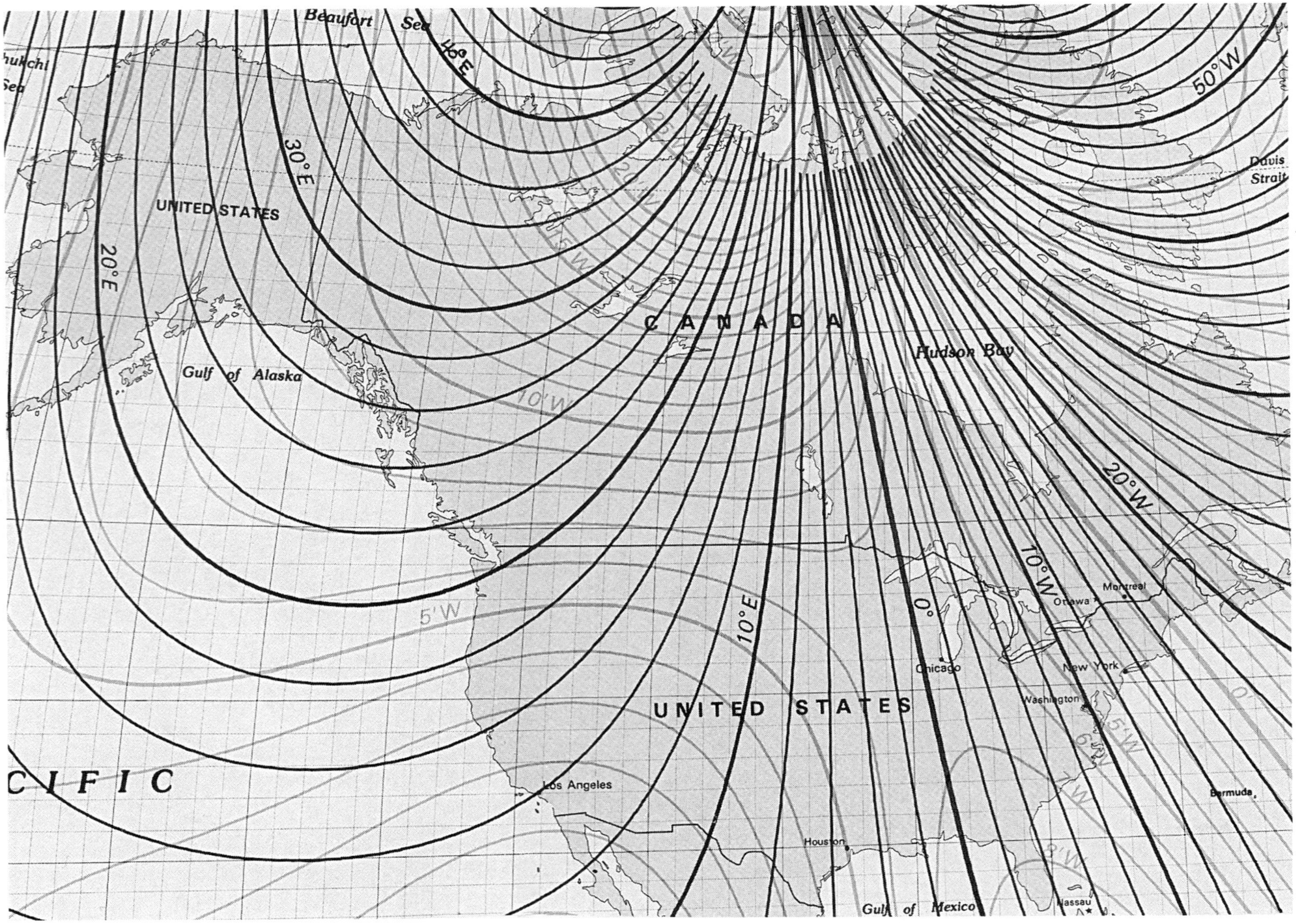

FIGURE 9-4 Magnetic declination in the United States and Canada from the map entitled "The Magnetic Declination of the Earth, 1990." (Courtesy U.S. Geological Survey.) Dark lines show declination in degrees while the lighter lines show estimated changes in declination in minutes/year.

9-7 DIRECTION ARROW CONVENTION

For property drawings a rather common arrow convention is used for showing the northerly direction. A full arrowhead is used for astronomic or geodetic directions, and a half arrowhead is used for magnetic directions. Should there be a western magnetic declination, the half arrowhead will be on the western side, and vice versa for an eastern declination (see Figure 9-5). It's not a bad practice to show this information in writing as well as with the types of arrowheads shown.

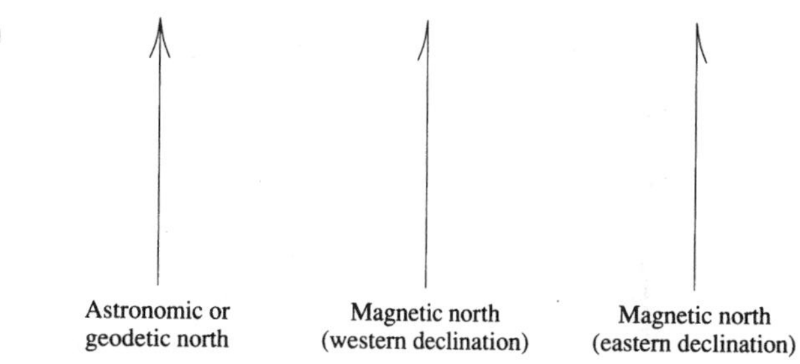

FIGURE 9-5 Direction arrow convention.

9-8 LOCAL ATTRACTION

The direction taken by a compass needle is affected by magnetic attractions other than that of the earth's magnetic field. Fences, underground pipes, reinforcing bars, passing cars, nearby buildings, iron ore deposits under the earth's surface, and other steel or iron objects may have a considerable effect on compass readings. In addition, the effect of power lines, particularly because of variations in voltage, may be so great that compasses are useless in some areas. Even the surveyor's wristwatch, pen, belt buckle, taping pins, or steel tape may have a distorting effect on compass readings.

On many occasions the surveyor may not realize that the magnetic bearings he or she reads with the compass have been affected by local attraction. To detect local attraction, he or she must read both forward and back bearings for each line to see if they correspond reasonably well. To read a forward bearing, the surveyor sights along a line in the direction of the next point. To read a back bearing for the same line, he or she moves to the next point and sights back to the preceding point. If the two readings vary significantly from each other, local attraction has probably been the cause. This topic is continued in Section 9-10.

9-9 READING BEARINGS WITH A COMPASS

Compass readings may be used for surveys in which speed is important and when only limited precision is required. In addition, they may be used for checking more precise surveys or for rerunning old property lines that were originally run with compasses. They may also be used for preliminary surveys, rough mapping, timber cruising, and checking angle measurements.

For many years the surveyor's compass (see Figure 9-6) was used for determining directions. This instrument was originally set up on a single leg called a Jacob staff. Later, tripods were used. The reader should particularly notice the folding upright sight vanes or peep sights that were used for alignment. These sights were fine slits running nearly the whole length of the vanes. This obsolete instrument (now a collector's item on the antique circuit) was formerly used for laying off old land boundaries. Today it is used occasionally when attempting to reestablish old property lines that were originally run with the same type of instrument.

FIGURE 9-6 Antique surveyor's compass. (Courtesy of Teledyne Gurley.)

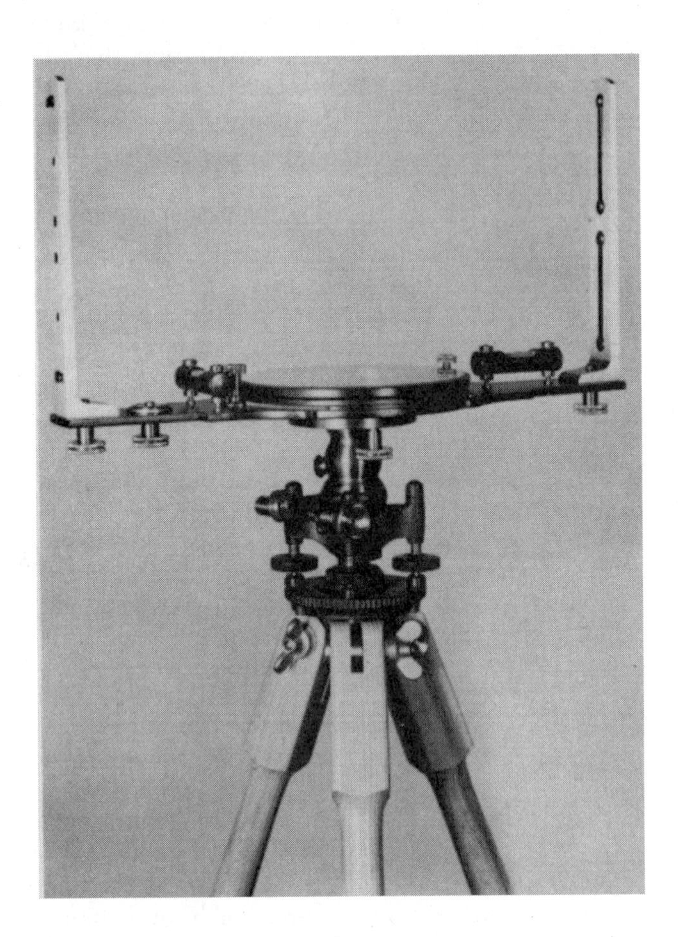

To read a bearing with a surveyor's compass or with the compass on a transit, the surveyor sets up the instrument at a point at one end of the line whose bearing is desired, releases the compass needle clamp, and sights down the line to a point at the other end of the line. He or she then reads the bearing on the circle at the north end of the needle and reclamps the compass needle. When the compass is unclamped, a jeweled bearing at the center of the needle rests on a sharp pivot point, allowing the needle to swing freely. It is important to take care of the point and keep it from becoming dull and causing a sluggish needle. This is avoided by reclamping the needle, that is, by raising the needle off the pivot when the compass is not in use to minimize wear.

The condition or sensitivity of a compass needle can easily be checked by drawing the needle out of position with a piece of magnetic material such as a knife blade. If when the magnetic material is removed the needle returns closely to its original position, it is in good condition. If the needle is sluggish and does not return closely to its original position, the pivot point can be resharpened with a stone. This is a difficult task and it is probably better to have the needle replaced.

In Figure 9-7 the instrument is assumed to be located at point A and the telescope is sighted toward point B so that the bearing AB can be determined. It will be noticed that whichever way the telescope is directed, the needle points toward magnetic north and bearings are read at the north end of the needle. For this to be possible, the positions of E and W must be reversed on the compass. Otherwise, when the telescope is turned to the NE quadrant, the needle would be between N and W on the compass. This fact can be seen in Figure 9-7.

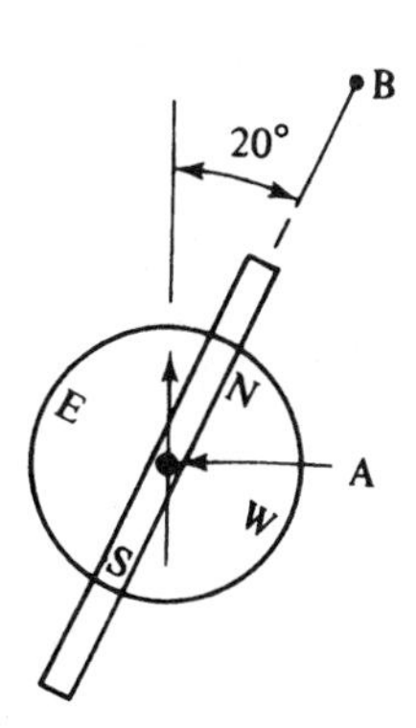

FIGURE 9-7 The telescope is over point *A* and sighted toward *B*. The bearing is N20°E. Notice that the scale turns with the telescope, but the needle continues to point toward magnetic north.

Sometimes in handling a compass, its glass face will become electrically charged because of the friction created by rubbing it with the hands or a piece of clothing. When this happens, the needle may be attracted to the undersurface of the glass by the electric charge. This charge may be removed from the glass either by breathing on it or by touching the glass at several points with wet fingers.

Many compasses (e.g., the old ones on transits) are equipped so that they may be adjusted for magnetic declinations; that is, the scale or circle under the compass which is graduated in degrees may be rotated in relation to the telescope by the amount of the declination. In this way the compass may be used to read directly approximate astronomic bearings even though declinations may be quite large.

In rerunning old surveys that were originally done with compasses, it is frequently necessary to try to reestablish some obliterated lines with the compass. In such cases it is absolutely necessary (if many years have elapsed) to estimate changes that have occurred in magnetic declinations. Such information may be obtained from old U.S.G.S. magnetic declination charts.

For new surveys it is usually desirable to work with astronomic bearings. This practice provides a permanence to surveys and greatly simplifies the work of future surveyors. An astronomic bearing may be determined for one line (probably from observations of the sun or Polaris), and the astronomic bearings for the other lines may be calculated from the angles measured with the total station to the nearest minute or closer. Of course, if there are a great many lines involved, it is well to recheck astronomic directions every so often during the survey. (Magnetic bearings may be used for present-day surveys by reading the magnetic bearing for one line and by computing the bearings for the other sides from the measured angles, but astronomic bearings are much preferred.)

Despite the increasing use of astronomic directions, it is not a bad practice to read magnetic bearings for each line if the instrument has a compass. Most new instruments do not. If mistakes are made in angle measurements, the magnetic bearings may frequently be used to determine where the mistakes occurred.

Although compass readings are estimated to the nearest 15′ to 30′, it is doubtful if the accuracy obtained is better than to the nearest degree. Transits, total stations, and theodolites may be read to the nearest minute or closer. In addition, compass readings may be decidedly affected by local magnetic attractions. As a result, *the precision obtained is quite limited, and compasses are becoming more obsolete for surveying purposes each year. Furthermore, modern total stations are not equipped with compasses.*

9-10 DETECTING LOCAL ATTRACTION

When magnetic bearings are read with a compass, local attraction can frequently be a problem; thus all readings should be carefully checked. This is normally accomplished by reading the bearing of each line from both its ends. If the forward and back bearings differ by 180°, there probably is no appreciable local attraction. If the bearings do not differ by 180°, local attraction is present and the problem is to discover which bearing is correct.

For this discussion, consider the traverse of Figure 9-8. It is assumed that the forward and back bearings were read for line *AB* as being S81°30′E and N83°15′W, respectively. Obviously, local attraction is present at one or both ends of the line.

FIGURE 9-8 Detecting local attraction in compass readings.

One method of determining the correct value is to read bearings from points *A* and *B* to a third point as perhaps *C* in the figure and then to move to that third point and read back bearings to *A* and *B*. If forward bearing *AC* agrees with back bearing *CA*, this shows that there is no appreciable local attraction at either *A* or *C*. Therefore, the local attraction is at *B* and the correct magnetic bearing from *A* to *B* is S81°30′E.

It should be noted that at a particular point, local attraction will draw the needle a certain amount from the magnetic meridian. Thus all readings taken from that point should have the same error, due to local attraction (assuming that the attraction is not a changing value as, say, voltage variations in a power line), and the angles computed from the bearings at that point should not be affected by the local attraction.[2]

9-11 TRAVERSE ANGLE DEFINITIONS

Before proceeding with a discussion of angle measurement, several important definitions concerning position and direction need to be introduced. In surveying, the relative positions of the various points (or stations) of the survey must be determined. Historically this was done by measuring the straight-line distances between the points and the angles between those lines. We will learn in Section 11-6, however, that another procedure (called radiation) is very commonly used today.

A *traverse* may be defined as a series of successive straight lines that are connected together. They may be *closed,* as are the boundary lines of a piece of land, or they may be *open,* as for highway, railroad, or other route surveys. Several types of angles are used in traversing, and these are defined in the following paragraphs.

An *interior angle* is one that is enclosed by the sides of a closed traverse (see Figure 9-9).

An *exterior angle* is one that is not enclosed by the sides of a closed traverse (see Figure 9-9).

[2]C. B. Breed and G. L. Hosmer, *Elementary Surveying,* 11th ed. (New York; John Wiley & Sons, Inc., 1977), pp. 27–28.

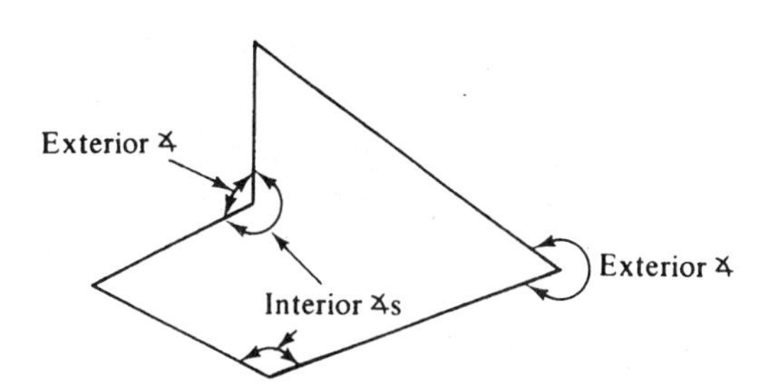

FIGURE 9-9 Several types of angles.

An *angle to the right* is the clockwise angle between the preceding line and the next line of a traverse. In Figure 9-10 it is assumed that the surveying party is proceeding along the traverse from *A* to *B* to *C*, and so on. At *C* the angle to the right is obtained by sighting back to *B* and measuring the clockwise angle to *D*.

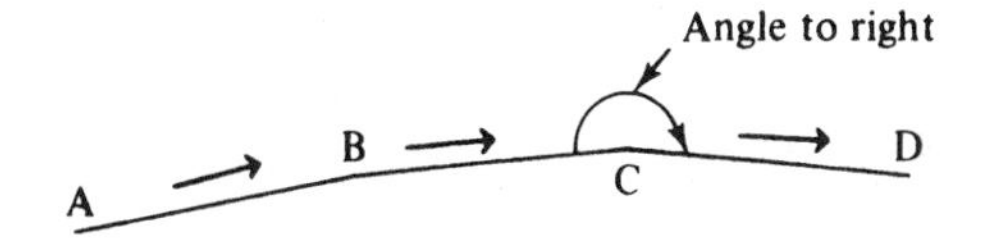

FIGURE 9-10 Angle to right.

A *deflection angle* is the angle between the extension of the preceding line and the present one. Two deflection angles are shown in Figure 9-11. The necessity for designating them as being either right (R) or left (L) should be noted. In each of these cases it is assumed that traversing is proceeding from *A* to *B* to *C*, and so on. The use of deflection angles permits easy visualization of traverses and facilitates their representation on paper. In addition, the calculation of successive bearings or azimuths is very simple (azimuths being easier to obtain).

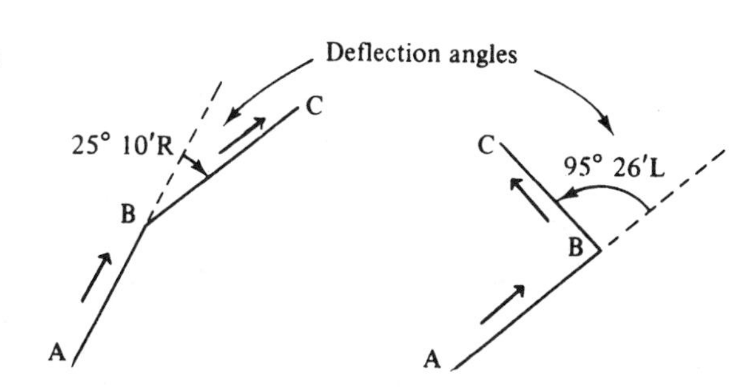

FIGURE 9-11 Deflection angles.

9-12 TRAVERSE COMPUTATIONS

For traverse computations the direction of at least one side must be known or assumed. The angles between the traverse sides are measured and with their values the directions of the other sides are computed. Actually, several possible methods of solving such a problem can be used, but regardless of which procedure is chosen, preparation of a careful sketch of the known data is required. Once the sketch is made, the required calculations are obvious.

One way to solve most of these problems is by making use of deflection angles. Example 9–1 illustrates the situation in which the bearing of one line and the angle to the next line are given and it is desired to find the azimuth and bearing of that second line.

Example 9–1

For the traverse shown in Figure 9-12, the bearing of side *AB* is given as well as the interior angles at *B* and *C*. Compute the north azimuths and the bearings of sides *BC* and *CD*.

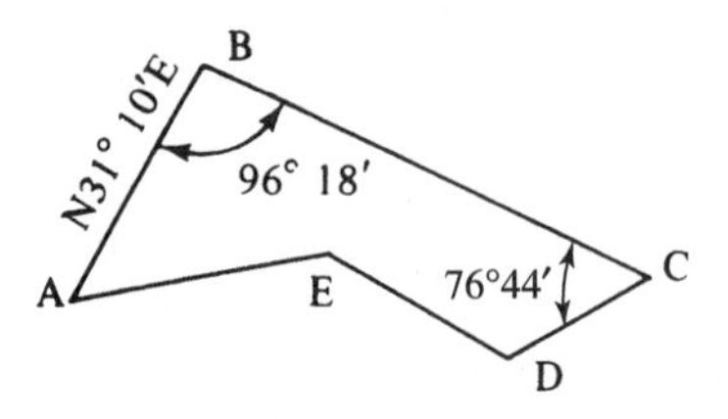

FIGURE 9-12 Computing azimuth and bearing for side *BC*.

Solution

A sketch (see Figure 9-13) is made of point *B* showing the north–south and east–west directions; the location of *AB* is extended past *B* (as shown by the dashed line), and the deflection angle at *B* is determined. With the value of this angle known, the azimuth and the bearing of side *BC* are obvious. A similar procedure is followed to calculate bearing *CD* as shown in Figure 9-14.

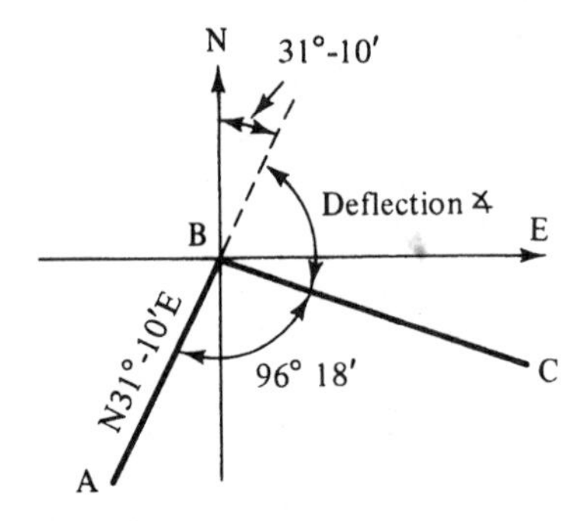

FIGURE 9-13 Deflection angle at B and bearing of line *BC*.

Deflection ∡ @ B = 180° − 96°18′ = 83°42′R
Azimuth of BC = 31°10′ + 83°42′
= **114°52′**
Bearing of BC = **S65°08′E**

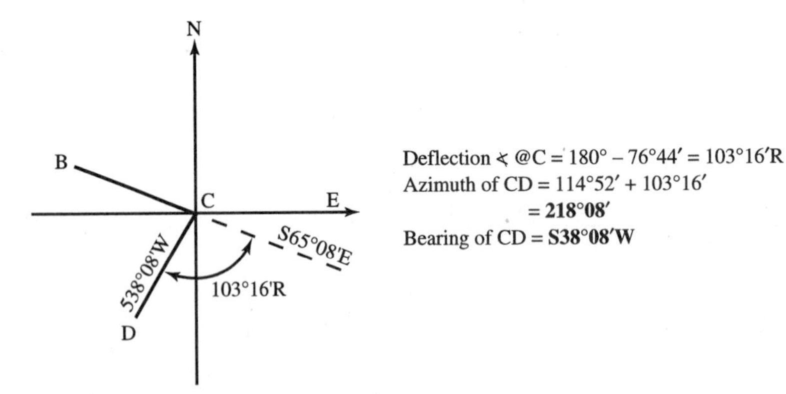

FIGURE 9-14 Deflection angle at *C* and bearing of line *CD*.

If the bearings of two successive lines are known, it is easy to compute the angles between them. A sketch is made, the deflection angle is calculated, and the value of any angle desired becomes evident. Example 9–2 presents the solution of this type of problem.

Example 9–2

For the traverse shown in Figure 9-15, the bearings of sides *AB, BC, CD,* and *DE* are given. Compute the interior angles at *B* and *D.*

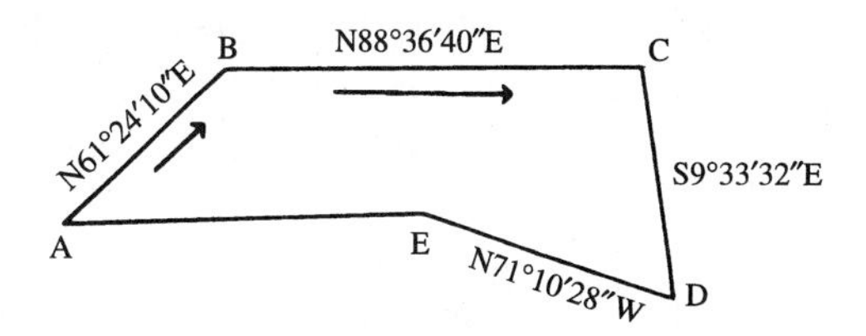

FIGURE 9-15 Computing interior angle at *B.*

Solution

A sketch is made (Figure 9-16), the deflection angle from the line *AB* extended to line *BC* is calculated, and any needed angle is immediately obvious. A similar procedure is shown in Figure 9-17 for the interior angle at *D.*

FIGURE 9-16 Interior angle calculation.

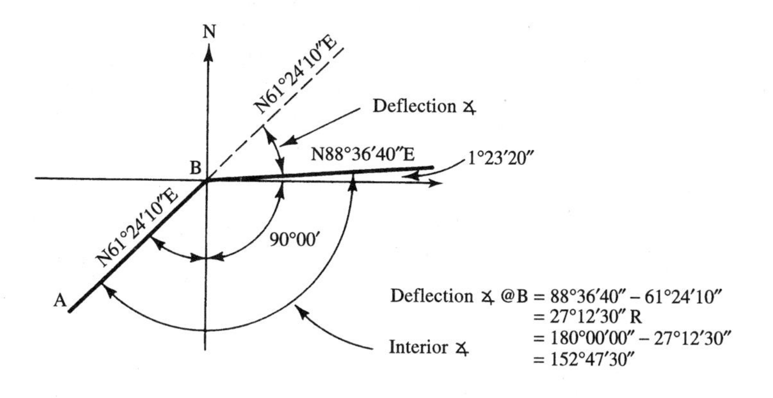

FIGURE 9-17 Interior angle calculation.

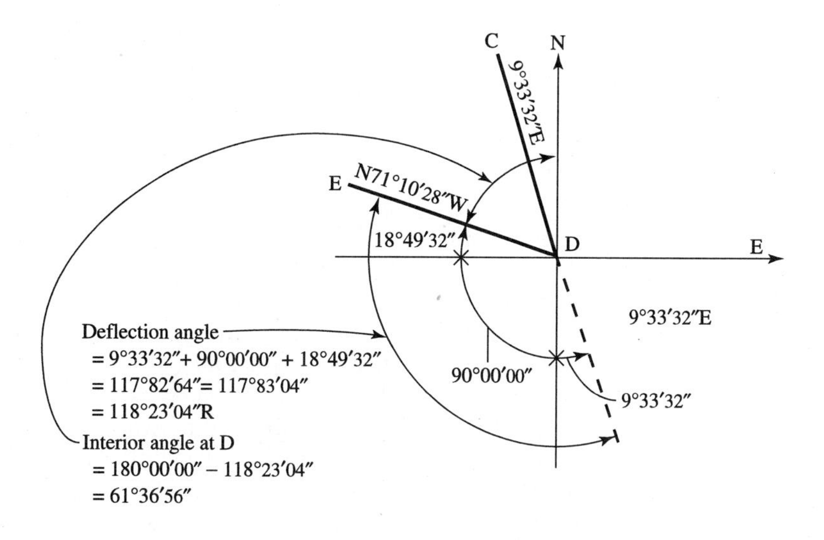

9-13 MAGNETIC DECLINATION PROBLEMS

The reader may become confused when first attempting to calculate the astronomic bearing of a line from its magnetic bearing for which the magnetic declination is known. A problem presenting similar difficulties is the calculation of the magnetic bearing of a line today when its magnetic bearing at some time many years ago is known, as are the entirely different magnetic declinations at the two times. These problems, however, may easily be handled if the student remembers one simple rule: *Make a careful sketch of the data given.* Example 9–3 presents the solution of a magnetic declination problem.

Example 9–3

The magnetic bearing of line *AB* was recorded as S43°30′E in 1888. If the magnetic declination was 2°00′E, what is the astronomic bearing of the line? If the declination is now 3°00′W, what is the magnetic bearing of the line today?

Solution

A sketch is made (Figure 9-18) in which the astronomic directions (N, S, E, W) are shown as solid lines and the magnetic directions are shown as dashed lines. Magnetic north is shown 2°00′E or clockwise of astronomic north and line *AB* is shown in its proper positions 43°30′E of magnetic south. Once the sketch is completed, the astronomic.bearing of the line is obvious.

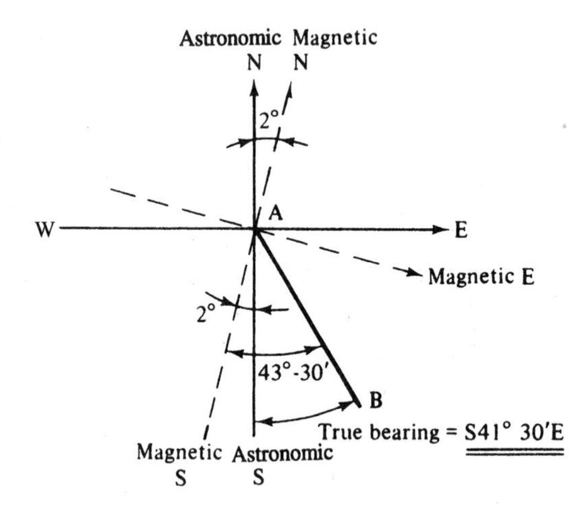

FIGURE 9-18 Astronomic bearing of side *AB*.

To determine the magnetic bearing of the line today another sketch is made (Figure 9-19), showing the astronomic bearing of the line and the present declination. From this completed sketch the magnetic bearing of the line today is obvious.

At the end of this chapter are several exercise problems dealing with angles, bearings, and azimuths. Before proceeding with the chapters that follow, the reader should be sure that he or she can handle all of these problems. The student should be able to compute azimuths of lines from their bearings and vice versa, calculate deflection angles from bearings, work from angles to bearings and bearings to angles, and be able to solve the magnetic declination problems discussed in Section 9-13.

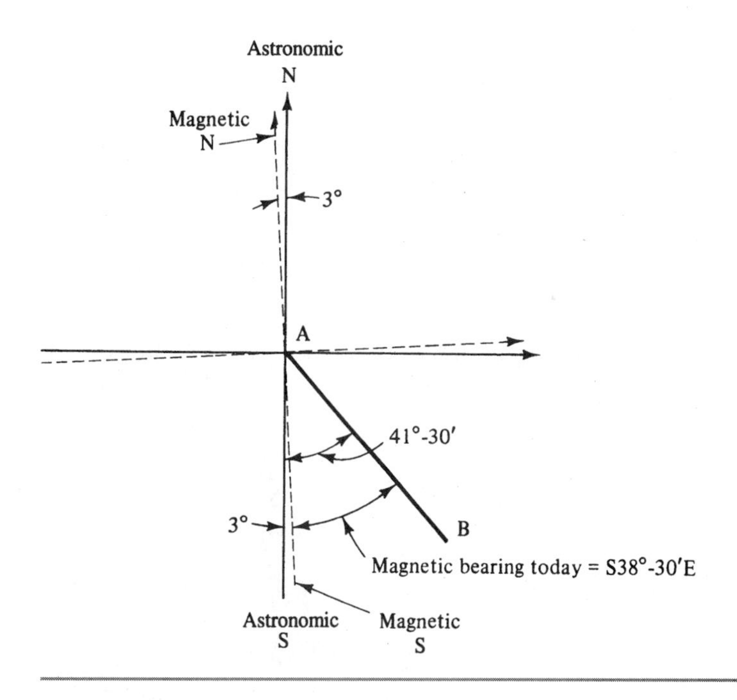

FIGURE 9-19 Magnetic bearing of side *AB*.

Problems

9-1. a Convert 63.2452^g to sexagesimal units. (Ans.: 56°55′14.4″)

b Convert 67°16′45″ to centesimal units (Ans.: 74.7546^g)

9-2. A circular arc has a radius of 560.00 ft and a central angle of 38°20′45″. Determine the central angle in radians and the arc length.

9-3. Three lines have the following north azimuths: 146°18′, 227°36′ and 332°48′. What are their bearings? (Ans.: S33°42′E, S47°36′W, N27°12′W)

9-4. Determine the north azimuths for sides *AB, BC,* and *CD* in the accompanying sketch for which the bearings are given.

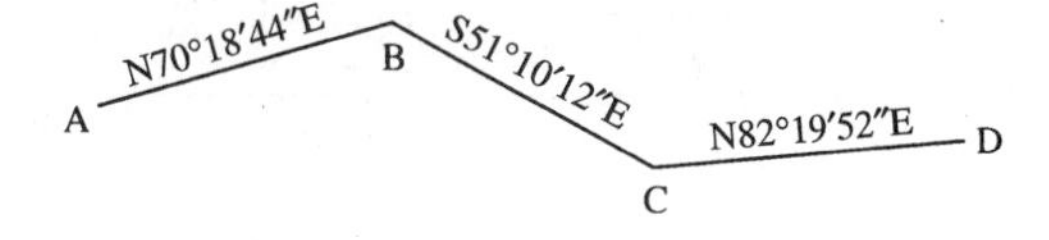

9-5. Calculate the north azimuth for sides *OA, OB, OC,* and *OD* in the accompanying figure. (Ans.: *OA* = 19°33′52″; *OB* = 105°13′54″)

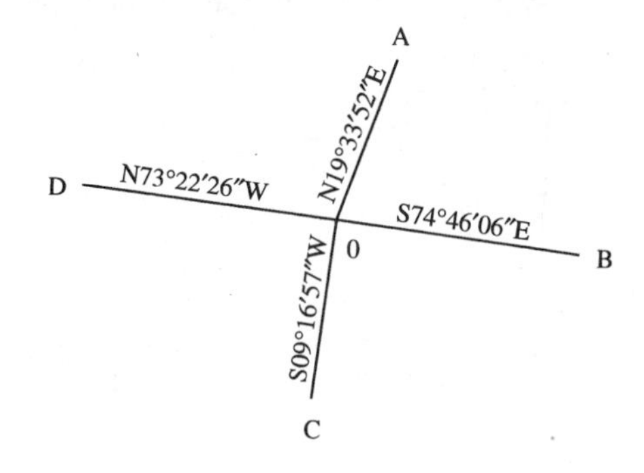

9-6. Find the bearings of sides *BC* and *CD* in the accompanying figure.

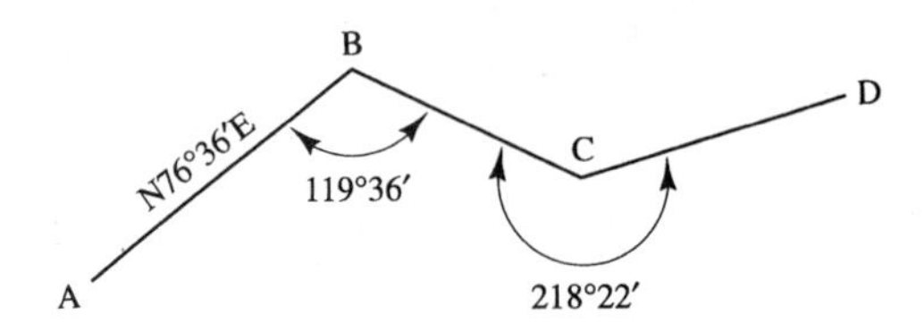

9-7. Compute the bearings of sides *BC* and *CD* in the accompanying figure.
(Ans.: *BC* = S26°36′E; *CD* = S64°58′W)

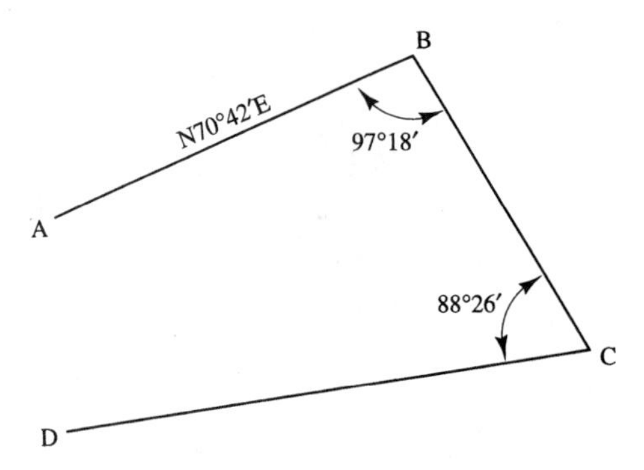

9-8. What are the bearings of sides *CD, DE, EA,* and *AB* in the accompanying figure?

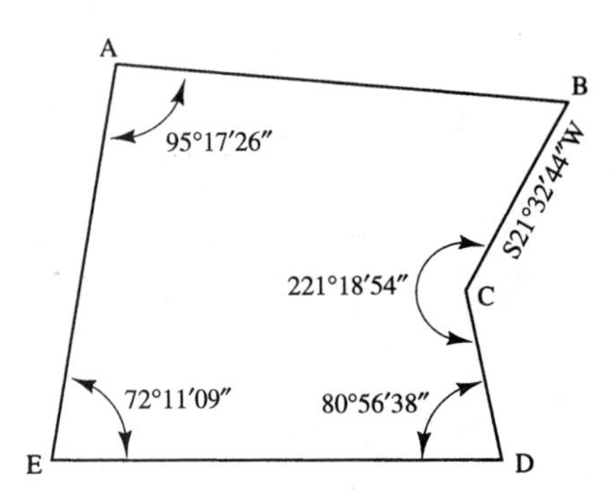

9-9. Determine the angles *AOB, BOC,* and *DOA* for the figure of Problem 9-5.
(Ans.: *AOB* = 85°40′02″; *BOC* = 84°03′03″)

9-10. Compute the value of the interior angles at *B* and *C* for the figure shown.

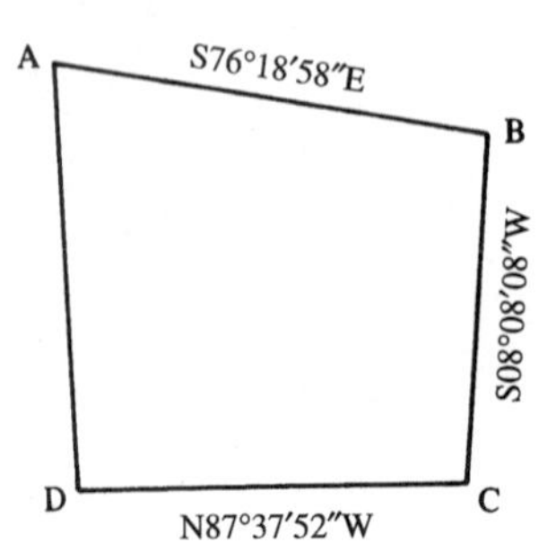

In Problems 9-11 to 9-14, compute all of the interior angles for each of the figures shown.

9-11. (Ans.: $A = 25°37'$; $C = 60°35'$)

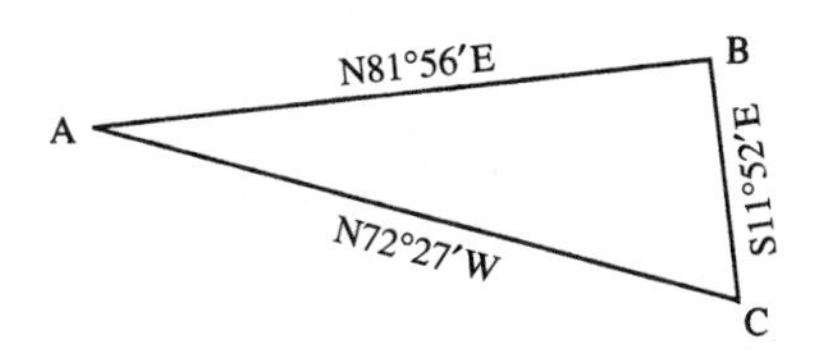

9-12.

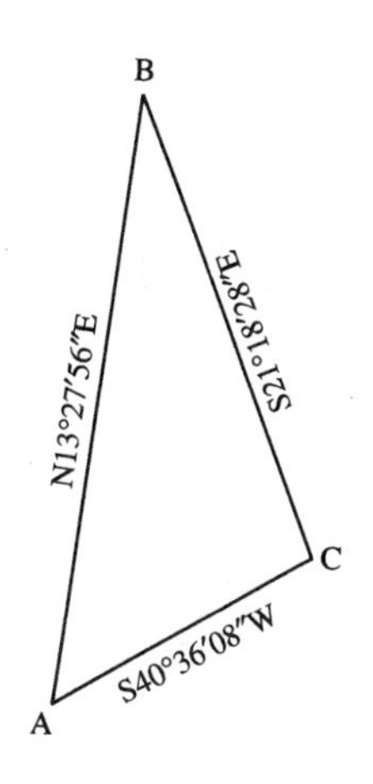

9-13. (Ans.: $B = 95°55'09''$; $D = 109°33'44''$)

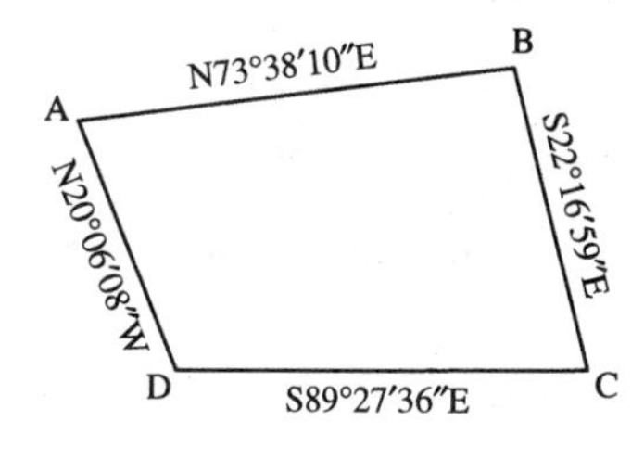

9-14.

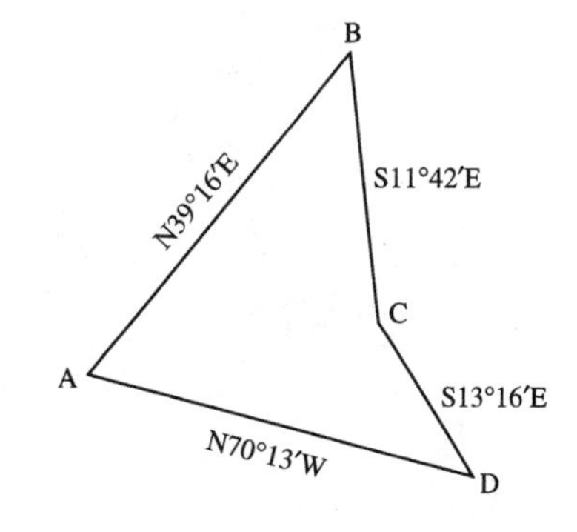

9-15. Compute the deflection angles for the traverse of Problem 9-12.
(Ans.: $B = 145°13'36''$R; $C = 61°54'36''$R)

9-16. Compute the deflection angles for the traverse of Problem 9-14.

9-17. From the data given, compute the missing bearings.

1–2 = _____

2–3 = _____

3–4 = N8°10′00″W

4–1 = _____

Interior ∢ at 1 = 51°16′00″

Interior ∢ at 2 = 36°22′00″

Interior ∢ at 4 = 221°37′56″

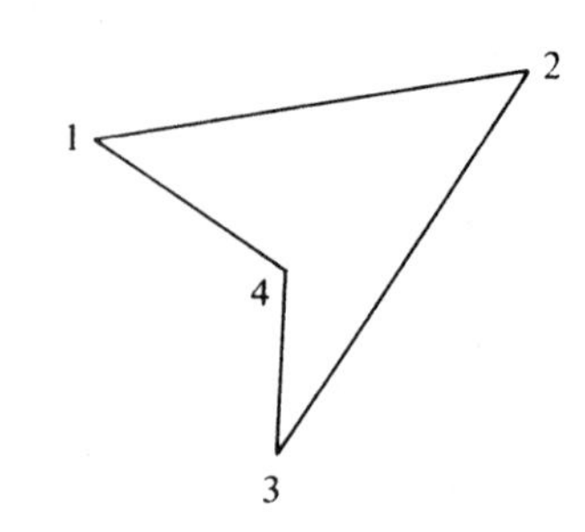

(Ans.: 1–2 = N78°56′04″E; 2–3 = S42°34′04″W)

9-18. From the data given, compute the missing bearings.

1–2 = N25°49′E

2–3 = _____

3–4 = S35°18′W

4–5 = _____

5–6 = N81°11′W

6–1 = _____

Interior ∢ at 1 = 114°06′

Interior ∢ at 2 = 79°54′

Interior ∢ at 4 = 242°38′

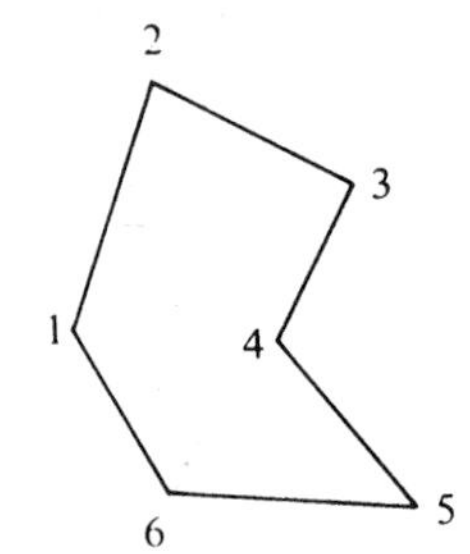

9-19. For the accompanying figure, compute the following:

a Deflection angle at *B*. (Ans.: 72°28′R)

b Interior angle at *B*. (Ans.: 107°32′)

c Bearing of the line *CD*. (Ans.: S26°40′E)

d North azimuth of *DA*. (Ans.: 289°18′)

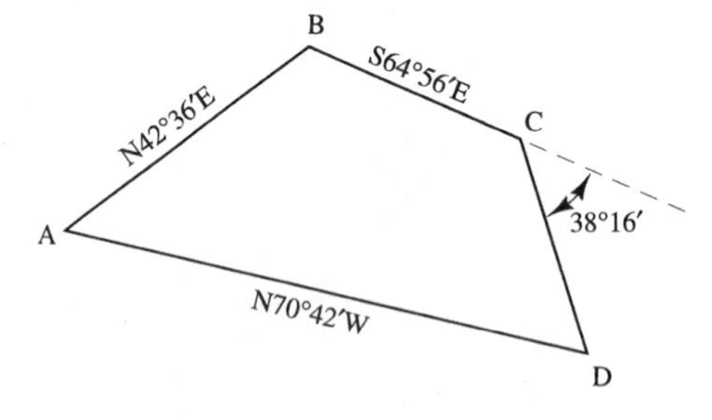

9-20. For the figure shown, compute the following:

a Bearing of line *AB*.

b Interior angle at *C*.

c North azimuth of line *DE*.

d Deflection angle at *B*.

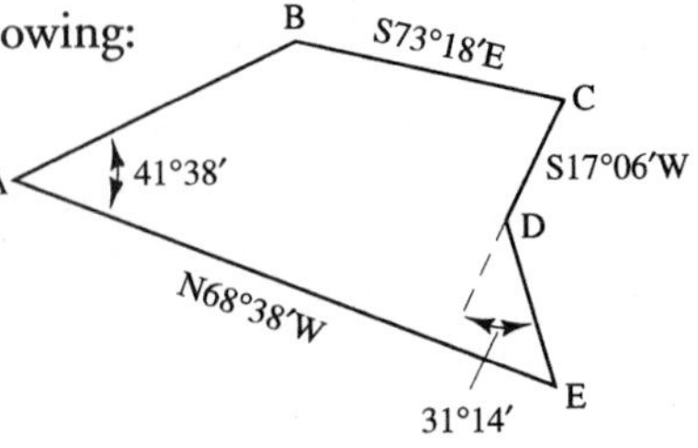

9-21. For the figure shown, compute the following:

a Deflection angle at *B*. (Ans.: 96°25′R)

b Bearing of *CD*. (Ans.: S60°50′W)

c North azimuth of *DE*. (Ans.: 159°54′)

d Interior angle at *E*. (Ans.: 58°12′)

e Exterior angle at *F*. (Ans.: 244°26′)

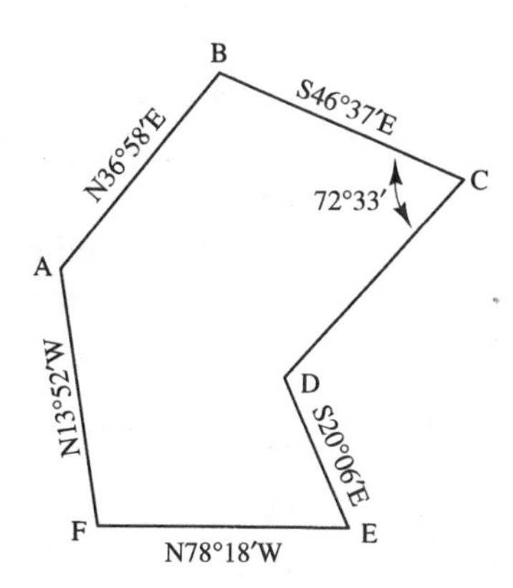

9-22. The following values are deflection angles for a closed traverse: *A* = 109°18′R, *B* = 84°27′R, *C* = 113°43′R, *D* = 80°22′L, and *E* = 132°54′R. If the bearing of side *CD* is S24°16′W, compute the bearings of the other sides.

9-23. The magnetic north azimuth of a line is 110°42′ while the magnetic declination is 8°30′E. What is the astronomic azimuth of the line? (Ans.: 119°12′)

9-24. At a given place the magnetic bearings of two lines are N42°41′E and S58°35′E. If the magnetic declination is 3°30′W, what are the astronomic bearings of the lines?

9-25. The magnetic bearings of two lines are N14°30′E and S85°30′E. If the magnetic declination is 4°45′E, what are the astronomic bearings of the lines? (Ans.: N19°15′E; S80°45′E)

9-26. The astronomic bearings of two lines are N85°44′E AND S43°38′W. Compute their magnetic bearings if the magnetic declination is 3°20′E.

9-27. Change the following astronomic bearings to magnetic bearings for a 3°45′W magnetic declination: N5°32′W, N16°32′E and S88°22′E. (Ans.: N1°47′W, N20°17′E, S84°37′E)

9-28. The magnetic north azimuth of a line was 134°30′ in 1890 when the magnetic declination was 7°00′E. If the magnetic declination is now 3°30′W determine the astronomic azimuth of the line and its magnetic azimuth today.

9-29. In 1860 the magnetic bearing of a line was S81°30′E and the magnetic declinations 3°15′W. Compute the magnetic bearing of this line today if the magnetic declination is now 4°30′E. What is the astronomic bearing of this line? (Ans.: Astronomic = S84°45′E, magnetic = S89°15′E)

From the information given in Problems 9-30 to 9-33, determine the astronomic bearing of each line and its magnetic bearing today.

	Magnetic bearing in 1905	**Magnetic declination in 1905**	**Magnetic declination today**
9-30.	N41°30′W	4°00′W	3°30′E
9-31.	S85°45′W	5°45′E	4°30′W
9-32.	N5°15′E	3°50′W	9°10′E
9-33.	N6°30′E	6°40′W	4°30′W

(Ans.: 9-31: Astronomic = N88°30′W; magnetic = S84°00′W)
(Ans.: 9-33: Astronomic = N0°10′W; magnetic = N4°20′W)

CHAPTER 10 Measuring Angles and Directions with Total Stations

10-1 TRANSITS AND THEODOLITES (OBSOLETE)

For many decades the instruments used for measuring horizontal and vertical angles were divided into two groups—transits and theodolites—but the distinction between the two was not clear. Originally, both of these instruments were called theodolites. The source of the term theodolite is not known with any certainty. In any case, transits and theodolites were both used for measuring horizontal and vertical angles. Originally the instruments that were manufactured with long telescopes and could not be inverted end for end were called *theodolites.* As time passed, however, some instruments were made with shorter telescopes that could be inverted or transited; these were called *transits.*

Eventually most instruments (whether transits or theodolites) were manufactured with telescopes which could be inverted; thus the original distinction between the two no longer applied and to a large extent what they were called was a matter of local usage. It seemed to be the convention that vernier-read instruments were called transits while the more precise and optically read instruments were called theodolites. The earlier theodolites had verniers and micrometer microscopes for reading angles. They then were manufactured with optical systems with which the user could read both horizontal and vertical angles through an eyepiece located near the telescope. The latest theodolites were made so the horizontal and vertical angles were shown digitally in a display window.

Transits and theodolites are now almost obsolete in American surveying practice. Their use and operation is briefly described in Appendix D of this textbook. They are only mentioned one or two other times times in the body of this textbook and there only with historical reference. The author refers to *transits* as being the American-style instruments which had four leveling screws and silvered horizontal and vertical scales. They were usually equipped with plumb bobs for centering over points. Figure 10-1 shows an old American transit. The term *theodolite* is used to refer to the instruments with three leveling screws and horizontal and vertical glass circles that could be read directly or with an optical micrometer and also to those instruments that provided digital displays of angle readings.

FIGURE 10-1 Transit. (Courtesy of Teledyne Gurley.)

10-2 INTRODUCTION TO TOTAL STATIONS

Although the use of global positioning systems for surveying purposes is becoming more common each year, the most commonly used surveying instrument today is the total station. It is a device that is a combination of a theodolite and an EDM together with an inboard computer or microprocessor, which has the capability of making various computations such as determining the horizontal and vertical components of slope distances, computing elevations and coordinates of sighted points. The original name for instruments of this type was *tacheometer* or *tachymeter* or *electronic tacheometer*, but Hewlett-Packard introduced the name total station over 30 years ago and the name immediately caught on with the profession.

In the mid 1970s it was rather common to use an electronic distance meter mounted above a theodolite as shown in Figure 10-2. Such an arrangement was a little cumbersome, but the surveying profession quickly saw the advantages of combined distance and direction measuring equipment. These early beginnings led to the magnificent total stations of today that are used for running traverses, for making topographic maps, for construction layout, and for many other types of surveys.

In the first half of the 1980s total stations were primarily used by surveying companies who had sufficient money for such investments. (Now these same companies probably have more money.)

With total stations the surveyor can handle all the tasks he or she formerly handled with transits or theodolites and do them better and quicker. Horizontal and vertical angles and slope distances are automatically read and with these measurements the microprocessor instantaneously computes horizontal and vertical components of slope distances as well as elevations and coordinates of sighted points. The values are provided in liquid crystal displays and may be stored in the microprocessor or transferred to external data collectors.

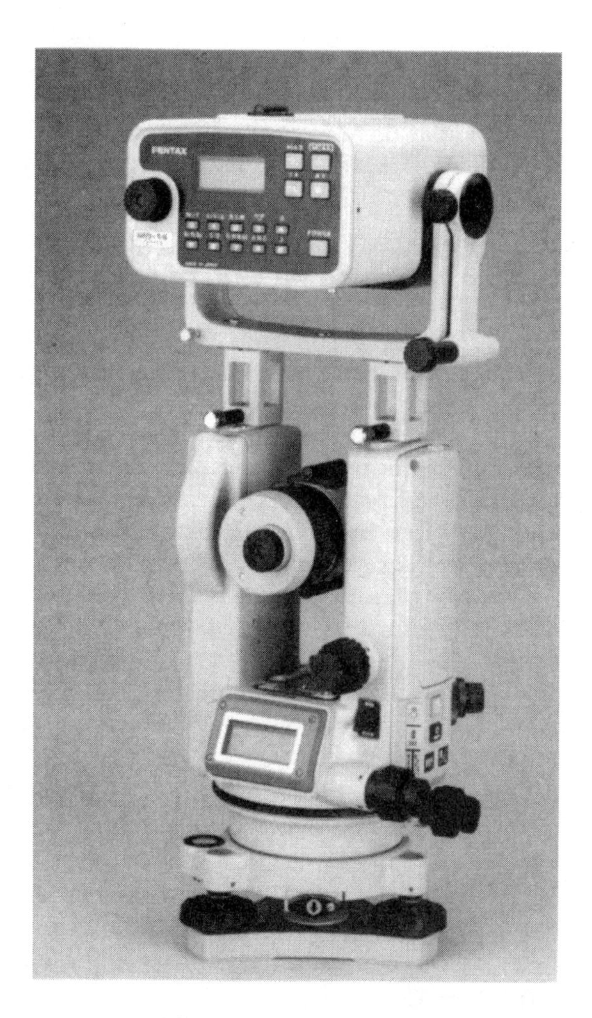

FIGURE 10-2 MD-14 Electronic distance meter mounted above EIOD theodolite. (Courtesy of Pentax Instruments.)

Not only can total stations be used for the tasks just described they can also be used in a tracking mode, which is very useful for construction staking. The desired distance (horizontal or slope) to a particular point can be input into the instrument with the keyboard and the telescope set along the correct line. The rodman estimates approximately where the stake should be located and places the prism there. A measurement is quickly made to the prism the instrument will instantly calculate and display the distance the prism needs to be moved forward or back to the correct location. The rodman estimates that next position where and the prism is placed and another measurement is taken. This process is quickly repeated as many times as necessary until the instrument reading indicates the correction is equal to zero. At that time the stake is driven.

10-3 TYPES OF TOTAL STATIONS

In the early days of total stations there were three classes of instruments available: manual, semiautomatic, and automatic.

Manual total stations. With these instruments it was necessary to read the horizontal and vertical angles manually, that is, by eye. The only values read electronically were the slope distances.

Semiautomatic total stations. For these instruments the user had to read the horizontal circle manually but the vertical circle readings were sensed electronically and the values shown digitally. Slope distances were measured electronically and the instruments could in most cases be used to reduce the values to horizontal and vertical components.

Automatic total stations. These are the common total stations used today. They sense electronically both the horizontal and vertical angles, measure slope distances, compute the horizontal and vertical components of those distances, and determine the coordinates of observed points. To perform this latter task it is necessary to properly orient the instrument as to direction. The coordinate information obtained can either be stored in the total station's computer or in an automatic data collector.

The first two types of total stations described above are sometimes described as antiques today but the reader should realize that many of them are still in use. Today's automatic total stations are sometimes referred to as *smart total stations.* Their prices range from \$6000 or \$7000 on up to \$40,000 or more. They may be used with external data collectors while the more expensive ones normally have self-contained data collectors.

Almost all total stations on the market use infrared as the carrier for distance measurements. The less expensive units can, with single prism reflectors, measure distances up to approximately 300 to 800 m. Those in the higher price range are satisfactory for measuring distances up to about 2000 m when single prisms are used. The accuracies of measurements with the less expensive instruments probably run about ±(5 mm + 5 ppm) while for the very expensive total stations they can run about ±(1 mm + 1 ppm.) Several total stations are shown in Figures 10-3 to 10-5. These are just a few examples of the approximately 100 different models on the market (2003).

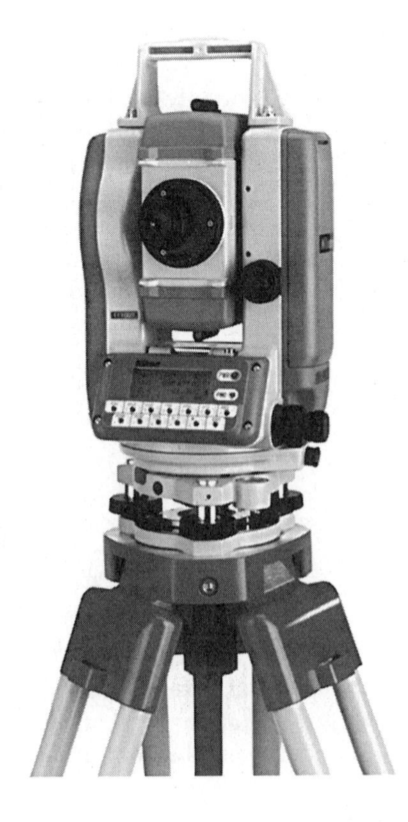

FIGURE 10-3 Top Gun DTM 400 Series Total Station. Provides slope, horizontal, and vertical distances; measures horizontal and vertical (zenith) angles; determines coordinates; has external data collector. (Courtesy of Nikon, Inc.)

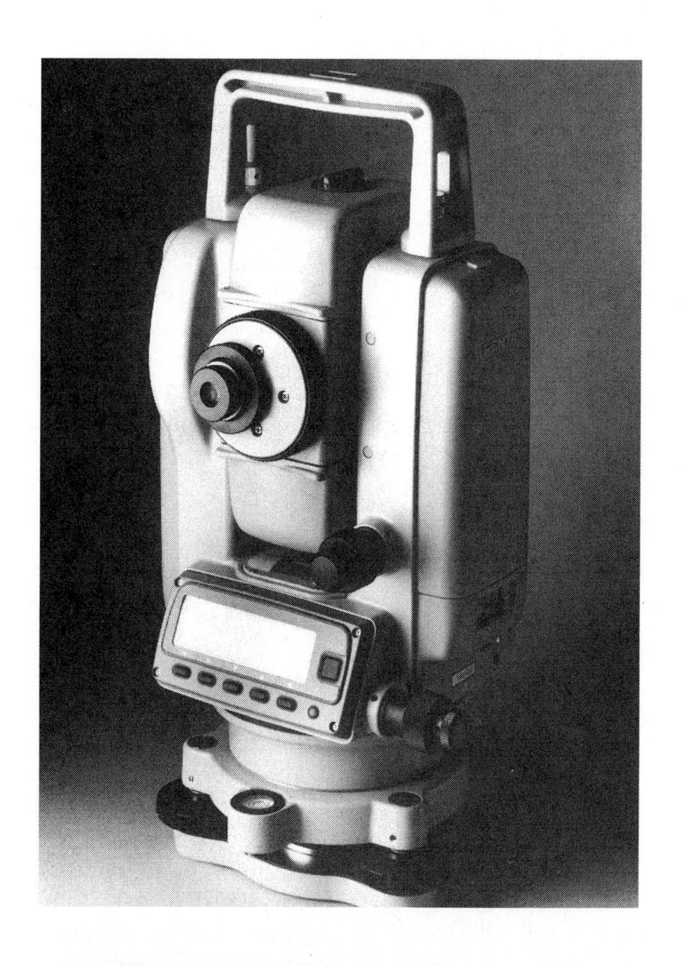

FIGURE 10-4 Pentax electronic total station PCS-215/PCS-225. (Courtesy of Pentax Corporation.)

FIGURE 10-5 Topcon total station in use. (Courtesy Topcon Corporation.)

A comment sometimes made by surveyors concerning total stations is "They're wonderful, but look at what's coming around the corner—the global positioning system (GPS)." The truth today, however, is that total stations and GPS may be used together. GPS is excellent and economical for setting up or expanding horizontal control networks. It should be noted that the surveyor is also a decision maker. He or she will need to apply the correct tool to a particular project (i.e., total station, GPS, etc.) to achieve the accuracy and precision needed while at the same time staying within the economical constraints of the job.

Total stations may be used to work from the control networks for boundary surveys, construction stake out, and for obtaining topographic data. As we will see in Chapter 15, GPS does have a visibility limitation. There must be a sufficient number of visible artificial satellites—a situation that may not be present when work is being done in forest areas or in metropolitan areas around tall buildings. Total stations can be used to take care of these types of situations.

10-4 DISADVANTAGES OF TOTAL STATIONS

Total stations do have a few disadvantages that should be clearly understood. Their use does not provide a hard set of field notes of the types we have studied in previous chapters. This means that it may be difficult for the surveyor to look over and check the work while in the field. For an overall check of the survey it will be necessary to return to the office and prepare the drawings (perhaps handled by computers).

Another disadvantage of total stations is that they should not be used for observations of the sun unless special filters are used, such as Roelof's prism. If this is not done, the EDM part of the instrument may be damaged. We will see another possible disadvantage of total stations which may occur in traversing a situation that is described in Chapter 11.

10-5 PARTS OF TOTAL STATIONS

A total station is made up of various parts including the tripod, tribrach, optical plummet, microprocessor, keyboard, display, and communications port. These parts are briefly discussed in this section.

To support the instrument it is necessary to have a good solid tripod so that accurate surveys can be made. The tripod legs may consist of wood or metal and they may be fixed in length or they may have adjustable lengths. Although adjustable lengths are convenient when rough sloping terrain is involved, fixed legs are a little more rigid and may help obtain slightly more accurate surveys.

The *tribrach,* previously described for levels in Chapter 5, contains three leveling screws, a circular or bull's eye level, and probably an optical plummet for centering the instrument over survey points. The tribrach is screwed down on the head of the tripod and the instrument is clamped to the tribrach with the tribrach screw.

The upper part of the total station is referred to as the *alidade* and includes the telescope, the graduated vertical and horizontal scales for measuring angles, and the other components that are involved in measuring angles and distances.

The short telescopes, which can be inverted end for end or transited, contain reticles or cross hair rings with the cross hairs etched in glass. On most telescopes there are two focusing devices. One is the objective lens that is used to sight to the point observed while the other one is the eyepiece control used to focus the crosshairs.

The optical plummet is a device that permits the surveyor to accurately center the instrument over a given point. The plummet may be a part of the alidade of the instrument but is more commonly a part of the tribrach. When it is a part of the tribrach, more accurate positioning is the result. The device provides a line of sight parallel and in line with the vertical axis of the total station. Its use is described in the next section of this chapter.

There are two graduated circles (horizontal and vertical). They are used to measure angles in mutually perpendicular planes. Leveling of the instruments will place the horizontal circle in a horizontal plane and the vertical circle in a vertical plane. Thus horizontal and vertical or zenith angles can be measured in their proper planes. Many of the earlier total stations had bubble tubes for leveling but most of the later ones make use of automatic compensators as previously described for levels in Chapter 6.

A total station is in actuality an electronic theodolite that contains an EDM and a microprocessor. The EDMs are actually rather small but they are nevertheless quite satisfactory for almost all surveying work. Using a single prism a total station can be used to measure distances up to 3 or 4 km; if triple prisms are used a total station can be used to measure distances roughly twice as long.

The angles and distances measured with a total station are entered into the built-in microprocessor. This device converts the measured slope distance to horizontal and vertical components (or elevation differences). If the elevation of the instrument center and the h.i. of the reflector are entered into the instrument, the elevation of the sighted point will be computed taking into account earth's curvature and atmospheric refraction. In addition, if the coordinates of the occupied station and a known direction or azimuth are available, the microprocessor will compute the coordinates of the sighted point.

10-6 SURVEYING WITH TOTAL STATIONS

A total station is roughly leveled using its three leveling screws. It is not necessary to turn the instrument around its vertical axis. Once it is roughly leveled the magnitude of any leveling errors present are received by the instrument's microprocessor, which will make appropriate corrections to the values of the measured horizontal and vertical angles. As a result instrumental errors, such as line of sight not perpendicular to the horizontal axis and horizontal axis not perpendicular to vertical axis, are considered by the microprocessor.

With older instruments (transits and theodolites) it was the practice to measure equal numbers of direct and reversed angles and average the results to compensate for these errors. This practice is not absolutely necessary with total stations. Nevertheless, some measurement errors creep into the work, and it is wise to take multiple readings and let the microprocessor average the measured values.

For this discussion it is assumed that rectangular coordinates of the initial point and the azimuth of a line are either known or assumed. The total station is set up, the coordinates are entered into the instrument, and the known azimuth is set on the horizontal circle display. Using the lower motion of the instrument a backsight is taken along the line whose azimuth is known. With the upper motion a sight is taken to the next point. Then the forward azimuth will be displayed and stored in the instrument's memory. This procedure is described in detail in Section 10-9.

The total station measures the slope distance from the instrument to the reflector along with the vertical and horizontal angles. The microprocessor in the instrument then computes the horizontal and vertical components of the slope distance. Furthermore the microprocessor, using these computed components and the azimuth of the line, determines by trigonometry the north–south and east–west components of the line and the coordinates of the new point. These new coordinates are stored in memory.

When the instrument is moved to the second point the procedure used at the first point is repeated, except that the back azimuth to the first point and the coordinates of the second point do not have to be entered. They are merely recalled from the instrument's memory after which the next point is sighted. This procedure is continued until the surveyor returns to the initial point or to some other point whose coordinates are known. The coordinates of this ending point are compared with those determined with the total station. If the difference or error of closure (to be discussed at length in Chapter 12) is within acceptable limits, proportional adjustments are made to the intermediate points (also described in Chapter 12) to produce their final coordinates.

As part of traversing the surveyor may want to obtain elevations of the points. This can easily be done with the total station. It is, of course, necessary to enter the height of the instrument and the height of the reflector. Then when the microprocessor computes the vertical components of the slope distance it will determine the elevation of the next point. As a part of this calculation a correction is made for earth's curvature and atmospheric refraction. When the last point is reached, the difference in its elevation and the one determined by the instrument (if within specified limits) is adjusted or spread out to the intermediate points. Elevations determined in this manner are not as accurate as those determined with levels as described previously in Chapter 7.

10-7 SETTING UP THE TOTAL STATION

To measure angles, directions, and elevations with a total station it is first necessary to set the instrument over a definite point such as an iron pin or a tack in a stake or a mark in the pavement. This process is described in this section.

Before the instrument is taken from its case, the tripod should be placed over the point to be occupied and the tripod legs firmly placed in the ground The total station is carefully removed from its carrying case by picking it up using its standards or its handles if such are provided. It is set on the tripod and the centering screw, which is located under the tripod head, is screwed up into the base of the instrument.

Although total stations are equipped with optical plummets, it is usually convenient to use plumb bobs for approximate centering before the plummets are used. As a part of the centering process it will probably be necessary to pick up and slightly move the tripod in one direction or the other one or more times until the instrument is roughly centered. Once this rough centering is completed the plumb bob will be removed and the fine centering done with the optical plummet.

The optical plummet is either a part of the tribrach or a part of the alidade of the total station. The line of sight of the plummet coincides with or is collinear with the vertical axis of the total station. You will thus note that the instrument must be leveled for the vertical axis of the instrument and the line of sight of the plummet to be truly vertical.

Once the line of sight of the plummet is very close to its desired position the instrument should be leveled using its plate bubble and level screws. Then the centering screw can be loosened and the instrument slid over to the desired position. It may then be necessary to repeat the steps of leveling the instrument, checking the optical plummet's line of sight, releasing the tribrach screw, sliding the instrument over a little, etc. For many total stations the optical plummet is in the tribrach. For such instruments it is wise to leave the total station in its carrying case until the last leveling and centering of the tribrach is completed.

After a total station has been leveled as much as possible with its tripod legs, it will be necessary to carefully finish the leveling process with the instrument's leveling screws. For an instrument that has a level vial the telescope is turned until it is parallel to a line through a pair of leveling screws. The bubble is then centered by turning that pair of screws. The screws are both turned in toward the center, , or in the opposite direction, . In each case as the screws are turned the bubble will follow the motion of the left thumb. The telescope is rotated 90° and the bubble is centered with the third screw.

Some total stations do not have a level vial (such as those with electronic leveling systems) and for them it is necessary for the surveyor to study the instrument instruction booklets to carry out the leveling.

It may seem to the reader that the setting up process just described is very lengthy. Perhaps that is true for the first two or three times a person goes through the process but after that he or she will be able to do the job very quickly.

10-8 POINTING THE INSTRUMENT

The surveyor should point at targets and not aim at them. Aiming, which is excessive pointing at a target, causes eyestrain and makes the target look as though it is moving. The most reliable sighting is usually the first trial when the target first appears to be aligned with the cross hairs. When aligning the cross hairs on the target with the tangent screw or slow motion screw, it is desirable that the last movement be in a clockwise direction. Such a procedure prevents the occurrence of small errors due to slack in the slow motion screw.

All pointings should be made with the target close to the intersection of the vertical and horizontal cross hairs. Figure 10-6 shows some of the common cross-hair arrangements for various instruments. The black dot in each case indicates the recommended position of the target. The first three arrangements (a), (b), and (c) are usually found on less precise transits. In (b) the top and bottom horizontal hairs are the stadia hairs. To avoid confusion with the middle hair they are sometimes shortened as shown in (c). Parts (d), (e), and (f) show double hairs. These permit the centering of distant targets instead of covering them with the hairs. They are usually used on more precise instruments.

FIGURE 10-6 Some cross-hair arrangements.

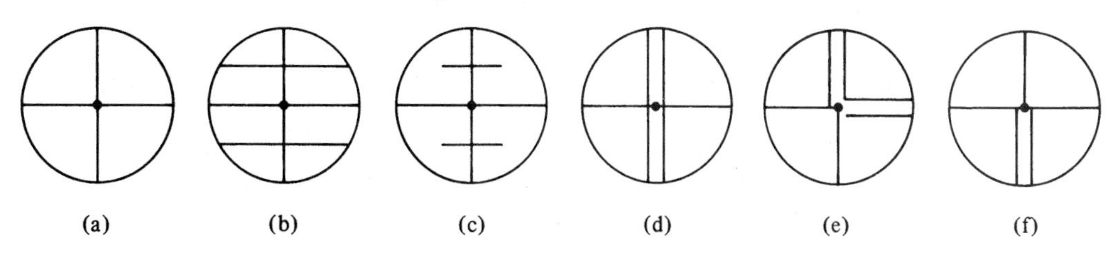

To measure an angle, the telescope is pointed at one target and then turned to another. If, as is often true, the targets in question are represented by tacks in wooden stakes, and if they are reasonably close to the instrument, it is often possible to leave the tacks sticking up where they may be seen with the telescope. If this is not feasible, a nail or taping pin may be held on top of the tack. When the stakes, iron pins, or whatever are too low for sighting, a plumb bob may be held over the point and the telescope lined up with the string. Such a procedure is very good for short sights because the narrow width of the string, as compared to the width of a range pole or prism pole, will enable the surveyor to obtain more accurate sights. Sometimes a card or paper is attached to the string so that the string will be more visible to the observer.

For a large percentage of work the surveyor will be sighting on a prism pole. A prism pole is kept vertical by means of a bull's eye level on the pole. When sighting on such a pole for angle or direction purposes the surveyor should line up the vertical cross hair of the instrument so that it bisects the pole just below the prism. Sights taken on the prism itself for measuring angles may contain significant errors particularly when short distances are involved.

10-9 MEASURING HORIZONTAL ANGLES

Depending upon the quality of the total station being used angles can be measured all the way from the nearest $\pm 1/2$ second (using very expensive equipment) to the nearest ± 20 seconds (using much less expensive instruments such as might be used for construction work). Some of the instruments will display the angle or direction in degrees, minutes and seconds as in 46° 22′10″, while other instruments will provide the value with a decimal such as 46.2210. The values may also be given in grads.

To facilitate the measurement of horizontal angles and directions a total station has a clamp screw and a tangent or slow-motion screw. When the clamp is released, the telescope can be freely turned about the vertical axis of the instrument. When the clamp screw is tightened, the telescope can only be moved slowly and in small angle increments by turning the tangent screw.

For this discussion reference is made to Figure 10-7 where it is desired to measure the angle ABC. The prism or reflector is usually mounted on an adjustable length prism pole at a height equal to the height of the total station (h.i.) above the set up point.

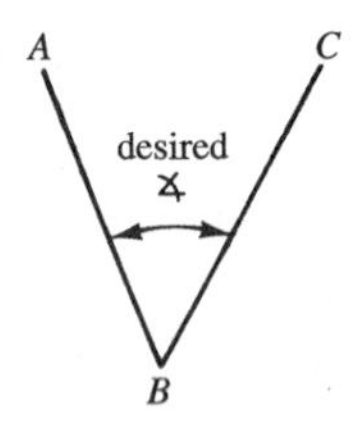

FIGURE 10-7 Measuring a horizontal angle.

The instrument is set up and centered over point *B* as previously described in Section 10-7 and a backsight is taken to Point *A*. This is done by releasing the horizontal clamp, sighting approximately on point *A*, tightening the clamp and sighting precisely on *A* with the horizontal tangent screw. Then an initial value equal to 0°00′00″, or any other desired value, is entered into the display. (To measure a vertical angle or a zenith angle as later described in Section 10-13, the user will go through a similar procedure with the vertical clamp and tangent screw.) To measure the horizontal angle the clamp is released, the telescope sighted close to point *C*, the clamp tightened, and the tangent screw used for fine sighting on point *B*. The value of the angle will automatically be displayed by the instrument. After the instrument is set up and a telescope sighting made, it usually takes no more than two to four seconds before the angle and distance is displayed. If, however, the tracking mode is used, the time is probably no more than one half a second. The values obtained with the first or normal method will be more precise than those obtained in the tracking mode because multiple measurements are taken and averaged by the instrument.

10-10 CLOSING THE HORIZON

Before the student attempts to use the total station to measure horizontal angles for an actual traverse, he or she probably needs a good practice session to be sure that the operation of the instrument is fully understood. One excellent way to do this is to set up and level the instrument at a convenient point and to drive or stick in the ground four or five wooden stakes or taping pins scattered around the instrument. If stakes are used, a tack should be placed in each stake so that it protrudes from the top of the stake.

The angle between each pair of taping pins or tacks in the stakes is measured with the value in the display set to equal zero before each angle is meassured. When all the angles around the instrument have been measured, they are added together to see if their total is equal to or very close to 360°. If each angle can be read to the nearest ±10 seconds with a particular instrument and there are five angles involved, their total as described in Section 2-11 should be within $\pm(\sqrt{5})(10'') = \pm 22''$ of 360° 00′″. Once the student is able to carry out this exercise correctly and with reasonable speed, he or she will clearly understand how to use the lock and tangent screws of the instrument and will be ready to measure other angles.

Sample notes for the type of exercise that is referred to as closing the horizon are shown in Figure 10-8. This general method is very useful for checking the measurement of any angle. If the surveyor has measured one or more angles at some position of the instrument, he or she can set the display back to zero and measure the angle necessary to complete the circle. If the angles are added together and the total is very close to 360°, the surveyor has a very good check on the work.

A good many total stations are able to automatically make corrections in measured angles for instrumental errors. For instance, with a specified calibration test the user can measure the magnitude of the index error for a vertical circle. If this value is stored in the microprocessor, the instrument will make that correction each time a vertical angle or zenith angle is measured.

10-11 MEASURING ANGLES BY REPETITION

All surveyors on more occasions than they would care to admit make mistakes in measuring angles. After they have measured an angle, they would like to be as sure as possible that no mistakes were made so they will not have to return and repeat the measurement. Usually it is much easier to prevent mistakes than it is to figure where they occurred at a later time. A method that nearly always enables the surveyor to know whether a mistake has been made is that of measuring angles by repetition.

FIGURE 10-8 Angle measurement—closing the horizon.

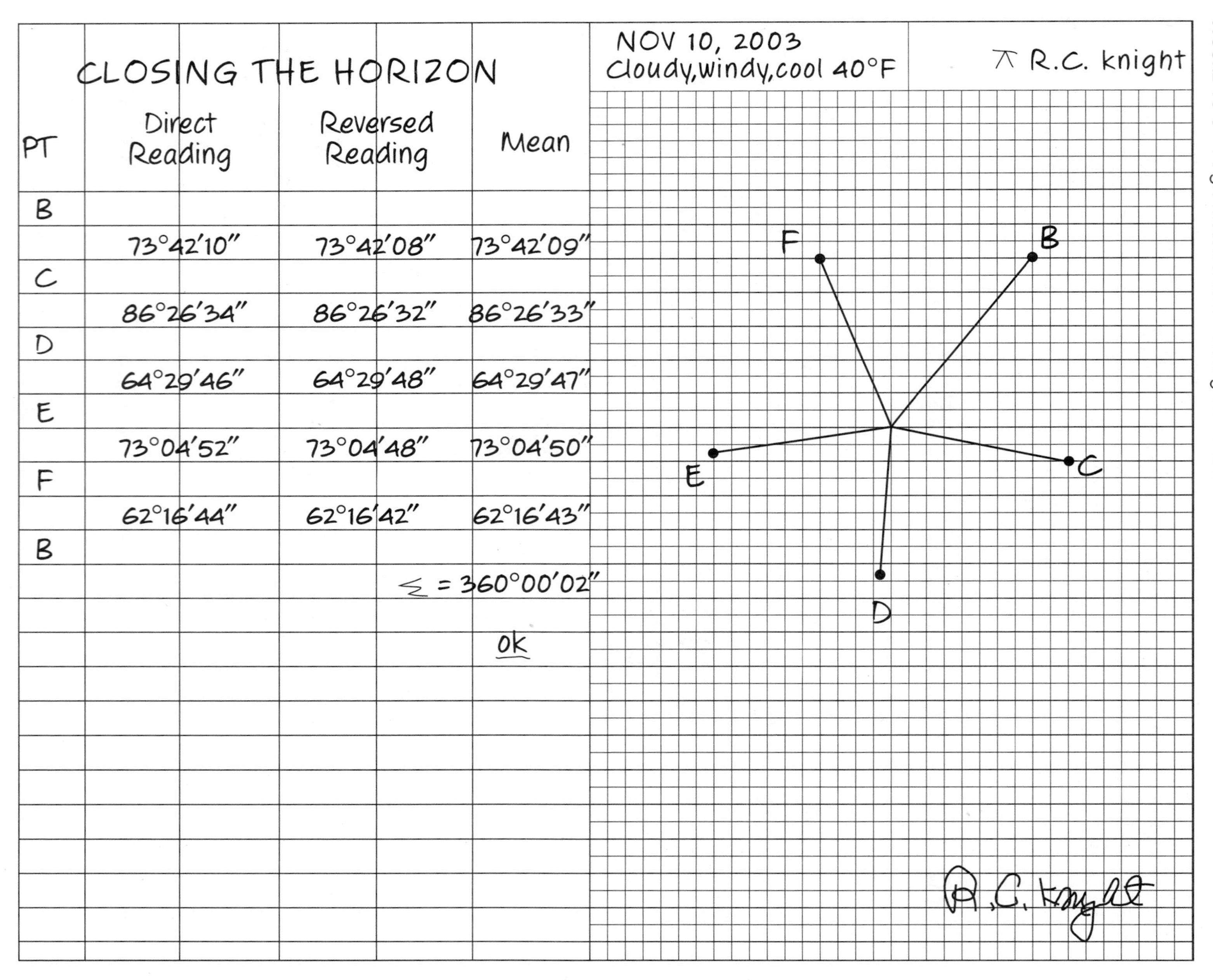

CLOSING THE HORIZON

NOV 10, 2003
Cloudy,windy,cool 40°F
⊼ R.C. knight

PT	Direct Reading	Reversed Reading	Mean
B			
	73°42'10"	73°42'08"	73°42'09"
C			
	86°26'34"	86°26'32"	86°26'33"
D			
	64°29'46"	64°29'48"	64°29'47"
E			
	73°04'52"	73°04'48"	73°04'50"
F			
	62°16'44"	62°16'42"	62°16'43"
B			
		Σ =	360°00'02"
			ok

FIGURE 10-9 Field notes for the measurement of the interior angles of a closed traverse.

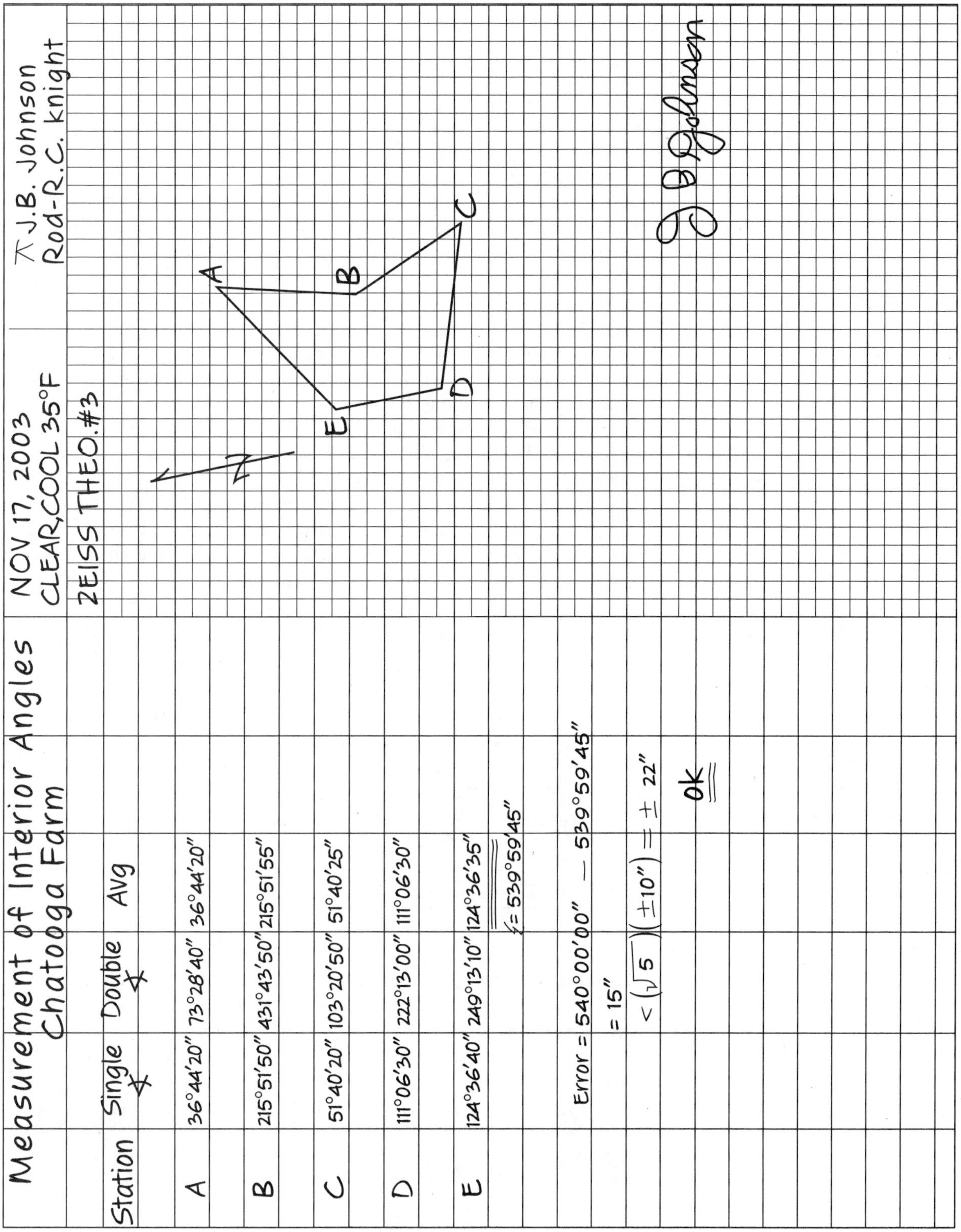

Measurement of Interior Angles
Chatooga Farm

NOV 17, 2003
CLEAR, COOL 35°F
ZEISS THEO. #3

⊼ J.B. Johnson
Rod-R.C. Knight

Station	Single ∡	Double ∡	Avg
A	36°44'20"	73°28'40"	36°44'20"
B	215°51'50"	431°43'50"	215°51'55"
C	51°40'20"	103°20'50"	51°40'25"
D	111°06'30"	222°13'00"	111°06'30"
E	124°36'40"	249°13'10"	124°36'35"
			Σ = 539°59'45"

Error = 540°00'00" − 539°59'45"
= 15"
< $(\sqrt{5})(\pm 10") = \pm 22"$
ok

After an angle is measured, its value is retained in the instrument display by pressing the appropriate button on the instrument keyboard. A backsight is taken to the first point using the horizontal clamp and tangent screws. The display is released and a foresight is taken to the second point. Two times the original value should be shown. If a mistake has been made, the entire procedure needs to be repeated. Assuming the angle has been measured correctly the repetition procedure can be repeated as many times as desired after which the whole procedure can be repeated with the telescope plunged or inverted.

Another example of measuring horizontal angles is presented in Figure 10-9. The interior angles of a Five sided closed traverse have been measured. As a check each of the angles was doubled and divided by two. Then the five angles were added together and their total judged to be sufficiently close to 540°00′00″.

For the example shown in Figure 10-10 the angle in question has been measured three times with the telescope in its normal position. The first measurement of the angle which is 52°16′12″ is used only as a check against the value obtained after the total value is divided by the number of measurements.

Measuring angles several times with the telescope in its normal position and in its reversed position and averaging the resulting values will result in improved precision. However, repeating an angle more than a total of six or eight times will not appreciably improve the measurement because of accidental errors in centering and pointing and instrumental errors in the total station.

Dividing the third reading in Figure 10-10 by three is quite easy because the numbers are all multiples of three and the answer is obviously 52°16′11″. Should the numbers be more complicated (or, that is, where they are not each multiples of three) the user may find it easier to put the value in decimal form, divide that value by three and then change the result back to degrees, minutes, and seconds. For instance for this case 156°48′33″ equals 156.8092° and when divided by 3 the result is 52.2697°.

FIGURE 10-10 Measuring angles by repetition.

MEASURING ANGLES BY REPETITION				
Angle	1st RDG	Third RDG	Avg	
A	52°16′12″	156°48′33″	52°16′11″	

Actually the procedures for measuring angles by repetition can vary somewhat with different instruments. As a result the surveyor should study his or her instrument's instruction manual before attempting to use this procedure.

10-12 DIRECTION METHOD FOR MEASURING HORIZONTAL ANGLES

In the last section of this chapter a method of measuring horizontal angles was presented in which a sight was taken to one point, an initial value 0°00′00″ entered into the display, and a sight taken to another point. Instead of using this procedure it is often convenient to use the so called direction procedure for determining horizontal angles. Such a procedure is particularly useful when multiple angles are to be measured from one instrument setup.

With the direction method horizontal circle readings are taken to various points and the angles between those sightings computed. When multiple angles are to be determined, the time the surveyor has to spend at a station is reduced as compared to the single angle procedure described in Sections 10-10 and 10-11. Just as with the previous procedure precision can be improved if multiple readings are taken in the direct and reversed positions and the results averaged.

10-13 MEASURING ZENITH ANGLES

A vertical angle is defined as the plus or minus angle from a horizontal plane to a line being sighted. Sometimes a plus angle is referred to as an *elevation angle* and a negative angle is called a *depression angle.* A *zenith angle* is the angle from a vertical line to the line in question. These angles are shown in Figure 10-11.

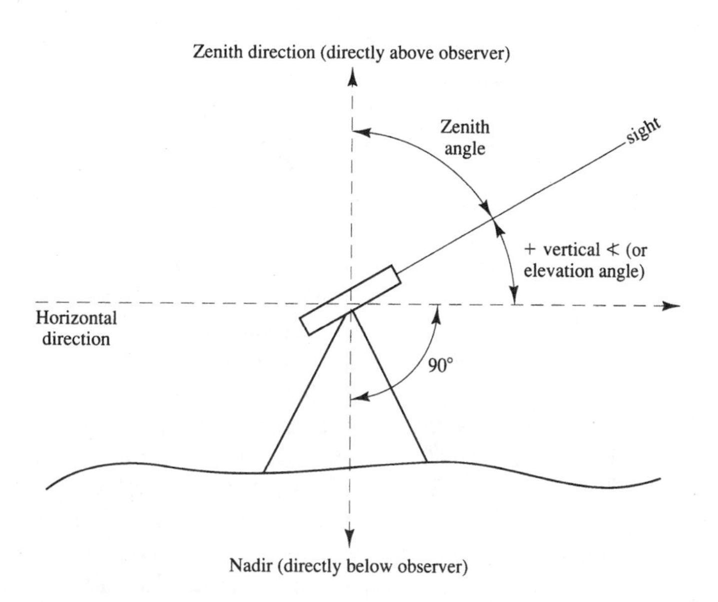

FIGURE 10-11 Measuring a zenith angle.

Today's total stations normally display zenith angles rather than vertical angles. A reading of 0° with one of these instruments will be obtained when the telescope is pointed vertically overhead. If the telescope is in its normal or direct mode and is horizontal, the zenith angle will be 90°. If the telescope is inverted, that is in its reversed mode, the zenith angle will be 270°. Notice the two readings will always add up to 360°. If a zenith angle of 70° is measured with the telescope in its direct mode, the zenith angle in the reversed mode will be 290°.

Should index errors be present in zenith angle readings they can be substantially reduced by taking an equal number of readings of each angle in the direct and reversed modes and computing the mean. Usually this is done with the instrument's microprocessor. Although index errors may be negligible with many instruments, it is always wise to assume the instrument is out of adjustment, measure the angles in the two modes, and average them together.

10-14 ROBOTIC TOTAL STATIONS

With robotic surveying the instrument is set up over a control point and left there while the surveyor carries the prism to the various points that are to be located. The instrument will track the prism and when the surveyor presses a button at his position, the instrument will quickly and accurately take and record the necessary readings.

Robotic surveying is now available with instruments such as the Geodimeter 4000, the Leica TCA 1103, Trimble 5603, Topcon GTS-800A/RC-2, and others. With this equipment surveyors can greatly improve their productivity and efficiency. On many occasions one person alone can perform complete surveys. Some devoted users of robotic equipment claim they can perform in one day an amount of work that would take them three or four days if they used regular total stations and they note that fewer people need to be employed to do the work. Users claim that robotics equipment will pay for itself within one to two years due to the speed of the work and the smaller payrolls.

The equipment list used for robotic surveying includes the instrument, a pole with a circular level, a reflector prism, telemetry equipment for communication with the robot, and the keyboard.

The instrument is set up over a control station and oriented in the usual manner. To do this it is necessary to enter the coordinates of the station where the instrument is located and take a backsight along a line with a known azimuth. After this setup is completed, the surveyor carries the remote positioning unit (RPU) to the points for which data is desired.

If the instrument is used as a robot, the surveyor carries out all measurements while located at the various target points using the RPU. When the RPU button on the instrument is activated, the instrument becomes a robot, permitting a one-person operation with data recording and data storage taking place at the instrument. The surveyor takes the RPU to the point to be sighted and then presses the search button. The robot searches for and locates the RPU. Furthermore, the robot will follow the RPU as it is moved from point to point. In fact, the robot will track the prism even if it is placed flat on the ground while the surveyor is driving a stake or performing some other task. Should the lock be lost the surveyor will press the search button again to renew the search and the instrument will lock onto the RPU. The maximum time required is usually only a few seconds.

The fact that the robotic instrument can remain unattended, and yet will continually track the reflector, cuts back greatly on the times required in normal surveying for sighting, focusing, and reading. To collect the data the only time required for each point is that necessary for the surveyor to walk to the point and plumb the rod. The robot will make and record the desired measurement within 2 or 3 seconds.

The Geodimeter 4000 uses infrared as its carrier and thus can be used during nighttime hours. This feature often results in much saving of time and money on construction projects as the layout surveying necessary for the next day's work can be done the night before. The next morning construction operations can begin immediately without having to wait for the surveyors. In a similar manner, surveying areas with high-volume daytime traffic may be performed when traffic volumes are lower during the night.

One concern of working with robotic instruments is that of protecting the instruments. There is the possibility of having the unmanned instruments stolen or run over by vehicles. Another problem may occur where heavy vehicular traffic is in the vicinity as it may cause so much interference that the robot may have difficulty in locking on the target.

10-15 USE OF DATA COLLECTORS WITH TOTAL STATIONS

The data collectors used with total stations may be handheld devices which are connected by cables to the total station or they may be built into the total station. Most data collectors are designed so that the data stored in them can be automatically downloaded into a computer and there used for various calculations and plotting. The capabilities of today's data collectors are phenomenal compared to what they were only a few years ago.

Total stations with data collectors are especially efficient for surveys that involve the location of a large number of points in the field. This is a situation that is particularly common to topographic mapping (a subject discussed in detail in Chapter 14). The total station is capable of handling many surveying tasks. One for which it is ideally suited is topographic surveying in which the x, y, and z coordinates of points are determined. (The z component is used here to refer to elevations.) The work can be done much faster (perhaps 2 or 3 times as fast) than it can be if a transit or theodolite and stadia are used for the work. This appreciably increases the area for which ground topographic surveys are competitive with those made with aerial photography.

With the data collector the field measurements are recorded and transferred or downloaded to a computer for final map preparations. In addition, complex projects can be set up on a computer in the office and transferred or uploaded to the data collector. The data collector is then taken to the field and connected to the total station. After this is done hundreds of points can rapidly be laid out in the field with the total station.

10-16 CARE OF INSTRUMENTS

Although the general rules given here for taking care of transits, theodolites, EDMs, and total stations are closely related to the ones previously given for levels (Section 6-11), the subject is of sufficient importance to warrant some repetition. The most important rule, as for levels, is "don't drop the instrument" because very serious damage will almost surely result. The following list presents some important items to remember in caring for these expensive instruments.

1. Dirt and water are problems for instruments and should be removed as soon as possible. After the instruments have been used dust should be removed with a cleaning brush and the instrument should be wiped off with a cloth. More resistant dirt and mud can be removed with the aid of mild household cleaners together with cotton balls and pipe cleaners.
2. If precipitation occurs, put the dustcap over the objective lens. In addition, it is a good idea to carry a waterproof silk bag with which to cover the instrument.
3. When the instrument is being transported in a vehicle it should be held in the lap, kept in its box, or cushioned in some other way to avoid sudden shocks.
4. When removing the instrument from its box, the box should be placed horizontally. The instrument should be held by its standards (for transits) or by its handle or yoke (for other instruments).

5. Place the tripod with its legs well apart and sunk firmly in the ground.
6. Do not place the instrument on a smooth, hard surface such as a concrete slab unless some provision (such as a triangular frame) is made to keep the tripod legs from slipping.
7. Do not turn its screws tightly. If more than fingertip force is needed to turn them, the instrument either needs cleaning or repairs. Overtightening may lead to appreciable damage to the instrument.
8. Never leave an instrument unattended because it may be upset by wind, vehicles, children, or farm animals, or it may be stolen.
9. Hold the tripod in your arms in a horizontal position with the instrument ahead when carrying it inside a building. This enables one to better avoid obstacles. Better yet, carry the instrument into the building in its box.
10. Optical glass is not very hard and is easily scratched. If the glasses get dirty, they should be brushed carefully with a clean camel's-hair brush. Fingers should not be placed on the lenses because skin oil holds dust. A clean lintless cotton cloth moistened with alcohol (or alcohol mixed with ether) is used to wipe the lens using a rotational motion working from the center outwards.
11. Total stations and EDMs must not be sighted on the sun unless filters are used because the internal components of the instruments may be damaged.
12. For high-precision work instruments should be shaded against direct sunlight. Furthermore, instruments should be protected against very high temperatures as well as against sudden changes in temperature.
13. The average surveyor should not try to disassemble or lubricate EDMs, theodolites, or total stations. The manufacturers should handle these tasks.

Problems

10-1. Distinguish between transits and theodolites.

10-2. Describe the leveling process for a total station.

10-3. What is an alidade? What are its parts?

10-4. List the types of surveying work that can be done with a total station.

10-5. What are the functions of the microprocessor in the total station?

10-6. What are the advantages of measuring angles by averaging an equal number of readings taken with the telescope in its normal and in its reversed position?

10-7. A horizontal angle was measured by repetition six times with a total station. If the initial display reading was 21°33′18″ and the final reading was 129°20′04″, determine the value of the angle to the nearest second.
(Ans. 21°33′21″)

10-8. A horizontal angle was measured by repetition six times with a total station. If the initial reading was 76°17′30″ and the final value was 97°44′48″, what is the value of the angle to the nearest second?

CHAPTER

Miscellaneous Angle Discussion

11-1 COMMON ERRORS IN ANGLE MEASUREMENT

Most of the errors commonly made in angle measurement are probably obvious enough, but they are nevertheless listed here together with comments on their magnitudes and methods of reducing them. These are divided into the usual categories: personal, instrumental, and natural.

Personal Errors

Most of the inaccuracy in angle measurement is caused by these errors. Personal errors are accidental in nature and cannot be eliminated. They can, however, be reduced substantially by following the suggestions made herein. Perhaps the largest personal errors occur in pointing and in setting up the instruments.

1. *Instrument not centered over point.* If the instrument is not centered exactly over a point, an error will be introduced into an angle measured from that position. Here it is necessary for a person to use his or her sense of proportion. If the points to be sighted are distant, errors caused by imperfect centering will be small. If, however, sight distances are very short, centering errors may be very serious. Should a sight be taken on a point 300 ft away and should the instrument be 1 in. off the theoretical line of sight, the angle will be in error by approximately 1′. The surveyor should periodically check to see that the instrument remains centered over the occupied point.
2. *Pointing errors.* If the vertical cross hair in the telescope is not perfectly centered on observed points, errors similar to those described for imperfect centering of the instrument will occur. The most important method of reducing errors in pointing is to keep the sight distances as long as possible. In fact, this is a basic principle of good surveying—*avoid short distances if at all possible.*

   If the points sighted are close to the instrument, the width of a range pole is an appreciable factor. Either a plumb bob may be held over the point with observations being made on the plumb bob string or it may actually be possible to sight on tacks in the stakes, and so on, as described in Section 10-8.

A good rule to follow in this regard is to sight only on vertical targets (range poles, taping pins, plumb-bob strings, etc.), which appear to be only a little wider than the vertical cross-hair thickness when looking through the telescope.

Another good practice is to initially use the longest sightings possible and then proceed to the shorter sights. As an example, the longest dimensions for a building project should be laid out prior to the shorter dimensions.

In using the telescope, the longer a person stares at the point the more difficult it will be to obtain a good reading because after a while the point will seem to move. As described previously, the surveyor should use the first clear sighting of the target, since that will in all probability be the most accurate.

3. *Unequal settling of tripod.* Tripod legs should be pushed firmly into the ground to provide solid support for the instrument. The instrumentman must be careful not to brush against the instrument and not to step too closely to the tripod legs if the ground is soft. A good practice, as in leveling, is to check the bubble tubes before and after readings to make sure that they are still centered. In very soft or swampy ground it may be necessary to provide special supports for the tripod legs, such as stakes driven into the ground. Many total stations contain sensors which provide the instrument operator with a warning when the leveling of the instrument becomes so far off as to appreciably affect the work.
4. *Improper focusing of telescope (parallax).* To minimize errors caused by improper focusing, the instrumentman should carefully focus the eyepiece until parallax disappears. Furthermore, objects being sighted should be placed as near to the center of the telescope's field of view as possible.
5. *Instrument not level.* Obviously the instrument must be level when readings are made. The surveyor should often check bubbles to see that they remain centered. If they are not centered they will need to be recentered—but only before a particular angle is measured or after its measurement is completed. Usually the total station or theodolite will display an error code or message if the instrument is not level. It is sometimes desirable to relevel transits, theodolites, and total stations between angle measurements, but this should never be done in the middle of a single angle measurement. If a backsight is taken, the instrument releveled, and then the foresight taken, the observations will have been made on different planes and as a result different elevations will be obtained.
6. *Placement and plumbing of rod.* Special care must be taken when the rod is placed behind a point to be sure that it is in line. Careless plumbing of the rod is another common error producer in angle or direction measurement. This is a particularly serious matter when the instrument operator can only sight on the top part of the rod due to intervening shrubs or other terrain features.

Instrumental Errors

Since no instrument is perfect, there will be instrumental errors. If the instruments are out of adjustment, the magnitudes of these errors will be increased, but they are greatly reduced by the process of double sighting, or double centering, in which the readings are taken both with the telescope in its normal position and in its inverted position. Then the results are averaged. The operation of rotating the telescope about its horizontal axis is called *plunging* or *inverting* the telescope.

Sometimes an optical plummet will get out of adjustment, causing the instrument to be set over erroneous points. A plummet should be periodically checked by comparing its centering with that of a plumb bob. Adjustment screws are available on the plummets for making corrections.

Natural Errors

In general, natural errors are not sufficiently large to affect work of ordinary precision. For more precise work some things can be done to reduce natural errors. These are included in the list that follows. Should weather conditions become unusually severe, work should be discontinued.

1. *Temperature changes.* Use an umbrella over the instrument or do the work at night.
2. *Horizontal refraction.* Try to keep sights away from items that radiate considerable heat such as pipes, tanks, buildings, and so on.
3. *Vertical refraction.* Read vertical or zenith angles from both ends of a line and average the readings—theoretically, the uphill angle will be too large by the amount of the refraction error and the downhill angle will be too small by the amount of the refraction error.
4. *Wind.* Shield instruments as much as possible and use an optical plummet for centering the instrument over the point. When using an instrument without an optical plummet on windy days an approximately 3-ft length of stove pipe can be placed around the plumb bob.

11-2 COMMON MISTAKES IN MEASURING ANGLES

A list of frequent mistakes in angle measurement is provided in this section. Many of these mistakes can be detected if the measurements are made by the repetition method.

1. Insufficient leveling of instruments.
2. Recording the wrong number, such as 131° instead of 113°.
3. Failing to center the bubble before measuring a zenith or vertical angle.
4. Recording the wrong algebraic sign for a vertical angle.
5. Sighting on the wrong target when measuring a horizontal angle.
6. Using the clamps and tangent screws incorrectly.
7. Leaning on tripod when pointing the instrument or making readings.

11-3 ANGLE-DISTANCE RELATIONSHIPS

For a particular survey it is logical for the angles and distances to be measured with comparable degrees of precision. It is not sensible to go to a great deal of effort to obtain a high degree of precision in distance measurements and not do the same with the angle measurements, or vice versa. If the distances are measured with a high degree of precision, the time and money spent have been partially wasted unless the angles are measured with a corresponding precision.

If an angle is in error by 1′, it will cause the line of sight to be out of position by 1 ft at a distance of approximately 3440 ft (i.e., 3440 times the tangent of 1′ = 1 ft). Therefore, an angle that is in error by 1′ is said to correspond to a precision of 1/3440. It should be noted that angles measured with a 1′ instrument are usually measured a little closer than 1′, so that reading them to the nearest 1′ (without repetition) probably corresponds to a precision of approximately 1/5000 in distance measurement.

A similar discussion may be made for the relative precision obtained for angles measured to the nearest 30″, 20″, and so on, or for angles measured with a 1′ instrument by repetition. Table 11-1 presents the angular errors that correspond to the degrees of precision described in this section.

TABLE 11-1 Precision of Angle Measurement

Angular error	Angular precision
5′	1/688
1′	1/3440
30″	1/6880
10″	1/20,600
1″	1/206,000

Should distance measurements with a precision of say 1/20,000 be made for a particular project it can be seen from the table that measurement of the angles should be made to about the nearest 10″ so that comparable precision is obtained.

The reader should realize that this is not the whole story on angle precisions. For instance, the trigonometric functions do not vary directly with angle sizes. In other words, the tangent of an angle of 1°11′ that is 1′ in error does not miss its correct value by the same amount as the tangent of an angle of 43°46′ that is 1′ in error. Nevertheless, the approximate relations given in Table 11-1 provide a satisfactory guide for almost all surveying work.

11-4 TRAVERSING

As described previously, a traverse consists of a series of successive straight lines that are connected together. The process of measuring the lengths and directions of the sides of a traverse is referred to as *traversing.* Its purposes are to find the positions of certain points.

Open traverses, which are normally used for exploratory purposes, have the disadvantage that arithmetic checks are not available. For this reason, extra care should be used in making their measurements. The angles should be measured by repetition, half with the telescope in its normal position and half with it inverted. Formerly a rough check was made for the bearings of traverse lines by reading magnetic bearings for each line with a compass and computing the angles between the lines from their bearings. Today's modern instrumentss probably don't have compasses and that method of checking is probably not available to us. The use of astronomic directions with occasional checks made by observations of the sun is a much better procedure. Distances should be measured both forward and back whether tapes or EDMs are used.

In Chapters 15 and 16 another method is presented for checking long open traverses. The positions of the end points and perhaps some intermediate points are established using the Global Positioning System (GPS). As described there closure can be checked using these points rather than by traversing back to the starting point along a different set of lines. A procedure similar to the GPS one can be used if there are nearby NGS monuments whose positions are known.

A *closed traverse* is one that begins and ends at the same point. A closed traverse could also be one which starts at a known point and ends at another known point, provided both points are on the same coordinate system. Whenever feasible a closed traverse is to be preferred over an open one because it offers simple checks for both angles and distances, as will be seen in Chapter 12.

11-5 OLDER METHODS OF TRAVERSING

For many decades traversing was performed for both open and closed figures by taping the distances and determining directions by measuring one of the following: deflection angles, angles to the right, interior angles, or azimuths. These methods of traversing, although perfectly satisfactory, are nevertheless somewhat obsolete. They are each briefly described in the following paragraphs and then the more modern procedures are discussed in the next section. Whichever method is used, it is wise to measure each angle two or more times.

Deflection Angle Traverse

A deflection angle, defined in Section 9-11, is the angle between the extension of the preceding line and the present one. To measure a deflection angle, the telescope is inverted and sighted on the preceding point, and the telescope is then reinverted and turned to the left or to the right as required to sight on the next point and the angle is read. This method permits easy visualization of traverses, facilitates their representation on paper, and makes the calculations of successive bearings or azimuths very simple. Deflection angles are sometimes used for route surveys, such as for highways, railroads, or transmission lines. Overall, their use has greatly decreased because of the frequency of mistakes in reading and recording angles as being right or left, especially when small angles are involved. The algebraic sum of all the deflection angles for a closed traverse (with no lines crossing) equals 360°. Note that right and left deflection angles must be given opposite signs in the summation or must be designated as being left (L) or right (R). Deflection angles should be measured by repetition as are the other types of angles.

Angle to Right Traverse

Perhaps the most common method of measuring the angles for a traverse in the past was the angle to the right method. With this method, which was partly supplanted by the deflection angle method as the years went by, the telescope is first sighted on the preceding traverse corner. Then it is turned in a clockwise direction until the next point is sighted and the angle is read.

Interior Angle Traverse

Obviously, interior angle measurement applies only to closed traverses. The telescope is sighted on the preceding corner and then is turned either in a clockwise or counterclockwise direction so that the interior angle may be measured. It will be noted that if the surveyor proceeds in a counterclockwise direction while traversing, he or she will always turn clockwise angles.

The sum of the interior angles of a closed polygon is given by the following expression, in which n is the number of sides in the figure:

$$\Sigma = (n - 2)(180^\circ)$$

Thus if all the interior angles of a closed traverse are measured and their total is very close to $(n - 2)(180^\circ)$, we are fairly sure that the angles were measured accurately. Of course, due to the errors that have been described (personal, instrumental, and natural), it will be normal for some difference to exist between the total and the correct value. The difference is called the *angle misclosure*. This value is eliminated by distributing it around the traverse by one of several methods described in Section 12-4.

Azimuth Traverse

An azimuth traverse may be used conveniently for a survey in which a large number of details are to be located, such as topographic mapping. If the instrument is set up so that azimuths can be read directly to each point, the work of plotting them on a map will be appreciably simplified. To accomplish this purpose the telescope is sighted to the preceding corner of the traverse with the instrument reading the back azimuth of that line. Then the telescope is sighted to as many points as desired from that instrument position and the forward azimuth to each of the points is read.

11-6 MODERN TRAVERSING WITH TOTAL STATIONS

Total stations are ideally suited for traversing as the work can be so quickly and efficiently accomplished. Distances and directions may be quickly measured to numerous points from each instrument station. Furthermore, the horizontal and vertical components of those distances and the elevations and coordinates of the sighted points are instantly computed and stored with the instrument's microprocessor.

The total station can be set up at successive traverse stations around the traverse. At each station the instrument is oriented by taking a backsight to a known station, setting the azimuth of that line to the known value along with the coordinates of the occupied position. Then readings (direction and distance) are taken to the next station.

Should the traversing proceed around the traverse and back to the original station or to some other point of known position the measured coordinates just obtained for that point are compared to the given or starting coordinates. If the differences in the two sets of values are within allowable values, they can be adjusted with the microprocessor and final coordinates established for the traverse points.

A large percentage of traverses are today handled with a very efficient procedure called *radiation*. With this method a convenient point or points are selected for instrument set ups from which sights can be taken to the corner stations of the traverse. Using this procedure most obstacles can be avoided and bush cutting kept to a minimum. From the selected points horizontal and vertical angles and distances are measured and the horizontal and vertical positions of the traverse corners computed. The instrument locations may be inside or outside of the traverse and in fact one or more of the traverse corners may be used.

It is assumed that the position of a nearby point is available, such as point *X* in Figure 11-1. This position may be given by coordinates as described in Chapter 12. The azimuth of a line such as *XY* in Figure 11-1 may be known from previous work or determined with an astronomic observation or with GPS.

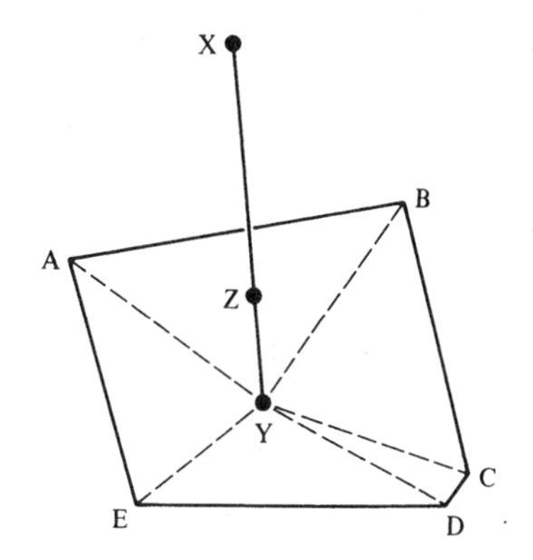

FIGURE 11-1 Radiation from a single point.

If the traverse is fairly small, a convenient point can probably be selected from which all of the corners of the traverse can be sighted. For instance, in Figure 11-1 point *Y* is assumed to be such a point. For this particular case it is assumed that the azimuth of line *XY* is known. The horizontal angles between all of the lines with respect to line *YX* are measured. In addition, the vertical angles (or the zenith angles for many instruments) and the distances to each corner are measured.

The azimuths of the lines are determined and with the vertical angles horizontal distances are computed. From these azimuths and lengths the positions of all the corners can be computed by a method called latitudes and departures, which is described in Chapter 12. With these values it is but a simple step to the determination of the lengths and azimuths of the side of the traverse. A numerical example of this type is presented in Section 13-9. (In Section 11-4 traversing was said to be the process of measuring the lengths and directions of the sides of a traverse. By that definition radiation is really not traversing because side observations are made to the traverse corners from a convenient point or points and the lengths and directions of the sides of the traverse computed.)

Although the calculations described here can be made with handheld calculators with reasonable simplicity, there are today a multitude of programs on the market for programmable calculators and computers with which the calculations can be done in minutes or even seconds once the data are input.

If a total station instrument is used, the measurements described herein for radiation and the subsequent calculations may be abbreviated even further. Furthermore, with an automatic total station the horizontal and vertical angles are both read electronically for use with the slope distances in internal computers or data collectors. With some total stations it is possible to have considerable data reduction in the field. Other instruments are constructed so that the data reduction and the plotting will be completed with office computers.

With an automatic total station, the coordinates (these may be assumed) of the position along with the reference line (*YX*) in Figure 11-1 can be input. A sight is taken along the reference line and sights are taken to each of the traverse corners. With the total station's microprocessor the distances, elevations, and coordinates are automatically computed and displayed.

A check can be made of the work by moving to a different point and repeating the measurements to all the corners and the calculations. In Figure 11-1 point *Z* is shown as the check point. It does not have to be along line *XY* as shown in this figure. If the values determined for coordinates, azimuths, and distances are in good agreement, the final values can be averaged.

Radiation does have one major weakness, in that a mistake may be made in one of the values. You may say "but we are using a total station instrument and the readings are recorded in electronic data collectors and thus there are no mistakes." Unfortunately, mistakes can be made with this modern equipment as with older equipment. A sight may be taken on the wrong point for the corner, the electronic signal may be reflected by something other than our reflectors (perhaps a nearby road sign or a piece of reflector tape on a parked vehicle), and so on.

If two points are close together in the traverse, such as *C* and *D* in Figure 11-1, a small error in the length of *YC* or *YD* may have a rather large effect in the direction of side *CD*.[1] For larger traverses it is probable that all of the traverse corners will not be visible from one instrument setup. Nevertheless, the same procedure may be followed except that there will be a need for more setups, as shown in Figure 11-2.

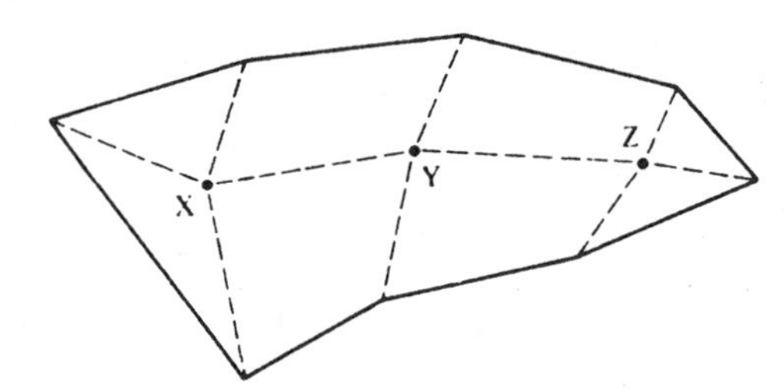

FIGURE 11-2 Radiation from several points.

Radiation is also very convenient for laying out construction projects. The coordinates of various points to be staked are determined from the design drawings. Then the angles and distances to those points from particular set up points of the total station are computed. The staking points are then laid out with the total station.

Auxiliary Closed Traverses

As we have described, it is often quite difficult to work directly on the property lines of a piece of land because of various obstacles. For such cases the surveyor may work

[1]F. H. Moffitt and H. Bouchard. *Surveying,* 8th ed. (New York: Harper & Row, Publishers, Inc., 1987), pp. 322–323.

from one or more points, as shown in Figures 11-1 and 11-2, or may have a complete closed auxiliary traverse, as shown in Figure 11-3. It is often feasible to establish an auxiliary traverse which itself can easily be traversed. The corners of such an auxiliary traverse are located in the vicinity of the property corners and located so that ties can easily be made to them by distances and directions. Such a situation is shown in Figure 11-3, where the outside traverse represents the land boundaries and the dashed lines represent the auxiliary traverse.

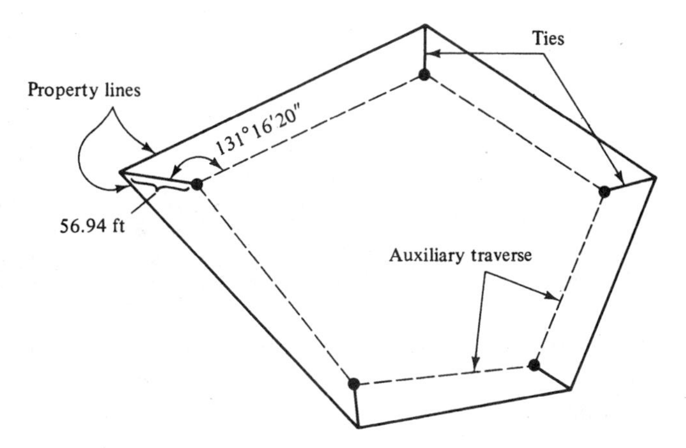

FIGURE 11-3 Using an auxiliary traverse.

The lengths and directions of the sides of the auxiliary traverse are determined. The lengths and directions of the ties are measured and from this information the location of the property corners are determined. As a last step the lengths and directions of the property lines will be computed. The reader will learn how to make these calculations in the next few chapters.

11-7 INTERSECTION OF TWO LINES

A frequent surveying problem involves the intersection of two lines. For this discussion, reference is made to Figure 11-4, in which points *A, B, C,* and *D* are established on the ground. It is desired to find the intersection of lines *AB* and *CD* (shown as point *x*).

If two instruments are available, the problem can be handled easily. One instrument can be set up at *A* and sighted toward *B,* and the other instrument can be set up at *C* and sighted toward *D.* A rodman is sent to the vicinity of the intersection point and is waved back and forth until his or her range pole (or plumb-bob string) is lined up with both instruments.

When only one instrument is available, the problem can still be handled with little difficulty with the aid of some stakes and tacks and a piece of string. It is assumed that the instrument is set up at *A* and sighted on *B.* Two stakes, *E* and *F,* are set on the line so that line *EF* straddles line *CD.* A tack is set sticking out of the top of each of these stakes (properly on the line) and a string is tied between them. After the position of the string is carefully checked, the instrument is moved to *C,* lined up with *D,* and sighted on the string. This is the desired point *x.* It should be obvious that the longer the sights used for this work, the more precise will be the measurements.

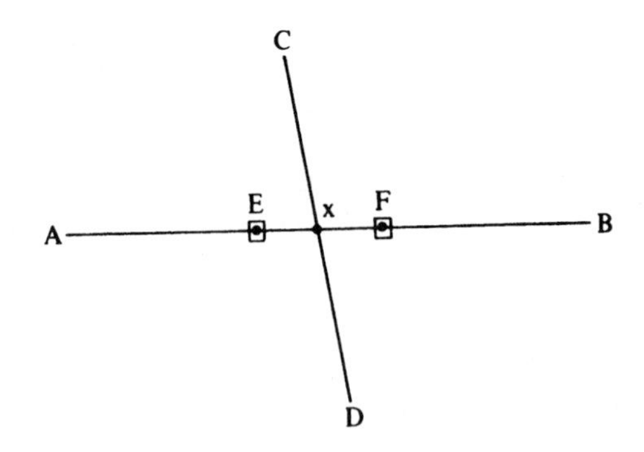

FIGURE 11-4 Intersecting two lines.

11-8 MEASURING AN ANGLE WHERE THE INSTRUMENT CANNOT BE SET UP

Another common problem faced by the surveyor is the one of measuring an angle at a point where the instrument cannot be set up. Such a situation occurs at the intersection of fences or between the walls of a building, as illustrated in Figure 11-5.

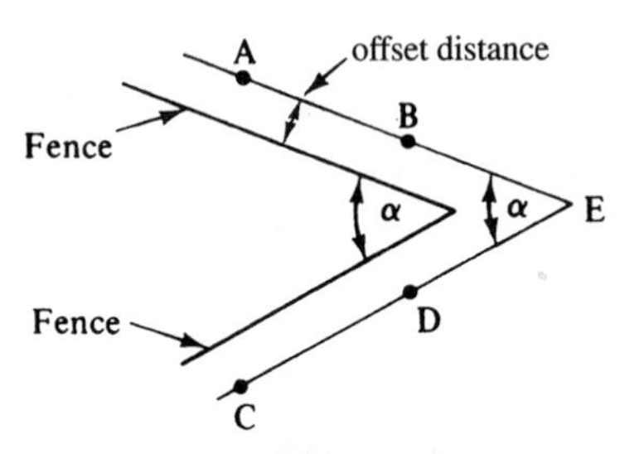

FIGURE 11-5 Measuring an angle between offset lines.

To handle this problem, line *AB* is established parallel to the upper fence, and line *CD* is established parallel to the lower fence. The lines are extended and intersected at *E*, as described in Section 11-7, and the desired angle α is measured with the instrument. In establishing line *AB*, point *A* is located at a convenient distance from the fence. The shortest distance to the fence is measured by swinging the tape in an arc with *A* as the center. The desired distance is the perpendicular distance. Similarly, point *B* is located the same distance from the fence. The same process is used to locate points *C* and *D* near the other fence in order to establish line *CD*.

A quicker and more precise method that is sometimes used involves the measurement of convenient distances up each fence line and the chord distances across from one fence to the other, as shown in Figure 11-6. The desired angle α may then be computed by trigonometry. If the distances measured up the fence lines are equal, the angle may be obtained from the equation shown in the figure. However, some type of radiation procedure such as the one previously described is better than the method illustrated in Figures 11-5 and 11-6.

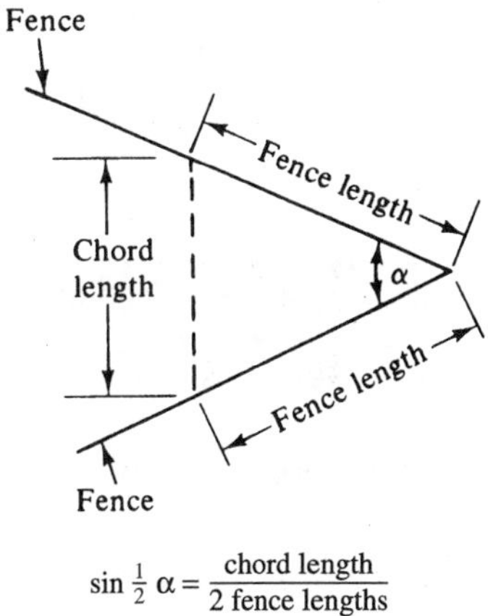

FIGURE 11-6 Determining an angle from distance measurements.

11-9 PROLONGING A STRAIGHT LINE BY DOUBLE CENTERING

An everyday problem of the surveyor is that of prolonging straight lines. Such work is quite common in route surveying, where straight lines may have to be prolonged for considerable distances over rough terrain. In Figure 11-7 it is assumed that line *AB* is the line to be extended beyond point *B*. The instrument is set up at *B* and backsighted on *A*. Then the telescope is plunged to set point *C′*. If the instrument is not in proper adjustment (if the line of sight of the instrument is not perpendicular to its horizontal axis), point *C′* will not fall on the desired straight line. For this reason *double centering* is the method chosen for this problem.

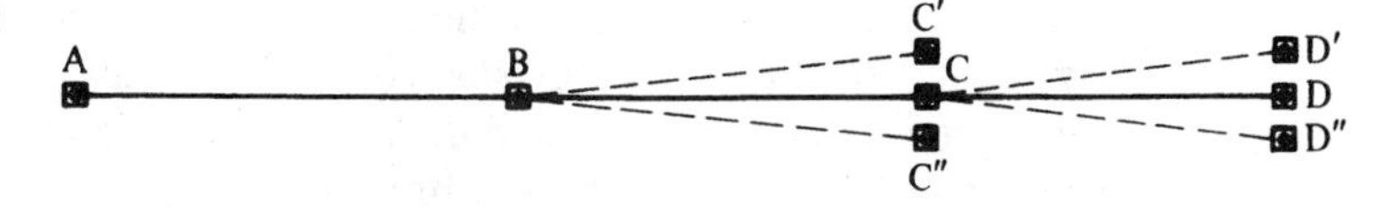

FIGURE 11-7 Prolonging a straight line by double centering.

With the double-centering method the telescope in its normal position is sighted on *A* and plunged to set point *C′*. The telescope is then rotated horizontally about its vertical axis until point *A* is sighted. The telescope is now inverted or upside down. It is once more plunged and point *C″* is set. The correct point *C* is halfway between the two points *C′* and *C″*. Of course, if the instrument is properly adjusted, points *C′* and *C″* should coincide if the distances are short. For long distances, however, there will be some displacement between the points, even for well-adjusted instruments. This procedure substantially reduces errors caused by instrument inadjustment and gives the instrumentman a check against the presence of other errors and mistakes. From point *C* the straight line is continued to point *D*, and so on.

The technique of double centering is a very important one for the surveyor and one that can be accomplished in a very short time. If one can learn to measure all angles (angles to the right, deflection angles, etc.) by double centering, the precision of the work will be appreciably improved and many blunders will be eliminated.

Double-centering techniques are also used for setting large building corners. When one is measuring angles by repetition, half of the measurements should be made with the telescope in the normal position and half with the telescope inverted.

11-10 ESTABLISHING POINTS ON A STRAIGHT LINE BETWEEN TWO GIVEN POINTS

Points Intervisible

If the entire line is visible between the two points, the surveyor has no problems. He or she can set up the instrument at one end, sight on the other end, and then establish any desired points in between. If a large vertical angle is involved in setting any of the points, the careful surveyor may very well set the points with the telescope in its normal position and then check them with the telescope inverted.

Balancing In

If the end points are not visible from each other but there is an area in between from which both the points may be seen (a surprisingly common situation), the process called *balancing in* or *wiggling in* may prove beneficial. In this procedure the surveyor sets up the instrument at a point that he or she believes to be on a straight line from *A* to *C* (see Figure 11-8). One point is sighted and the telescope plunged to see if the other point is in line. If not, the instrument is moved laterally to another position and the steps are repeated. Since it is very difficult to estimate closely the first time, the instrument may have to be moved several times, with the final adjustment probably made by loosening the leveling screws and shifting the head of the instrument. Once the instrument is in line, any desired intermediate points from *A* to *C* may be set. When balancing in is possible, it may very well save the time involved in running trial lines, as described in the following paragraph, and the delay required to make the necessary calculations to finish the problem.

FIGURE 11-8 Balancing in or wiggling in.

Two Points Not Visible from Each Other

When it is desired to establish intermediate points between two known points that cannot be seen from each other and cannot be seen from any point in between, the surveyor may very well use a random line procedure. This is very common in surveying when hills, trees, and great distances are involved. The surveyor can run a trial straight line (measuring the distances involved) in the general direction from *A* toward *E* (Figure 11-9). He or she sets points *B*, *C*, and *D* (preferably by double centering), and when near *E*, measures the distance *DE* and the interior angle at *D*. From this information the desired interior angle at *A* is computed, the straight line from *A* toward *E* is run, and points are established in between. If point *E* is missed slightly, the intermediate points are readjusted proportionately.

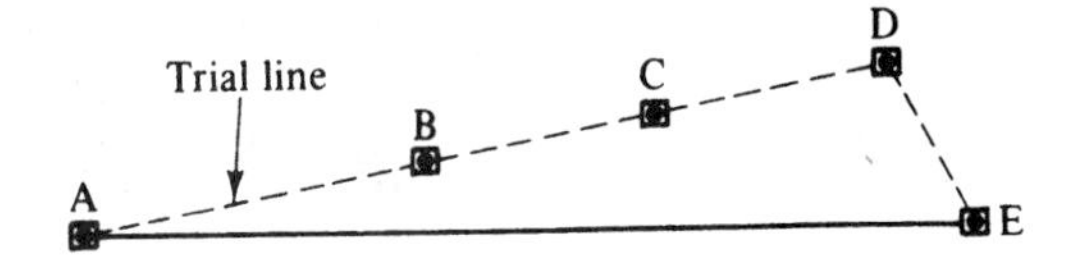

FIGURE 11-9 Random line procedure.

Another random line method which may be a little more practical will become obvious after study of Chapter 12 is completed. The surveyor runs a random set of straight lines from *A* to *E*, as shown in Figure 11-10, and then computes the desired length and direction of line *AE*. Then he or she can return to *A* and run the straight line *AE* and compute the distances required to set points on the desired line by measuring over from the random trial lines and working from the ends *A* and *E*.

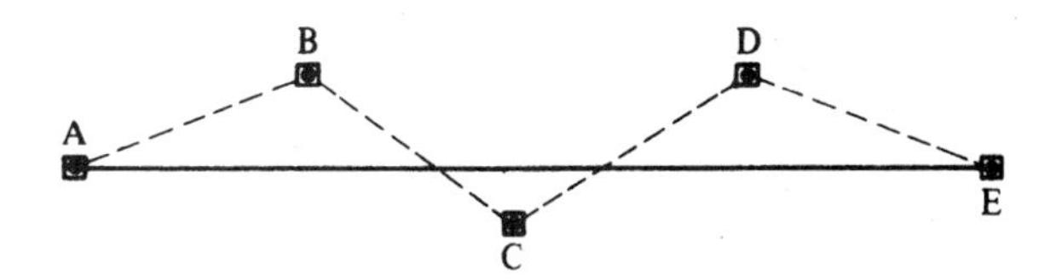

FIGURE 11-10 Another random line procedure.

11-11 CLEANING SURVEYING EQUIPMENT

Equipment will frequently become wet or dirty, and to ensure its precise operation and to lengthen its life, the equipment must be cleaned at regular intervals and at times when it has been subject to unusually severe conditions. The equipment can be sent to a professional repair service for thorough cleaning and lubrication (a good practice every few years), but the surveyor should continually do several things to keep equipment in good condition.

Professional help may be needed for internal cleaning and lubrication of instruments, but external dust, dirt, and water should be removed immediately. Water and dirt can be removed from instruments with cotton balls and pipe cleaners. Dust can be removed from lenses with camel's-hair brushes. If the lenses are streaked, it will be necessary to use an optical glass cleaner and lint-free cloth. Do not use strong cleaners nor rub too firmly.

Problems

11-1. List four errors that frequently occur in angle measurement.

11-2. Why is it desirable in angle measurement to avoid short sight distances if at all possible?

11-3. If angles are measured to the nearest 5″, to what precision does it correspond? (Ans.: 1/41,300)

11-4. It is desired to measure angles with a precision of 1/100,000. How close should the angles be measured to correspond to such a precision?

11-5. Describe how a deflection angle traverse is run.

11-6. What are the advantages of traversing by the radiation method?

11-7. What is the disadvantage of traversing by the radiation method?

11-8. Describe two methods for measuring the angle at a fence corner where the instrument cannot be set up.

11-9. What is meant by "balancing in"?

CHAPTER

Traverse Adjustment and Area Computation

12-1 INTRODUCTION

Although this chapter is concerned primarily with the calculation of land areas, several other important topics are included, such as the precision of field work, the balancing of errors, and the use of coordinates in surveying.

It may seem to the reader that the first part of this chapter is concerned only with situations where the surveyor measures distances and angles around the perimeter of a land tract. However, the author is attempting to prepare a foundation upon which all types of traverses may be handled, such as the common radiation surveys of today.

12-2 COMPUTATIONS

Almost all surveying measurements require some calculations to reduce them to a more useful form for determining distance, earthwork volumes, land areas, and so on. This chapter is devoted to the calculation of land areas. Perhaps the most common need for area calculations arises in connection with the transfer of land titles, but area calculations are also needed for the planning and design of construction projects. Some obvious examples are the laying out of subdivisions, the construction of dams, and the consideration of watershed areas for designing culverts and bridges.

12-3 METHODS OF CALCULATING AREAS

Land areas may be calculated by several different methods. A very crude approach that should be used only for rough estimating purposes is a graphical method in which the traverse is plotted to scale on a sheet of graph paper and the number of squares inside the traverse are counted. The area of each square can be determined from the scale used in drawing the figure, and thus the area is roughly estimated.

A similar method, which yields appreciably better results but is again satisfactory only for estimating purposes, involves the use of a device called a planimeter (see Section 12-13). The traverse is carefully drawn to scale and a planimeter is used to measure the traverse area on the paper. From this value the land area can be computed from the scale of the drawing. It is probable that with careful work, areas may be estimated by this method within a range of from 1/2% to 1% of the correct values.

A very useful and accurate method of computing the areas of traverses that have only a few sides is the triangle method. The traverse is divided into triangles and the areas of the triangles are computed separately. The necessary formulas appear in Figure 12-1, in which several traverses are shown. The area of a triangle is considered in part (a) while the area of a four-sided figure is considered in part (b). If the traverse has more than four sides, it is necessary [as shown in part (c) of the figure] to obtain the values of additional angles and distances either by field measurements or by lengthy office computations. For this five-sided figure, the triangle areas *ABE* and *CDE* can be obtained as before, but additional information is necessary to determine the area *BCE*. The problem is amplified for traverses that have more than five sides. For such cases as these, the surveyor is probably better advised to use one of the methods described later in this chapter.

Other methods of computing areas within closed traverses are the double meridian distance, double parallel distance, and coordinate methods discussed in Sections 12-8 through 12-12. In addition, several methods are presented in Section 12-13 for computing the areas of land within irregular boundaries. Computer methods of handling the same problems are discussed in Chapter 13.

FIGURE 12-1 Computing areas from triangles.

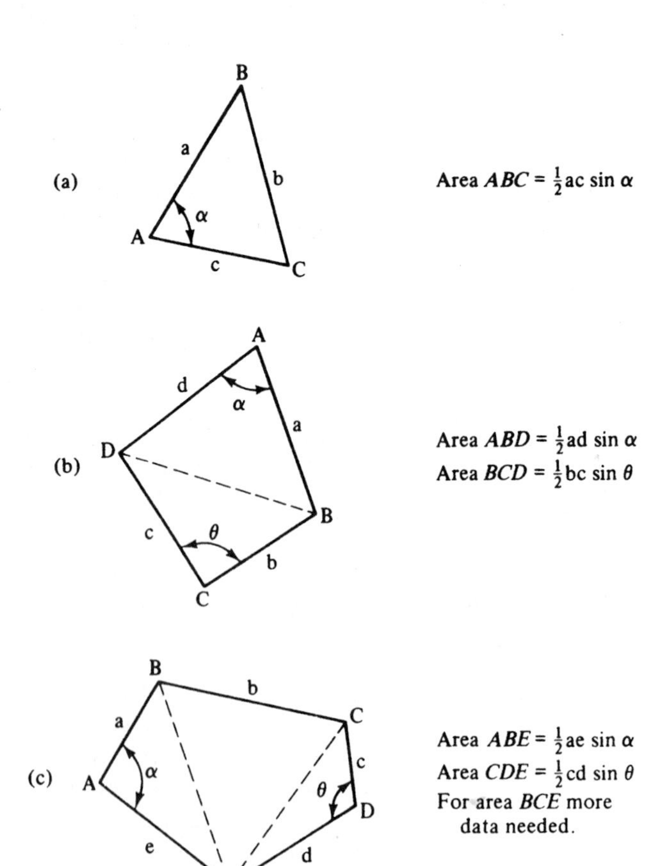

12-4 BALANCING ANGLES

Before the area of a piece of land can be computed, it is necessary to have a closed traverse. The first step in obtaining a closed figure is to balance the angles. The interior angles of a closed traverse should total (n -2)(180°), where n is the number of sides of the traverse. It is unlikely that the angles will add up perfectly to this value, but they should be very close. The usual rule for average work is that the total should not vary from the correct value by more than approximately the square root of the number of angles measured times the least division readable with the instrument (see Section 2-10). For an eight-sided traverse and a 10″ total station, the maximum error should not exceed

$$\pm 10'' \sqrt{8} = \pm 28.3'' \text{ say, } \pm 28''$$

It is customary for the instrumentman to check the sum of the angles for the traverse before leaving the field. If the discrepancies are unreasonable, he or she must remeasure the angles one by one until the source of the trouble is found and corrected.

If the angles do not close by a reasonable amount one or more mistakes have been made. If a mistake has been made in only one angle, that angle can often be identified by plotting the lengths and directions of the sides of the traverse to scale. If this is done the plotted traverse will have an error of closure such as the one shown by the dotted line in Figure 12-2.Then if a line is drawn perpendicular to the error of closure, it will often point to the angle where the mistake was made. It can be seen in Figure 12-2 that if the angle containing the mistake is reduced, it will tend to cause the error of closure to be reduced.

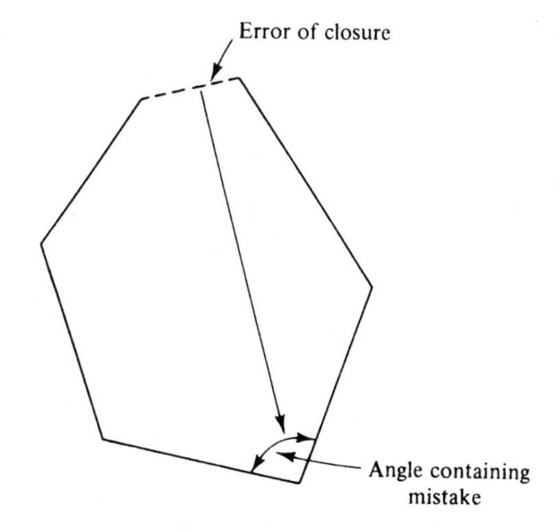

FIGURE 12-2 Trying to find "busted" angle.

When the angular errors for a traverse have been reduced to reasonable values, they are distributed among the angles so that their sum will be exactly (n –2)(180°). Each angle may be corrected by the same amount: only certain angles may be corrected because of difficult field conditions; or an arbitrary rule may be used to make the corrections. For instance if there is an angle that is suspect in the surveyor's mind (that is one in which obstructions, short sides, or other problems were involved), a large part or all of the total correction may be applied to that angle. For the traverse considered in this chapter all the angles were corrected by an equal amount.

The very sensible thought may occur to the reader that all the angles in the traverse should be corrected by an equal amount. For a 3′ error in 12 angles, each correction would be $3'/12 = 0.25' = 15''$. It is probably not advisable to make the adjustments in units smaller than the least value readable with the instrument. For the example given, if the angles were measured to the nearest 15″ or less, this probably is a reasonable procedure. If the angles were measured to the nearest 1′, however, it does not seem very reasonable to correct them to the nearest 15″. After the angles are balanced, the bearings of the sides of the traverse are computed. The initial bearing is preferably an astronomic one, but a magnetic or assumed bearing may be used and the other bearings computed from the balanced traverse angles.

The interior angles for the traverse of Figure 10-9 in Chapter 10 are added together in Table 12-1 and total 539°59′45″ which is less than the correct total of 540°00′00″ for the interior angles of a closed five-sided figure. The error was corrected as shown in the table by adding 3″ to each of the five angles. An astronomic bearing[1] was obtained for side *AB* and the bearings of the other sides were calculated with the balanced angles and shown in the table.

TABLE 12-1 Balanced Angles and Bearings

Point	Measured ∡	Corrected ∡	Calculated bearings from corrected ∡'s
A	36°44′20″	36°44′23″	
			S6°15′00″W
B	215°51′55″	215°51′58″	
			S29°36′58″E
C	51°40′25″	51°40′28″	
			N81°17′26″W
D	111°06′30″	111°06′33″	
			N12°23′59″W
E	124°36′35″	124°36′38″	
			N42°59′23″E
	Σ = 539°59′45″	Σ = 540°00′00″	

12-5 LATITUDES AND DEPARTURES

The closure of a traverse is checked by computing the latitudes and departures of each of its sides. The *latitude of a line* is its projection on the north-south meridian and equals its length times the cosine of its bearing. In the same manner, the *departure of a line* is its projection on the east-west line (sometimes called the *reference parallel*) and equals its length times the sine of its bearing. These terms (illustrated in Figure 12-3) merely describe the *x* and *y* components of the lines.

For the calculations used in this chapter, the latitudes of lines with northerly bearings are designated as being north or plus. Those in a southerly direction are designated as south or negative. Departures are positive for lines having easterly bearings and negative for lines having westerly bearings. For example, line *AB* in Figure 12-3(a) has a northeasterly bearing and thus a + latitude and a + departure. Line *CD* in Figure 12-3(b) has a southeasterly bearing and thus a – latitude and a + departure. Calculations of latitudes and departures are illustrated in the next section.

[1] J.C. McCormac, *Surveying Fundamentals*, 2nd ed. (Englewood Cliffs, N.J., Prentice-Hall, 1991) pp. 330–352.

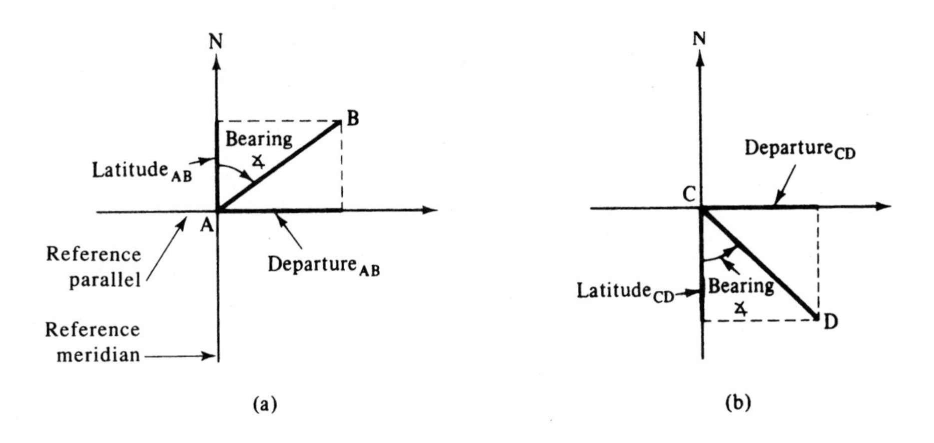

FIGURE 12-3 Latitudes and departures.

12-6 ERROR OF CLOSURE

If you start at one corner of a closed traverse and walk along its lines until you return to your starting point, you will have walked as far north as you have walked south and as far east as you have walked west. This is the same thing as saying that for a closed traverse the sum of the latitudes should equal zero and the sum of the departures should equal zero. When the latitudes and departures are calculated and summed, they will never be exactly equal to zero (except by accidental results of canceling errors).

When the latitudes are added together, the resulting error is referred to as the *error in latitude* (E_L); the error that occurs when the departures are added is referred to as the *error in departure* (E_D). If the measured bearings and distances of the traverse in Figure 12-4 are plotted exactly on a sheet of paper, the figure will not close because of E_L and E_D. The usual magnitude of these errors is greatly exaggerated in this figure.

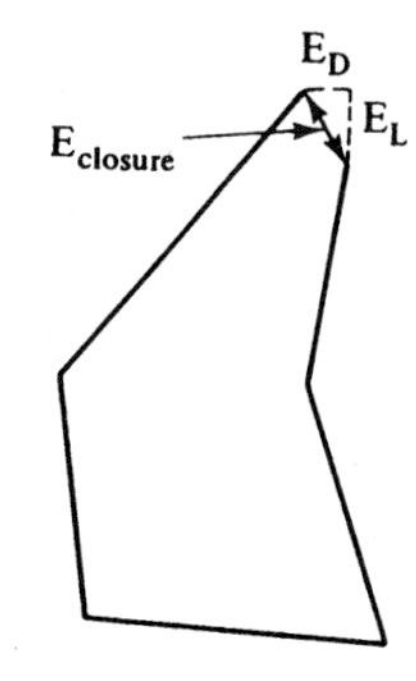

FIGURE 12-4 Error of closure.

The error of closure can be easily calculated as follows:

$$E_{\text{closure}} = \sqrt{(E_L)^2 + (E_D)^2}$$

and the precision of the measurements can be obtained by the expression

$$\text{precision} = \frac{E_{\text{closure}}}{\text{perimeter}}$$

After the precision is determined, the surveyor will make a decision as to whether the work has been done satisfactorily. In most areas of the United States minimum acceptable precisions are provided by law for various kinds of surveys. Typical values are 1/5000 for rural land, 1/7500 for suburban land, and 1/10,000 for urban land. If the precision is satisfactory, the surveyor will proceed to balance the errors in latitudes and departures and compute the area of the traverse as described in the next few sections of this chapter.

Should the precision obtained be unsatisfactory for the purposes of the survey, it will be necessary to recheck the work. Of course, as a first step, the math should be checked carefully for mistakes, and if mistakes cannot be found, the field measurements must be checked. Mistakes must not be included in the data reduction.

In Table 12-2 the values of E_L, E_D, $E_{closure}$, and the precision are computed for the traverse previously considered in this book. The lengths are those shown in Figure 4-1 while the bearings are the balanced ones of Table 12-1. Usually, the degree of precision is rounded to the nearest 100. If the precision is unsatisfactory, the surveyor should study the calculations carefully before rechecking the field measurements. If the angles balance, a mistake in distance is to be suspected. Usually, it will have occurred in a side roughly parallel to the direction of the error of closure line. For instance, if a traverse has a large error in latitude but a small error in departure, one might look for a side running primarily in the north-south direction. If such a side exists, it would be logical to remeasure the distance for that side first. In a similar manner, if the error in departure is twice the error in latitude, the surveyor would want to check the distances for any sides whose calculated departures and latitudes were approximately in that proportion.

TABLE 12-2 Computing Latitudes and Departures

			Latitude = Length x cosine of bearing		Departure = Length x sine of bearing	
Side	**Bearing**	**Length**	**+N**	**–S**	**+E**	**–W**
AB	S6°15′00″W	189.53	—	188.403	—	20.634
BC	S29°36′58″E	175.18	—	152.293	86.571	—
CD	N81°17′26″W	197.78	29.949	—	—	195.499
DE	N12°23′59″W	142.39	139.068	—	—	30.575
EA	N42°59′23″E	234.58	171.590	—	159.952	—
		Σ = 939.46	+340.607	–340.696	+246.523	–246.708
			E_L = –0.089 or too much south		E_D = –0.185 or too much west	

$$E_{\text{closure}} = \sqrt{(0.089)^2 + (0.185)^2} = 0.205 \text{ ft}$$

$$\text{Precision} = \frac{0.205}{939.46} = \frac{1}{4583} \text{ say } \frac{1}{4600}$$

Despite these ideas, small mistakes may not be located easily and the surveyor may have to remeasure carefully all or nearly all of the sides of the traverse. Once the errors of closure are reduced to reasonable values, they will be adjusted so that the traverse will close perfectly (to the number of places being used), as described in the next section.

12-7 BALANCING LATITUDES AND DEPARTURES

The purpose of balancing the latitudes and departures of a traverse is to attempt to obtain more probable values for the locations of the corners of the traverse. If, as described in the preceding section, a reasonable precision is obtained for the type of work being done, the errors are balanced out in order to close the traverse. This usually is accomplished by making slight changes in the latitudes and departures of each side so that their respective algebraic sums total zero.

Theoretically, it is desirable to distribute the errors in a systematic fashion to the various sides, but practically, the surveyor may often use a simpler procedure. He or she may decide to make large corrections to one or two sides where the most difficulties were encountered in making the measurements. The surveyor may look at the magnitude of E_L and E_D and decide, after studying the lengths of the various sides and the sines and cosines of their bearings, that by changing the length of such and such a side or sides the traverse can just about be balanced.

The practical balancing methods just described may not seem to the reader to give very good results. However, they may be as satisfactory as the results obtained with the more theoretical rules described in the following paragraphs, since the latter methods are based on assumptions that are not altogether true and may, in fact, be far from being true.

Very often the surveyor may have no idea which side should get the correction and, in fact, the person performing the calculations may be someone other than the person who made the measurements. For such cases a systematic balancing method such as the compass rule, the transit rule, or one of several others is desirable.

A very popular rule for balancing the errors is the *compass* or *Bowditch rule,* named after the distinguished American navigator Nathaniel Bowditch (1773–1838), who is given credit for its development. It is based on the assumption that the quality of distance and angular measurements is approximately the same. It is particularly applicable to surveys made with EDMs and theodolites or with total stations. It is further assumed that the errors in the work are accidental and thus that the total error in a particular side is directly proportional to its length. The rule states that *the error in latitude (departure) in a particular side is to the total error in latitude (departure) as the length of that side is to the perimeter of the traverse.* For the example traverse, the correction for the latitude of side *AB* is calculated below.

$$\frac{\text{Correction in lat}_{\text{AB}}}{E_L} = \frac{L_{\text{AB}}}{\text{perimeter}}$$

$$\text{Correction in lat}_{\text{AB}} = E_L \frac{L_{\text{AB}}}{\text{perimeter}} = \frac{(+0.089)(189.53)}{939.46} = +0.018 \text{ ft}$$

If the sign of the error is +, the correction will be minus. Actually, the sign of the corrections can be determined by observing what is needed to balance the numbers to zero. The author finds it much easier not to worry with recording signs for these corrections as he has done in Table 12-3. In Table 12-2 the sum of the south latitudes is larger than the sum of the north latitudes. Thus to make the corrections he makes the north latitudes larger and the south ones smaller. Similarly the sum of the west departures is larger than the sum of the east departures. Therefore the corrections used are to make the east values larger and the west ones smaller.

In Table 12-3 the latitudes and departures for the traverse are balanced by the compass rule.

TABLE 12-3 Balancing Latitudes and Departures

	Latitude Correction		Departure Correction		Balanced Latitudes		Balanced Departures	
	N	S	E	W	N	S	E	W
AB	—	+0.018	—	+0.037	—	188.385	—	20.597
BC	—	+0.017	+0.035	—	—	152.276	86.606	—
CD	+0.019	—	—	+0.039	29.968	—	—	195.460
DE	+0.013	—	—	+0.028	139.081	—	—	30.547
EA	+0.022	—	+0.046	—	171.612	—	159.998	—
					340.661	340.661	246.604	246.604

A rule that is occasionally used is the *transit rule.* It is based on the assumption that errors are accidental and that the angle measurements are more precise than the length measurements, such as for a stadia traverse (obsolete). The computations involve corrections to the latitudes and departures in such a manner that the lengths of the sides are changed but not their directions. In this method *the correction in the latitude (departure) for a particular side is to the total correction in latitude (departure) as the latitude (departure) of that side is to the sum of all the latitudes (departures).*

$$\frac{\text{Correction in lat}_{AB}}{E_L} = \frac{\text{lat}_{AB}}{\sum \text{latitudes}}$$

Computer programs are readily available for solving all the problems (from balancing to area calculations) described in this chapter. These programs use one of the systematic methods of balancing, such as the compass rule or the transit rule described here, or other methods, such as the Crandall method or the least squares method.

The *Crandall method* provides another systematic adjustment procedure, which is rather similar in application to the transit rule. It is particularly applicable to traverses where the angles have been measured more precisely than the distances, stadia surveys again being an example. With this method the angles are corrected equally while a weighted least squares adjustment is made to the lengths.

The *least squares method* is usually the best method available for adjusting surveying data. It is difficult to apply, however, unless a computer is being used. In each of the other methods (compass, transit, estimation, and Crandall) some type of systematic correction is made to errors that are accidental or random in nature. The least squares method, developed from probability theory, provides the most probable results for the traverses. With this method the angle and distance measurements are adjusted so as to make the sum of the squares of the residuals the least possible.[2]

In the previous few paragraphs of this section various methods were introduced for balancing the latitudes and departures of the sides of closed traverses. Such balancing will cause the lengths and bearings of the traverse sides to change. The new

[2]P. R. Wolf and R. C. Brinker, *Elementary Surveying,* 9th ed. (New York: Harper Collins College Publishers, 1994). p. 249.

length of each side can be determined by computing the square root of the sum of the squares of its adjusted latitude and departure ($l_{AB} = \sqrt{(\text{dep}AB)^2 + (\text{lat}AB)^2}$. Its adjusted bearing may be determined by trigonometry. For example,

$$\tan \text{bearing} \measuredangle = \frac{\text{adjusted departure}}{\text{adjusted latitude}}$$

Space is not taken here to show these calculations for the example traverse considered in these sections. The computer programs commonly used today to make precision, balancing, and area computations (which are discussed in Chapter 13) will usually provide the final adjusted lengths and bearings of the sides.

12-8 DOUBLE MERIDIAN DISTANCES

The best-known procedure for calculating land areas with handheld calculators is the *double meridian distance* (DMD) method. The *meridian distance* of a line is the distance (parallel to the east-west direction) from the midpoint of the line to the reference meridian. Obviously, the *double meridian distance* (DMD) of a line equals twice its meridian distance. In this section it will be proved that if the DMD of each side of a closed traverse is multiplied by its balanced latitude and if the algebraic sum of these values is determined, the result will equal two times the area enclosed by the traverse.

Meridian distances are considered positive if the midpoint of the line is east of the reference meridian and negative if it is to the west. In Figure 12-5, the plus meridian distance of side *EA* is shown by the dashed horizontal line. For sign convenience, the reference meridian is usually assumed to pass through the most westerly or the most easterly corner of the traverse.

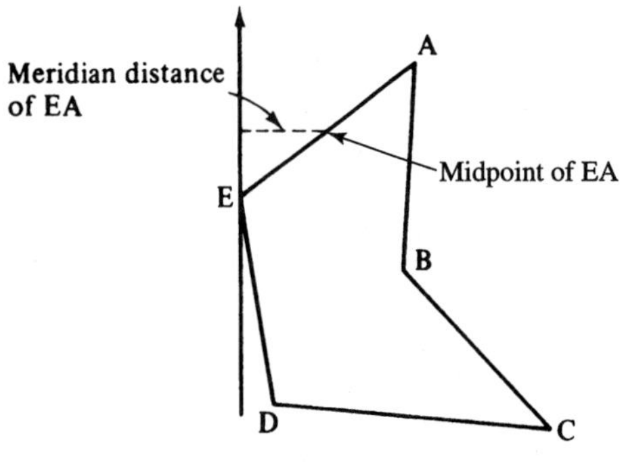

FIGURE 12-5 Meridian distance.

If the surveyor has difficulty in determining the most westerly or the most easterly corners, he or she can solve the problem quickly by making a freehand sketch. He or she starts at any corner of the traverse and plots the departures successively to the east or west for each of the lines until returning to the starting corner. The location of the desired points will be obvious from the sketch. In this manner the most westerly and easterly points of the sample traverse used in this chapter are located in Figure 12-6. The departures shown were taken from Table 12-3.

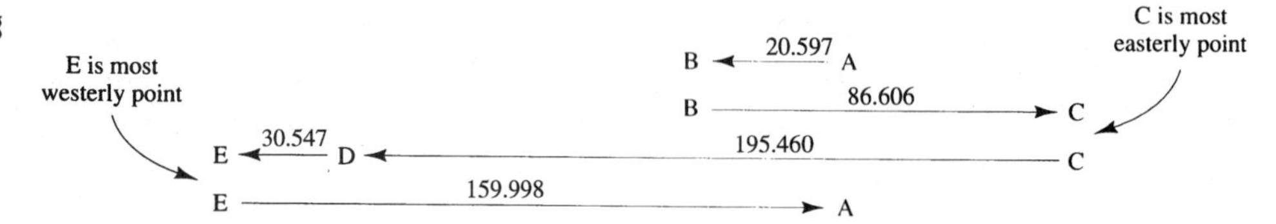

FIGURE 12-6 Locating the most westerly and easterly points of a traverse.

In Figure 12-5 it can be seen that the DMD of side *EA* equals twice its meridian distance or equals its departure. The DMD of side *AB* equals two times the departure of *EA* plus two times one-half the departure of *AB*. In this manner the DMD of any side may be determined. By studying this process, however, the reader will develop the following rule for DMDs that will simplify the calculations: *The DMD of any side equals the DMD of the last side plus the departure of the last side plus the departure of the present side.* The signs of the departures must be used and it will be noticed that the DMD of the last side (*DE* in Figure 12-5) must equal the departure of that side, but it will of necessity be of opposite sign.

To see why the DMD method for area calculation works, we refer to Figure 12-7. In this discussion, north latitudes are considered plus and south latitudes are considered minus. If the DMD of side *AB*, which equals *B′B*, is multiplied by its latitude *B′A*, the result will be plus two times the area of the triangle *B′BA*, which is outside the traverse. If the DMD of side *BC* is multiplied by its latitude *B′C′*, which is minus, the result will be minus twice the trapezoidal area *B′BCC′*, which is inside and outside the traverse. Finally, the DMD of side *CA* times its latitude *C′A* equals plus two times the area *ACC′*, which is outside the traverse. If these three values are added together, the total will equal two times the area inside the traverse, because the area outside the traverse will be canceled. This same method can be used to prove that the DMD method will work for computing the area inside any closed traverse that consists of straight lines.

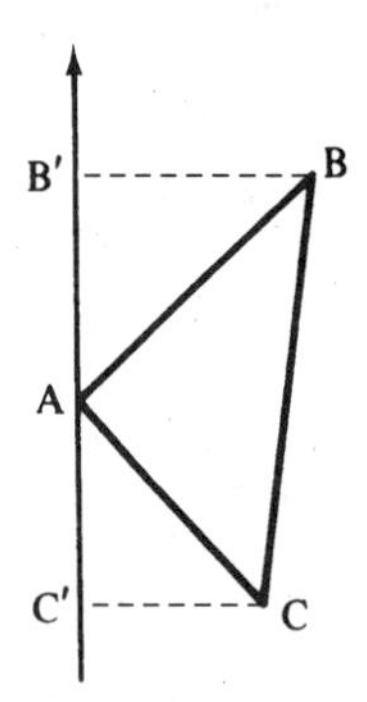

FIGURE 12-7 Explaining why the DMD method works.

Table 12-4 shows the area calculations by the DMD method for the example traverse used in this chapter. Because the signs of the latitudes must be used in the multiplications, the table provides one side for positive values under the north column for north latitudes and one for negative values under the south column for south latitudes. The traverse area equals one-half the algebraic sum of the two columns. *It does not matter whether the final value is positive or negative.* The resulting area may be quickly checked by moving the reference meridian to another corner and repeating the calculations. If the refererence meridian has been assumed at the

TABLE 12-4 Computing Area Using DMDs

Side	Departures E	W	DMD	Latitudes N	S	Double Areas +N	–S
AB	—	20.597	299.399	—	188.385	—	56,402
BC	86.606	—	365.408	—	152.276	—	55,643
CD	—	195.460	256.554	29.968	—	7,688	—
DE	—	30.547	30.547	139.081	—	4,249	—
EA	159.999	—	159.998	171.612	—	27,458	—

$\Sigma = +39{,}395 \; -112{,}045$

$2A = +\,39{,}395 - 112{,}045 = -72{,}650$

$A = -36{,}325$ sq ft $= 0.834$ acre

most westerly corner, it will probably be moved to the most easterly corner. This same problem is solved with a great saving in time and effort in Section 13-4 using a computer and the computer disk enclosed with this book.

When lengths are in meters the area is expressed in square meters (m^2). The SI system does not specify a particular unit for land areas, but the *hectare,* which equals 10,000 m^2, is used by many countries. One hectare equals 2.47104 acres.

12-9 DOUBLE PARALLEL DISTANCES

The same procedure as that used for DMDs may be used if double parallel distances (DPDs) are multiplied by the balanced departures for each side. The final areas will be the same. The *parallel distance* of a line is the distance (parallel to the north-south direction) from the midpoint of the line to the reference parallel or east-west line. The parallel is probably drawn through the most northerly or the most southerly corner of the traverse.

12-10 RECTANGULAR COORDINATES

Rectangular coordinates are the most convenient and probably the most used method available for describing the horizontal positions of survey points. In the world of computers, just about everyone uses coordinates to define the positions of such points. In courthouses coordinate systems are frequently used to describe property corner locations. Dams, highways, industrial plants, and mass-transit systems are located, planned, designed, and constructed on the basis of computerized information, which includes coordinates as well as other information concerning topography, geology, drainage, population, and so on. As a result of these facts, it is absolutely necessary for the surveyor to be familiar with, and able to use, coordinates. The coordinates of a particular point are defined as the distances measured to that point from a pair of mutually perpendicular axes. The axes are usually labeled X and Y, the perpendicular distance from the Y axis to a point is called the x coordinate, and the perpendicular distance from the X axis to the point is called the y coordinate.

In the United States it is common to have the *X* axis coincide with the east-west direction and the *Y* axis with the north-south direction. Just the opposite system is used in some countries, particularly in Europe. In Figure 12-8 the coordinates for two points *A* and *B* at the ends of a line are shown. The positive directions are indicated by the arrows on the axes. Point *B* is below the *X* axis and thus has a negative *Y* value.

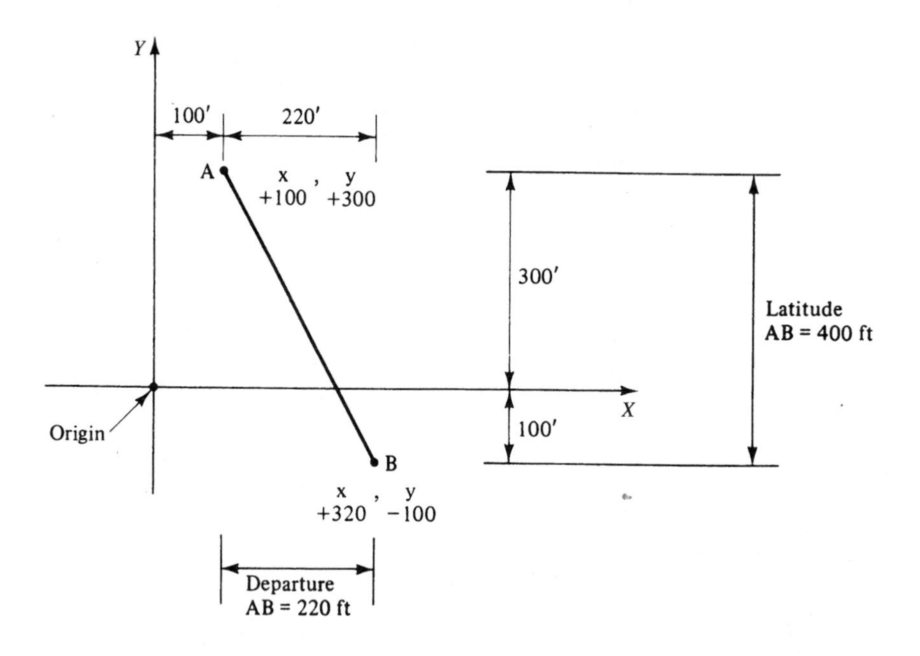

FIGURE 12-8 Rectangular coordinates.

Example 12-1

With reference to Figure 12-9, suppose that the coordinates have been assumed or calculated for point *A* at the left end of the line *AC*. It is desired to calculate the coordinates for point *C* at the other end.

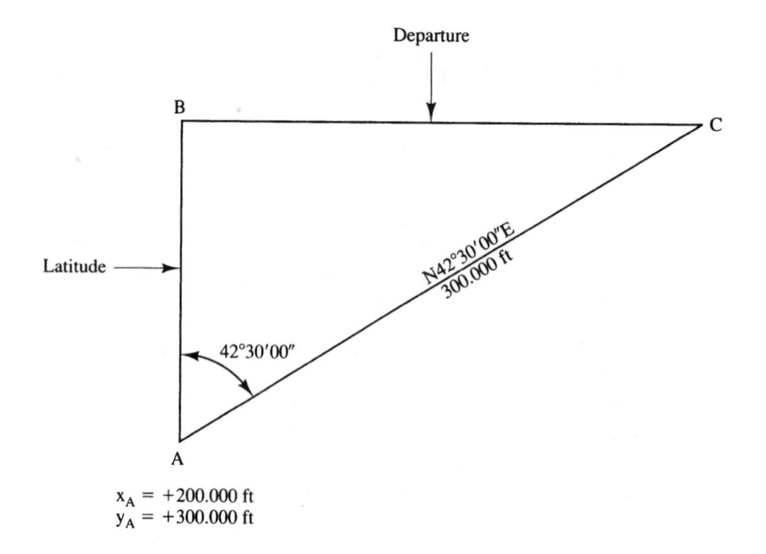

FIGURE 12-9 Computations for coordinates.

Solution

First the x or horizontal component of the line is calculated. It equals the departure of the line.

$$\text{Departure } BC = (\text{length } AC)(\sin 42°30'00'') = 202.678 \text{ ft}$$

Then the y or vertical component of the line is calculated. It equals the latitude of the line.

$$\text{Latitude } BC = (\text{length } AC)(\sin 42°30'00'') = 221.183 \text{ ft}$$

Finally, the coordinates of point C are as follows:

$$x_c = 200.000 + 202.678 = \mathbf{402.678 \text{ ft}}$$

$$y_c = 300.000 + 221.183 = \mathbf{521.183 \text{ ft}}$$

If the latitudes and departures of the sides of a traverse have been computed, a coordinate system can easily be established. For illustration the balanced latitudes and departures of Table 12-4 are used to compute the coordinates of the corners of the traverse considered earlier in the chapter.

A convenient location is selected for the origin. It may be placed at one of the corners of the traverse or at some other convenient point. It is often located at such a point that the entire survey will fall within the first or northeast quadrant. If this is the case, there will be no negative coordinates.

For the balanced latitudes and departures of Table 12-3, the most westerly point, E, is assumed to fall on the Y axis. As a result, the x coordinates of all the points will be positive. They are determined as follows:

$$E = 0.000$$

$$A = 0.000 + 159.998 = 159.998$$

$$B = 159.998 - 20.597 = 139.401$$

$$C = 139.401 + 86.606 = 226.007$$

$$D = 226.007 - 195.460 = 30.547$$

$$E = 30.547 - 30.547 = 0.000$$

The most southerly point, C, is assumed to fall on the X axis, so that all y coordinates are positive. They are computed as follows:

$$C = 0.000$$

$$D = 0.000 + 29.968 = 29.968$$

$$E = 29.968 + 139.081 = 169.049$$

$$A = 169.049 + 171.612 = 340.661$$

$$B = 340.661 - 188.385 = 152.276$$

$$C = 152.276 - 152.276 = 0.000$$

The x and y coordinates for each of the corners of the traverse are shown in Figure 12-10. The use of coordinates has become standard practice for many types of surveys. One of its earliest surveying applications was in mine surveys. Coordinates are also quite useful for plotting maps and for computing land areas.

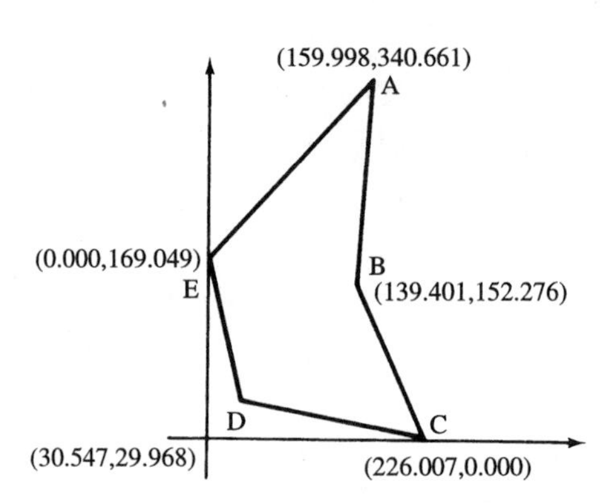

FIGURE 12-10 Coordinates for sample traverse.

12-11 AREAS COMPUTED BY COORDINATES

Another useful method for computing land areas is the method of coordinates. Today almost all area calculations are made with computers and the large percentage of the programs used are written using the coordinate method. Furthermore, with total station instruments commonly used for surveying, the coordinates of the corners of traverses are computed and displayed by the instruments. As a result it is quite easy to move from those values into the coordinate area calculation method. Some surveyors like this method better than the DMD method because they feel that there is less change of making mathematical mistakes. The amount of work is approximately the same for both methods. Instead of computing the DMDs for each side, the coordinates of each traverse corner are computed and then the coordinate rule is applied. This rule is derived in Figure 12-11 in much the same way that the DMD rule was derived in Section 12-8.

From this figure it can be seen that to determine the area of a traverse, each y coordinate is multiplied by the difference in the two adjacent x coordinates (using a consistent sign convention such as minus the following plus the preceding). The sum of these values is taken and the results equal two times the area. The math can be checked quickly by taking each x coordinate times the difference in the two adjacent y coordinates.

The coordinates of the corners of the example traverse were determined previously and shown in Figure 12-10. Using these values the coordinate rule is used to compute the area of the figure and the results shown below. The answer is exactly the same as that determined earlier with the DMD method.

$$2A = (0.000)(-340.661 + 29.007) + (159.998)(-152.276 + 169.049)$$

$$+ (139.401)(-0.000 + 340.661) + (226.007)\ (-29.968 + 152.276)$$

$$+ (30.547)(-169.049 + 0.000)$$

$$= 72.650$$

$$A = 36,325 \text{ sq ft} = 0.834 \text{ acre}$$

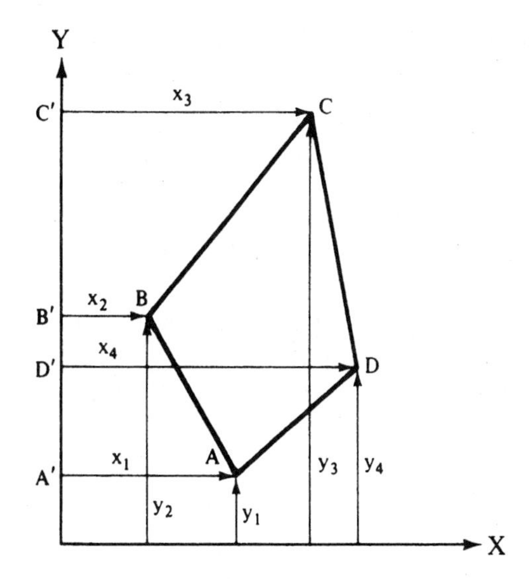

FIGURE 12-11 Derivation of coordinate rule.

Area *ABCD* = area *C'CDD'* + area *D'DAA'* – area *C'CBB'* – area *B'BAA'*

$$\text{Area ABCD} = (\tfrac{1}{2})(x_3 + x_4)(y_3 - y_4) + (\tfrac{1}{2})(x_4 + x_1)(y_4 - y_1) - (\tfrac{1}{2})(x_3 + x_2)(y_3 - y_2) - (\tfrac{1}{2})(x_2 + x_1)(y_2 - y_1)$$

Multiplying these values and rearranging the results yields

$$2 \text{ area} = y_1(-x_2 + x_4) + y_2(-x_3 + x_1) + y_3(-x_4 + x_2) + y_4(-x_1 + x_3)$$

12-12 ALTERNATIVE COORDINATE METHOD

There is available a very simple variation of the coordinate method for area computations which is a little easier to remember and apply. For this discussion reference is made to Figure 12-12, where the *x* and *y* coordinates of the corners of a traverse are shown. The formula that was presented in the last line of Figure 12-11 may be rewritten in the following form:

$$2A = x_1y_2 + x_2y_3 + x_3y_4 + x_4y_1 - y_1x_2 - y_2x_3 - y_3x_4 - y_4x_1$$

With this equation it is possible to quickly compute the area within a traverse by following the steps:

1. For each of the corners of the figure a fraction is written with *x* as the numerator and *y* as the denominator. These are listed on a horizontal line *and the fraction for the first or starting corner is repeated at the end of the line.* Next, a solid diagonal line is drawn from x_1 to y_2, from x_2 to y_3, and so on. Then a dashed diagonal line is drawn from y_1 to x_2, from y_2 to x_3, and so on.

$$\frac{x_1}{y_1} \quad \frac{x_2}{y_2} \quad \frac{x_3}{y_3} \quad \frac{x_4}{y_4} \quad \frac{x_1}{y_1}$$

2. The summation of the products of the coordinates joined by the solid lines minus the summation of the products of the coordinates joined by the dashed lines equals twice the area within the traverse. This is exactly a statement of the formula given earlier in this section.

$2A$ = summation of solid-line products
minus the summation of the dashed-line products

The area within the traverse of Figure 12-12 is determined by this alternative coordinate method as follows:

$$\frac{100}{400} \times \frac{600}{800} \times \frac{900}{300} \times \frac{500}{100} \times \frac{100}{400}$$

$$2A = (100)(800) + (600)(300) + (900)(100) + (500)(400) - (400)(600)$$
$$- (800)(900) - (300)(500) - (100)(100) = -570{,}000$$
$$A = 285{,}000 \text{ sq ft}$$

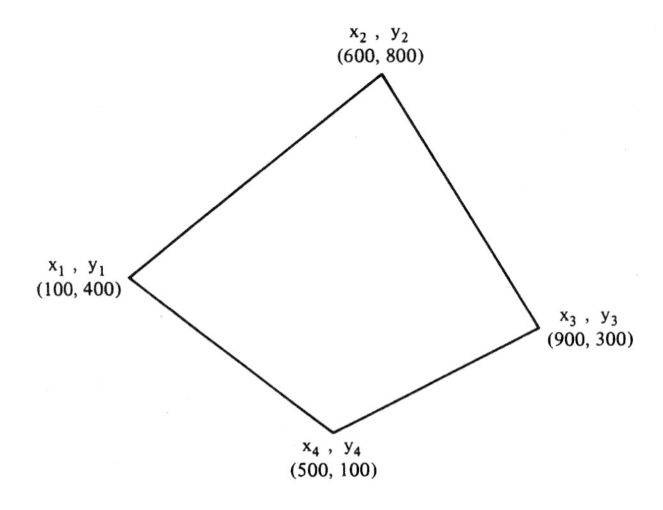

FIGURE 12-12 Alternative coordinate method.

12-13 AREAS WITHIN IRREGULAR BOUNDARIES

Very often, property boundaries are represented by irregular lines, for example, the center line of a creek or the edge or center line of a curving road. For such cases as these, it is often not feasible to run the traverse along the exact boundary line. Instead, it may be practical to run it a convenient distance from the boundary and locate the position of the boundary by measuring offset distances from the traverse line, as shown in Figure 12-13. The offsets may be taken at regular intervals if the boundary does not change direction suddenly, but when it does change suddenly, offsets are taken at irregular intervals, as shown by *ab* and *cd* in the figure.

The area inside the closed traverse may be computed by one of the methods described previously, and the area between the traverse line and the irregular boundary may be determined separately and added to the other value. If the land in question between the traverse line and the irregular boundary is carefully plotted to scale, the area may be determined satisfactorily with a planimeter. Other methods commonly used are the trapezoidal rule, Simpson's one-third rule, and another coordinate method involving offsets. All of these methods are described in this section, but it should be recognized that they probably do not yield results that are any more satisfactory than those obtained with a planimeter, because of the irregular nature of the border between the measured offsets.

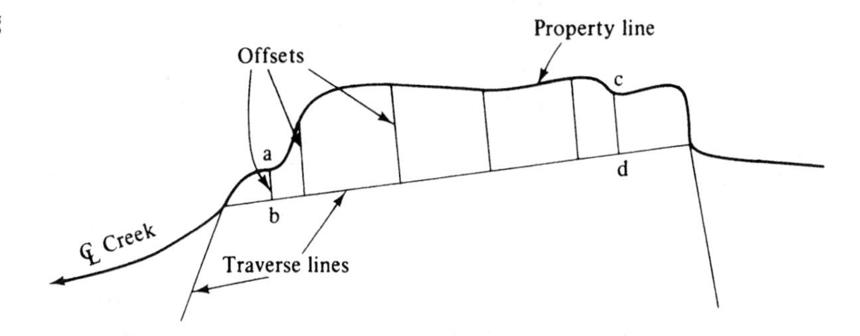

FIGURE 12-13 Locating boundary line with offsets.

Planimeter

A polar *planimeter* (see Figure 12-14) is a device that can be used to measure the area of a figure on paper by tracing the boundary of the figure with a tracing point. As the point is moved over the figure the area within is mechanically integrated and recorded on a drum and disk. An excellent mathematical proof of the workings of the planimeter is presented in the book by Davis, Foote, and Kelly.[3] When a planimeter is used to determine an area, it is not necessary to compute latitudes,

FIGURE 12-14 Tamaya digital planimeter. (Courtesy of the Leitz Company.)

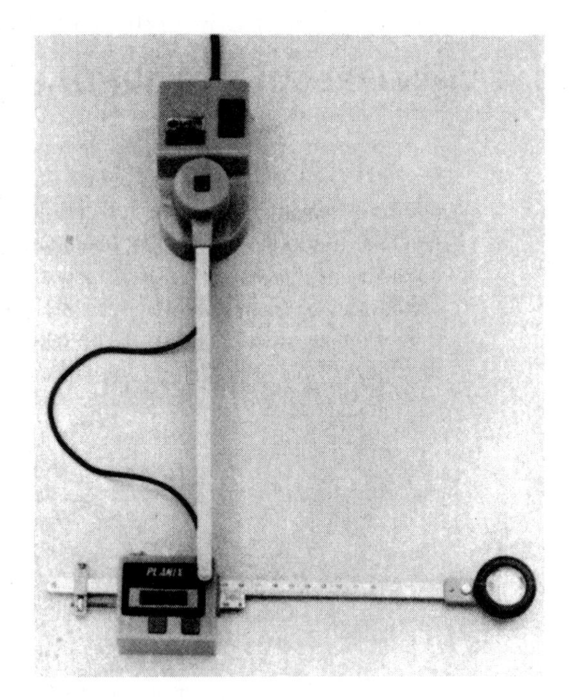

[3]R. E. Davis, F. S. Foote, and J. W. Kelly, *Surveying Theory and Practice,* 5th ed. (New York: McGraw-Hill Book Company, 1966), pp. 67–69.

departures, or coordinates. It is not even necessary to have figures consisting of straight lines as is required with the DMD and coordinate methods. It is, however, necessary to plot the figures carefully to scale before the planimeter is used.

The planimeter is particularly useful for measuring the areas of irregular pieces of land as well as the areas of cross sections. If the operator is careful, he or she can obtain results within 1% or better, depending on the accuracy of the plotted figures, the types of paper used, and the carefulness with which the figures are traced.

The major parts of a planimeter are the tracing point, the anchor arm with its weight and post, the scale bar, and the graduated drum and disk. When an area is to be traced, the drawing is stretched out flat so that there are no wrinkles and the anchor point is pushed down into the paper at a convenient location so that the operator can trace all or a large part of the area desired at that one location. The tracing point is set at a distinct or marked point on the drawing and the graduated drum is read (or set to zero). Then the perimeter is carefully traced until the tracing point is returned to the starting point, at which time a final drum reading is taken. In traversing around the figure the operator must be very careful to note the number of times the drum reading passes zero. Counters are available with some planimeters with which the number of revolutions is automatically recorded. The initial reading is subtracted from the final reading and represents to a certain scale the area within the figure. If the perimeter is tracked clockwise, the final reading will be larger than the initial reading, but it will be smaller if traversed counterclockwise.

If the anchor point is placed outside the area to be traversed, the area of the figure will equal

$$A = Cn$$

where C is a constant and n is the difference from the initial to the final readings on the drum. The constant C is usually given on the top of the tracing arm or on the instrument box. It is equal to 10.00 sq in. for many planimeters. If the user is not sure of C, he or she can easily construct a figure of known area (e.g., a 5 in. by 5 in. square), run the tracing point around the area, and determine what C would have to be to give the correct area when multiplied by the net reading on the drum. For some instruments the value of C is given in SI units.

If the anchor point is placed inside the area to be traversed (as it will often be for large areas), it will be necessary to make a correction to the area computed. It is possible to hold the tracing arm in such a position that the tracing point can be moved completely around 360° without changing the drum reading. The area of this circle, called the *zero circle* or the *circle of correction,* must be added to Cn if a figure is traversed in a clockwise direction with the anchor point on the inside. It will be noted that if the area of the figure is less than that of the zero circle, the change in the drum reading will be minus for a clockwise traverse. Example 12-2 illustrates the use of the planimeter for determining areas of figures in a drawing.

EXAMPLE 12-2

Upon calibration it was found that a given planimeter traversed 10 sq in. for each revolution of its drum.

(a) If a given map area is traversed with the anchor point outside the area and a net reading of 16.242 revolutions is obtained, what is the map area traversed?

(b) If the same area is traversed with the anchor point located inside the area and the net reading is 8.346 revolutions, what is the area of the zero circle?

(c) Another map area is traversed with the anchor point placed inside the area and a net reading of 23.628 revolutions is obtained. If this area were plotted on the map with a scale of 1 in. = 20 ft, what would be the actual area on the ground?

Solution

(a) Area = (10)(16.242) = **162.42 sq in.** on the map

(b) Area of zero circle = (10)(16.242 - 8.346) = **78.96 sq in.**

(c) Area on map = (10)(23.628) + 78.96 = **315.24 sq in.**
Area on ground = (20 × 20)(315.24) = **126,096 sq ft**

Electronic polar planimeters are available that provide digital readouts in large, bright numbers. These devices are simple to read and may easily be set to zero. They may also be used to handle cumulative adding and subtracting of areas. When using a planimeter it is well to measure each area several times and average the results. Modern planimeters will usually compute the average for you.

Trapezoidal Rule

When the offsets are fairly close together, an assumption that the boundary is straight between those offsets is satisfactory and the trapezoidal rule may be applied. With reference to Figure 12-15, the offsets are assumed to be located at regular intervals and the area inside the figure equals the areas of the enclosed trapezoids, or

$$A = d\left(\frac{h_1 + h_2}{2}\right) + d\left(\frac{h_2 + h_3}{2}\right) + \bullet\bullet\bullet + d\left(\frac{h_{n-1} + h_n}{2}\right)$$

from which

$$A = d\left(\frac{h_1 + h_n}{2} + h_2 + h_3 + \bullet\bullet\bullet + h_{n-1}\right)$$

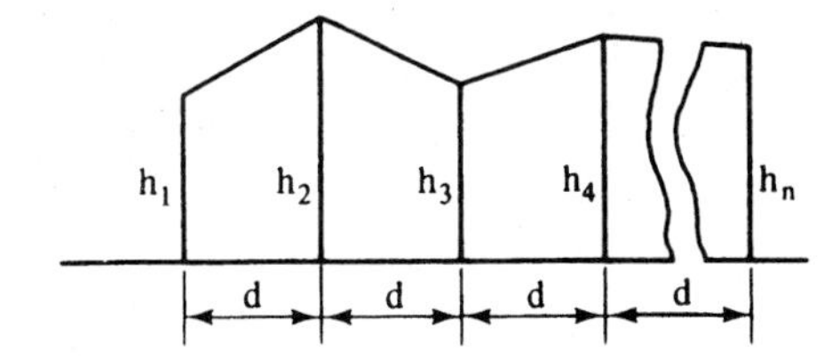

FIGURE 12-15 Figure used for developing the trapezoidal rule.

Simpson's One-Third Rule

If the boundaries are found to be curved, Simpson's one-third rule (which is based on the assumption that the boundary lines are parabolic in shape) is considered better to use than the trapezoidal rule. Again, it is assumed that the offsets are evenly spaced. *The rule is applicable to areas that have an odd number of offsets. If there is an even number of offsets, the area of all but the part between the last two offsets (or the first two) may be determined with the rule.* That remaining area is determined separately, ordinarily assuming it to be a trapezoid.

With reference to Figure 12-16, Simpson's rule is written as follows:

$$A = (2d)\left(\frac{h_1 + h_3}{2}\right) + \left(\frac{2}{3}\right)(2d)\left(h_2 - \frac{h_1 + h_3}{2}\right) + (2d)\left(\frac{h_3 + h_5}{2}\right)$$
$$+ \left(\frac{2}{3}\right)(2d)\left(h_4 - \frac{h_3 + h_5}{2}\right), etc.$$

This reduces to

$$A = \frac{d}{3}\left[(h_1 + h_n)\right] + 2(h_3 + h_5 + \bullet\bullet\bullet + h_{n-2}) + 4(h_2 + h_4 + \bullet\bullet\bullet + h_{n-1})$$

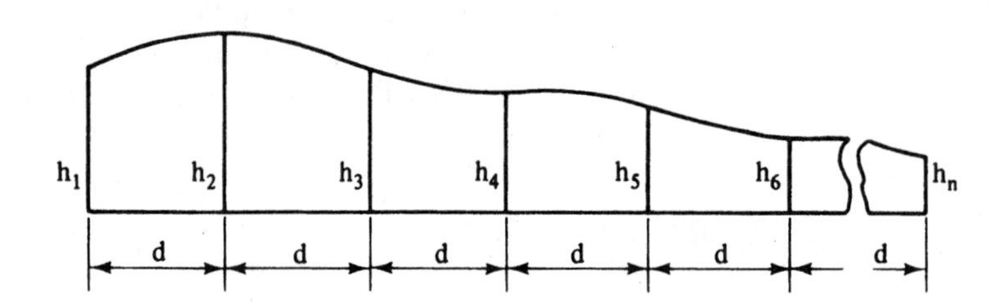

FIGURE 12-16 Figure used for developing Simpson's one-third rule.

The trapezoidal rule and Simpson's one-third rule are used in Example 12-3 to compute the irregular land area shown in Figure 12-17.

EXAMPLE 12-3

Compute the area of the land shown in Figure 12-17:

(a) Using the trapezoidal rule.

(b) Using Simpson's one-third rule.

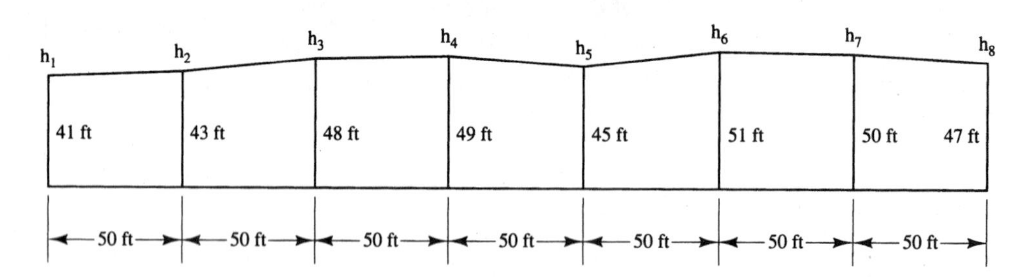

FIGURE 12-17 Area by trapezoidal rule and Simpson's one-third rule.

Solution

(a) Area by trapezoidal rule:

$$A = 50\left(\frac{41 + 47}{2} + 43 + 48 + 49 + 45 + 51 + 50\right) = \mathbf{16,500\ sq\ ft}$$

(b) Area by Simpson's one-third rule: Since there are an even number of offsets the area between the last two offsets on the right end is computed separately as a trapezoid.

$$A = \frac{50}{3}\left[\left(41 + 50\right) + 2\left(48 + 45\right) + 4\left(43 + 49 + 51\right)\right] + \left(\frac{50 + 47}{2}\right)\left(50\right)$$

$$= \mathbf{16,575\ sq\ ft}$$

Coordinate Rule for Irregular Areas

When offsets are taken at irregular intervals, the area of each figure between pairs of adjacent offsets may be computed and the values totaled. In addition, the planimeter method is particularly satisfactory here. There are other methods of handling the problem, for example, *the coordinate rule for irregular spacing of offsets, which says that twice the area is obtained if each offset is multiplied by the distance to the preceding offset plus the distance to the following offset.* It will be noted that for the end offsets the same rule is followed, but there will be only one distance between offsets because the outside one does not exist. An application of this coordinate rule is given in Example 12-4.

EXAMPLE 12-4

Using the coordinate rule for irregular areas, determine the area contained in the irregular tract shown in Figure 12-18.

Solution

$$2A = \left(32\right)\left(50\right) + \left(46\right)\left(50 + 30\right) + \left(38\right)\left(30 + 40\right) + \left(40\right)\left(40 + 20\right) + \left(39\right)\left(20 + 30\right)$$
$$+ \left(36\right)\left(30 + 30\right) + \left(33\right)\left(30\right)$$

$$A = \mathbf{7720\ sq\ ft}$$

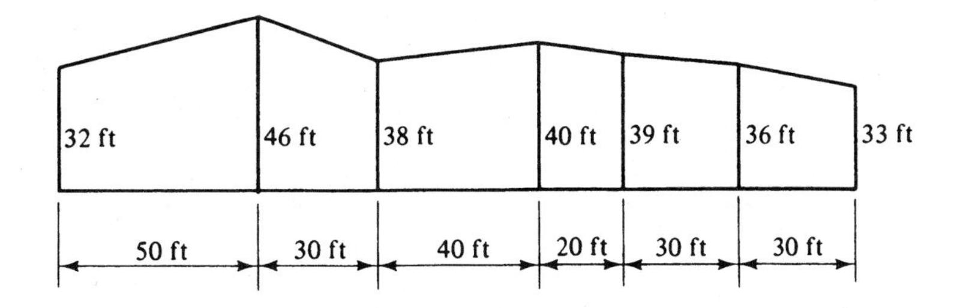

FIGURE 12-18 Area by coordinates.

Area of Segment of Circle

If we have a tract of land that has a circular horizontal curve for one of its boundaries (as is often the case where the tract is adjacent to a highway or railroad), we can compute the area of the land in question with little trouble, as described herein. For this discussion the traverse *ABCDEFA* shown in Figure 12-19 is considered. One method of determining the total area within the figure is to separate the traverse into two parts: *ABCDEA* and the segment of the circle *AEFA.* In the figure, *X* is the center of the circle, *R* is the radius whose arc is *EFA,* and *I* is the angle between the two radii shown.

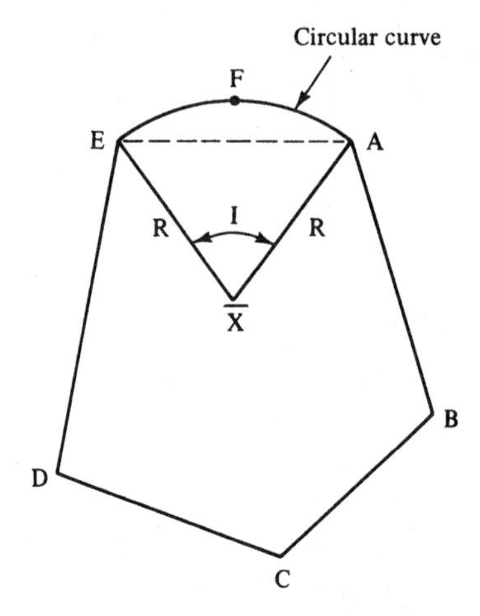

FIGURE 12-19 Area of segment of a circle.

The area *ABCDEA* can be determined by one of the methods previously described and added to the area of the segment of the curve:

$$\text{area of segment} = \text{area } A\overline{X}EFA - \text{area } A\overline{X}EA = R^2\left(\frac{\pi I^\circ}{360^\circ} - \frac{\sin I}{2}\right)$$

An alternative solution is to calculate the area $ABCDE\overline{X}A$ and add it to the area *AXEFA* calculated as follows:

$$\text{area } A\overline{X}EFA = \frac{I^\circ}{360^\circ}\left(\pi R^2\right)$$

Problems

In Problems 12-1 to 12-4, compute the latitudes and departures for the sides of the traverses shown in the accompanying figures. Determine the error of closure and precision for each of the traverses.

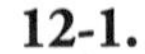

12-1. (Ans.: $E_C = 0.133$ ft, precision = 1/6432)

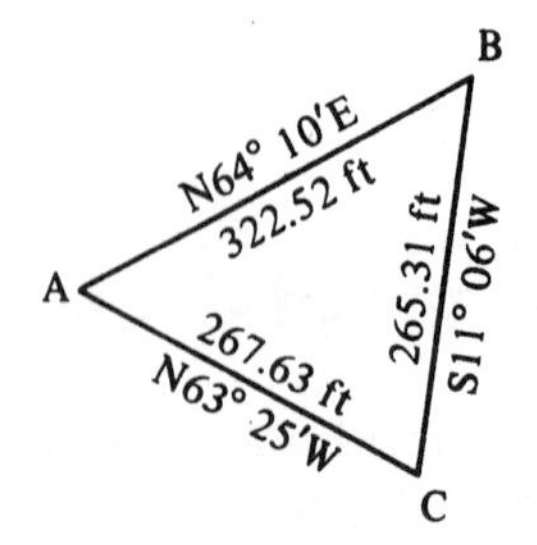

12-2.

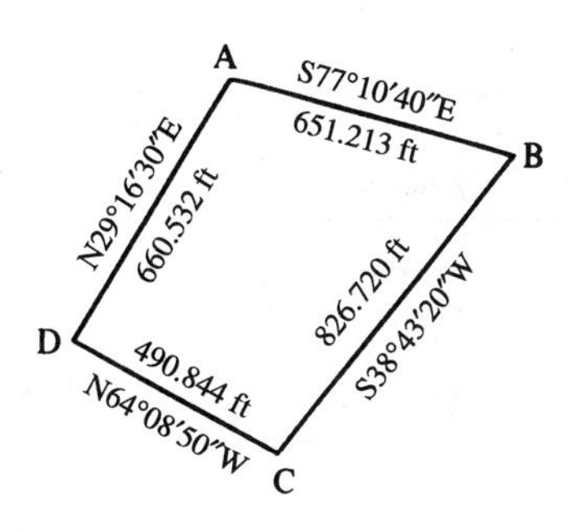

12-3. (Ans: E_C = 1.44 ft, precision = 1/4411)

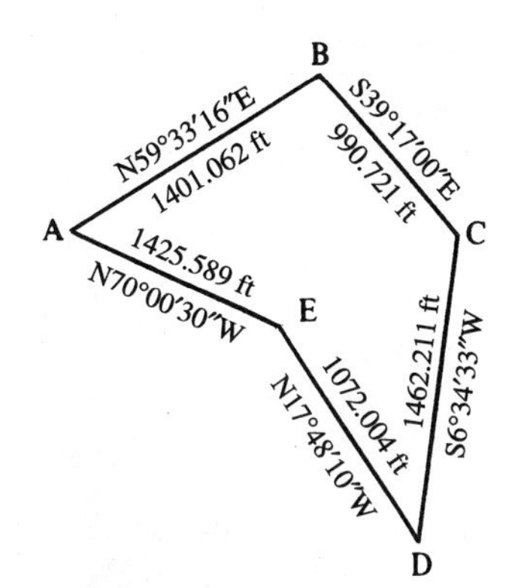

12-4.

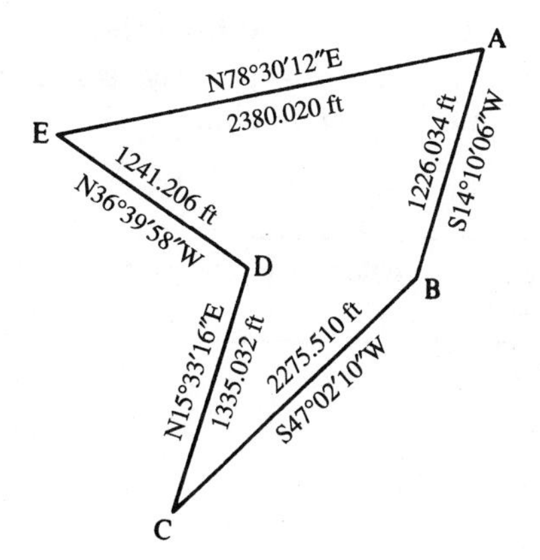

In Problems 12-5 to 12-7, balance each of the sets of latitudes and departures given by the compass rule and give the results to the nearest 0.01 ft.

12-5.

		Latitudes		Departures	
Side	**Length**	**N**	**S**	**E**	**W**
AB	400.00	320.00		245.00	
BC	300.00		180.00	235.36	
CA	500.00		140.24		480.00

(Ans.: 320.08; 179.94; 140.14; 244.88; 235.27; 480.15)

12-6.

Side	Length	Latitudes N	S	Departures E	W
AB	600.00	450.00		339.00	
BC	450.00		285.00	259.50	
CA	750.00		164.46		599.22

12-7.

Side	Length	Latitudes N	S	Departures E	W
AB	220.40	185.99		118.26	
BC	287.10		234.94	165.02	
CD	277.20		181.25		209.73
DE	200.10	187.99		68.55	
EA	147.90	42.01			141.81

(Ans.: *AB* 186.03, 118.20; *EA* 42.04, 141.85)

In Problems 12-8 to 12-10 balance by the compass rule the latitudes and departures computed for each of the traverses of Problems 12-1 to 12-3.

(Ans.: Problem 12-9 for *AB* 144.692, 635.196, for *CD* 645.216, 516.869)

In Problems 12-11 to 12-13, from the given sets of balanced latitudes and departures, calculate the areas of the traverses in acres using DMDs with the meridians through the most westerly points.

12-11.

Side	Balanced Latitudes N	S	Balanced Departures E	W
AB	600		200	
BC	100		400	
CD	0	0	100	
DE		400		300
EA		300		400

(Ans.: 4.13 acres)

12-12.

Side	Balanced Latitudes N	S	Balanced Departures E	W
AB	100			200
BC	100		200	
CD	200		100	
DE		700	100	
EA	300			200

12-13.

Side	Balanced Latitudes N	S	Balanced Departures E	W
AB	200		100	
BC	100		200	
CD		150	150	
DE		50		200
EA		100		250

(Ans.: 1.29 acres)

For Problems 12-14 to 12-17, with the latitudes and departures balanced with the compass rule, compute the areas in acres using DMDs with the meridians through the most westerly points.

12-14. Problem 12-1

12-15. Problem 12-2 (Ans.: 9.28 acres)

12-16. Problem 12-3

12-17. Problem 12-4 (Ans.: 65.5 acres)

12-18. Repeat Problem 12-11 with the meridian passing through the most easterly point.

12-19. Repeat Problem 12-12 using double parallel distances (DPDs) with the reference parallel passing through the most northerly point.
(Ans.: 1.95 acres)

12-20. Repeat Problem 12-13 using DPDs with the reference parallel passing through the most southerly point.

In Problems 12-21 and 12-22, for the given coordinates compute the length and bearing of each side.

12-21.

Point	*x*(ft)	*y*(ft)
A	0	0
B	+300	+200
C	+500	−100
D	+400	−300

(Ans.: $AB = 360.555$ ft, N56°18′36″E)

12-22.

Point	*x*(m)	*y*(m)
A	+200	+300
B	+350	−400
C	+150	0

In Problems 12-23 to 12-25, compute the area in acres by the method of coordinates for each of the traverses whose corners have the coordinates given.

12-23.

Point	x(ft)	y(ft)
A	+100	+300
B	+500	–100
C	+200	–400
D	–200	–250
E	–150	+200

(Ans.: 7.03 acres)

12-24.

Point	x(m)	y(m)
A	0	0
B	+300	+150
C	+600	–300
D	+200	–250
E	+100	–300

12-25.

Point	x(ft)	y(ft)
A	+150	+250
B	+350	+400
C	+400	–250
D	–200	–500
E	–300	–300
F	+200	–100

(Ans.: 5.28 acres)

12-26. Repeat Problem 12-11 using the coordinate method.

12-27. Repeat Problem 12-12 using the coordinate method. (Ans.: 1.95 acres)

12-28. Repeat Problem 12-13 using the coordinate method.

12-29. Determine the area of the traverse of Problem 12-21 in acres using the coordinate method. (Ans.: 2.75 acres)

12-30. Using the coordinate method compute the area in hectares (1 ha = 10,000 m^2) for the traverse of Problem 12-22.

12-31. A given planimeter has a constant of 2; that is, one revolution of the drum equals 2 sq in.

(a) If a map area is traversed with the anchor point outside the area and a net reading of 18.247 revolutions is obtained, what is the map area? (Ans.: 36.494 sq in.)

(b) If the same area is traversed, with the anchor point inside the area and the net reading is 7.695 revolutions, what is the area of the zero circle? (Ans.: 21.104 sq in.)

12-32. Upon calibration, it was found that a given planimeter traversed 10 sq in. for every revolution of its drum and the area of the zero circle was 72.00 sq in. With the anchor point inside the area the reading before traversing an unknown area 5.162 revolutions and after traversing the area was 9.034 revolutions.

(a) What is the area in square inches?

(b) If the map scale is 1 in. = 50 ft, what is the area on the ground?

12-33. A given planimeter has a constant of 8. A map area is traversed with the anchor point outside the area and a net reading of 15.896 revolutions is obtained. The same area is traversed with the anchor point inside the area and a net reading of 8.124 revolutions is obtained. The scale of the map on which the area is being measured is 1 in. = 50 ft. Determine the following:

(a) Map area traversed. (Ans.: 127.168 sq in.)

(b) Area of the zero circle. (Ans. 62.176 sq in.)

(c) Actual ground area. (Ans. 7.30 acres)

In Problems 12-34 and 12-35, compute the area (in square feet) of the irregular tracts of land shown using the trapezoidal rule.

12-34.

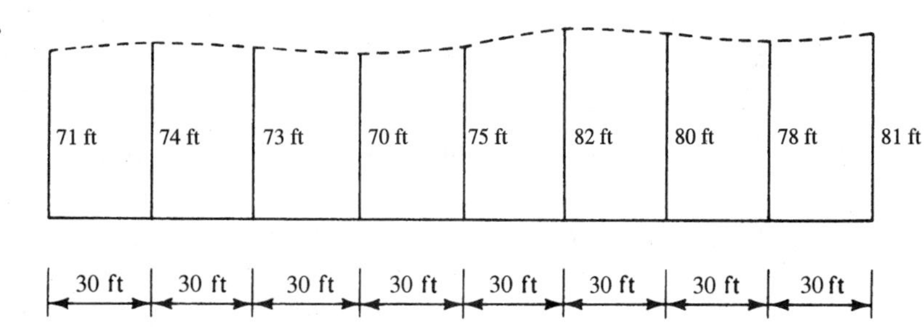

12-35.

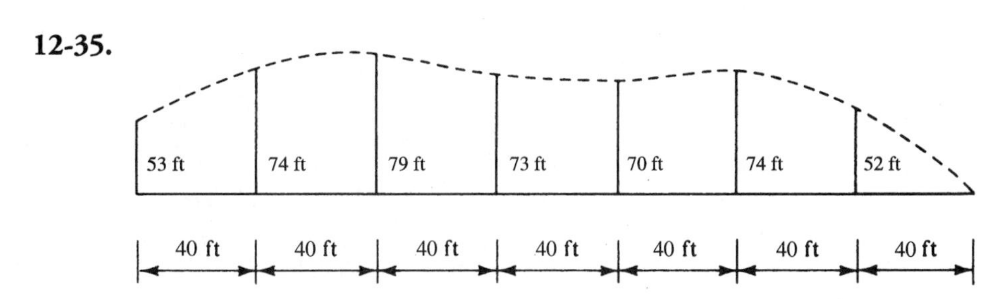

(Ans.: 17,940 sq ft)

12-36. Repeat Problem 12-34 using Simpson's one-third rule.

12-37. Repeat Problem 12-35 using Simpson's one-third rule.

(Ans.: 18,200 or 18,167 sq ft)

12-38. To determine the area between a base line *AB* and the edge of a lake the following offset distances were measured at 40-ft intervals. Compute the area (in square feet) of the tract by using Simpson's one-third rule. Offset distances: 32 ft, 41 ft, 49 ft, 60 ft, 73 ft, 68 ft, 60 ft, 55 ft, 41 ft, 37 ft.

For Problems 12-39 to 12-41, plot to the scale of 1 in. = 50 ft the referenced information and determine the areas (in square feet) using a planimeter.

12-39. Problem 12-34. (Ans.: 18,300 sq ft)

12-40. Problem 12-35.

12-41. Problem 12-38. (Ans.: 19,200 sq ft)

For Problems 12-42 to 12-44, the offsets are taken at irregular intervals. Determine the area (in square feet) between the traverse line and the boundary for each case. Use the coordinate method. The distances given are measured from the origin.

Problem 12-42.		Problem 12-43.		Problem 12-44.	
Distance	Offset	Distance	Offset	Distance	Offset
0	30	0	18	0	17
30	50	20	22	10	30
50	35	40	28	20	45
80	45	60	34	30	56
90	58	70	46	50	50
100	50	75	58	100	62
120	68	80	64	150	71
150	60	85	66	200	78
		100	60	210	63
		120	55	230	52
				250	50

(Ans.: 4905 sq ft)

12-45. Determine the area of the figure shown below. (Ans.: 285,882 sq ft)

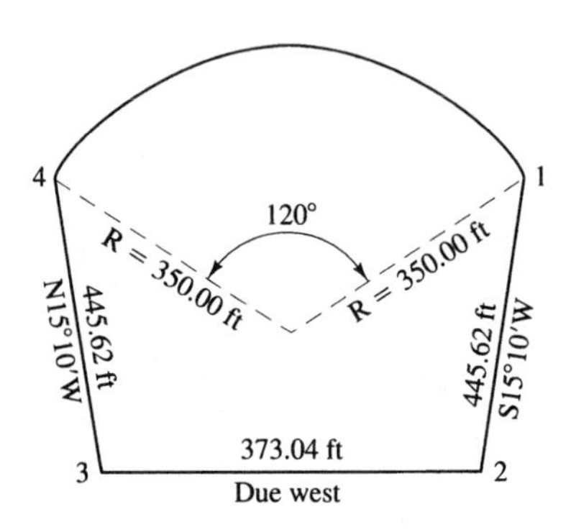

CHAPTER 13

Computer Calculations and Omitted Measurements

13–1 COMPUTERS

The types of calculations presented in the preceding chapter and in this one are almost "everyday" problems for the practicing surveyor. Until several decades ago these calculations were made laboriously using trigonometric tables and logarithms. As the years passed, large mechanical calculators and then smaller and smaller calculators were developed with which the numbers were handled. Since the 1960s, however, handheld calculators and digital computers have made the other equipment obsolete. The latter devices, which have built-in trigonometric tables, have appreciably simplified the computational work of the surveyor, both in the office and in the field.

In field work the surveyor often makes some measurements and then has to stop and make what are sometimes rather involved calculations (frequently of the types discussed in Chapter 12) before he or she can proceed with more field work. A great advantage of small programmable calculators (and small portable computers) is that they may be used conveniently in the field. Thus the necessary calculations can be made quickly without the delays and inconvenience involved in returning to the office, making calculations, and returning to the field.

13-2 PROGRAMS

There are today an increasing number of software programs that the surveyor may use. To prepare a list of all the companies having such programs available is an almost impossible task and the list would be obsolete before this chapter could be printed. It is sufficient to say that a surveyor in any part of the United States can find suitable programs from one organization or another for almost any problem that he or she faces.

Regardless of where a program is obtained, almost any surveyor would want to take a short problem or two that he or she has previously solved and check out the correctness of the program. There are few surveyors who would be willing to stake their professional reputations blindly on the claims made by the representative of a software company or other organization without making a preliminary check.

Of particular interest to the surveyor is the fact that many computer programs are available that will not only make the computations necessary for precision, balancing, areas, and so on, but will also print out a drawing or plat of the traverse and show the directions and lengths of its sides. Other programs are available that subdivide property into lots according to given limitations and print out a drawing of the results.

Computers and calculators are only "as good" as the person using them. The surveyor/technician must have a good understanding of the principles behind the programs being used. Furthermore, he or she must be capable of looking at the output and detecting gross mistakes that may be the result of incorrect key strokes, etc.

13-3 APPLICATION OF THE COMPUTER PROGRAM SURVEY

In this section the reader is introduced to a set of computer programs called SURVEY, which were prepared for use on Windows-based computers. These programs, which are written in Microsoft® Windows 95 using Visual Basic 5.0, are furnished on the CD-ROM enclosed with this book.

Although an interested user may obtain a listing of the various steps involved in the programs, SURVEY is designed to operate as a "black box" that is, the user need only supply the specified input data and the software will perform the calculations and supply the appropriate results automatically. *Even though the reader may have had little or even no previous computer experience, he or she can learn to apply these programs in a very short time (minutes, not hours).*

In this section and the next, the very simple bits of information needed to apply SURVEY to a problem of the type handled in Chapter 12 (precision, balancing, and area calculations) are presented. Later in this chapter the same CD-ROM is used for an omitted-line problem, while in Chapters 20 and 21, computer calculations are made for horizontal and vertical curves. Although SURVEY is very easy to use and the screen prompts are quite clear as to the steps to be taken, the procedure is described in some detail in the next few paragraphs.

The user will need to put the CD-ROM into the CD-drive, use Explorer or My Computer to find the file "setup.exe" on the CD-ROM, and double-click that file to start the Set-up Wizard. The Wizard will install a file called "survey.exe." in a folder (by default it goes into C:\Program Files\Survey, but the location can be changed during set-up). Once installed, use Explorer or My Computer to locate the file and double-click to start the program.

13-4 COMPUTER EXAMPLE

Later in this section we present an example that illustrates the reworking of the example traverse problem of Chapter 12 using SURVEY. It is assumed that any error in the angles have been balanced throughout the traverse as described in Section 12-4 and the azimuth or bearing of each side determined.

The user will input the number of sides of the traverse and then will select coordinates for the starting point. Any set of values may be used unless coordinates have been determined from previous work, you may like to input large positive numbers so that all points will have positive values. (Should you not type in values for coordinates, the values $x = +10{,}000$ and $y = +10{,}000$ will be used automatically.)

The user is asked to specify the length units in feet or meters by inputting the letter f or m. If feet are used, the resulting area will be given in square feet and in acres (1 acre = 43,560 square feet). If meters are specified, the computed area will be in square meters and in hectares (1 hectare = 10,000 square meters).

After the errors in latitudes and departures are computed, they are balanced with the compass or Bowdich rule. (Many computer programs use the somewhat more accurate method of least squares for balancing these errors. The author, however, uses the compass rule in the program so that the results will be consistent with the calculations presented in Chapter 12.)

Example 13-1

Using the program SURVEY, determine the errors of closure and the precision for the traverse of Figure 13-1. Then balance the closure errors and determine the area of the figure. In addition it is desired to determine the coordinates of the corners of the traverse from the balanced latitudes and departures as well as the balanced lengths and directions of the sides. This is the same traverse that we considered in Chapter 12.

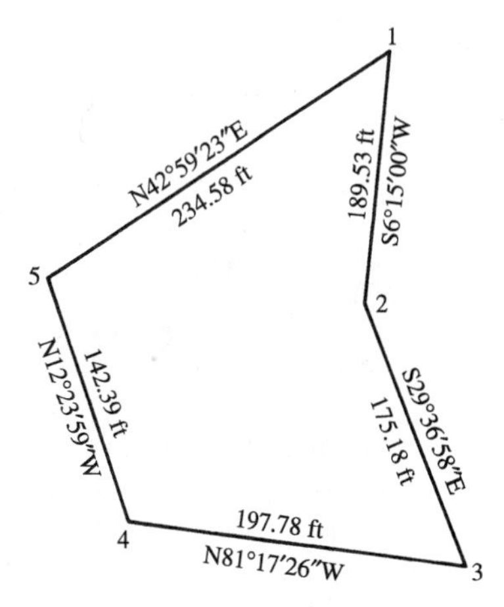

FIGURE 13-1 Traverse used for computer program.

Solution

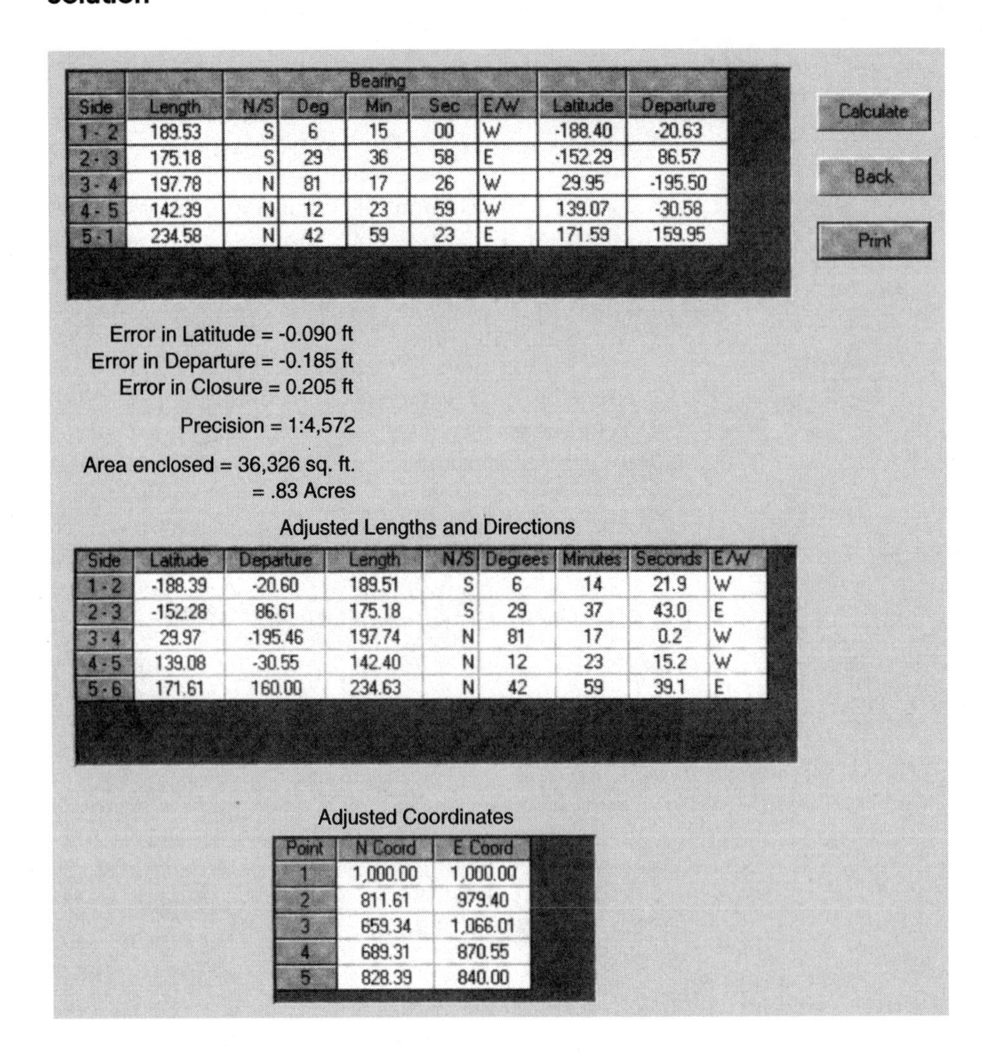

Side	Length	Bearing N/S	Deg	Min	Sec	E/W	Latitude	Departure
1 - 2	189.53	S	6	15	00	W	-188.40	-20.63
2 - 3	175.18	S	29	36	58	E	-152.29	86.57
3 - 4	197.78	N	81	17	26	W	29.95	-195.50
4 - 5	142.39	N	12	23	59	W	139.07	-30.58
5 - 1	234.58	N	42	59	23	E	171.59	159.95

Error in Latitude = -0.090 ft
Error in Departure = -0.185 ft
Error in Closure = 0.205 ft

Precision = 1:4,572

Area enclosed = 36,326 sq. ft.
= .83 Acres

Adjusted Lengths and Directions

Side	Latitude	Departure	Length	N/S	Degrees	Minutes	Seconds	E/W
1 - 2	-188.39	-20.60	189.51	S	6	14	21.9	W
2 - 3	-152.28	86.61	175.18	S	29	37	43.0	E
3 - 4	29.97	-195.46	197.74	N	81	17	0.2	W
4 - 5	139.08	-30.55	142.40	N	12	23	15.2	W
5 - 6	171.61	160.00	234.63	N	42	59	39.1	E

Adjusted Coordinates

Point	N Coord	E Coord
1	1,000.00	1,000.00
2	811.61	979.40
3	659.34	1,066.01
4	689.31	870.55
5	828.39	840.00

FIGURE 13-2
SURVEY printout.

13-5 A POTENTIAL WARNING: DANGER IN COMPUTER USE

The reader can almost instantly become an "office surveying expert" by studying the few paragraphs contained herein which pertain to the CD-ROM SURVEY and by using the program to solve a few of the example problems. This person will then be able to handle many of the problems presented in the text, whether or not he or she knows anything about the material. *To the author, this is a potentially dangerous situation.*

The danger is that inexperienced persons will use programs such as SURVEY to handle various problems with which they have little or no experience. Computers by themselves do not extend a person's knowledge of surveying. Programs such as

SURVEY will provide correct answers *if the correct input is used.* But if poor data are supplied, will the user be able to see if the results are reasonable?

In practice, it is absolutely necessary that an experienced surveyor review computer work for suspicious results. Ultimately, the surveyor is responsible for his or her work, regardless of what kind of software is used.

Concerning the use of the programs on the enclosed CD-ROM, the author would like to make the usual disclaimer: *The reader may use the SURVEY programs in any way that he or she sees fit, but neither the author nor the publisher take responsibility for any difficulties arising as a result.* It is almost impossible for a programmer to predict all the ways that users will attempt to apply their programs. As a result, the author would be grateful to receive any comments, criticisms, or suggestions from the users of SURVEY.

13-6 OMITTED MEASUREMENTS

Occasionally, one or more angles and/or lengths are not measured in the field and their values are computed later in the office. There could be several reasons for not completing the field measurements, such as difficult terrain, obstacles, hostile landowners, lack of time, sudden severe weather conditions, and so on.

Should all the measurements for a closed traverse be completed and an acceptable precision obtained, it is perfectly permissible to compute bearings and distances within that traverse. For instance, in Figure 13-3 it is assumed that a group of lots (1 to 4) are being laid out in the field. The perimeter of the lots, represented by the solid lines, has been run with an acceptable precision and the errors of closure balanced. For such a case it is perfectly permissible to compute the lengths and bearings of the missing interior lot lines, which are shown dashed in the figure. Furthermore, a great deal of time may be saved where there is an appreciable amount of underbrush along the lines and/or where the ends of the lines are not visible from each other due to hills, trees, and so on.

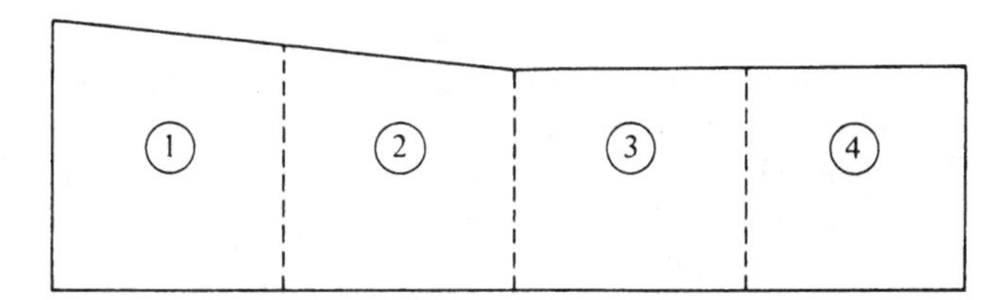

FIGURE 13-3 Computing lengths and bearings of dashed lot lines.

The omission of some of the measurements for one or more sides of a closed traverse is a very undesirable situation. Even though it is possible to calculate the values of up to two missing lengths or two missing bearings (that is the same as three angles missing) or one length and one bearing (two angles), the situation should be avoided. The trouble with such calculations is that there is no way to calculate the precision of the field measurements that were made, since the figure was not closed. The measurements which were taken are assumed to be "perfect" in order that the missing quantities may be calculated. As a result, severe blunders can be made in the field, causing the computed values to be meaningless.

From this discussion it is evident that if the measurements of any lengths or angles that are part of a closed traverse are omitted in the field, it is advisable and almost essential to use some kinds of approximate checks on the computed values. Such things as distance readings, angle measurements or compass bearings, and even eyeball estimates can be critical to the success of a survey. The most commonly omitted measurement problem is the one where the length and bearing of one side are missing. An example problem of this type is presented in the following section.

13-7 LENGTH AND BEARING OF ONE SIDE MISSING

For the traverse of Figure 13-4, the lengths and bearings of sides *AB, BC,* and *CD* are known but the length and bearing of side *DA* are unknown. This is the same thing as saying that the angles at *D* and *A* are missing as well as the length of *DA*. From the assumption that the measurements for the three known sides are perfect, it is easily possible to compute the missing information for side *DA*. The problem is exactly the same as the one handled in Table 12-2, where the error of closure of a traverse was determined.

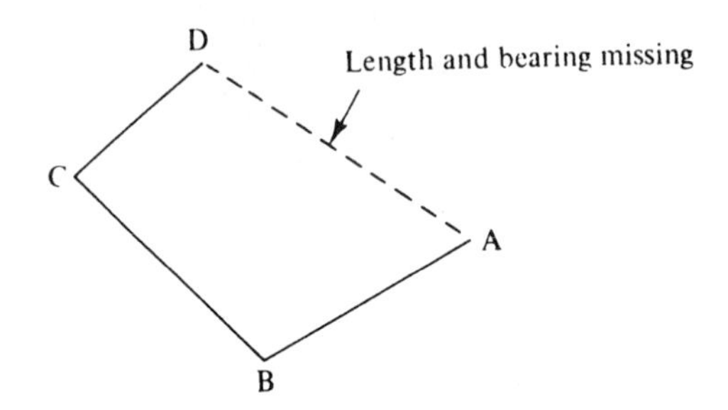

FIGURE 13-4 Length and bearing of one side missing.

The latitudes and departures of the three known sides are computed, summed, and the "error of closure" represented by the dashed line *DA* in the figure is determined. This is the length of the missing side; the tangent of its bearing angle equals the horizontal component of its distance divided by its vertical component, and may be written as follows:

$$\tan \text{bearing} \measuredangle = \frac{\text{error of departure}}{\text{error of latitude}}$$

From this expression the bearing can be obtained as illustrated in Example 13-2. A simple way to determine the direction of the missing side is to examine the latitude and departure calculations. To close the figure of this example, the sum of the south latitudes needs to be increased, as does the sum of the east departures. Thus the bearing of the line is southeast.

Since everyone is subject to making mathematical errors, it is well to compute the latitude and departure for this newly calculated line to determine if the traverse closes.

Example 13-2

Determine the length and bearing of side *BC* of the traverse shown in Figure 13-5. The lengths and bearing of the other sides are known.

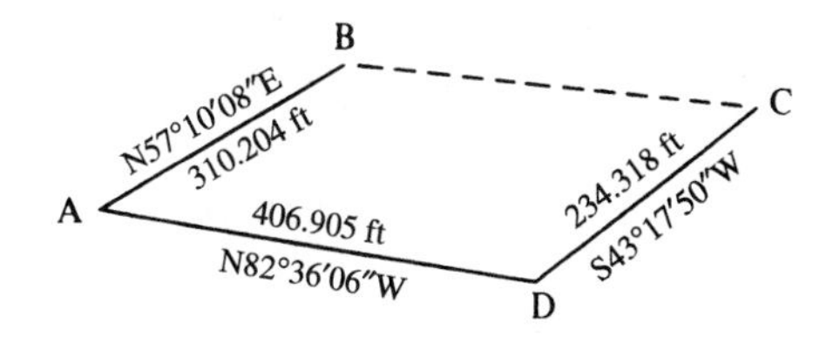

FIGURE 13-5 Computing length and bearing of side *BC*.

Solution

Compute the latitudes and departures of the known sides.

Side	Bearing	Length	Cosine	Sine	Latitudes N	Latitudes S	Departures E	Departures W
AB	N57°10′08″E	310.204	0.542164551	0.840272337	168.182	—	260.656	—
BC	—	—	—	—	—	—	—	—
CD	S43°17′50″W	234.318	0.727806006	0.685783069	—	170.538	—	160.691
DA	N82°36′06″W	406.905	0.12876675	0.991674908	52.396	—	—	403.517
					$\Sigma = 220.578$	$\Sigma = 170.538$	$\Sigma = 260.656$	$\Sigma = 564.208$
					$E_L = 50.040$		$E_D = 303.552$	

Next, compute the length and bearing of the missing side.

$$L_{BC} = \sqrt{(50.040)^2 + (303.552)^2} = 307.649 \text{ ft}$$

$$\tan \text{bearing} \measuredangle = \frac{E_D}{E_L} = \frac{303.552}{50.040} = 6.06618705$$

$$\text{Bearing} = \text{S}80^\circ 38' 21'' \text{E}$$

13-8 USING SURVEY TO DETERMINE THE LENGTH AND BEARING OF A MISSING SIDE

In this section Example 13-2 is reworked using SURVEY. The user in answer to the screen prompt selects the specified program desired, which in this case is 2 (*omitted-line computation*). The RETURN key is pressed and the directions given on the screen followed. The computer will then determine the length and bearing of a line from the ending point to the beginning point.

Notice that in answer to the screen prompt asking for the number of sides in the traverse we supply the number of sides for which we have data, thus not counting the missing side.

Example 13-3

Using the computer program SURVEY, repeat Example 13-2.

Solution

FIGURE 13-6 SURVEY printout.

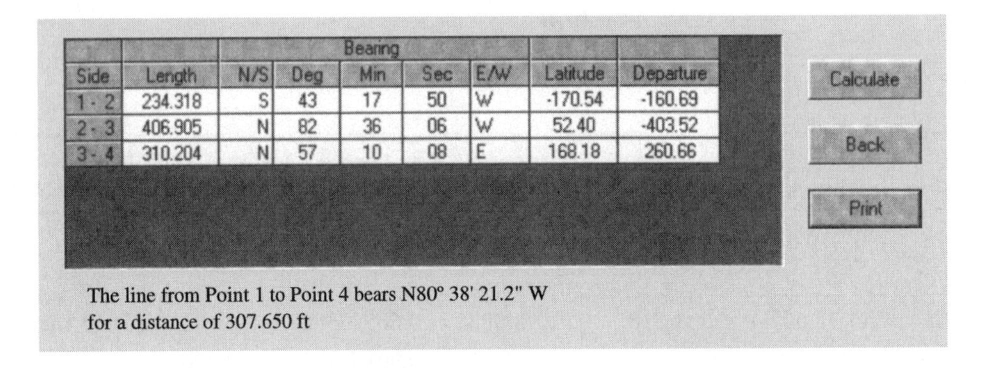

		Bearing						
Side	Length	N/S	Deg	Min	Sec	E/W	Latitude	Departure
1 - 2	234.318	S	43	17	50	W	-170.54	-160.69
2 - 3	406.905	N	82	36	06	W	52.40	-403.52
3 - 4	310.204	N	57	10	08	E	168.18	260.66

Calculate

Back

Print

The line from Point 1 to Point 4 bears N80° 38' 21.2" W
for a distance of 307.650 ft

13-9 EXAMPLE RADIATION PROBLEM

The radiation method of traversing was discussed briefly in Section 11-6. Now that latitudes, departures, and coordinates have been introduced, the mathematics of radiation can easily be handled.

With reference to Figure 13-7, it is assumed that the instrument is set up at a convenient location where all of the corners of the traverse can be sighted. Coordinates can be assumed for point 1 or they can be determined by measuring angles and distances from another point whose coordinates are known.

The bearing of line *AB* is assumed, or it is established by measuring angles from other points and lines (with known bearings), or it is determined by measuring from

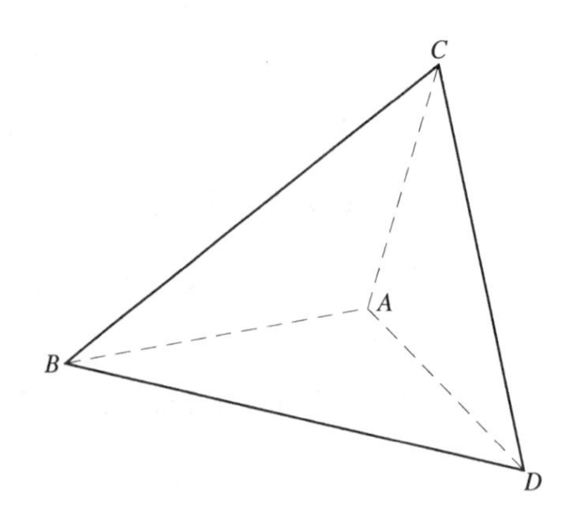

FIGURE 13-7 Traversing by radiation.

astronomical observations. Next the distances *AB*, *AC*, *AD* are measured (by EDM or taping) and the angles at point *A* are determined. The latitude and departure for each line is computed, thus enabling us to determine the coordinates of the traverse corners. Once these coordinates are available, the lengths and bearings of the traverse sides are computed. For an example, the length of side *BC* equals the square root of the sum of the squares of its latitude and departure.

$$L_{BC} = \sqrt{\left(X_C - X_B\right)^2 + \left(Y_C - Y_B\right)^2}$$

The tangent of its bearing angle is as follows:

$$\text{tangent of bearing } BC = \frac{\text{departure } BC}{\text{latitude } BC}$$

Example 13-4 illustrates the calculations involved for a traverse that has been run by radiation.

Example 13-4

The three-sided traverse of Figure 13-8 has been traversed by radiation. The bearing of line *AB* has been assumed or determined previously, as were the coordinates given in the figure for point *A*. The measured angles and distances are shown in the figure.

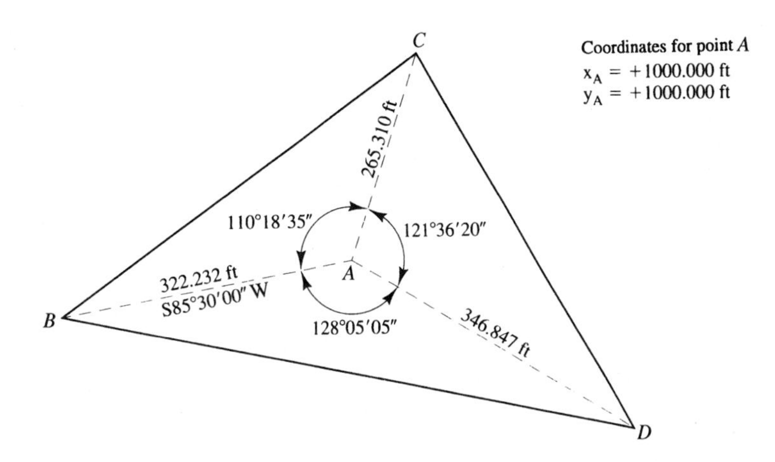

FIGURE 13-8 Radiation example.

Solution

The bearings of lines *AC* and *AD* are determined first, then the latitudes and departures of lines *AB*, *AC*, and *AD* are computed. Notice that the lengths used in the calculations must be horizontal.

Rays	Bearing	Length (ft)	Cosine	Sine	Latitudes (ft) N	Latitudes (ft) S	Departures (ft) E	Departures (ft) W
AB	S85°30′00″W	322.232	0.0784591	0.9969173	×	25.282	×	321.239
AC	N15°48′35″E	265.310	0.9621718	0.2724435	255.274	×	72.282	×
AD	S42°35′05″E	346.847	0.7362775	0.6766797	×	255.376	234.704	×

The coordinates of points *A*, *B*, and *C* are determined from the latitudes and departures just calculated.

Corner	x (ft)	y (ft)
B	678.761	974.718
C	1072.282	1255.274
D	1234.704	744.624

Finally, the lengths and bearings of the traverse sides are determined and shown in the table that follows. A simple calculation for side *BC* is shown.

$$L_{BC} = \sqrt{(1072.282 - 678.761)^2 + (1255.274 - 974.718)^2}$$

$$= 483.291\,ft$$

$$\text{Tan bearing departure } BC = \frac{\text{departure } BC}{\text{latitude } BC} = \frac{1072.282 - 678.761}{1255.274 - 974.718} = \frac{393.521}{280.556} = 1.4026469$$

$$\text{Bearing latitude } BC = \text{N}54^\circ 30' 49'' \text{E}$$

Traverse Side	Latitudes (ft) N	Latitudes (ft) S	Departures (ft) E	Departures (ft) W	Length (ft)	Bearing
BC	280.556	X	393.521	X	483.291	N54°30′49″E
CD	X	510.650	162.422	X	535.858	S17°38′39″E
DB	230.094	X	X	555.943	601.678	N67°30′59″W
	Σ = 510.650	510.650	555.943	555.943		

When radiation is used for traversing, the calculations are obviously a little long and tedious. However, computer programs are readily available to handle this type of problem.

Alternative Solutions

The author used latitudes and departures to obtain the coordinates of the traverse corners and then from those values computed the lengths and bearings of the traverse sides. There are obviously other methods for making the same calculations.

With reference to Figure 13-8, note that we have measured one angle and the lengths of two sides for each of the triangles. With this information the missing angles and the length of the third side can easily be determined with standard triangle formulas (see Tables C-1 and C-2 of Appendix C).

With many total stations the coordinates of the corners can be obtained directly and the resulting data fed directly into computers for calculations of the traverse lengths and directions desired.

13-10 COMPUTER SOLUTION FOR RADIATION PROBLEM

Example 13-5 illustrates the solution to the problem of Example 13-4 using the enclosed SURVEY program. If horizontal distances have not been determined previously, it will be necessary to input the inclined distances and their vertical angles and the program will determine the horizontal values. If horizontal distances are already available, we pass over the inclined distance and vertical angle spaces and put in the horizontal values.

Example 13-5

Repeat Example 13-4 using the enclosed SURVEY program.

Solution

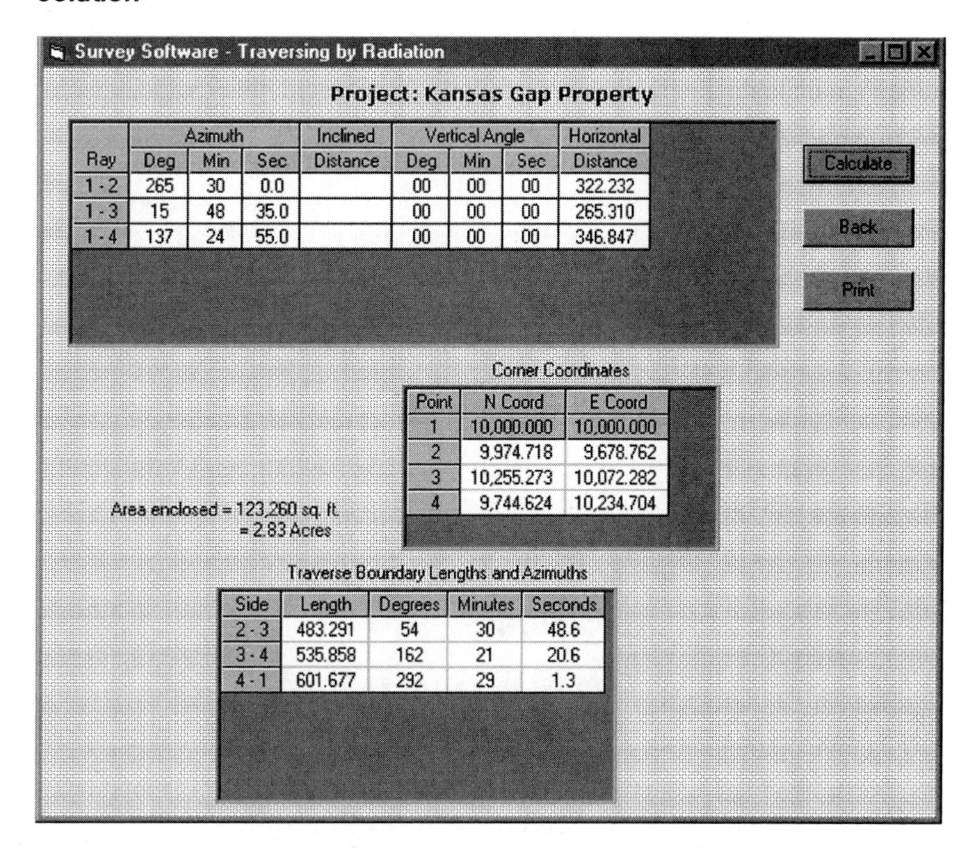

Ray	Azimuth Deg	Azimuth Min	Azimuth Sec	Inclined Distance	Vertical Angle Deg	Vertical Angle Min	Vertical Angle Sec	Horizontal Distance
1 - 2	265	30	0.0		00	00	00	322.232
1 - 3	15	48	35.0		00	00	00	265.310
1 - 4	137	24	55.0		00	00	00	346.847

Point	N Coord	E Coord
1	10,000.000	10,000.000
2	9,974.718	9,678.762
3	10,255.273	10,072.282
4	9,744.624	10,234.704

Side	Length	Degrees	Minutes	Seconds
2 - 3	483.291	54	30	48.6
3 - 4	535.858	162	21	20.6
4 - 1	601.677	292	29	1.3

FIGURE 13-9 SURVEY printout.

13-11 RESECTION

A very useful procedure with total station instruments for many situations is *resection.* This is a method sometimes called *free stationing* by which the location of an unknown point can be determined by setting up the instrument at the point in question and taking sights on at least three points of known location. The position of the desired point is determined by measuring the two angles between the points as shown in Figure 13-10 and by solving the triangles involved. The detailed trigonometric equations needed are given in the book footnoted here.[1]

FIGURE 13-10
Resection.

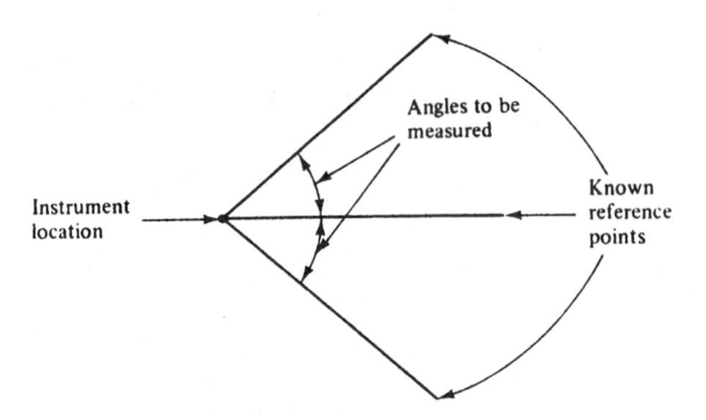

The instrument can be set up at some convenient point where there is excellent visibility for other work and then sight on at least three points whose positions have previously been determined. Then with a total station that is programmed for this situation the coordinates of the instrument position can be determined. An accuracy check can be made by making a sight to a fourth point whose position is known. If the surveyor sights on just three known points and is careless with the work a bad solution will be the result. Obviously, it is desirable to use another point when feasible.

Once the unknown position is determined sights can be made to other points and their positions determined by measuring directions and distances. Resection is a particularly useful procedure for building layout work where the instrument can be set up at any convenient position, perhaps out on the building itself. After the coordinates of the instrument are determined the coordinates of desired layout points can be input into the total station and its programs will compute the lengths and directions of ties to those points.

Problems

For Problems 13-1 to 13-3, determine the precision and land areas using the program SURVEY provided with this book.

13-1. Problem 12-1. (Ans.: $E_c = 0.133$ ft, precision $\frac{1}{6414}$)

13-2. Problem 12-2.

13-3. Problem 12-3. (Ans.: $E_c = 1.456$ ft, precision $\frac{1}{4363}$)

[1] R. E. Davis, F. S. Foote, and J. W. Kelley, *Surveying Theory and Practice,* 5th ed. (New York: McGraw-Hill Book Company, 1966), pp. 413-415.

13-4. Compute the lengths and bearings of sides *BG* and *CF* for the accompanying closed traverse. The latitudes and departures for the outer traverse (in solid lines) have been balanced.

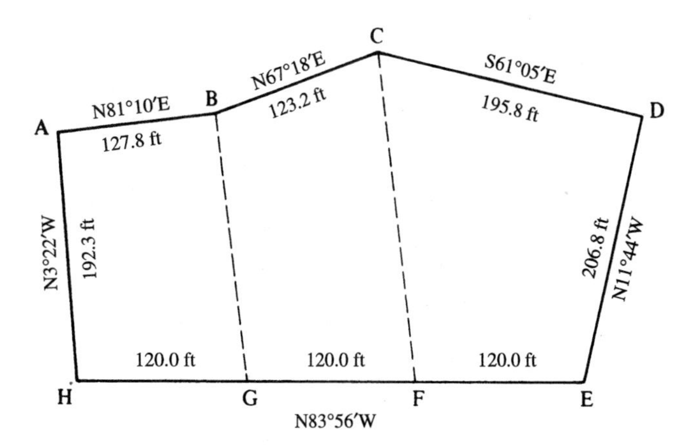

13-5. A random traverse is run from points *A* to *F*, as shown in the accompanying figure. Compute the length and bearing of a straight line from *F* to *A*.

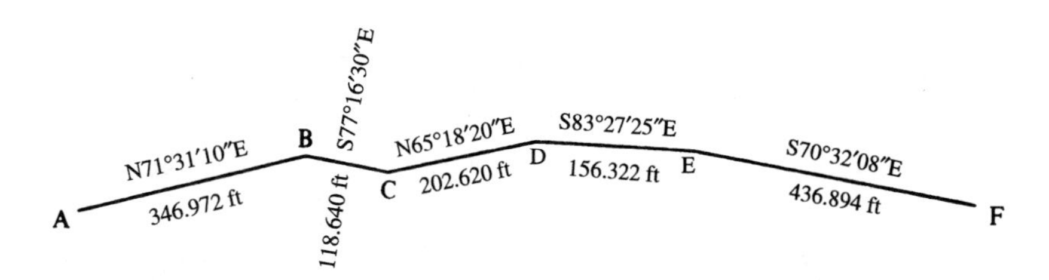

(Ans.: S89°45′19.5″W, 1196.135 ft)

In Problems 13-6 to 13-8, for each of the closed traverses given in the tables, the length and bearing of one side are missing. Compute the latitudes and departures of the given sides and determine the length and bearing of the missing side.

13-6.

Side	Bearing	Length (ft)
AB	N45°30′E	280.00
BC	S31°30′E	330.00
CD	S41°42′W	218.00
DE		

13-7.

Side	Bearing	Length (ft)
AB	N63°19′12″E	346.880
BC	S18°03′20″W	294.844
CA		

(Ans: N60°19′6.3″E, 251.572 ft.)

13-8.

Side	Bearing	Length (ft)
AB	N48°50′10″E	256.252
BC	S42°48′30″E	240.396
CD	S18°10′18″W	241.129
DE		
EA	N20°32′20″W	321.664

For Problems 13-9 to 13-12, repeat the problems given using the SURVEY program provided with this book.

13-9. Problem 13-5. (Ans.: S89°45′19.5″W, 1196.135 ft)

13-10. Problem 13-6.

13-11. Problem 13-7. (Ans.: N60°19′6.3″W, 251.572 ft)

13-12. Problem 13-8.

13-13. Using SURVEY determine the length and direction of the missing side. Azimuths and lengths are provided for the other sides.

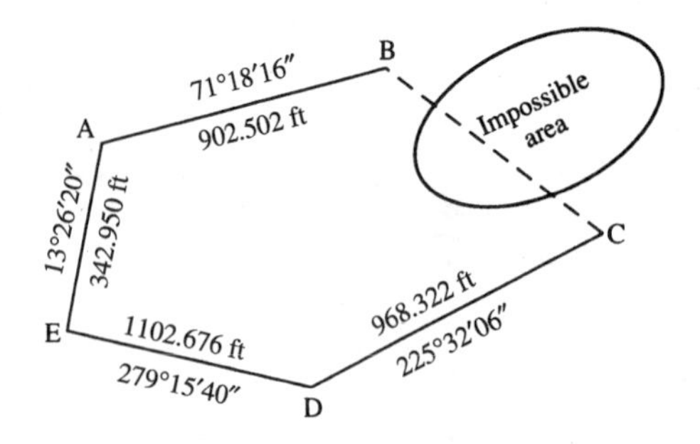

(Ans.: 278°13′8.3″, 853.554 ft)

For Problems 13-14 and 13-15, the figures shown were traversed by radiation. The distances given are horizontal. With the information given, determine the lengths and bearings of the sides of the traverses.

13-14.

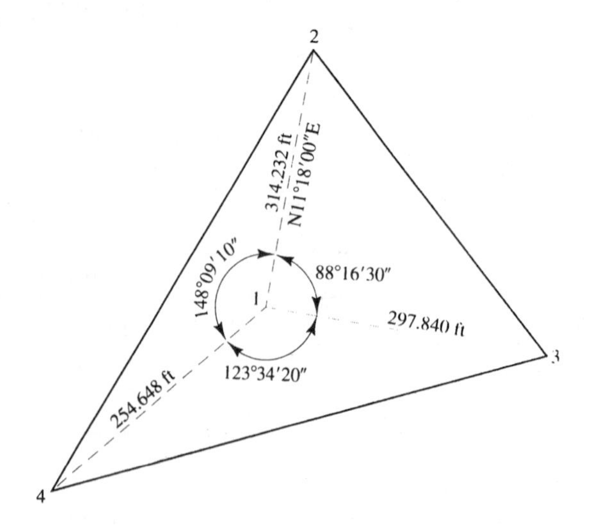

13-15.

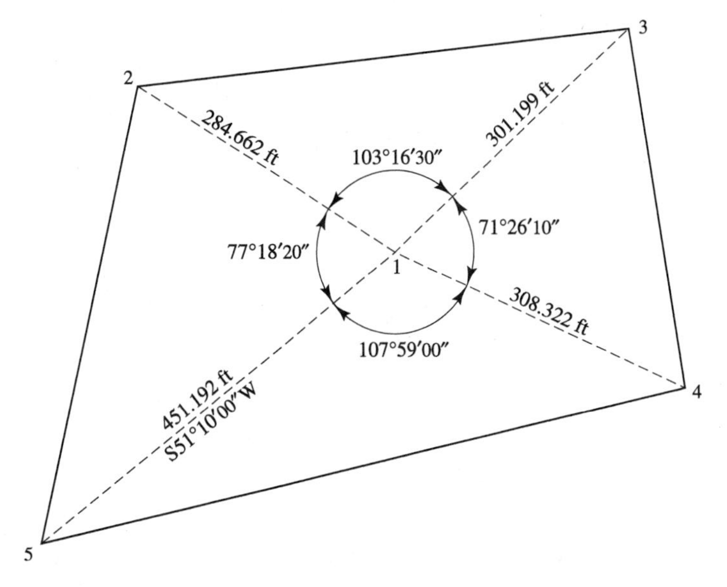

(Ans: 2–3, N88°49′47″E, 459.489 ft, 4–5, S79°23′26″W, 620.156 ft)

For Problems 13-16 to 13-18 the figures given were traversed by radiation. Using the SURVEY program, determine the lengths and bearings of the sides of the traverses.

13-16. Problem 13-14.

13-17. Problem 13-15.

(Ans.: Side 4–5, 620.110 ft, N79°23′25.4″W;
Side 5–2, 477.659 ft, N15°37′7.2″E)

13-18. The traverse shown in the accompanying figure, for which horizontal distances are given.

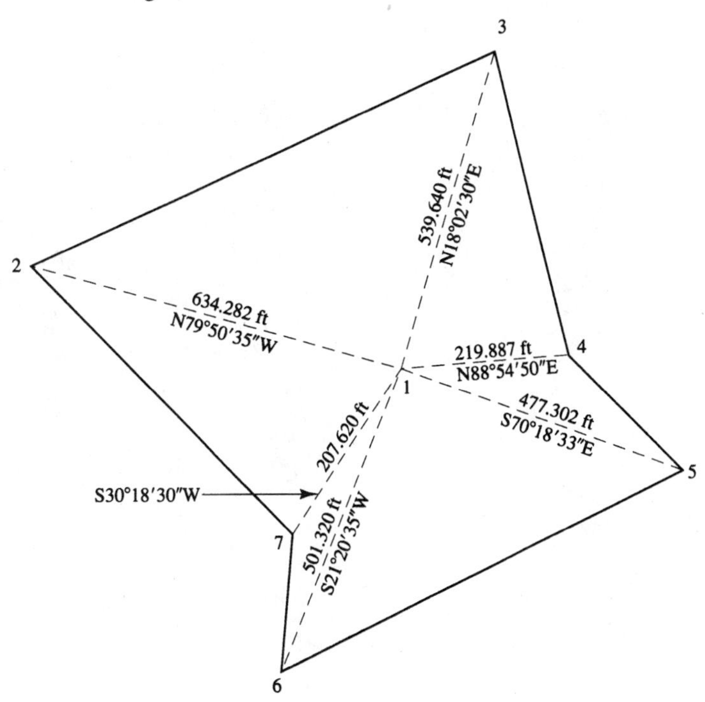

CHAPTER

14 Topographic Surveying

14-1 INTRODUCTION

Topography can be variously defined as the shape, configuration, relief, roughness, or three-dimensional quality of the earth's surface. Topographic maps are made to show this information, together with the location of artificial and natural features of the earth, including buildings, highways, streams, lakes, forests, and so on. Obviously, the topography of a particular area is of the greatest importance in planning large projects such as buildings, highways, dams, or pipelines. In addition, unless the reader lives in fairly flat country, he or she would probably want to have a topographic map prepared for the land before locating and planning the building of a house. Topography is also important for soil conservation projects, forestry plans, geological maps, and so on.

The ability to use maps is very important to people in many professions other than surveying, for example, engineering, forestry, geology, agriculture, climatology, and military science. In October 1793, at the age of 24, Napoleon Bonaparte received his first promotion because of his ability to make and use maps when he was placed in command of the artillery at the siege of Toulon.

Since the preparation of maps is quite expensive, the surveyor should learn what maps have previously been made for the area in question before beginning a new one. For instance, the U.S. Geological Survey has prepared topographic maps for a large part of the United States. These maps are readily available at very reasonable prices. A large percentage of their old maps were published with a scale of 1:62,500 (1 in. = almost 1 mile). These maps, however, are being phased out and the present goal of the National Mapping Program is to provide maps with a scale of 1:24,000 (1 in. = 2000 ft) for all of the United States except Alaska, where a scale of 1:25,000 has been used on some recent metric maps. The USGS maps have contour intervals (a term defined in Section 14-2) of 10 ft in relatively flat country and up to 100 ft in mountainous country.

Many other state and federal government agencies have prepared maps of parts of our country and copies of them are easily obtained. In fact, more than 30 different federal agencies are engaged in some phase of surveying and mapping. To obtain information about the maps available for a particular area, one may write to the National Cartographic Information Center, 507 National Center, 12001 Sunrise Valley Drive, Reston, VA 22092. If the surveyor cannot find a map from these sources that has a sufficiently large scale for the purpose at hand, he or she should nevertheless be able to find one that will give general information and serve as a guide for the work.

14-2 CONTOURS

The most common method of representing the topography of a particular area is to use contour lines. It is thought that contours were first introduced in 1729 by the Dutch surveyor Cruquius in connection with depth soundings of the sea. Laplace used contours for terrain representation in 1816.[1] A contour is an imaginary level line that connects points of equal elevation (see Figure 14-1). If it were possible to take a large knife and slice off the top of a hill with level slices at uniform elevation intervals, the edges of the cut lines around the hill would be contour lines (see Figure 14-2). Similarly, the edge of a still lake is a line of equal elevation or a contour line. If the water in that lake is lowered or raised, the edge of its new position would represent another contour line.

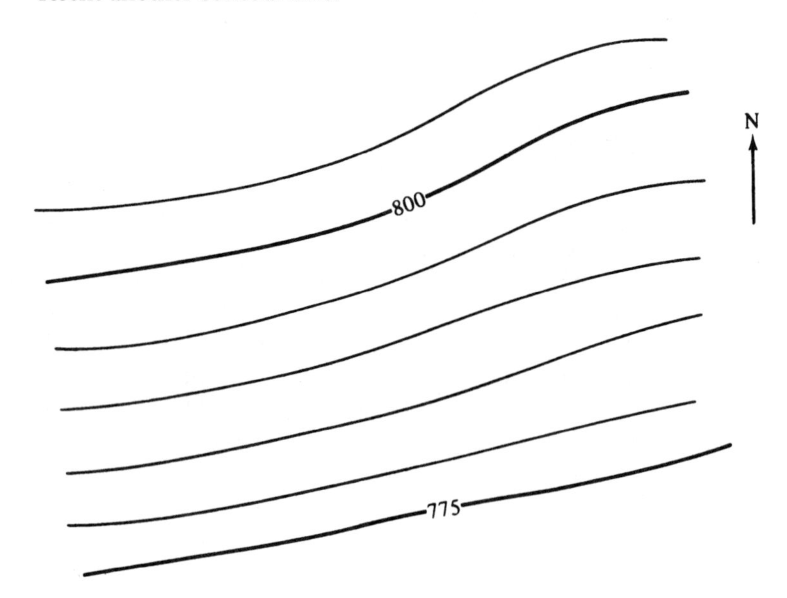

FIGURE 14-1 Typical contours (5-ft contour interval). Notice that the slope is fairly uniform, for the lines are almost equally spaced.

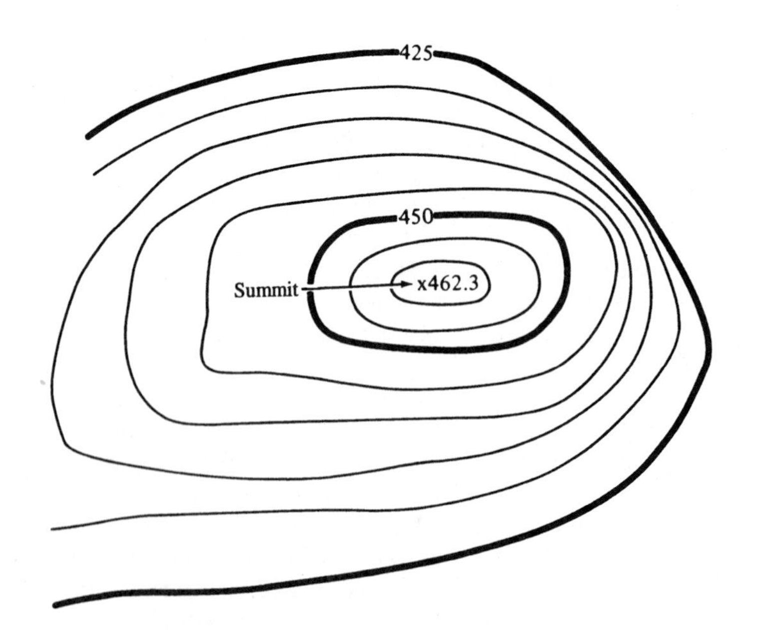

FIGURE 14-2 Notice that × 462.3 ft indicates the top of the hill or summit and its highest point. The close spacing of the contours on the upper right indicates a steep slope.

[1]M. O. Schmidt and K. W. Wong, *Fundamentals of Surveying,* 3rd ed. (Boston: PWS Engineering, 1985), p. 350.

The *contour interval* of a map is the vertical distance between contour lines. The interval is determined by the purpose of the map and by the terrain being mapped (hilly or gently sloping). For normal maps, the interval varies from 2 to 20 ft but it may be as small as 1/2 ft for relatively flat country and as large as 50 to 100 ft for mountainous country.

The selection of the contour interval is an important topic. The interval must be sufficiently small so that the map will serve its desired purpose while being as large as possible, to keep costs at a minimum. When the maps are to be used for earthwork estimates, a 5-ft interval is usually satisfactory unless very shallow cuts and fills are to be made. For such cases a 2-ft contour interval is probably needed. When the maps are to be used for planning water storage projects, it is usually necessary to use 1- or 2-ft intervals.

A sufficient number of contour lines are numbered so that there is no confusion regarding the elevation of a particular contour. Usually, each fifth contour line is numbered and shown as a wider and heavier line. Such a line is called an *index contour.* If any uncertainty remains, other contour lines may be numbered. Figures 14-1, 14-2, and 14-3 present introductory examples of contour lines together with descriptive notes.

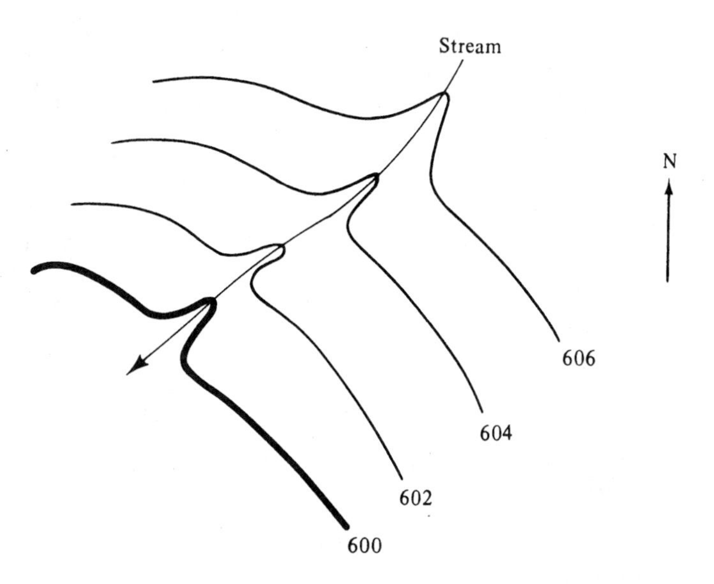

FIGURE 14-3 Notice how the contour lines bend or point upstream in crossing streams or rivers.

In Figure 14-4 the author has drawn a set of contour lines that illustrate quite a few of the situations that may be encountered in preparing a topographic map. Various notes are shown on the drawing to identify different features of the land. Note in this figure that the term *saddle* is used to describe the contours, which show two closely spaced high points or *summits.*

Other methods that may be used to show elevation differences are relief models, shading, and hachures (defined later). Relief models are a most effective means of showing topography. They are made from wax, clay, plastic, or other materials and are shaped to agree with the actual terrain. The Army Map Service once produced and distributed beautifully colored molded plastic relief maps for those parts of the country that have significant elevation differences. These maps are now commonly available in many stores around the U.S.A.—particularly those which specialize in outdoor activities.

FIGURE 14-4 Various contour map features.

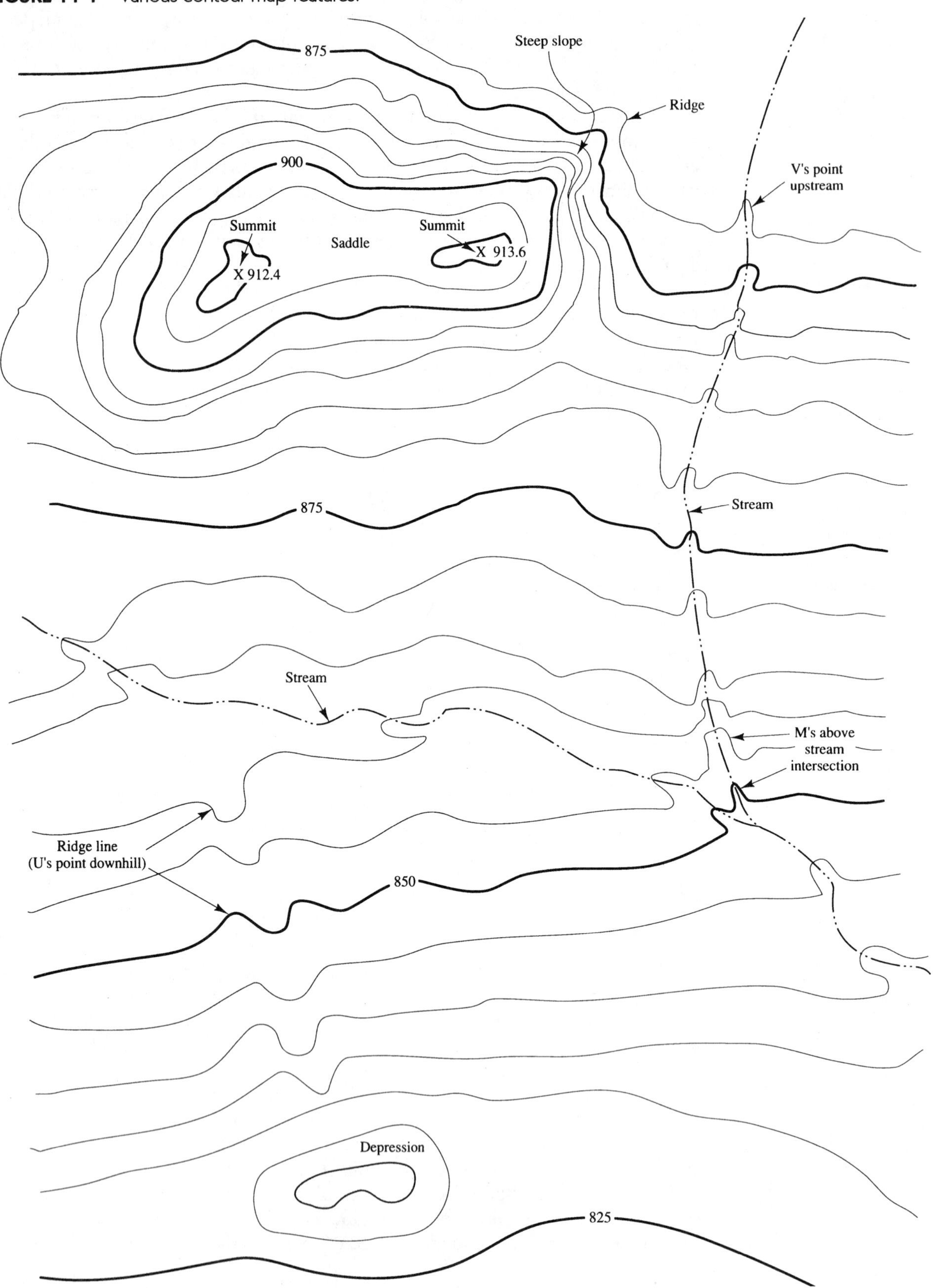

An old method used to show relative elevations on maps was shading. In this method an attempt was made to shade in the various areas as they might appear to a person in an airplane. For instance, the steeper surfaces might be in shadows and would be darkly shaded on the drawings. Less steep slopes would be indicated with lighter shading, and so on.

A few decades ago it was common practice to represent topography by means of hachures, but their use today is more infrequent, and they have been replaced by contour lines. *Hachures* are short lines of varying widths drawn in the direction of the steepest slopes. They give only an indication of actual elevations and do not provide numerical values. By their spacing and widths, these lines produce an effect similar to shading, but they are perhaps a little more effective.

14-3 PLOTTING OF TOPOGRAPHIC MAPS

With a topographic survey information is obtained concerning the elevations and positions of a large number of points for a particular land area. In addition, data is obtained as to the character and position of various man-made and natural features.

In this section a brief description is given of the steps involved in hand plotting a topographic map using the data obtained in a topographic survey. First the elevations of various points around a tract of land are plotted as shown in Figure 14-5 along with a control traverse for the area. This traverse is shown with the solid straight lines in the figure.

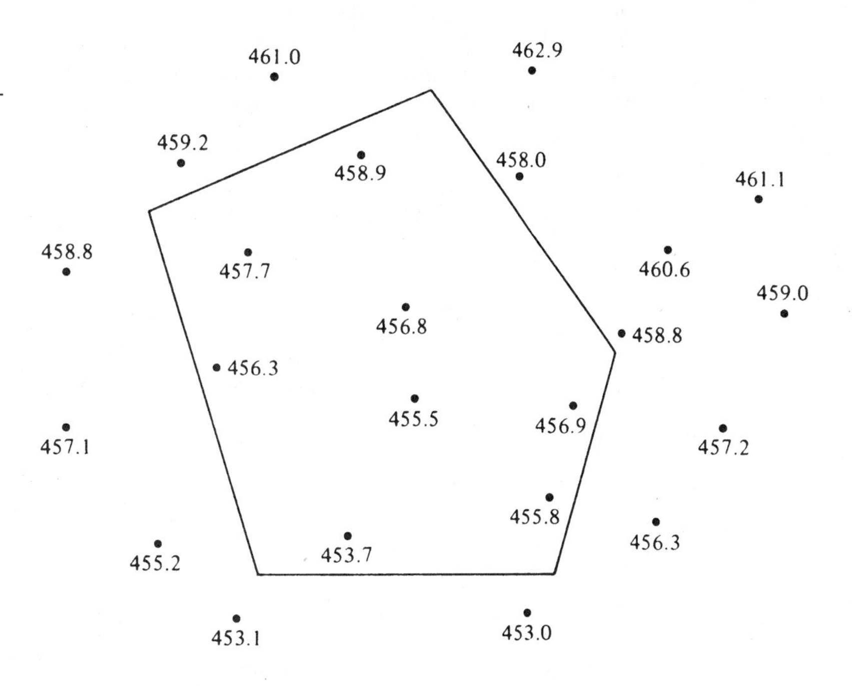

FIGURE 14-5 Elevations of a group of points determined by stadia.

In Figure 14-6, 2-ft interval contours are drawn from the elevations that were plotted in Figure 14-5. These lines were sketched in freehand by "eyeball" interpolation of the proportionate distances between elevation points. For very precise maps, the interpolation may be made with a calculator with the proportionate distances scaled between the points.

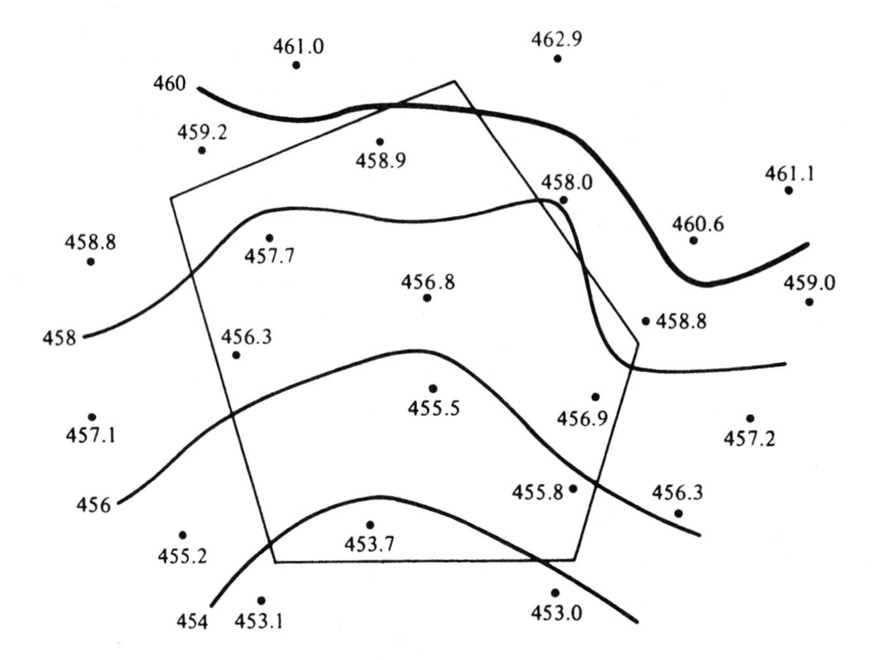

FIGURE 14-6 Contour lines plotted for the elevations shown in Figure 14-5.

Finally, in Figure 14-7 the control traverse is removed from the drawing, and the locations of important objects, such as roads, buildings, and power lines, are plotted. On many topographic maps, for example, those prepared for house lots or building lots, the property lines and corners are so important for the project in mind that they are included on the map. Final details such as the legend and title box are not shown in this figure.

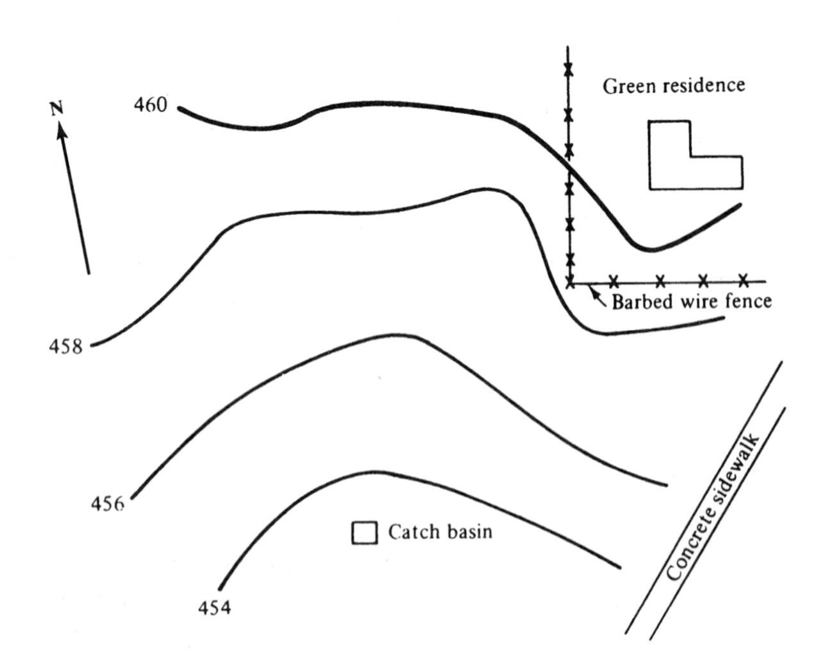

FIGURE 14-7 Control traverse removed and more details added to Figure 14-6.

There is no doubt but that the work of plotting topographic maps by hand is a very tedious and time-consuming process. Within the last few decades much work has been done in the area of plotting such maps with computers. Originally, computer-aided contouring was available only on mainframe computers, but today programs are available to perform the necessary calculations and plot the contour lines and other details quickly and accurately with personal computers. The information can be inputted into the program from the field notes, or a total station instrument can be used. The instrument will simultaneously compute the x, y, and z coordinates of the points and record them in an electronic field book. Back in the office the map is plotted by the computer and the contour map can be viewed on the screen. Furthermore, the contour lines can be smoothed or rounded off. Today, computer-generated graphics are a part of a large number of surveying projects in North America. A sample set of computer plotted contour lines is shown in Figure 14-8 (a). With such computer programs it is also possible to print out a three-dimensional picture of the area looking at it from any angle desired. In part (b) of Figure 14-8, such a view is shown looking at the land area of part (a) from the upper end toward the lower side.

FIGURE 14-8 Computer-plotted maps: (a) contour lines for sample land area; (b) three-dimensional view of land of part (a) looking at it from the lower left side.

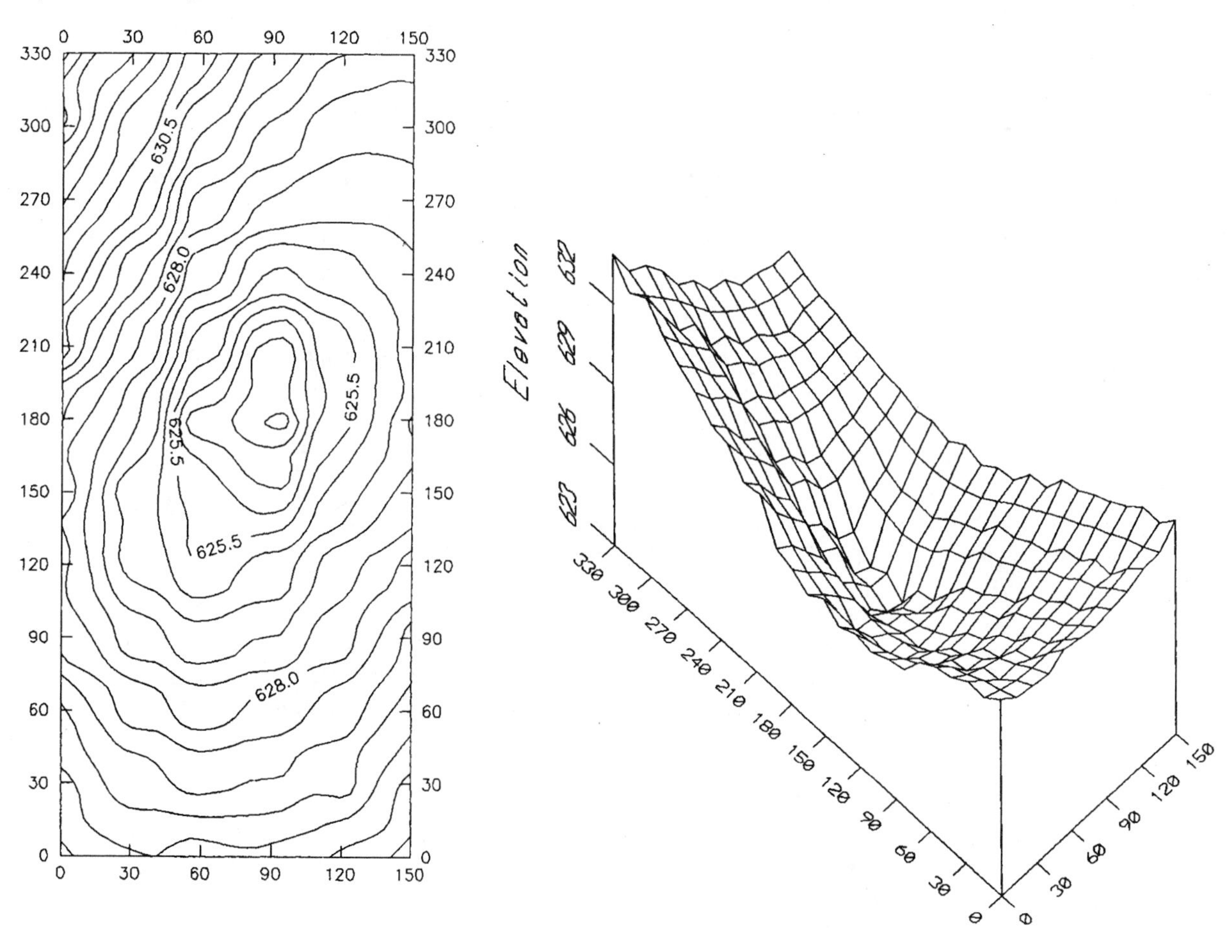

One serious problem may occur with the computer drafting of topographic maps if we have a point whose elevation is completely wrong. The computer will proceed to draw contour lines indicating a hill or hole in the ground where none exists. If the map is drawn by hand, we are more likely to discover and discard the erroneous value.

14-4 SUMMARY OF CONTOUR CHARACTERISTICS

To aid the student in learning to read and prepare topographic maps, a summary of contour line characteristics follows:

1. Contour lines are evenly spaced when the ground surfaces are uniformly sloping.
2. For steep slopes contours will be close together, while for relatively flat areas they will be widely spaced.
3. When the ground surface is rough and irregular, contours will be irregular.
4. Contours are drawn perpendicular to the directions of the ground slopes and thus water flow will be perpendicular to the contours.
5. Contour lines crossing streams or ditches point upstream in approximate ⌒ or ∧ shapes.
6. Contours connect points of equal elevation and thus cannot cross each other except in caves or similar situations.
7. The tops of hills and ground depressions look the same as to contour lines and may not easily be distinguished. For this reason spot elevations may be shown as at the low or high point, some extra contour lines may be numbered, and shading or hachure lines may be used.
8. Contour lines are stopped at the edges of buildings.
9. Contour lines just above stream intersections and between the streams usually have ⌒‿⌒ shapes.

14-5 MAP SYMBOLS

It is quite convenient to represent various objects on a map with symbols that are easily understood by the users. Such a practice saves a great deal of space (compared to the lengthy explanatory notes that might otherwise be necessary) and results in neater drawings. Many symbols are standard across the United States, whereas others vary somewhat from place to place. In Figure 14-9 a sample set of rather standard symbols are presented. Very detailed symbol lists are available from the U.S. Geological Survey, these being the ones they use for their maps. Occasionally, the surveyor encounters some feature for which he or she does not have a standard symbol. For such a situation a new symbol may be made up, but it must be described in the map's legend.

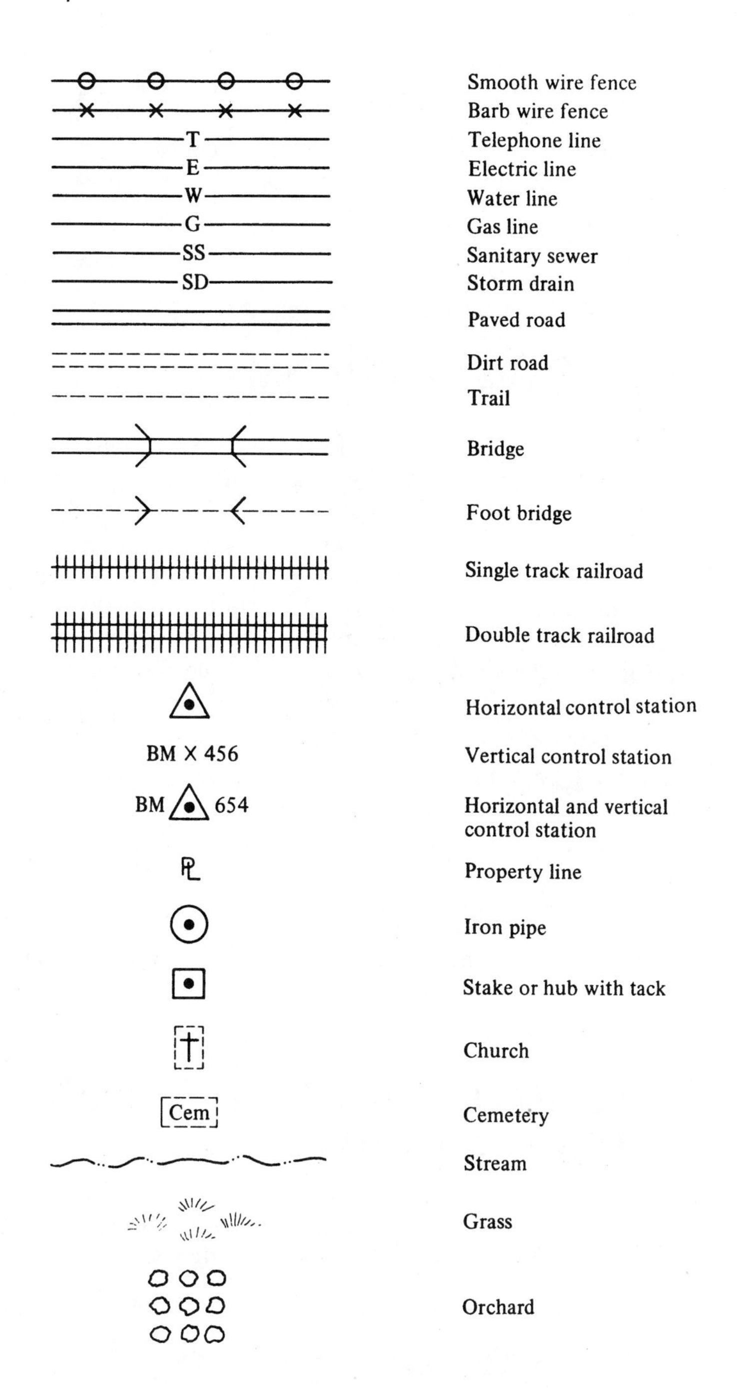

FIGURE 14-9 Typical map symbols.

14-6 COMPLETING THE MAP

The appearance of the completed map is a matter of the greatest importance. It should be well arranged and neatly drawn to a scale suiting the purpose for which it is to be used. It should include a *title, scale, legend,* and *meridian arrow.* The lettering should be neatly and carefully performed or perhaps lettering sets or press-on letters should be used.

The title box may be placed where it looks best on the sheet but is frequently placed in the lower right-hand corner of the map. It should contain the title of the map, the name of the client, location, name of the project, date, scale, and the name of the person who drew the map and the name of the surveyor.

14-7 SPECIFICATIONS FOR TOPOGRAPHIC MAPS

To describe the accuracy of the horizontal dimensions for a particular topographic map the usual procedure is to say that the average error that results in scaling a distance between two plotted points should not exceed a certain value as compared to the measured value on the ground. The accuracy of contour lines may be described by saying that the error in any elevation taken from the map may not exceed a certain value, such as perhaps one-half of the contour interval used on the map.

The federal agencies involved in mapping have agreed upon certain minimum requirements that when met enable agencies to print on their maps the following statement: "This map complies with the National Map Accuracy Standards." These requirements are summarized in the paragraphs to follow.

Horizontal Accuracy. For maps with scales larger than 1:20,000, no more than 10% of the positions of well-defined points tested may be in error by more than 1/30 of an inch. For maps with scales of 1:20,000 or smaller, the value becomes 1/50 of an inch.

Vertical Accuracy. It is specified that no more than 10% of the points tested can be in error by more than one-half of the contour interval.

14-8 METHODS OF OBTAINING TOPOGRAPHY

There are several methods available for obtaining topography. These are listed and briefly discussed in the following paragraphs.

Obsolete Methods

1. *Transit-stadia or theodolite-stadia.* With this, the classical or historic method of gathering topographic data, the necessary measurements were made in the field, recorded in the field book, and then plotted on paper in the office. The author feels that an introductory study of this procedure, although now made obsolete by modern equipment, will provide the student with an excellent background for the present day methods of handling topography. As a result the transit-stadia method is briefly introduced in Section 14-9 of this chapter.

2. *Plane table.* With the plane table procedure the measurements were made in the same way as those in the preceding method, but the data was plotted in the field on paper that was attached to a drawing board mounted on a tripod. This device is called a plane table and its use is described in Section 14-10.

Modern Methods

1. *Total stations.* A very large percentage of topographic surveys today are made with total stations. These instruments, which are used for almost all topographic surveys of small areas, are discussed in Section 14-11.

2. *Photogrammetry.* Today topographic maps for small areas up to 20 to 40 acres are usually prepared using total stations. Photogrammetry is used for larger areas. The dividing line between the use of the two methods is not based on area alone but also on the terrain conditions. For instance, if there is dense growth which will require a large amount of bush cutting or if other features make sighting with a total station difficult, photogrammetry will be economical for areas as small as 20 acres or even less. On the other hand, where topographical features can easily be sighted the use of total stations is economical for somewhat larger areas.

Total station surveys and those made by the transit-stadia and plane table procedures can be made during almost all weather and vegetation conditions. On the other hand, photogrammetric surveys need to be conducted on clear days and for situations where snow or dense vegetation do not obscure the ground surface. These conditions can decidedly reduce the accuracy of photogrammetric surveys and may actually prevent their use.

3. *Global positioning system (GPS).* This system, which is discussed in Chapters 15 and 16, is being increasingly used for all types of surveys, including topographic ones. Perhaps someday it will become the predominant method used not only for topographic but also for many other types of surveys. GPS systems have advanced mapping to such a state that now people with little background experience with just a little training can map almost anything.

14-9 TRANSIT-STADIA METHOD OF MAPPING

In this section the author attempts to describe quite briefly the stadia method that was used for many years to gather topographic data for mapping. He feels this will provide the student with a very good background and understanding of how mapping was done until the last few decades thus enabling the student to do much better work with the presently used methods.

The equipment used for stadia surveying consisted of an instrument whose telescope was provided with stadia hairs (as previously described in Section 3-4 of this book), and a regular level rod or perhaps a stadia rod (described later in this section). The telescope stadia hairs were mounted on the cross-hair ring, one a distance above the center horizontal hair and one an equal distance below. A regular level rod was satisfactory for stadia readings of up to 200 or 300 ft, but for greater distances, stadia rods similar to the ones shown in Figure 14-10 were used. A study of this figure will show how easily good stadia rods could be made, and for this reason there were numerous types of rods in use, many of which were homemade.

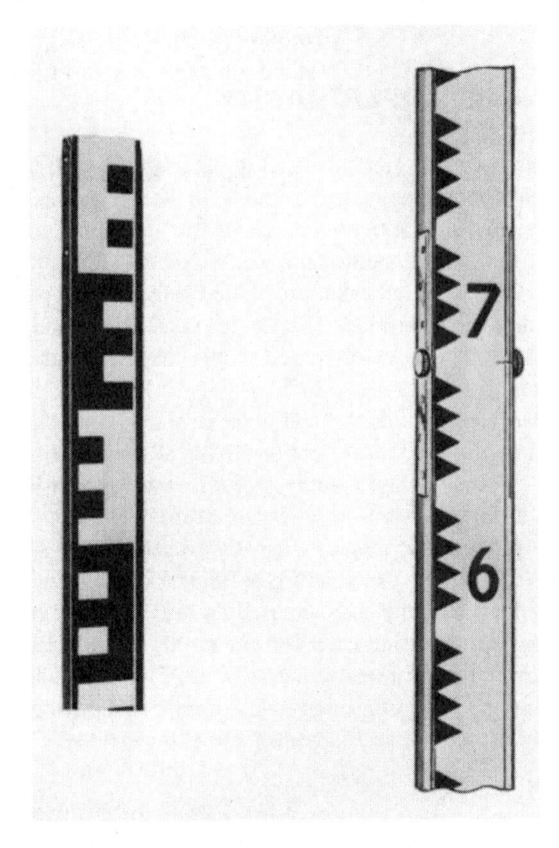

FIGURE 14-10 Stadia rods which can be read up to 1500 or 2000 ft. (Courtesy of Keuffel & Esser, a Kratos Company.)

Horizontal Measurements

Transits were manufactured so that if the telescope was horizontal and the level rod vertical, the distance D from the center of the instrument to the rod equaled the stadia multiplier or *stadia interval factor* K times the interval from the rod reading at the top stadia hair to the rod reading at the bottom stadia hair. This interval between the stadia hairs is referred to as the *rod intercept* and is represented herein by the letter s, as shown in Figure 14-11. The stadia multiplier was 100 for almost all instruments.

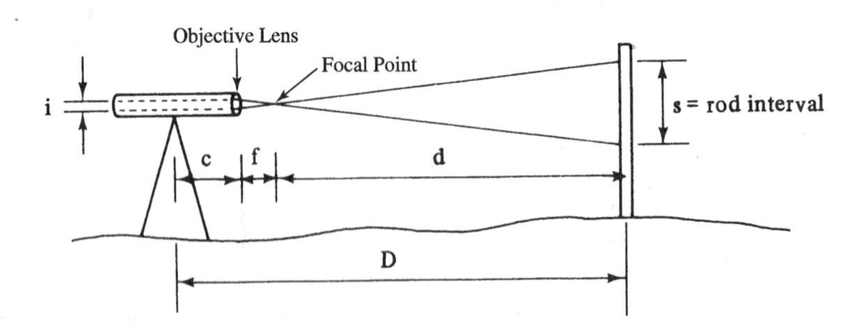

FIGURE 14-11 Stadia constant $c + f$.

The earlier transits were externally focusing. Thus their focal points were located out in front of the objective lens. Referring to Figure 14-11 you can see that the distance from the center of the instrument to the focal point equals the sum of the distances c and f as shown in the figure. We need to determine the distance D from the center of the instrument to the level rod. This value, which equals c + f + d, is labeled D in the figure.

If we let the distance from the top stadia hair to the bottom stadia hair be *i*, it is possible to write the following expression and from it determine the distance *D:*

$$\frac{i}{f} = \frac{s}{d}$$

$$d = \frac{f}{i}s$$

$$D = \frac{f}{i}s + \left(c + f\right)$$

In nearly all old instruments the value of f/i or K (stadia interval factor) is 100 and the value of $c + f$, called the *stadia constant,* varies from approximately 0.8 to 1.2 ft with an average value of 1 ft. Manufacturers showed on their instrument boxes more precise values for the stadia constant for each instrument. In general, when the telescope was in a horizontal position, the horizontal distance from the center of the instrument to the center of the rod was

$$H = Ks + 1$$

The stadia intervals usually measured were probably a little too large because of unequal refraction and because of unintentional inclination of the rod by the rodman. To offset these systematic errors, the stadia constant was neglected in most surveys.

For the internally focusing instruments the distance from the center of the telescope to the focal point was almost zero and was ignored. The objective lens of one of these instruments remained fixed in position while the direction of the light rays is changed by means of a movable focusing lens located between the plane of the cross hairs and the objective lens.

Inclined Measurements

Since stadia readings are seldom horizontal, it is necessary to consider the theory of inclined sights. For such situations it is necessary to read vertical angles as well as the s values or rod intercepts and from these values to compute elevation differences and horizontal components of distance.

If it were feasible to hold the rod perpendicular to the line of sight, as shown in Figure 14-12, the horizontal and vertical components of distance would be as follows (again, the stadia interval factor is neglected):

$$H = D\cos\alpha = Ks\cos\alpha$$

$$V = D\sin\alpha = Ks\sin\alpha$$

Obviously, it is not practical to hold the rod perpendicular to the line of sight, that is, at an angle α from the vertical; therefore, it is held vertically. This means that the rod is held at an angle α from being perpendicular to the line of sight. As can be seen in Figure 14-12, the rod intercept reading is too large if the rod is not perpendicular to the line of sight. If the readings are taken with the rod plumbed, they can be corrected to what they should be if the rod had been perpendicular to the line of

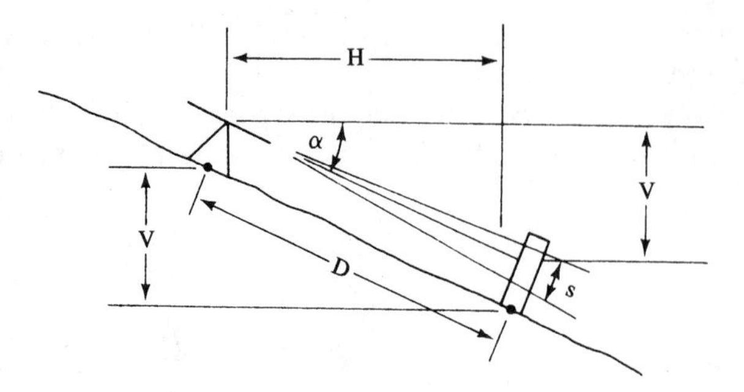

FIGURE 14-12 Inclined sights with level rod held perpendicular to telescope line of sight *(impractical).*

sight by multiplying them by the cosine of α. The resulting values for *H* and *V* may then be written as follows:

$$H = (Ks \cos \alpha)(\cos \alpha)$$

$$H = Ks \cos^2 \alpha$$

$$V = (Ks \sin \alpha)(\cos \alpha)$$

$$V = \frac{1}{2} Ks \sin 2\alpha$$

As a practical matter, if the vertical angle α is less than 3°, the diagonal distance may be assumed to equal the horizontal distance, but for such cases the vertical differences in elevation (the *V* values) definitely cannot be neglected.

A typical set of stadia notes for the nearly obsolete transit-stadia method is shown in Figure 14-13. The first four columns in these notes identify the points by number and show the three measurements taken for each. The last three columns are used to record computed values. The remaining space on the right-hand page is used to describe the points carefully. Obviously, these descriptions are very important to the map plotter back in the office. In addition, the recorder should make any necessary explanatory sketches in the field book.

FIGURE 14-13 Typical stadia notes.

STADIA SURVEY OF WILD HOG FARM

Dec. 1, 1998
Clear, mild, 60°F
K+E Transit # 31621

J. B. Johnson ⊼, N
N. C. Hanson ⌀
A. N. Nelson ⌀

Sta.	Azimuth	Rod. Int.	Vert. ∡	Elev. Diff.	Horiz. Dist.	Elev.	
⊼	@ C,	Elev =	207.30,	h.i. =	4.7, BS	on D	
1	27°08′	1.45	−11°05′	−27.4	140	179.9	℄ of creek
2	51°56′	2.60	−3°06′	−14.0	259	193.3	Keller's front corner far side of road
3	77°05′	1.30	−14°20′	−31.2	122	176.1	Pt. under power line
4	95°00′	1.70	−10°50′	−31.4	164	175.9	" " " "
5	127°58′	2.97	−3°41′	−19.0	296	188.3	Keller's back corner
6	161°20′	0.75	−4°50′	−6.3	74	201.0	Edge of cleared field

An example of the calculations needed to complete the final three columns in the notes is presented at the end of this paragraph. For this work a pocket calculator was used to solve the just presented formulas for H and V.

$$H = Ks \cos^2\alpha = (100)(1.45)(\cos^2 11°05') = 1396 \text{ ft rounded to 140 ft in the notes}$$

$$V = {}^1/{}_2 Ks \sin 2\alpha = ({}^1/{}_2)(100)(1.45)(\sin 22°10') = 27.354 \text{ ft rounded to 27.4 ft in the notes}$$

$$\text{Elevation of point } 1 = 207.30 - 27.4 = 179.9 \text{ ft}$$

When the transit-stadia method was in common use for topographic surveys, various tables, slide rules, and computer programs were developed to perform stadia reduction calculations. The reader should realize that in a typical topographic survey observations are made for hundreds or even thousands of different points and these reduction devices saved a great deal of time. One old computer program that was formerly used for transit-stadia work to perform the above calculations has been left in the computer program SURVEY, which accompanies this book.

14-10 PLANE TABLE SURVEYS

A variation of the transit-stadia procedure for obtaining map data involved the use of the plane table (See Figure 14-14). An alidade was used to take stadia measurements and the results were plotted on drawing paper attached to a portable drafting board or plane table. The table was mounted on a tripod so that it could be rotated and leveled. Map details were plotted on the paper as soon as they were measured. The contour lines were drawn as the work proceeded. The great advantage of plane table mapping was that the plotter could compare or correlate his or her drawing with the actual terrain as the work progressed, thus greatly reducing the chances of making mistakes.

The alidade consisted of a telescope mounted on a brass ruler approximately 2 1/2 in. by 15 in. The ruler was aligned with the line of sight of the telescope, which enabled its user to take readings and scale them off in the proper direction along the ruler's edge. The telescope was similar to those used on transits and had the usual stadia hairs. In addition, the alidade had a trough compass mounted in a small metal box, a bubble tube mounted on the straight edge for leveling purposes, and a vertical arc that permitted the measurement of vertical angles needed for determining elevation differences. If the surveyor had his or her position located on the drawing and if the drawing was oriented in the proper direction, he or she could sight on a desired point and determine the distance and elevation difference to that point by stadia and immediately plot the information on the paper.

Plane table paper was subject to rough field conditions, particularly dirt and moisture. Consequently, a special tough paper was used that minimized expansion

FIGURE 14-14
Plane table and alidade—obsolete. (Courtesy of Wild Heerbrugg Instruments.)

and contraction. This paper had been subjected to alternating drying and wetting cycles to make it more resistant to moisture changes. Sharp, hard pencils (6H to 9H) were used for plotting details, and great care was needed in handling the alidade to reduce smearing; it was lifted and moved rather than slid on the paper.

14-11 TOPOGRAPHIC DETAILS OBTAINED WITH TOTAL STATIONS

Today total stations provide the most common ground method for obtaining topographic details. With these instruments the surveyor can simultaneously measure directions and slope distances and have the instruments compute the horizontal and vertical components of the distances as well as the elevations and *X-Y* components for the points being sighted.

Using total stations the surveyor is extremely efficient in gathering topographic data—in fact, several times as efficient as when traditional stadia methods are used. An experienced surveyor using a total station can probably gather the data needed for 600 to 1000 points in a day. This is two or three times the number of points that can usually be handled using the classical stadia procedure. The situation with total stations is even better than described if a data collector and a computerized surveying system is used. With such a system the data can be collected, downloaded to the computer, adjustments made to the measurements, if necessary, and the map automatically plotted. This increase in productivity means that ground topographic surveying is more economically competitive with photogrammetry for larger areas than when the transit-stadia method was used.

As with all of the methods described in this chapter for obtaining topographic details, considerable thought must be given to the positions where the instrument is to be set up so the area in question can be covered from as few positions as possible. As an illustration it may be possible to set up the total station on a hill where quite a large area can be covered.

In Figure 14-15 a sample set of topographic notes is presented as they might be recorded in a field book by a surveyor using a total station. For this case it was necessary for the surveyor to take the information back to the office for map preparation, probably by computer.

FIGURE 14-15 Topographic data obtained with total station.

TOPOGRAPHIC SURVEY FOR GIGNILLIAT FARM — Dec. 1, 2003; Clear, mild, 60°F; Topcon total station #3106 — J.B. Johnson ⊼, N; N.C. Hanson, Ø

Pt	Azimuth	Horiz. Distance	Elev. Diff.	Elev.	
⊼ @B, Elev. = 654.32, h.i. = 5.0, BS on A					
1	0°00′00″	100.15	−3.26	651.06	P.O.G. (pt. on ground)
2	10°16′15″	210.16	+7.45	661.77	P.O.G.
3	20°30′30″	154.13	+3.25	657.57	Top of creek bank
4	25°25′35″	161.67	−1.01	653.31	¢ dirt road

The economics of using total stations can be much improved if an automatic data collector is used. The total station is set up at a known point and oriented by backsighting to another point with the azimuth of the line between the points and the coordinates of the points entered with the keyboard. Then each of the points selected for topography is sighted and the azimuth, zenith angle, and distance measured. The total station will compute the horizontal and vertical components of distance as well as the coordinates and elevations of each point. It is necessary to provide for each point an identification number as well as other descriptive details (edge of lake, northeast corner of building, and so on).

Many modern data collectors are so constructed that the surveyor can recall or scroll the recorded data and have it displayed on the screen. Then the surveyor can check the values and supply additional information as needed.

At various intervals (perhaps lunch time and quitting time) the information stored in the data collector may be transferred to another instrument or perhaps downloaded to a computer back in the office. Then the computer may be used to generate the desired topographical maps. (When working far away from the office the data obtained may be sent back to the office using telephone lines.)

14-12 SELECTION OF POINTS FOR TOPOGRAPHIC MAPPING

For mapping purposes it is necessary to determine the elevations of a sufficient number of points in order to picture the contour or relief of the area accurately. Several approaches are possible in selecting these points. One is the so-called *checkerboard method,* in which the desired area is divided into squares or rectangles and the elevation determined at the corners of each of these figures. In addition, elevations are determined at points where slopes suddenly change, as, for example, at valley or ridge lines. This method is best suited to small areas that have little relief or roughness.

The most common method of selecting the points is the *controlling point method.* Elevations are determined for key or controlling points and the contour lines interpolated between them. The controlling points are usually thought of as those points between which the slope of the ground is approximately uniform. They are such points as the tops of hills, the bottoms of valleys, and tops and bottoms of the sides of ditches, and other points where important changes in slope occur. If there is a uniform slope for a long distance, it is theoretically necessary to obtain only one elevation at the top of the slope and one at the bottom, for the map plotter will merely interpolate between those points in order to obtain contour lines. Practically speaking, however, since the eye is easily fooled by slopes, it is well to take an occasional intermediate point, even though the slope seems constant.

A third procedure, which was occasionally used when the transit-stadia method was common, was called the *tracing contour method.* With this procedure, which simplified the drawing of contour lines, a number of points were located whose elevations were equal to those of a desired contour. Those points were plotted on the map and connected together to form the desired contour line. The telescope of the transit or theodolite was leveled, its height of instrument (H.I.) determined and from that value it was determined what the FS rod reading should be when the rodman was on the desired contour. For instance, if the H.I. was 457.2 ft, the rodman would be on the 450 ft contour if the FS reading was 7.2 ft. Therefore, he or she would be waved up or down the slope until the rod reading was 7.2 ft. Then the distance and direction to that point would be measured. After a sufficient number of points were located for the 450-ft contour line, the surveying party would work on say the 448-ft contour line by positioning the rodman at points where the rod readings were 9.2 ft.

14-13 PROFILES FROM CONTOUR MAPS

One of the important advantages of contour maps is the fact that their users can quickly plot the profiles for lines running across the map in any direction. This is illustrated in Figure 14-16, where the profile for line *AB* is plotted. Such a profile presents information that can conveniently be used for establishing grades for various projects and estimating earthwork quantities.

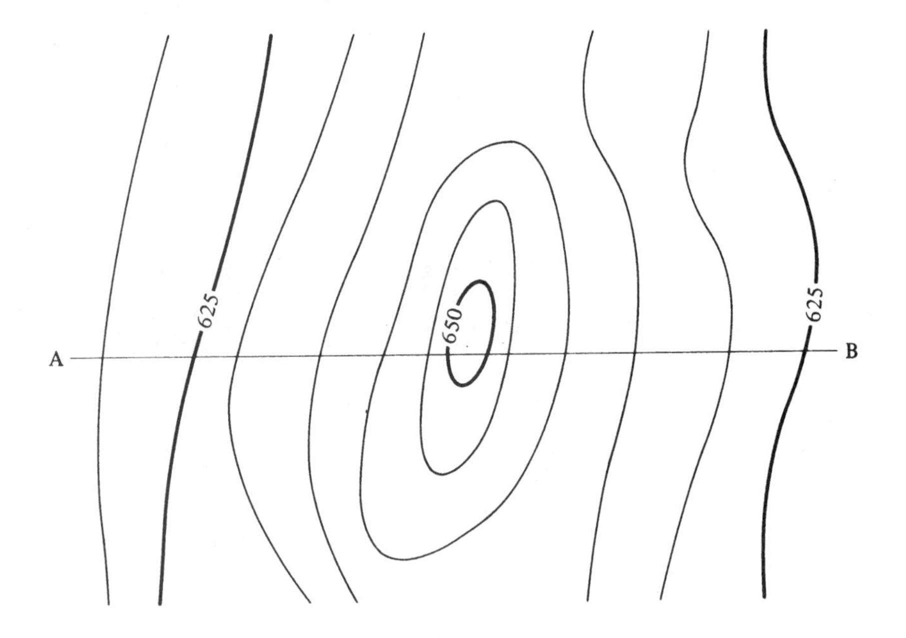

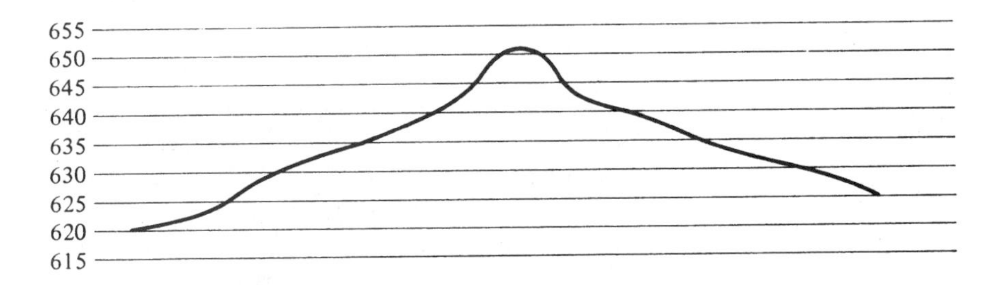

FIGURE 14-16 Contour map and profile for line *AB* plotted from the contour map.

14-14 CHECKLIST OF ITEMS TO BE INCLUDED ON A TOPOGRAPHIC MAP

A checklist of items that may be needed for a topographic survey is presented in this section. This list should be helpful for surveyors in the field as well as for persons drawing maps in the office. The most important item is for all parties to be aware of the purpose of the map. If a surveyor were to imagine that he or she were preparing a map of his or her own property for some type of development and wanted to list the items that would be of significance, the result would probably be a list similar to the following:

1. Location of property (as to state, county, town, highways, etc.).
2. Direction of north meridian.
3. Available access to property (from highways, railroads, etc.).
4. Information concerning property corners and monuments as well as the lengths and directions of property lines and land area.
5. Sufficient elevations to plot contours and show ridges and valleys.
6. Locations, sizes, and descriptions of buildings on property.
7. Locations and sizes of any roads (in use or abandoned) on or near property. Similar information concerning railroads.
8. Location of power lines, water lines, sewer lines, and other utilities on or near property.
9. Location and sizes of springs, streams, lakes, wells, and drainage ditches—also culverts, bridges, and fences. This information would include land areas subject to flooding.
10. Positions and areas of forest land, cleared land, cultivated land, and so on.
11. Description and location of any horizontal or vertical control monuments on or near the land.
12. Other significant features that the surveyor might think are of importance to the owner. Included in this list might be information concerning the characteristics and development of neighboring property.
13. Title, scale, legend, and names of surveyor and draftsperson.

Problems

14-1. What is a contour line?

14-2. Define the term "contour interval."

14-3. What factors should be considered in selecting the contour interval to be used for a particular map?

14-4. What is a saddle on a topographic map?

14-5. Where would a contour line have an M shape?

14-6. When would parts of contour lines be straight?

14-7. Describe how shading and hachure lines might be used to show relative elevations on maps?

14-8. The following values are the elevations in feet at the corners of 50-ft coordinate squares.

Draw 2-ft interval contours.

84.6	81.2	76.6	65.3	61.8
80.7	76.1	69.6	66.2	60.5
78.7	74.0	70.4	64.6	58.7
75.0	71.4	69.2	62.8	58.5

14-9. Repeat Problem 14-8 using 1-ft contour intervals and assuming that a fifth line of elevations is added at the bottom as follows:

70.3	67.3	64.9	62.4	56.6

14-10. The total station method is the primary ground procedure used today for obtaining topographic information. Why is this true?

14-11. Complete the following transit-stadia topographic notes if the elevation of the instrument station is 559.76 ft and the h.i. is 5.1 ft.

Point	Azimuth ⊼ @ B	Rod intercept (ft)	Vertical angle	Elevation difference	Horizontal distance	Elevation
1	29°16′	3.64	–8°22′			
2	35°18′	5.12	–4°46′			
3	51°33′	3.96	+3°16′			
4	77°37′	6.42	+5°38′			

14-12. When the transit-stadia method was used for obtaining topographic details, what were the advantages and disadvantages of the "checkerboard method"?

14-13. Describe the "controlling point" method of obtaining topographic details.

14-14. Describe the "tracing contour method" for obtaining topographic details.

CHAPTER 15

The Global Positioning System (GPS)

15-1 INTRODUCTION

Humankind has looked for an accurate system for locating points on the earth's surface since almost the beginning of recorded history. Such a system is now available—it is the global positioning system (GPS). This system may turn out to be the greatest surveying tool ever developed. With GPS points can be quickly and accurately located on earth by measuring distances to artificial satellites. You may be quite surprised to learn that locations of points on the earth and the distances between those points, whether short or long, can be determined as accurately or more accurately by measuring distances to satellites located thousands of miles out in space rather than by using conventional techniques right on the earth where the points are located.

In 1978 the Department of Defense (DOD) began to launch satellites into space (Figure 15-1) with the goal of being able to quickly and accurately locate positions on the earth. This system was kept secret for five years. Today the DOD has a network of 24 of these orbiting satellites with at least three spares. As the estimated life of each satellite is approximately seven years it is necessary to periodically launch replacements.

The gravitational pull of the mass of the earth keeps the satellites in orbit. Each satellite tries to fly by the earth in a straight line at about 8650 mph but the gravitational force of the earth pulls it down. As a result the satellite falls vertically in a path that is parallel to the earth's curved surface. If the satellite's velocity were decreased, it would fall to the earth, if increased it would leave the earth and move further out into space. The GPS satellites are approximately 12,600 miles above the earth's surface, where their orbits are completely free of the earth's atmosphere. They are located in six different circular orbital planes inclined 55° to the equator and spaced every 60° in longitude. They are manufactured by Rockwell International, weigh 1900 lbs each, and have a 17-ft span when their solar panels are extended.[1]

The system is referred to as the Navigation Satellite Timing and Ranging (NAVSTAR) Global Positioning System (GPS). Actually the designations NAVSTAR and GPS are used interchangeably.

[1]B. F. Kavanagh and S. J. G. Bird, *Surveying Principles and Applications* (Englewood Cliffs, N.J.: Prentice-Hall, 1996), p. 453.

FIGURE 15-1 GPS satellite. (Courtesy of Rockwell Semiconductor Systems.)

The original purpose of the satellite system was to enable planes, ships, and other military groups to quickly determine their geodetic positions. Although the system was developed for military purposes, it is of tremendous benefit to other groups, such as the National Geodetic Survey, the private surveying profession, and much of the general public, as will be discussed later. With perhaps a few exceptions (as for surveys in places where it is difficult or impossible to receive radio signals from the satellites, such as in mining work, where there are nearby tall buildings, and in deep forests), the GPS system can be used to accomplish anything that can be done with conventional surveying techniques.

A particularly useful feature of the satellites is that their radio signals are available free of charge to users anywhere in the world at any time of day or night and during rain, snow, fog, heat of day, or any other weather condition. Over 10 billion dollars has been spent by the DOD in establishing the system.

15-2 MONITORING STATIONS

In addition to the satellites in space the GPS system includes five monitoring stations on earth. These are located in Hawaii, Ascension Island (in the middle of the Atlantic Ocean between South America and Africa), Kwajalein (in the Pacific Ocean northeast of New Guinea), Diego Garcia (in the Indian Ocean), and in Colorado Springs, Colorado. The master control station is located in Colorado Springs.

The satellites travel around the earth in 11 hours and 58 minutes. As a result they each pass over one of the monitoring stations twice a day. Their altitude, speed, and position are carefully measured and the information is transmitted to the master station every few hours. Although the satellites are launched into very precise orbits they do tend to drift a little. These variations in positions (which are generally quite small) are due to gravitational pulls from the sun and moon and the pressure of solar radiation. The DOD relays the information concerning satellites positions as often as three times per day to the satellites and each satellite broadcasts the necessary corrections along with its signals.

15-3 USES OF GPS

In this section a brief summary is presented of the list of present and anticipated uses of the GPS system by the military, by other government organizations, by private surveying groups, and by the general public.

Military. The GPS system was originally designed for military positioning situations. Such items as the navigation of aircraft, ships, and land vehicles were included. Other military applications are photo reconnaissance, missile guidance, the monitoring of nuclear tests, and the positioning and navigation of other space vehicles.

There is a Russian GPS as well called GLONASS (Global naya Navigatsionnaya Sputnikova System or the Global Navigation Satellite System).[2] A California company, Ashtech, Inc., introduced the "GG Surveyor" in 1996. It is the first receiver to utilize both GPS and GLONASS. At one time GLONASS had 24 working satellites but the system is deteriorating and in July, 2002 only seven of their satellites were operational according to their web site.

The radio signals from both GPS and GLONASS are transmitted from the satellites to the receivers and not the other way around. This is very important from the military standpoint because they can establish the positions of their units without transmitting any signals of their own which would reveal their positions to hostile forces.

Several GPS receivers on the market today are capable of tracking both GPS and GLONASS satellites. As a result observers may be able to do their work more quickly and perhaps with greater accuracy utilizing both systems. In the spring of 2002 the European Union decided to proceed with a new constellation of satellites of their own for global positioning, a system called Galileo. The combination of GPS, GLONASS, and Galileo is called the Global Navigation Satellite System (GNSS).

The deployment of Galileo which will be complete in about 2009 will provide considerable benefit for civilian GPS users around the world. There will be increased availability of satellites at all times and the result will be better accuracy for observations. Furthermore, the new system will particularly benefit users who are working in places where the satellite signals may be blocked as in urban areas with tall buildings in the vicinity or in places where there are nearby hills or tall trees. For these situations the availability of more observable satellites may very well prove to be of considerable advantage.

Other Government Surveys. The National Geodetic Survey has roughly 240,000 control monuments located at intervals of 6 to 8 miles in most localities. In the past, to establish first–order triangulation points they built towers, ran theodolite lines at night, and so on. Today the costs of such work would probably run at least $6000 to

[2]R. B. Langley, "GLONASS: Review and Update," *GPS World,* July 1997, vol. 8, no. 7, pp. 46–51.

$8000 per point. With the satellite system it may eventually be possible to establish these monuments for as low as a few hundred dollars each. Furthermore, it may be cheaper not to establish these monuments any more. With satellites it is possible to quickly and cheaply determine the positions of any points desired without the use of fixed monuments.[3]

Private Surveyors. There are a great many situations where GPS can be useful to private surveyors. These include the following:

1. The location of points for long traverses, particularly where there are no previously positioned end points.
2. The measurement of very long distances which can't be handled with EDMs. Notice that with GPS the cost of measuring a 20-mile distance is about the same as that required for a 500-ft distance.
3. Surveys over rough country with difficult access.
4. Measurements across property to which the owners will not allow access.
5. The location of roads, power lines, water lines, and shore lines of bodies of water, the mapping of wet lands, and so on.

General Public. The uses of the GPS system for the general public will probably exceed by far the uses of all the preceding organizations just mentioned.

Approximately nine million GPS units were built in the world in the year 2001. Trying to predict all of the uses that will be made of these devices is extremely difficult and probably impossible. The public already uses receivers for:

1. Keeping track of fleets of trucks, railroad cars, rental cars, and taxis.
2. Navigation for ships and planes.
3. Uses by hikers, hunters, and bikers.
4. Locations for fishing and recreational boats.
5. Law enforcement and emergency vehicles.
6. Units in cars that display vehicle locations together with navigation help to desired destinations.

A common saying today and probably a true one is that "in the near future no one will ever again have to stop their vehicle and ask directions". In other words if their vehicle has a GPS on board they will be able to obtain excellent and immediate instructions to whatever destination they are seeking.

An increasing number of surveyors are using GPS to determine coordinates for the control of various types of construction and route projects. GPS is extremely well suited for aerial survey control, route surveys, and other projects that have rather isolated points. The method works exceptionally well for surveys that are to be referenced to a coordinate system. Software is readily available to convert GPS data to various types of coordinate systems. Perhaps someday GPS will be used for establishing accurate coordinates for every piece of land in the United States. Surveyors will be able to establish coordinates for every corner of a piece of property, regardless of how far it may be to a reference monument.[4]

[3]J. Collins, "A Satellite Solution to Surveying," *Professional Surveyor,* Nov./Dec. 1982, vol. 2, no. 6, pp. 13–17.

[4]J. Collins, "Shooting for a Full GPS Constellation," *Professional Surveyor,* July/Aug. 1989, vol. 9, no. 4, pp. 22–23, 40–41.

Satellites are being used today for the location of control points for important government surveys. They will undoubtedly be used much more frequently in the future to provide the location of natural and manmade objects. The information obtained will be used by engineers, tax offices, fire departments, and so on, to better manage their resources.[5]

15-4 BASIC THEORY

The measurement of distances from a position on earth to artificial satellites is called *satellite ranging.* Distances are determined by measuring the time required for radio signals sent from satellites travelling at 186,282 miles per second (299,792,458 meters per second) to reach our position. Of course the speed of light and radio signals are only constant in a vacuum. The actual satellite radio signals are not in a perfect vacuum and are delayed by interference with the charged particles in the ionosphere. These particles are held in place in the ionosphere by the earth's magnetic field.

Notice the extraordinary accuracy that must be used in determining the times. If we miss a time by 0.1 second, we miss the distance by (0.1)(186,282) = 18,628 miles. Today's best GPS receivers can measure time to perhaps the nearest nanosecond (0.000000001 second). If we are off that much in time, our satellite distance will be off by

$$(0.000000001)(186{,}282) = 0.000186282 \text{ miles or } 0.984 \text{ ft}$$

The time required for a signal to travel from a satellite to a receiver is measured. Then this time is multiplied by the speed of light and the distance obtained. The resulting value is referred to as a *pseudo-range* (Figure 15-2) where the prefix *pseudo* means "false" because there are some inherent errors in the time difference measurement. GPS consists of measuring distances from points whose location is unknown to satellites whose positions are known at the time the measurements are taken.

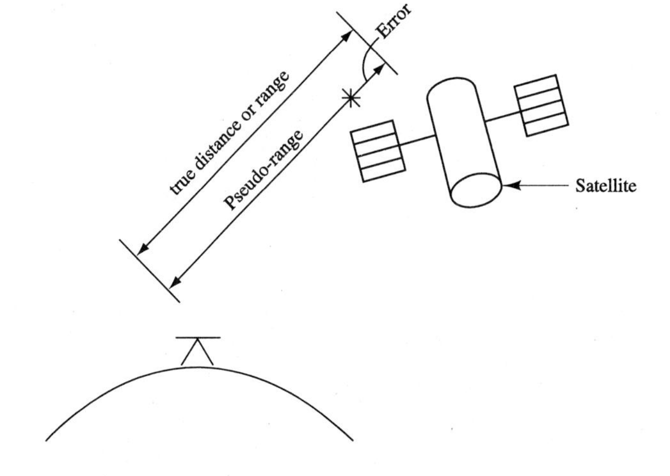

FIGURE 15-2 Satellite ranging.

If we can determine the distances to four different satellites whose positions are known, we will have four simultaneous equations that can be solved to obtain our

[5]Hurn, J., *GPS: A Guide to the Next Utility* (Sunnyvale, Calif.: Trimble Navigation, 1989), pp. 1–76.

position in the x, y, and z directions. For example, let's assume that we have received a signal from one satellite, which we calculate is 15,000 miles away. Our position is obviously located somewhere on the surface of an imaginary sphere of radius 15,000 miles whose center is at the satellite, as shown in Figure 15-3.

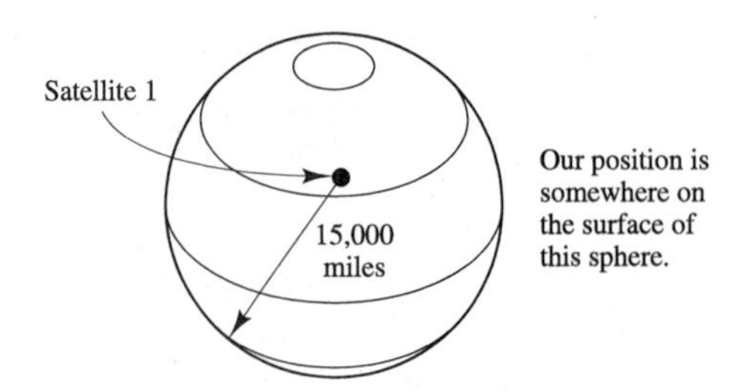

FIGURE 15-3 Distance measurement to one satellite.

Next, it is assumed that we are receiving radio signals from two satellites at the same time. The distances to the two satellites are calculated to be 15,000 and 12,000 miles, respectively. Our position obviously falls somewhere on the circle where the two spheres (of radii 15,000 and 12,000 miles) intersect, as shown in Figure 15-4.

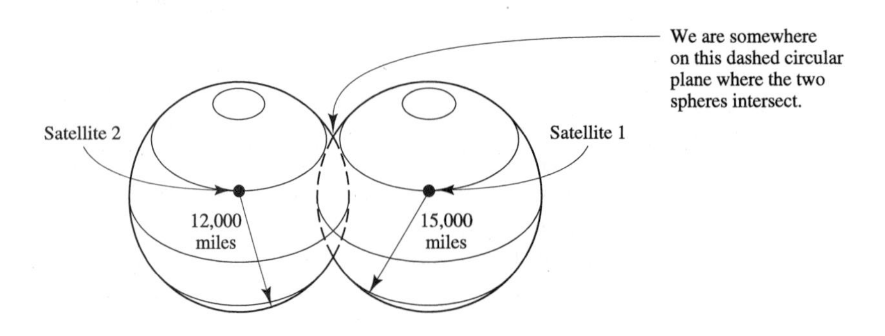

FIGURE 15-4 Distance measurements to two satellites.

Then let's assume that we are receiving signals from three satellites simultaneously and that the distances measured to them are 15,000, 12,000, and 16,000 miles, respectively. Now our position may only be located at one of two possible points in the universe. These are the two points where the three spheres intersect, labeled A and B in Figure 15-5.

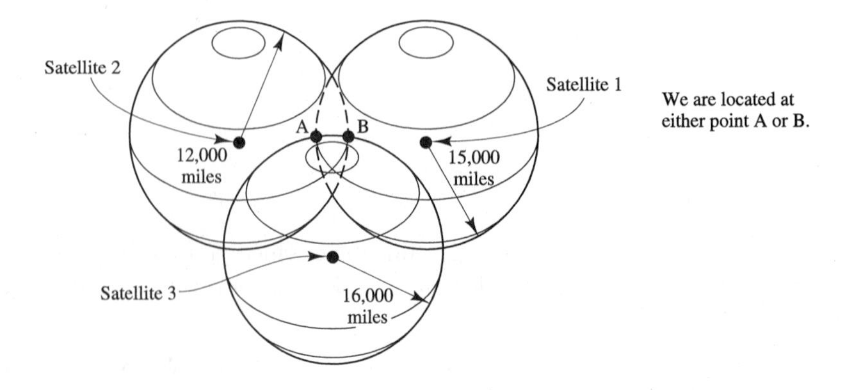

FIGURE 15-5 Distance measurements to three satellites.

Usually, one of these two points is obviously not possible (it is probably way out in space), and thus our position is known. Furthermore, the computers in GPS equipment are programmed to select the correct or feasible position. Notice also that if we know very accurately the elevation of our position above sea level, we have another sphere of known radius.

If we could make perfect distance measurements we could locate our position perfectly. Unfortunately, our time measurements (and thus our distance calculations) are somewhat in error. For this reason measurements are made to four or more satellites and we will find that this will enable us to eliminate the error in timing if that error is the same for all the measurements.

If we take measurements to three satellites, we will find that the spheres intersect at the two points A and B in Figure 15-5. If, however, the timing is in error, the points of intersection are wrong because the sphere radii are wrong. Now if we take a measurement to a fourth satellite (assuming that the measurements for all four satellites have the same error), the four spheres will not intersect at one point. As a result, the computers in the receivers are set to keep adding or subtracting small time increments for all four measurement times until one intersection solution is obtained for all the spheres.[6] Actually the computer does not search aimlessly for the clock offset. Using four simultaneous equations, which are presented in section 15-5, the necessary correction can be determined.

15-5 HOW CAN THE TRAVEL TIME OF A SATELLITE SIGNAL BE MEASURED?

Suppose you and a friend are standing about a block away from each other and you both yell "Hi" at the same instant. Your friend's "Hi" will not reach you until a little while after you have finished yelling. The time interval is the time required for sound to travel from your friend to you. You can multiply this interval by the speed of sound (1129 ft/second @70°F) and estimate the distance between the two of you.

We will see that this is the general idea on which GPS distance measurements are made. Each satellite transmits a unique coded signal every millisecond. The receiver (Figure 15-6) is synchronized with the satellite so that it generates the same code at the same time. After the satellite's signal is received we can go back and determine how much time has elapsed since the same code was generated by the receiver. Once that information is available the pseudo-range can be computed by multiplying the elapsed time by the speed of light. The various errors occurring in these measurements are discussed in Sections 15-6 to 15-8, along with methods for making corrections to them.

The codes transmitted by the satellites and generated by the receivers each consist of a string of bits (these are digital ones and zeros) arranged in a rather complex pattern. The code is referred to as the PRN-code (where PRN stands for pseudo-random noise). Each signal looks something like the one shown in Figure 15-7.

In the memory of each receiver is a replica of each of the twenty-odd satellite codes being broadcast. When the receiver picks up a satellite signal it immediately detects which satellite is sending the signal. The receiver compares the signal it is receiving with the same code it has internally generated. The pattern generated by the receiver will not match in position that of the incoming signal because of the time elapsed. This is the situation shown in Figure 15-8.

[6] Hurn, J., *GPS: A Guide to the Next Utility* (Sunnyvale, Calif.: Trimble Navigation, 1989), pp. 29–30.

FIGURE 15-6 GIR 1000 System with 12 parallel receiver channels for GPS and differential GPS surveys. (Courtesy of Sokkia Corporation.)

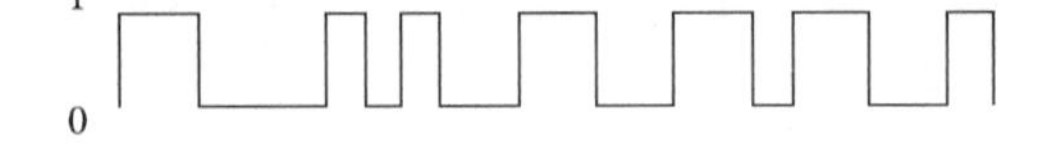

FIGURE 15-7 A typical satellite signal.

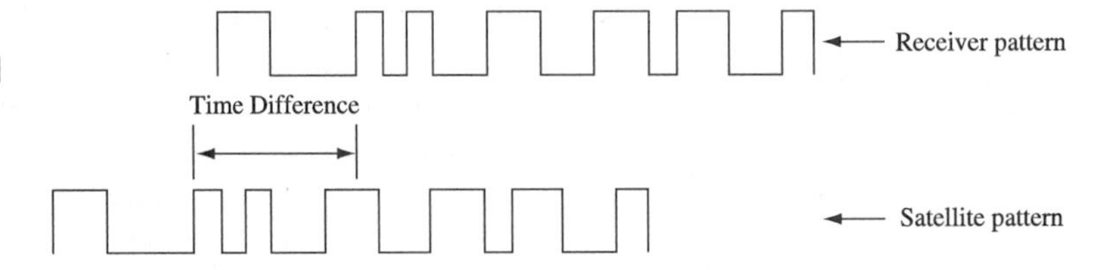

FIGURE 15-8 Time required for satellite signal to reach receiver.

The receiver slides its pattern along in time until it lines up with the satellite's pattern. The travel time of the signal equals the time which the receiver must slide its pattern along the satellite's pattern to line them up. Of course, this cannot be done with absolute perfection and the result is a few meters of error.

You may wonder how we are going to do all this wonderful timing we've been talking about. Well, four enormously expensive and precise clocks costing about $100,000 each are used in each satellite. (These are often referred to as atomic clocks even though they don't run on atomic energy. Their name comes from the fact that the oscillations of a particular atom are used in the metronome.) Four clocks are used to be sure that at least one of them is always working. Each receiver gets by with one relatively economical but very good quartz clock. (If each receiver had a clock equivalent to those used in the satellites hardly anyone, except perhaps the federal government, could afford to have receivers and use the GPS system.) Theoretically, measurements to only three satellites are required to establish a position. However, as described in the last section, a measurement to a fourth satellite is needed so that corrections can be made for timing errors. Among those errors are the ones caused by the less precise clocks used in the receivers.[7]

15-6 CLOCK BIAS

If we know the correct distances or ranges from our receiver to three satellites whose positions x_1, y_1, z_1: x_2, y_2, z_2, etc, are known we can determine our position on earth (U_x, U_y and U_z) by solving the following three simultaneous equations in which Eu_{U1}, E_{U2} and E_{U3} are the pseudo-ranges determined as described in the preceding section.

$$E_{U1} = \sqrt{(x_1 - U_x)^2 + (y_1 - U_y)^2 + (z_1 - U_z)^2}$$

$$E_{U2} = \sqrt{(x_2 - U_x)^2 + (y_2 - U_y)^2 + (z_2 - U_z)^2}$$

$$E_{U3} = \sqrt{(x_3 - U_x)^2 + (y_3 - U_y)^2 + (z_3 - U_z)^2}$$

To determine the distances from our position to a set of satellites correctly it is necessary to have the satellite and receiver clocks closely synchronized. As this cannot be done perfectly there will be a clock bias or timing error in our work. The result will be large errors in the distances and coordinates obtained.

If, however, the clock bias is assumed to be constant for each of the satellite observations, the coordinates of our position can be obtained by making an observation to a fourth satellite. To accomplish this objective it is necessary to solve the following four simultaneous equations in which c represents the velocity of the radio signals (186,282 mps) and dT is the timing error or clock bias.

$$E_{U1} = \sqrt{(x_1 - U_x)^2 + (y_1 - U_y)^2 + (z_1 - U_z)^2} + c(dT)$$

$$E_{U2} = \sqrt{(x_2 - U_x)^2 + (y_2 - U_y)^2 + (z_2 - U_z)^2} + c(dT)$$

$$E_{U3} = \sqrt{(x_3 - U_x)^2 + (y_3 - U_y)^2 + (z_3 - U_z)^2} + c(dT)$$

$$E_{U4} = \sqrt{(x_4 - U_x)^2 + (y_4 - U_y)^2 + (z_4 - U_z)^2} + c(dT)$$

In practice it is customary to keep a receiver at a station for a sufficient time to make observations to more than four satellites. Each additional pseudo-range (beyond four) provides an additional equation for determining dT and the coordinates of the position. These extra or redundant equations are solved by the receiver with improving accuracy for each additional satellite observed.

[7]Hurn, J., *GPS: A Guide to the Next Utility* (Sunnyvale, Calif.: Trimble Navigation, 1989), pp. 18–25.

15-7 GPS ERRORS IN DETAIL

Several types of errors occur in GPS surveying. These include errors due to atmospheric conditions, equipment imperfections, and others. The errors are briefly described in the next few paragraphs and in more detail in Chapter 16. As we will discover in Section 16-8 most of these errors can be greatly reduced by a process called *differencing*.

Atmospheric Errors. We learned in school that the speed of light was a constant 186,000 miles per second (it is actually 186,282 mps) at least in a vacuum as supposedly exists in space. When light passes through denser mediums, however, such as the heavily charged particles in the ionosphere (the outer part of the earth's atmosphere) and the water vapor in the troposphere (the part of the earth's atmosphere next to the earth's surface) it slows down. The more expensive GPS receivers add in correction factors for these items. Such corrections are helpful but are not perfect. Furthermore, the atmosphere is constantly changing.

Multipath Errors. When transmitted signals arrive at the surface of the earth they may be reflected from other objects before they reach the receiver, thus causing time values to be slightly large. Such errors are called multipath errors because the signals come to our receiver from more than one path. An illustration of this situation is shown in Figure 15-9. We have all seen examples of this phenomena while watching TV when a multiple image appears on the screen (referred to as "ghosting"). The signal from the TV station has taken more than one path and the result is overlapping images.

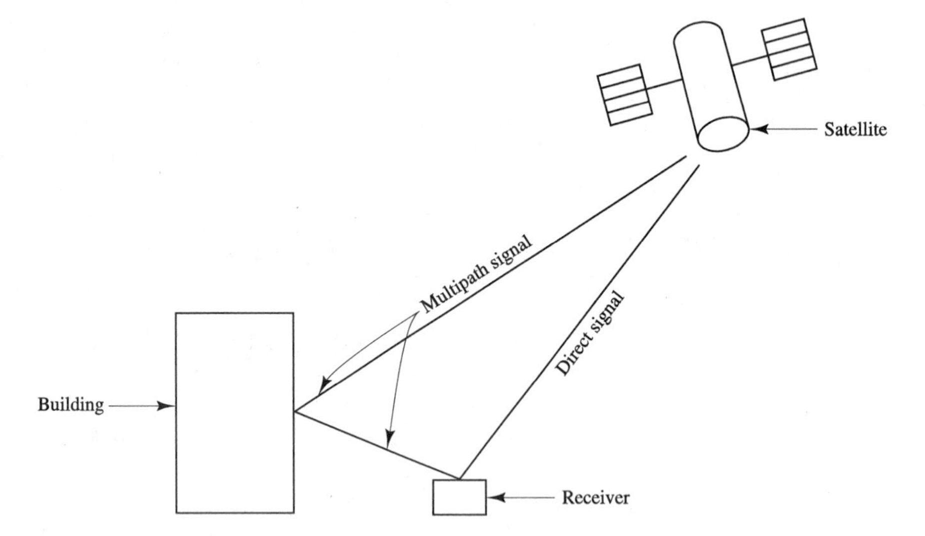

FIGURE 15-9 Multipath error.

Satellite Errors. Obviously timing is an extraordinarily important part of GPS work. Even though satellites are equipped with very accurate atomic clocks they aren't perfect, and slight errors in their measurements cause large errors in our distance calculations.

Other errors are caused by the fact that satellites drift a little from their predicted orbits due to the gravitational forces from the moon and sun and due to the pressure of solar radiation. The results are satellite ephemeris errors that affect our predicted satellite locations. An *ephemeris* (plural ephemerides) is an astronomical almanac, which provides the positions of various bodies in space such as the sun, moon, planets, and other stars at certain intervals. Here the term is used to refer to an almanac that provides information concerning the position of the artificial satellites circling the earth. As previously mentioned, the DOD sends corrections from their monitoring stations to the satellites and these corrections are broadcast by the satellites along with their timing signals.

Receiver Errors. As is the case for all instruments the receivers are not perfect, with the result that errors are caused. These usually are due to the imperfections of the clocks and due to the presence of internal noise. (Noise is referred to as anything that is not the radio signal. It comes from random electrical disturbances.)

Setup Errors. These errors are caused by imperfect centering of the receiver over points and by imperfect measurement of the h.i. of the equipment. Centering errors can be greatly reduced by careful use of a properly adjusted optical plummet. The h.i. errors can be minimized if equipment is used that has built in or accessory equipment for measuring those values.

Selective Availability. Several types of errors caused by our instruments have been described in the last few paragraphs. Until May 1, 2000, there was another type of error and it was deliberately caused by the DOD. This error was referred to as *selective availability*. The satellite system was conceived, developed, and paid for by the military establishment. They did not wish to furnish this asset to our enemies. Although they have control of the system, they have stated that they have no plans to completely deny use to the general public except in time of emergency. It is still their intent to retain control of GPS so that the United States and its allies will be able to exploit the system to their advantage in time of war.

Selective availability (SA) was caused intentionally by the DOD by feeding the satellites with wrong information—primarily clock and ephemeris errors. In other words the DOD can alter the satellite positions or radio broadcast messages from the satellites and in this way causes errors in measurements, and those errors can be varied in size and magnitude.

This "denial of accuracy" did not really affect the work of surveyors with two GPS receivers that were working in a differential survey mode. In a differential survey mode, one receiver is placed at a point whose coordinates have previously been located, and the other receiver is placed at a point whose coordinates are desired. It is possible in this way to determine the differences in coordinates between the two points to great accuracy (about ±2 ppm).

It is still the intent of the military to retain control of GPS so that the United States and its allies will be able to exploit the system to their advantage in case of war.

15-8 MINIMIZING ERRORS

It is possible to roughly estimate the errors caused by the speed of light as it passes through the atmosphere and to apply those errors as corrections to the travel time of the signals. Unfortunately, atmospheric conditions change from day to day as do the errors involved. A better method for determining errors is readily available, however, because of the relatively large distances to the satellites as compared to the short distances between the reference and roving receivers on earth. These latter distances are probably not more than a few score, or at most a few hundred miles.

The reference receiver is placed at a point which has been accurately located. This instrument receives the same signals as do the roving receivers. As the position of a particular satellite is known from the ephemeris the distance from that satellite to the reference receiver can be computed. This distance is theoretically equal to the measured time of travel multiplied by 186,282 miles per second. The difference between the known distance and the theoretical distance is the error of measurement. It can be divided by 186,282 to obtain the timing error.

As the distance from the reference station to a particular satellite is roughly equal to the distances from the roving receivers to that satellite it seems logical to assume that the timing errors are the same. Thus the reference receiver sends this correction to the roving receivers. Of course, the errors for the signals of different satellites will be different because their distances from the reference receiver will vary widely. As a result, the reference receiver will have to identify the satellites from which signals are received and provide different corrections for each. The information is used by the roving receivers to compute their positions. You should realize that the magnitude of these timing errors are in the order of a few nanoseconds.

It is important to understand that the errors described can't just be determined once and used to correct the measurements made all that day. They are constantly changing and there must be two receivers working simultaneously at all times to do the job.

15-9 POINT POSITIONING AND RELATIVE POSITIONING

There are two GPS methods used for locating points. These are referred to as *point positioning* and *relative positioning*. With the first of these methods a GPS receiver is placed at a point whose position is desired. With observations made on the satellites with a single receiver the three coordinates (X, Y, and Z) of the point can be determined with an accuracy varying from a few meters up to 20 or more meters. The accuracy obtained depends on the time spent in making the observations, the quality of the receiver, and on the accuracy with which the positions of the satellites are known.

When greater accuracy is needed, as for surveying applications, it is necessary to use the relative positioning or differential positioning method. With relative positioning two receivers are simultaneously used to receive the signals from the satellites. One of the receivers is placed at a point whose position has previously been determined with great accuracy while the other receiver is placed at a point whose position is desired. With appropriate software the computer determines the X, Y, and Z differences between the two points. Then the position of the new point with respect to the fixed point can be determined with an accuracy of a few millimeters and the precision of the line measurement between the two points can approach or exceed 1/1,000,000.[8] Of course the procedure described here for differential positioning can be repeated for as many other points as desired. Furthermore, we may be using several roving receivers simultaneously placed at different points.

We can obtain the best accuracy in the shortest period of time by tracking five or more satellites. Excellent results can be obtained with four satellites but more time will be required.

[8]Reilly, J. P., *Practical Surveying with GPS* (Canton, MI: P. O. B. Publishing Co., 1992), pp. 1–4 to 1–7.

15-10 RECEIVERS

The first GPS commercial receiver was put on the market around 1980. It was the Standard Telecommunications ST1 5010 and its price was several hundred thousand dollars. Today there are several hundred different receivers on the market (Figure 15-10) and their prices vary tremendously depending on their capabilities. Some small receivers can be purchased for prices under $200 (2003). In the $1500 to $5000 price range better receivers can be obtained—ones that can provide much better accuracy and ones that can store files for post-processing with base station files. For survey grade work receivers (which provide results in the mm accuracy range) prices can range up to as high as $30,000 or $40,000 with average prices in the $12,000 to $15,000 range. It is to be remembered that the surveyor normally needs at least two of these receivers to carry out his work. Thus, for survey grade work and two receivers we are talking about a minimum investment of at least $30,000 to $50,000.

FIGURE 15-10 Topcon Odyssey receiver with 20 channels. (Courtesy of Topcon Positioning Systems, Inc.)

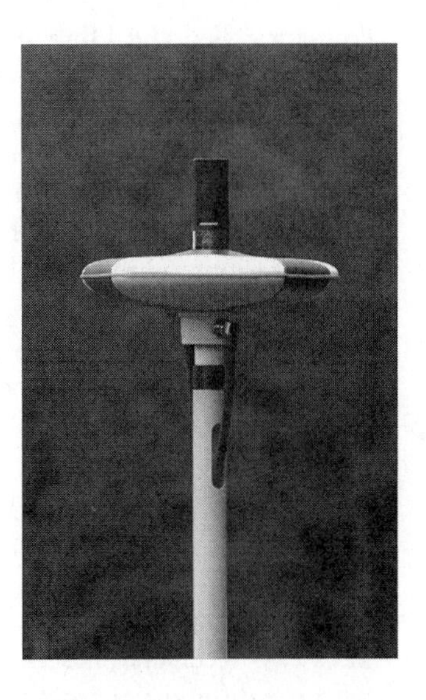

For mapping and Geographic Information System or GIS work (see Chapters 17 and 18) receivers priced in the $3,000 range (with accuracy in approximately the 1 meter or less range) will probably be satisfactory. For marine navigation receivers costing about $1000 each (with accuracy in roughly the few meter to 100 meter range) are normally sufficient. Then there are the receivers that cost about $100 to $300 and are used by hikers, hunters, and perhaps by persons making low accuracy maps.

As we have earlier learned it is necessary to observe at least four satellites to establish an accurate position on earth. Receivers are divided into two large groups: those being the ones that can simultaneously track four or more satellites at a time and those which cannot and must sequence between the satellites.

One and Two Channel Receivers. Receivers that must sequence between satellites are either one or two channel receivers. The single channel ones must move that channel from one satellite to another to gather data. Such sequencing can interrupt the positioning and cause a considerable reduction in accuracy. However, for hiking, recreational boating, hunting, fishing, and the like they are very satisfactory and can be used anywhere on earth.

The addition of a second channel to receivers greatly improves their capability as they can pick up signals under more adverse conditions and can track satellites near the horizon. For such units one channel is used to continually monitor the positioning data while the other channel is focusing on the next satellite.[9]

Continuous Receivers. For surveying and other scientific purposes receivers that can track four or more satellites simultaneously and provide instantaneous positions and velocity are necessary. Many of these units are manufactured so they can simultaneously track all of the visible satellites.

15-11 HARN

The preceding sections of this chapter have clearly shown the accuracy that can be obtained with differential positioning and the need for reference stations. The NGS and various state groups have established a system of such stations around the country called the High Accuracy Reference Network (HARN). The positions of these stations were themselves established using GPS methods.

The station locations are referenced to the coordinate system established by the NGS called the North American 1983 Datum (NAD 83). The NGS is committed to maintaining about 1400 of these stations.

There are now available to GPS surveyors in their vicinity some thousands of other stations maintained by the various states. Information concerning the locations and accuracy of these points can be obtained from the NGS. Notice that when using a HARN station it is necessary to have one receiver placed at the station and another one placed at the point whose position is desired.

15-12 CORS

The preceding sections of this chapter have seemed to indicate that it is necessary to have two receivers (and thus two people) working simultaneously to locate points with the differential positioning method. It is,however, advantageous on many occasions to have a system with which there is a fixed station which does not require the attendance and presence of a receiver and operator.

Today there is in the United States a network of several hundred stations called the Continuously Operating Reference Stations (CORS). At each of the stations of this network, which is sometimes called the National CORS Network, there is a continuously operating receiver and transmitter. These transmitters provide positioning information for many private and government agencies involved in mapping, navigation, GIS work, and so on.

If a surveyor is within range of the CORS (say 100 to 150 km), he or she will be able to locate points vertically and horizontally to the nearest 1 or 2 centimeters with the error becoming larger as the distance from the CORS station increases (about 1 mm/km). Work done with CORS will not be up to survey grade work (mm) as are the positions obtained from the HARN stations. It will be noted that the information from a CORS station will need to be used in the post processing work before the positions of the desired points can be obtained.

[9]Hurn, J., *GPS: A Guide to the Next Utility* (Sunnyvale, Calif.: Trimble Navigation, 1989), pp. 62–69.

There are some very slight differences in positions established using the CORS and HARN systems but these differences are being adjusted. The differences are due to the fact that the CORS stations are located within ±1 to 2 centimeters while the HARN positions were established from the NAD 83 coordinate system, which in general has accuracies of ±5 to 10 centimeters.[10]

15-13 GPS SIGNALS

The satellites broadcast their signals on two L–band frequencies. These are the L1 signal with a frequency of 1575.42 MHz and the L2 signal at 1227.60 MHz. The L1 signal is modulated with two codes—the course acquisition (C/A) code, which is available to the public, and the precise or protected (P) code, which is designed for military use. Supposedly more accurate measurements can be made by the military with the P code.

These signals are very low (those from our local radio and television stations being several thousand times stronger) and thus have very little penetrating power. For instance, solid metal surfaces completely stop them, as do tree trunks and limbs only a few centimeters thick. If the signals pass through as little as 1 centimeter of water they fall off so much as to be unusable. You can see that the leaves on deciduous trees will have a very detrimental effect on the signals and imagine what would happen if the trees are covered with snow or ice. On the other hand, glass or thin plastic has little effect on the signals.

15-14 GPS ON THE WORLD WIDE WEB

Due to work in the Department of Geography at the University of Texas at Austin an overview of GPS is available on the Web. Included in a group of Internet teaching modules is not only information on GPS but also overviews of Coordinate Systems, Map Projection, and the Geodetic Datum. This information is available to a worldwide audience.[11]

Among the users of this resource are not only surveyors and related professionals but also a large number of recreational users of GPS. The Uniform Resource Locator, URL or address, for the GPS Overview is as follows:

http://www.utexas.edu./depts/grg/gcraft/notes/gps/gps.html

Problems

15-1. What is satellite ranging?

15-2. What is a pseudo-range?

15-3. Why is it desirable to make measurements to four or more satellites when determining positions on the earth?

[10]Cheeves, M. "CORS Update" *Professional Surveyor*, March 1997, vol. 17, no.2, pp. 34–35.
[11]Dana, P. H. "Home on the Web: The University of Texas Builds a GPS Site," *GPS World*, November 1995, vol. 6, no. 11, pp. 44-51.

15-4. What is GLONASS?

15-5. Why is it so important to the military that the radio signals be broadcast from the satellites and not from the observer's positions?

15-6. What are multipath errors in relation to the satellite signals?

15-7. What is the meaning of the term *selective availability?*

15-8. What is an ephemeris?

15-9. Is GPS satisfactory for all field situations? If not, explain where it is not satisfactory.

15-10. What is CORS? HARN?

15-11. In speaking of GPS, what is meant by the term *kinematic surveying?*

CHAPTER 16

GPS Field Applications

16-1 GEOID AND ELLIPSOID

The earth is not quite a perfect sphere. It is flattened at its poles and its polar semi-axis is approximately 21 km less than its equatorial semi-axis. Its surface is roughly that of an *ellipsoid*, also called a spheroid. An ellipsoid is a curved surface that approximates the shape and size of the earth. (An ellipsoid is formed when an ellipse is rotated about its minor axis.)

A *geoid* is defined as a hypothetical figure representing the spheroidal shape of the earth but with its surface represented as mean sea level (MSL). The surface of this geoid is said to be equipotential in that the potential due to gravity is equal at all points on its surface. The surface is perpendicular to the direction of gravity at every point. The geoid is a surface that can vary as much as 100 m from the spheroid. Parts of a geoid, an ellipsoid, and the earth's surface are shown in Figure 16-1.

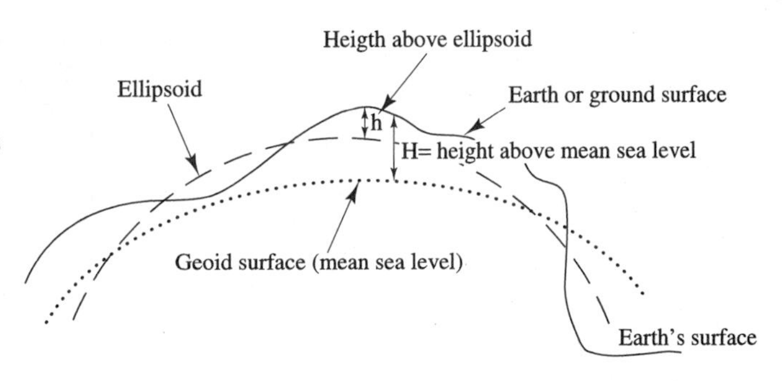

FIGURE 16-1 Three types of surfaces: geoidal, ellipsoidal, and earth's surface (vertical dimensions greatly exaggerated).

Surveyors have been accustomed to using spirit levels to measure vertical distances from the the mean sea level of the earth, or from the geoid. Generally, they have carried out plane surveys for small areas of the earth. Surveyors have in effect ignored the implications of *geodesy* (the effect that the curved or ellipsoidal shape of the earth had upon their work). The distances that they measured were relatively short and thus the global implications were negligible.

Today, however, the surveyor needs to realize that the elevation of a particular point (h) given by GPS work is actually the height to a point from the surface of the ellipsoid to the point in question. The surveyor, however, needs the height (H) above or below MSL. (See Figure 16-1.)

The ellipsoid used to picture the earth for GPS work is referred to as WGS 84 (where WGS means World Geodetic System). Satellite observations in the past few decades have enabled us to better our estimates of the size, shape, and mass of the earth. This ellipsoid is consistent with the ellipsoid called GRS 80 (where GRS represents the Geodetic Reference System of the International Union of Geodesy and Geophysics).

For the ellipsoid of the earth a Cartesian coordinate system is used where the center of the mass of the earth has the zero coordinates for *x*, *y* and *z*. The *x*-axis runs from this center of mass to a point on the ellipsoid's equator where it strikes the Greenwich meridian. The *y*-axis is perpendicular to the axis and the *z*-axis is perpendicular to the *xy* plane.

In effect, GPS observations enable surveyors to obtain the Cartesian coordinates for all points. These values can then be changed to latitude and longitude and ellipsoidal height. Finally, these latter values can be converted to state plane or other coordinates.

16-2 FIELD APPLICATIONS

GPS receivers (Figure 16-2) can be used to locate either stationary or moving objects. If a receiver is kept in one place, it will be possible to make repeated observations on the satellites. The receivers can make such a position calculation in less than one second.

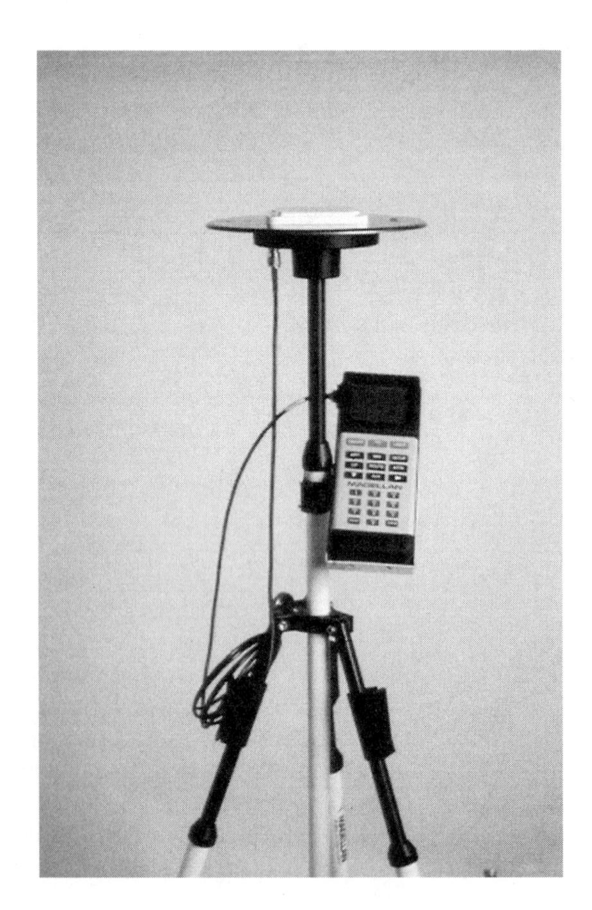

FIGURE 16-2 The Pro Mark X-CM with 10 parallel receiver channels for GPS and differential GPS surveys. (Courtesy of Magellan.)

The receivers can be quite small and easily carried. In fact, they are frequently small enough to be carried in one hand. The latest generation of receivers consume much less power than their predecessors, thus permitting the use of smaller and lighter batteries. Furthermore, the new receivers require only small antennas to receive satellite signals, even though the signals are rather faint.

GPS surveying is far less affected by weather than are most other types of surveying. Since it makes use of microwaves the work can be done in snow, rain, or fog. Even a buildup of snow on the antenna does not affect the accuracy of the work.

To be able to use GPS to locate points, it is not necessary to be able to sight between adjacent points, but the system does not work if there are obstructions blocking the satellite signals from the antenna on the equipment. It is necessary to have unobstructed views of the satellites for vertical angles greater than about 15° to 20° above the horizon. (Pocket compasses and clinometers are quite useful for locating satisfactory positions for observations. With a compass one can determine the directions to various obstacles and with the clinometer can measure the vertical angles from proposed points to the top of surrounding trees, hills, and buildings.)

Not all points are going to be located by the GPS; rather, certain control points can be located and traditional surveying measurements may be used to determine the location of other points with respect to these control points. One possibility may be to bring in a firm that specializes in GPS work to establish certain control points and then use conventional traversing methods for the remainder of the survey.

The satellite signals cannot penetrate water, soils, walls, or other obstacles. Thus they cannot be used for underground positioning or for underwater navigation. Furthermore, there can be problems in large cities with many tall buildings.

It is rather difficult to make GPS observations in forested areas. There needs to be a fairly clear view of the sky and suitable open areas may not be found. For such situations it may be necessary to cut some trees or raise the antenna above the tops of the trees. There are various lightweight portable towers on the market for this purpose. When using a tower remember to keep the antenna located directly over the point in question by plumbing and to carefully measure the height of the antenna above the point.

A very important use of GPS observations is with open traverses. If a long open traverse is run along a highway, power line, or railroad by the usual methods, a check is not available upon reaching the other end. Thus it is usually necessary to tie into some NGS monuments (if they happen to be reasonably close) or more likely to run a new set of lines back to the starting point and check the closure and precision of the resulting closed figure. Such a procedure is very time consuming and there are often precision problems if the lines are run back along approximately the same route as the initial ones. That is because the end angles can be very small, and slight errors in their values can greatly distort the traverse (thus resulting in weak measurements).

With GPS the end points can be located and perhaps a few intermediate ones along the route and the traverse can be tied into them. The results will be considerable saving in time and probably appreciable increases in accuracy.

16-3 STATIC GPS SURVEYS

If one receiver is placed at a point whose coordinates are known (from previous surveying) and the other receiver is placed over a point whose coordinates are desired, the process is referred to as *static GPS*. A surveyor may use more than two receivers and simultaneously collect data for multiple points perhaps many kilometers apart. The most accurate GPS observations are made with the static procedure.

With static GPS it is possible to observe repeated ranges from the satellites to desired points. With each new observation an improvement is made to the previously determined positions With first class equipment the relative positions of two points that may be many score miles apart can be determined closer than one centimeter.

Static GPS is generally used by the surveyor for control surveys where accuracy is extremely important. A disadvantage of the procedure is the length of the observations which may last 30 minutes to one hour or more depending on the receiver, the accuracy required, atmospheric conditions, the configuration of the observed satellites, and the distances from the base receiver to the roving receivers. As a result of the considerable time required for these observations, surveying crews jokingly refer to the system as "Getting Paid to Sit" (GPS).

The least accurate part of the GPS measurements is in the area of elevations, but if static surveys are planned and carried out carefully, reliable values may be obtained, and thus the need for conventional leveling may be eliminated or at least drastically reduced.

16-4 KINEMATIC SURVEYING

The development which has had the greatest impact on surveying in recent years is real-time kinematic surveying done with GPS to centimeter accuracy. It has changed GPS from a method used for control surveys into one that can be used for the collection of data for large-scale maps as well as one which can be used for the layout of engineering projects. The term *kinematic surveying* refers to differential positioning where one GPS receiver is located at a base station while another receiver is moved to other points. The roving receiver can be walked around a lake shore or along a road, or mounted on a bicycle, boat, or aircraft, and continuous positions obtained. Thus the trajectory of a moving object, such as a car, ship, or plane, can be obtained.

Kinematic surveying enables the surveyor to rapidly obtain topographic information in areas where the satellites can be observed (that is, away from forests or high-rise buildings). The accuracies achieved are approximately equal to those obtained when the roving unit is kept at each point for a short period of time. The values obtained are not sufficiently accurate for vertical control.

For kinematic surveying it is necessary for the roving receiver to be able to lock onto at least four satellites, while five or more are preferable. The lock must be continuously maintained, which means the roving receiver can't be carried through forests, between tall buildings, under overpasses, and so on. Thus, as a result of these equipment limitations, kinematic surveying is not a "cure-all" for all surveying.

With static GPS the points or objects being observed are stationary and observation times are long. On the other hand with kinematic GPS, it is necessary to locate moving objects or to locate a large number of points quickly as needed for topographic mapping or construction surveys. For instance, kinematic GPS is used for keeping track of the location of trucks, cars, ships, and planes. In these latter regards, kinematic GPS is very useful for establishing the location of ships when ships are making depth soundings or of planes when photos are taken for photogrammetric surveys.

16-5 DILUTION OF PRECISION (DOP)

It is to be remembered that the GPS satellites are in near circular orbits about 12,600 miles above the earth. Furthermore, they make one complete revolution of the earth every 11 hours and 58 minutes. As such, each satellite is visible at a particular locality for about 4 to 6 hours during each of its revolutions. The constellation is so arranged that no less than four satellites are visible at anytime anywhere on the face of the earth.

The strength of measurements is a very important topic in GPS as it is in all phases of surveying. This term refers to the effect of errors in measurement on the accuracy with which surveying is accomplished. If small errors in measurement affect final results very little, then the measurements are said to be strong. One illustration is presented here. If signals are being received from a group of satellites that are rather close to each other, the spheres shown in Figure 15-5 will not intersect each other at sharp angles as their surfaces will almost be parallel. Consequently, small errors in time measurement will result in large errors in the observer's computed position. The measurements are said to be weak.

If the angles between the incoming radio signals are small, the geometry will be weak and anticipated errors will be larger. If the angles are large, we will anticipate better results. Figure 16-3 shows examples of weak and strong geometry for GPS observations.

FIGURE 16-3 Satellite geometry.

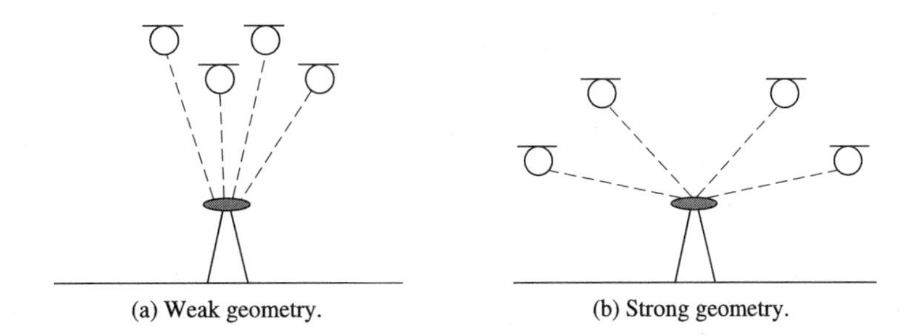

When GPS is used to determine positions on the earth, the accuracy obtained is very much dependent on the geometry or, that is, the position of the visible satellites at the time of the observation. The effect of the satellite configuration is expressed by the Dilution of Precision (DOP) factor. This factor is computed by the GPS unit using a least squares method to determine the effect of the satellite geometry on the measurement accuracy. The value obtained is called the Geometric Dilution of Precision (GDOP). The Positional Dilution of Precision (PDOP) is equal to GDOP corrected for time measurement errors.

The values of GDOP and PDOP are determined from the observations of several satellites. These values change as the satellites move along in their orbits. It is the objective to make observations when GDOP and PDOP are as small as possible. The more important DOP to consider is PDOP.

Should satellites be located at or near the horizon their signals will have to pass through more of the atmosphere than will the signals coming from satellites with larger vertical angles or elevations. As a result observers normally do not use signals coming from satellites with threshold or vertical angles less than arbitrarily assumed values of roughly 10 to 20 degrees above the horizon because of the magnitude of ionosphere errors. The assumed cutoff angle is called the *mask angle.*

Before a survey is made, the number and positions of the visible satellites at the proposed time and site of the observations is determined and the DOP value obtained. The weaker the geometric configuration of the satellites the larger will the DOP number become. It is actually the numerical ratio of the positional accuracy obtained to the measurement accuracy of the equipment.

As previously mentioned the measure of the geometrical strength of the array of satellites at a particular time and place is referred to as the Geometrical Dilution of Precision (GDOP). This value is due to geometry only and does not take into account obstacles that block or weaken the line of sight from the observer's position to the satellites. Although a good (low number) GDOP has been computed for a certain area and time, the observations made may still be poor if they have been adversely affected by obstacles such as trees, buildings, or hills. Other DOP factors include:

(a) HDOP - horizontal dilution of precision in position (latitude and longitude)
(b) VDOP - vertical dilution of precision in position (height)
(c) TDOP - time dilution of position

The predicted accuracy of a particular measurement equals the PDOP multiplied by the square root of the sum of the squares of the anticipated errors (clock, ephemeris, receiver, ionosphere, etc.). Under very good conditions the PDOP will be about 2.0. Under perfect conditions it will equal 1.0. The PDOP value should be 5 or less when observations are made.

Should the PDOP be greater than 5.0 the surveyor should delay his or her observations until a more favorable array or arrangement of satellites is available. When the PDOP is 1.0 the situation is as good as it is going to get. Such a value might occur when one satellite is directly overhead and three others are close to the horizon and separated from each other by 120°.[1]

16-6 PLANNING

As previously mentioned in Chapter 2 (Section 2-15) no other phase of surveying is more important than good planning. This statement is particularly true for GPS surveys. For such work 2, 3, 4, or more receivers may be used simultaneously to collect data from various points and those points may be some kilometers apart. Obviously sound planning to coordinate the various phases of the work is essential.

The goals of planning are first to obtain accurate measurements and secondly to do the work economically. With good planning surveying can be done more easily, more accurately, more economically, and more quickly. In this section several steps are suggested to do just that.

In the Office

1. Determine the accuracy required for the survey in question and select the GPS receiver or receivers to be used (if such a choice is available).
2. Obtain or prepare a map of the area involved. A USGS quad map or a GIS map will probably do admirably.
3. Examine the map to see if there are any previously established, carefully located government monuments. One is particularly interested in any points that have been located with accuracies equal to or exceeding those desired for the survey. Such points may serve very well as control monuments.

[1]Kennedy, M. *The Global Positioning System and GIS* (Chelsea, Michigan: Ann Arbor Press, Inc.) 1996), pp. 62–63.

4. Tentatively select a time period for doing field work (a particular day or days, a particular afternoon or morning etc.). Then with computer software plot the paths of the satellites for the times considered. Suppose that we are thinking of doing the fieldwork tomorrow afternoon. With our software we can plot the paths of the satellites that will be visible during that several hour period. Next, the GDOP values can be plotted for the same hours and a specific time period selected, which we think is best for the observations. A period in which the GDOP is 5 or below is usually considered to be satisfactory.
5. Prepare a list of the equipment needed to perform field reconnaissance and carry out the survey. In addition to the receivers, it is wise to have cell telephones to coordinate the work and perhaps two of those instruments previously declared in this book to be almost obsolete—clinometers (See Figure 3-17) and a compass (See Chapter 9).

Visit Job Site

For GPS surveys it is essential to make a preliminary visit to the job site. Some of the items one needs to note there include:

1. Search for existing monuments on the ground. GPS receivers may very well help in this endeavor. It is well to make some careful notes as to these locations so that the field crews will not waste time in finding the points.
2. Decide where to establish new control points. For such points and any previously located points, make notes of obstructions to observations such as buildings, trees, hills, etc. In this regard, the clinometer and compass will often prove quite useful. Carefully note the azimuths and elevations of the obstructions thus enabling one to later enter them into software for computer display.

16-7 EXAMPLE PROBLEM

In this section the determination of the latitude and longitude of a particular point is described. A static differential procedure is used with reference made to a nearby base station whose position is accurately known. The steps involved in making the measurement are also described in this section. Although the points involved are very close to each other, the same procedure could have been used if the points had been many kilometers apart. If one neglects transportation between the two points, the times and costs of measurement would have been equivalent.

A very important step in a GPS survey is the selection of the time period during which the observations will be made. In this case we have decided to try to make the observations during the middle of the day on October 9, 2002. Our first step is to make use of computer software that provides information as to the availability of visible satellites, their orbits, and the values of PDOP from 11:00 hours local time until 13:00 hours local time for the date in question. Software from several companies may be downloaded from the internet to determine this information.

As previously mentioned the orbits of the satellites vary just a little as time goes by due to solar radiation and gravitational pull from the moon and sun. The corrected orbits can be downloaded at any time from the internet. Although it is easy to download these corrections at frequent intervals, a few weeks old almanac is almost as satisfactory as an updated one because the changes are so very small.

Next, a mask angle is input into the computer (15° was selected for this case) and the best time is determined to conduct the survey from the satellite availability standpoint. To do this the software selected was used to print out the information shown in Figure 16-4 entitled "Number SVs and PDOP". The top part of the figure shows the number of visible satellites-SV (remembering a mask angle of 15° is being used) for the time period in question. An examination of the figure shows that the number of SVs varies from four to six during this time.

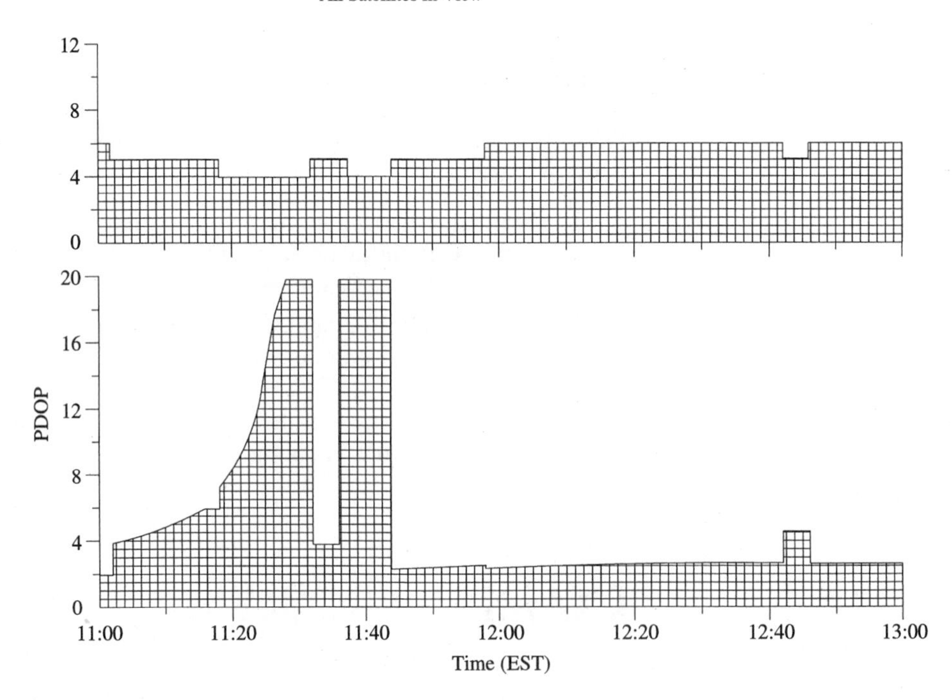

FIGURE 16-4 Mission planning satellite availability and PDOP.

At the bottom of the figure, the values of PDOP are shown for this same time interval. You can see that PDOP is extremely high and unsatisfactory at two times near 11:20 and 11:40. If you look at the upper part of the figure, you will see that these poor times occurred when the number of satellites went from five to four.

Plots can also be made with the software for the same time period for the azimuths to the visible satellites and for their elevations or vertical angles. These two plots can be combined into one plot called the *sky plot.* In this chapter the author includes only the elevation and sky plots. Also available from the software are plots of VDOP, HDOP and TDOP. On some rare occasion it may be possible to have high VDOPs and PDOPs but satisfactory HDOPs. Should this happen, and if one is only interested in *x*- and *y*-coordinates it is satisfactory to proceed with the measurements.

The elevation or vertical angle to each of the visible satellites is shown in Figure 16-5 for the 11:00 to 13:00 time period of the date in question. It is rather interesting to study this figure and compare it to the information given in Figure 16-4 as to the number of satellites visible above the mask angle and the PDOP values. For instance, in the first of these figures notice that the number of visible satellites changed from 6 to 4 during the period from about 11:18 to 11:38 and PDOP increased tremendously. Figure 16-5 shows that two satellites went below the mask angle during that time. Notice that after 11:45 the PDOP looks good as two more satellites have now moved above the mask angle.

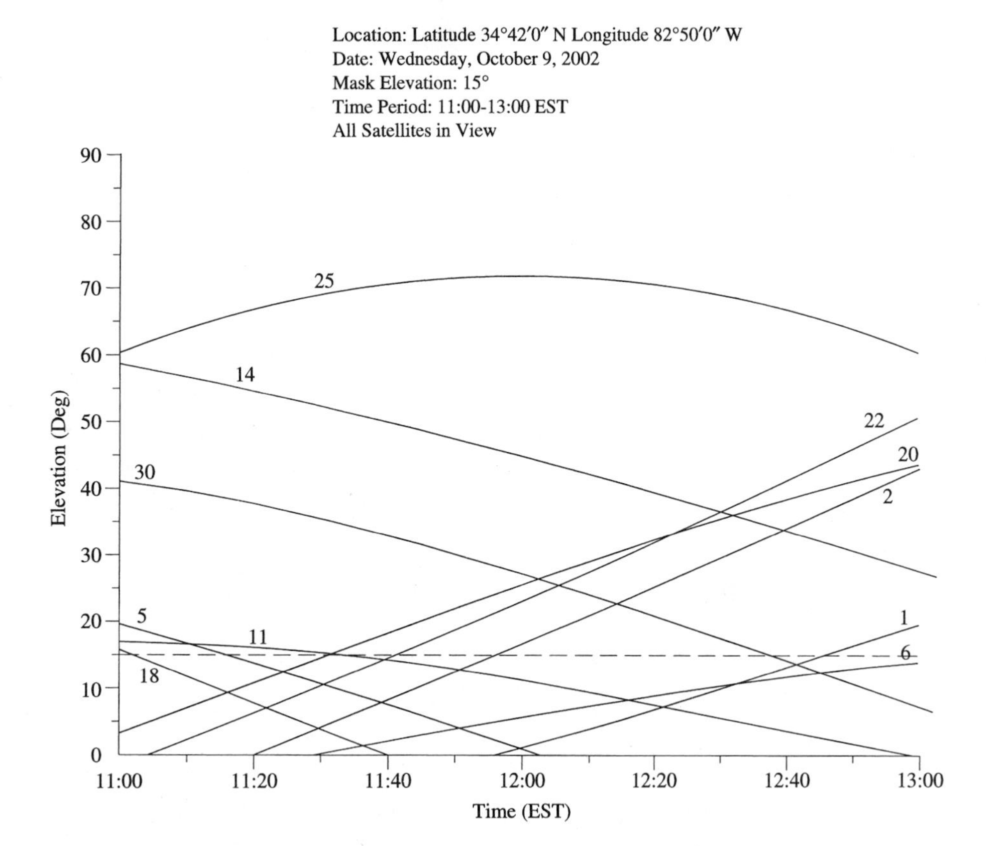

FIGURE 16-5 Mission planning elevation plot.

Next, a sky plot is presented in Figure 16-6. This figure shows the orbits of the visible satellites from 11:00 hours to 13:00 hours for the date in question. The satellites in this case are moving from their unnumbered ends toward their numbered ends. The satellite elevations, or their vertical angles from the horizon, can be determined from the concentric circles shown. The outside circle represents a 0° vertical angle while the next circle represents a 30° vertical angle. The next circle is 60° and the intersection of the NS and EW axes represents a 90° vertical angle. Notice also that the dotted circle at 15° represents the mask angle. Suppose we wish to determine the vertical angle to satellite #14 from the sky plot at 12:00 hours. As 12:00 is halfway between 11:00 and 13:00 we move halfway along the orbit shown for that satellite and see that it is about two-thirds of the distance from the 30° circle to the 60° circle (say 50°).

In the field the tripod is carefully set over the point whose coordinates are desired. If there is much wind or a nearby highway or railroad with heavy traffic, it is not a bad idea to use sandbags to steady the tripod legs from vibration. The

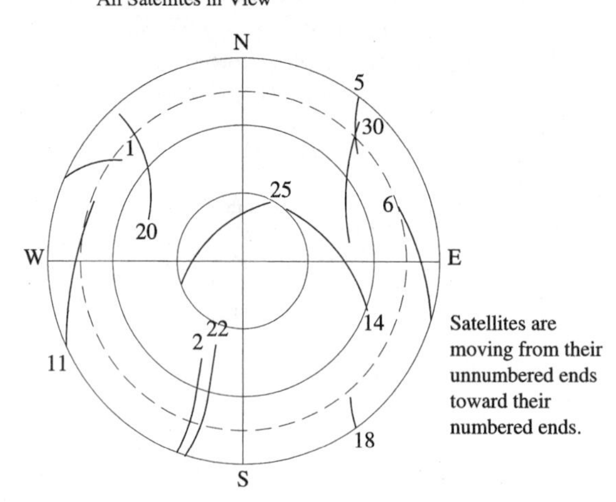

FIGURE 16-6 Mission planning sky plot.

antenna is attached to the top of the tripod and its cable attached to the receiver and to the antenna. The antenna height is measured very carefully and recorded. As significant errors often occur in such measurements the use of a fixed height tripod makes good sense.[2]

With a figure like 16-4 we obtained the best possible PDOP for our area. When an observer goes into the field and sets up a receiver, PDOP may be much higher because of reduced visibility caused by nearby buildings, trees, or hills. As one makes the observations the receiver will indicate the actual PDOP and one should adjust work accordingly.

In the field the receiver will be continually making observations of the satellites. If the observer limits the number of those observations to one every so many seconds or the data storage requirements will be kept within reason. The time interval set is referred to as an *epoch of time.* Later back in the office computer software will provide the latitude and longitude obtained with each observation and may be used to average the measured values if so desired.

16-8 DIFFERENCING

Differential positioning involves the use of two or more receivers to simultaneously measure the same satellite signals and to compute the values of the vectors (or baselines) between the receivers. Better accuracy can be obtained because of the correlation between the measurements. It's rather like leveling where the elevation of a particular point is desired and where that elevation is measured by leveling to the point in question over several different routes. Then its most probable value is computed from the various measured values as described in Section 8-1 of this text.

[2]Martin, D.J. "Static GPS: Field Procedure" *Professional Surveyor,* February 2002, vol. 22, no. 2, pp. 38–40.

Under the heading of differencing we have single differencing, double differencing, and triple differencing. These methods are described in the paragraphs to follow.

Single Differencing

With this procedure two receivers are simultaneously used to observe a single satellite. After the measurements are taken, different linear combinations of the equations can be written and as a result most of the satellite clock (bias) errors, orbit errors, and the delay errors due to the ionosphere and the troposphere (the lowest part of the atmosphere) can be eliminated. (See Figure 16-7).

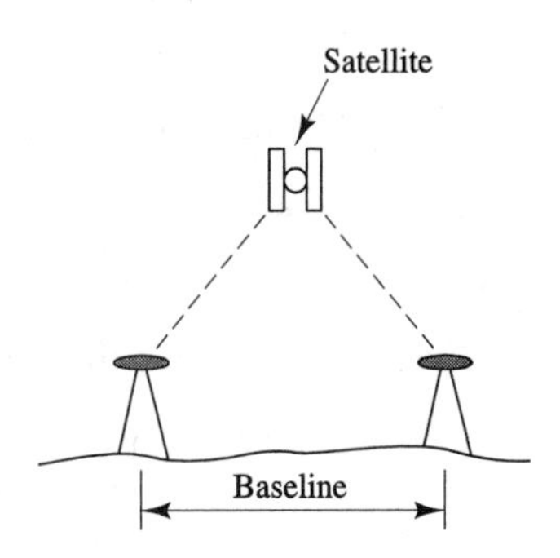

FIGURE 16-7 Single differencing.

Double Differencing

If, as mentioned, one receiver is used to observe two or more satellites it is possible to greatly reduce clock errors and atmospheric delay errors. Should two receivers be used to sight on two satellites as shown in Figure 16-8 a double difference is obtained and clock errors, atmospheric delay errors, and orbit errors are greatly lessened.

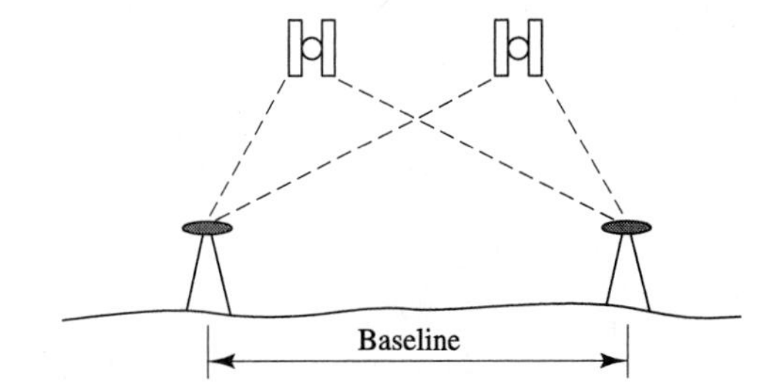

FIGURE 16-8 Double differencing.

Triple Differencing

Triple differencing involves the comparison of double differencing over two or more epochs of time. (An epoch of time is a moment in time.) In other words, double differences are compared over two or more successive epochs. (See Figure 16-9.) With triple differencing it is possible to find and minimize the effect of cycle slips.

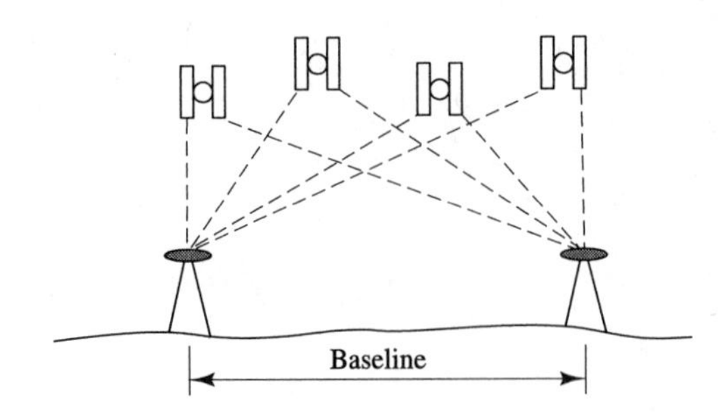

FIGURE 16-9 Triple differencing.

Problems

16-1. What is the ellipsoid or spheroid of the earth?

16-2. What is the geoid of the earth?

16-3. Distinguish between static and kinematic GPS surveys.

16-4. What is meant by the term "strength of measurements"?

16-5. What is GDOP?

16-6. What is the difference between GDOP and PDOP?

16-7. Why can the PDOP obtained with computer software for the general area where one is working be so different from the PDOP the receiver provides in the field?

16-8. What is a mask angle? What is its purpose?

16-9. What is a sky plot? What does it show?

16-10. What is the purpose of differencing in GPS work?

CHAPTER

Introduction to Geographic Information Systems (GIS)

17-1 DEFINITION OF GEOGRAPHIC INFORMATION SYSTEMS

Defining a geographic information system (GIS) is quite difficult because it is such a broad, complex, and rapidly changing technology. Even the experts do not agree on all the basic points of the definition. Nevertheless, the author believes very strongly that his first duty in this chapter is to implant the best definition of the topic that he can in the student's mind. To attempt to achieve such a goal he first defines an information system. Then he expands that system in both scope and definition in order to derive the definition of a GIS.

Information Systems

Table 17-1 contains some housing data concerning part of the state of South Carolina. The table includes names, areas, number of single family dwellings, multi-family dwellings, and mobile homes for some zip code areas in the state.

The data stored in this table can be processed to derive some rather useful information. For instance, for the zip code areas given one can quickly determine the number of single family dwellings, the number of multi-family dwellings, the number of mobile homes, the percentage of all housing units that are single family dwellings, and so on.

The entire process involves the collection of data, the storing of that data in a spreadsheet or table, and the analysis of that data to obtain some information from which various decisions can be made.

From the preceding paragraphs we can define an information system as being *the series of activities that includes the planning of the observations, the collection of data, the storage and analysis of the data, and finally, the use of the derived information in some decision-making process.*

TABLE 17-1 Partial Housing Data of South Carolina by Zip Codes

Zip	Name	Area sq km	Single family dwellings	Multi-family dwellings	Mobile homes	Total H. units
29001	ALCOLU, SC	205.393	520	24	246	790
29003	BAMBERG, SC	400.991	2032	248	464	2744
29006	BATESBURG, SC	415.594	2909	164	740	3812
29009	BETHUNE, SC	153.594	514	9	150	672
29010	BISHOPVILLE, SC	598.547	3282	345	1030	4656
29014	BLACKSTOCK, SC	178.573	184	4	70	259
29015	BLAIR, SC	51.779	63	4	19	86
29016	BLYTHEWOOD, SC	201.925	1735	46	782	2563
29018	BOWMAN, SC	320.442	1015	37	409	1460
29020	CAMDEN, SC	560.224	6731	880	1254	8865
29030	CAMERON, SC	237.835	734	29	362	1126
29031	CARLISLE, SC	349.396	614	15	252	880
29032	CASSATT, SC	109.417	163	3	77	243
29033	CAYCE, SC	23.092	3841	729	265	4833
29036	CHAPIN, SC	155.630	2534	37	448	3019

Geographic Information Systems

A geographic information system, on the other hand, enables someone not only to respond to queries that could be handled by an information system but also to handle spatial queries. The term *spatial* is used here to mean reference to particular positions on the earth's surface. For instance, in addition to the data presented in Table 17-1, a GIS would include information concerning the geographic location of all zip code areas in terms of their latitudes and longitudes. Given such information we can determine the total land area for all of the zip codes listed, the centroid (center of the mass of an area having equal density) of each zip code area, the number of single family dwellings within 5 miles of the centroid of zip code 29030, and so on.

With a GIS it is possible to handle all queries that can be handled with a regular information system and also to perform or answer spatial queries. There is also the capability of plotting maps since all data is geographically referenced. For this discussion consider the map shown in Figure 17-1. This particular map is the result of a query in which the density of single family dwellings in each zip code of South Carolina was derived.

This definition clearly distinguishes a GIS from an information system as one that has the capabilities of performing queries on spatially referenced data. In further establishing a definition of GIS it is important to outline the relationship of GIS with computer-aided design (CAD), computer cartography, database management systems (DBMC), and remote sensing information systems. As is evident in Figure 17-2 a GIS can be termed as a subset of the four listed technologies. While a GIS is not capable of replacing any one of those four technologies entirely, it shares some common capabilities with each of them.

A uniform referencing system that enables users to link data within a system to other related data is the basic fundamental of a GIS. A true GIS can be distinguished

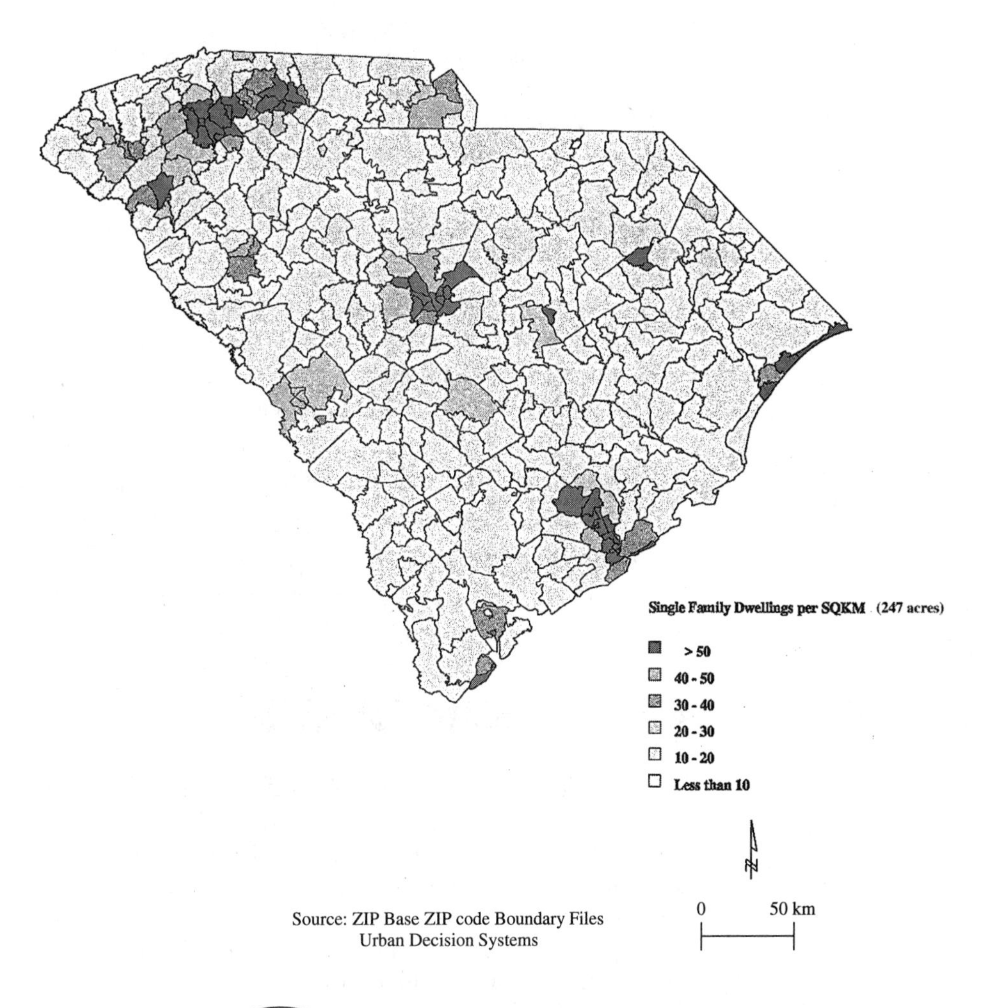

FIGURE 17-1
Density of Single Family Dwellings in SC by zip code.

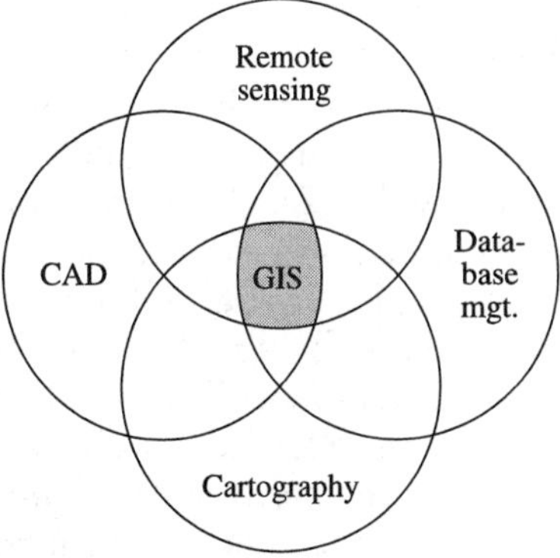

FIGURE 17-2
Relationship of GIS to four other systems.

from other systems by the fact that it can be used to conduct special searches and overlays, which actually generate new information.

17-2 WHY GIS?

Many organizations are spending large amounts of money on geographic information systems and on developing huge geographic databases. Predictions suggest that billions of dollars will be spent on these items over the next decade.

Rapidly declining computer hardware costs have made GIS affordable to an increasingly wider audience. More importantly, we have come to realize that geography (and the data describing it) is a part of our everyday world—almost every decision we make is constrained, influenced, or dictated by some fact of geography. A few examples follow:

1. Emergency vehicles are routed by the fastest available routes (not necessarily the shortest).
2. The federal government often awards grants to local governments based on population.
3. Many studies are made of various diseases. A large proportion of these are concerned with areas of prevalence of the diseases and the rates of spreading.
4. The impact of hurricanes can be predicted with the help of housing distribution, their dollar value, and the profile of the wind speed.
5. Routes are selected for highways, pipelines, transmission lines, and so on.
6. The geographic distribution of crimes and accidents are prepared.
7. Environmental impact of various projects are studied.

17-3 EVOLUTION OF GIS

The early beginnings of GIS were in the middle of the 18th century when the first accurate base maps began to be produced. Up until that time it was not feasible to accurately show the spatial attributes of points on the earth's surface. In the years that followed, maps that showed specific information about various ground features were introduced. These features included contour lines as well as the location of various items such as streams, buildings, and roads. This led to more variations of maps such as those used for military campaigns as well as those prepared for agriculture, forestry, and medicine.

Throughout the history of the development of GIS, the primary goal has been the desire to take raw data and transform it into new information that can support the decision-making process. The idea of recording different layers of data on a series of similar base maps was established soon afterwards. This process immediately opened up many applications in land management, agriculture, and military operations.

At the same time advances in physical and social sciences provided geographers with the intellectual tools needed for spatial analysis. Statistical techniques, number theory, and advanced mathematics were all beginning to be developed. The first geological map of Paris appeared in 1811 and it was soon followed by a geological map of London in 1815. The British census of 1825 produced enormous amounts of data to be analyzed. The science of demography (study of population characteristics) soon evolved. In 1854 Dr. John Snow overlaid a map of the city of London with the locations of the water wells in the city and the areas in the city where cholera deaths were particularly prevalent.[1] This enabled the city to find and close the dangerous wells. With steps such as these humankind began to record different types of data on different maps of the same areas.

Early modern systems of GIS received a great boost from the advent of the electronic computer in the 1940s. By 1952 London was processing its 1950 census data.

[1] P.R. Wolf and C.D.Ghilani, *Elementary Surveying: An Introduction to Geomatics*, 10th ed. (Englewood Cliffs, N.J.: Prentice-Hall, 2002), p. 832.

This process involving electronic calculations opened new research possibilities based on the massive manipulation of huge data files. Intricate models and alternatives could be generated to simulate future events. By the 1960s urban data banks were being developed in the United States to help implement data processing for state and urban governments.

From 1945 to 1965 the tools for innovative geographic processing advanced greatly opening up new applications in the areas of transportation, urban modeling, and environmental inventory and management. In 1955 the Detroit Metropolitan Area commissioned a traffic study to plan for their future needs. In this study Detroit was divided into a grid of one-quarter square-mile cells and each cell was inventoried as to traffic flow. Subsequently, by using statistical analysis, future traffic volumes were predicted. This study was used to prioritize future highway development. Although no computer output in this project was in a geographically ordered format, it was perhaps the earliest computer-based GIS system developed. Apparently, it was not until the 1960s that the term Geographic Information Systems was first used.

While many advances were made in the development of GIS techniques, it still was an expensive proposition. To own an average GIS system would cost hundreds of thousands of dollars. While government agencies were the predominant users of GIS during this period, cost-conscious commercial users did not adopt them. However, by the 1980s the situation had changed tremendously. There were incredible advances in the areas of computer hardware and software and the cost to own a GIS system had plummeted from hundreds of thousands of dollars to a few thousand dollars. Many capabilities were added to the GIS software generating new applications across many disciplines.

17-4 THEMATIC LAYERING

Prior to the 1980s, a town that decided to build a new school building, courthouse, park, or some other public facility would have had a long and tedious job of gathering information. Some of the items they would have needed include the following:

1. Location of suitable pieces of land as to size, location, and price
2. Access to property
3. Locations of roads and rights of way
4. Locations of water, sewer, and electrical lines
5. Locations of telephone lines and TV cables
6. Political boundaries
7. Fire and police jurisdiction
8. Topography of land
9. Property boundaries
10. Soil types
11. Flood zones
12. Characteristics of surrounding property
13. Zoning regulations
14. Census data
15. Tax maps

After this information was finally gathered, it would have been necessary to put the information together in some usable form (not an easy task). Would you, the reader, care to estimate how long it would have taken them to assemble all of this information from the courthouse, fire department, police department, county registry of deeds, various public utilities, and many other organizations? The correct answer might be several months.

An increasingly large number of local governments in the United States are now placing data of this type into a computer format. This enables them to store, recall, and display natural and man-made features of the earth. The result is a *geographic information system (GIS). A GIS is a computer-based system used to store and manipulate information about geographic features.*

With a GIS all of the relevant information for a particular area can be called up on the computer screen. The system can further be used to call up maps with the needed data displayed graphically, perhaps using different colors. (The use of different colors is very effective in GIS maps.) Furthermore, all or parts of the data from various maps can be overlayed, enabling users to examine different sets of data simultaneously.

If a town has a good GIS and is asked how long it would take to obtain the information needed for planning a facility, the answer would probably be a few hours at most.

Figure 17-3 shows a few of the typical items that might be digitally stored for a particular piece of land in a rural setting. The items are really not "layered" within

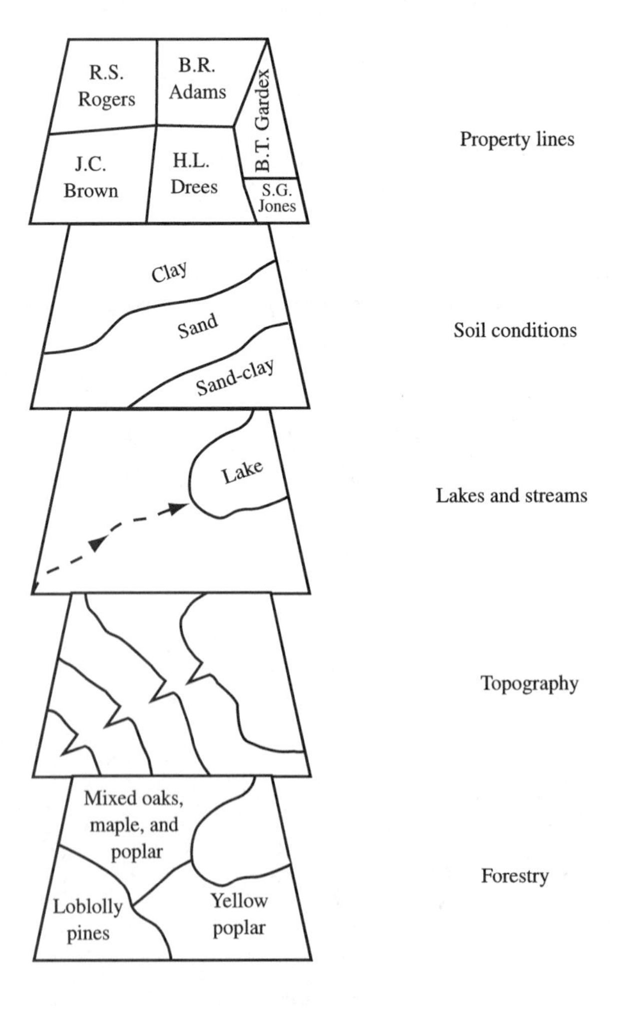

FIGURE 17-3 Thematic layering of a few of the typical items needed for a rural GIS.

the computer as shown in this figure, but this type of figure is commonly used to list the types of data that are stored. (The term *thematic layering* is often used in relation to a GIS, where the word *thematic* means relating to or consisting of a theme or themes.)

17-5 LEVELS OF USE OF A GIS

There are three levels at which a geographic information system might be used. These levels can be classified as data management, analysis, and prediction of "What if?" investigations.

Data Management. With this the lowest level of application, the GIS is used to input and store data, to retrieve that data through spatial and conditional queries, and to display the results. For the data management type of application, the GIS is merely used as an inventory system with the purpose of storing and displaying information about spatial features. These features are things like the width, number of lanes, and traffic count for a particular highway.

Analysis. The second level of GIS application is the analysis one and it is here that the user makes use of the spatial analysis capability of the system. Examples include determining the shortest path between two locations, grouping of areas of land into larger ones depending on certain criteria, and so on.

Prediction. The very highest application level of GIS falls into the prediction of "What if?" category. It is here that the data management and analysis capabilities of a GIS are combined into a modeling operation such as predicting the effect on traffic on a certain highway when certain land areas are developed in a certain way, predicting the effect of a hurricane, or predicting the effect of a certain disaster on air quality.

17-6 USES OF GEOGRAPHIC INFORMATION SYSTEMS

Geographic information systems can be used for many purposes. They may be used to determine optimum locations for roads, railroads, airports, utilities, subdivisions, retail market outlets, and hazardous waste facilities. They help both government and industry to efficiently manage their infrastructure such as water, gas, electric, telephone, and sewer lines.

They can be used to make maps, to establish the most efficient routes for emergency vehicles and school buses, to locate fire hydrants, to plan snow removal, and to appraise real estate. Several dozen federal agencies are presently making use of these systems in one way or another.

Business geographics is the latest trend in GIS. It is probable that at least 80% of business information is tied to geography.[2] (Think of the costs involved for a company that chooses a poor site for their business.) Today over 98% of the Fortune 500 companies in the United States use GIS for the following:

1. Market analysis
2. Customer analysis

[2]G. H. Castle, "The Bigger Picture," *Business Geographics,* February 1997, vol. 5, no. 2, p. 16.

3. Competitor analysis
4. Site selection
5. Studies for moving goods from warehouses to various customers

Some other applications of GIS are as follows:

1. Emergency response (911 calls)
2. Study of natural hazards such as earthquakes, hurricanes, and tornadoes
3. Transportation
4. Economic development

Many different fields or disciplines are involved in GIS activities. Some of the fields are *cartography* (the art or science of making maps), geography, civil engineering, computer science, environmental engineering, land use planning, surveying, photogrammetry, *geodesy* (the science of studying the form and dimensions of the earth), *remote sensing,* and many others. (The term *remote sensing* is used to identify a broad area of application in which images produced by electronic sensing devices or aerial or space photographs are studied.)

17-7 OBJECTIVES OF A GIS

An important objective of a GIS is to reduce the time- and money-consuming activities of handling, recording, researching, etc., the mountains of data we have accumulated and are generating relating to the land features mentioned herein. (Perhaps 75% or more of the information gathered by various government agencies can be easily recorded in a GIS.)

Around the United States numerous organizations are spending hundreds of millions of dollars in developing large geographic databases. The rapidly declining costs of computer hardware and software necessary to handle the jobs is making GIS available to a wider audience. The cost of storing data in computers has decreased dramatically in recent years. Computer memory has increased from kilobytes to gigabytes. (The bit is the smallest storable unit in a computer's memory. It is codable with one binary unit. A byte is eight consecutive bits.)

The primary goal of GIS is to take raw data and transform it by overlap and by various analytical calculations into new information that can help us make decisions. It is extremely important to understand that a GIS is not built for just one or two specific purposes. A GIS is a problem-solving tool; we may not know what we will be using it for next year or even next week. Will it be to plan a sewer system, an airport, a golf course, a shopping mall, a park, a zoo, a transportation system, or something else? A GIS provises an organization with the capability of applying geographic analysis methods to designated geographic areas to solve various problems.

17-8 ADVANTAGES OF A GIS

A GIS enables us to make spatial queries. Its users can visualize the desired information with the aid of maps. The oft-repeated story of two girls and their oranges illustrates well one of the great advantages of a GIS. One girl would peel an orange, eat the inside or pulp, and throw the peel away. The other girl would take her orange, peel it, grind the peel for use in a cake, and throw the pulp away. After they got together and each girl found what the other one was doing, one orange was sufficient instead of two.[3]

Most of the GIS data used to develop the database is public domain. Examples are the TIGER (Topographically Integrated and Geographic Encoding and Reference) files, census data, meteorological records, data from USGS, etc. Hence they are all shared. However, when someone has added other data to this public domain data we will probably have to pay for it.

Another advantage of GIS is that people from various groups or organizations put their information into a common database. The results are hopefully better cooperation between the various groups, less overlap in gathering information, efficiency in retrieving information, and perhaps the learning of new things when the various data is put together in GIS form and the whole picture is made available.

Figures 17-4 and 17-5 concern damages caused by Hurricane Hugo in South Carolina in 1989. Figure 17-4 shows the estimated maximum gust speeds occurring during the storm. Figure 17-1 showed the density of single-family dwellings

FIGURE 17-4 Estimated maximum gust wind speeds in South Carolina during Hurricane Hugo.

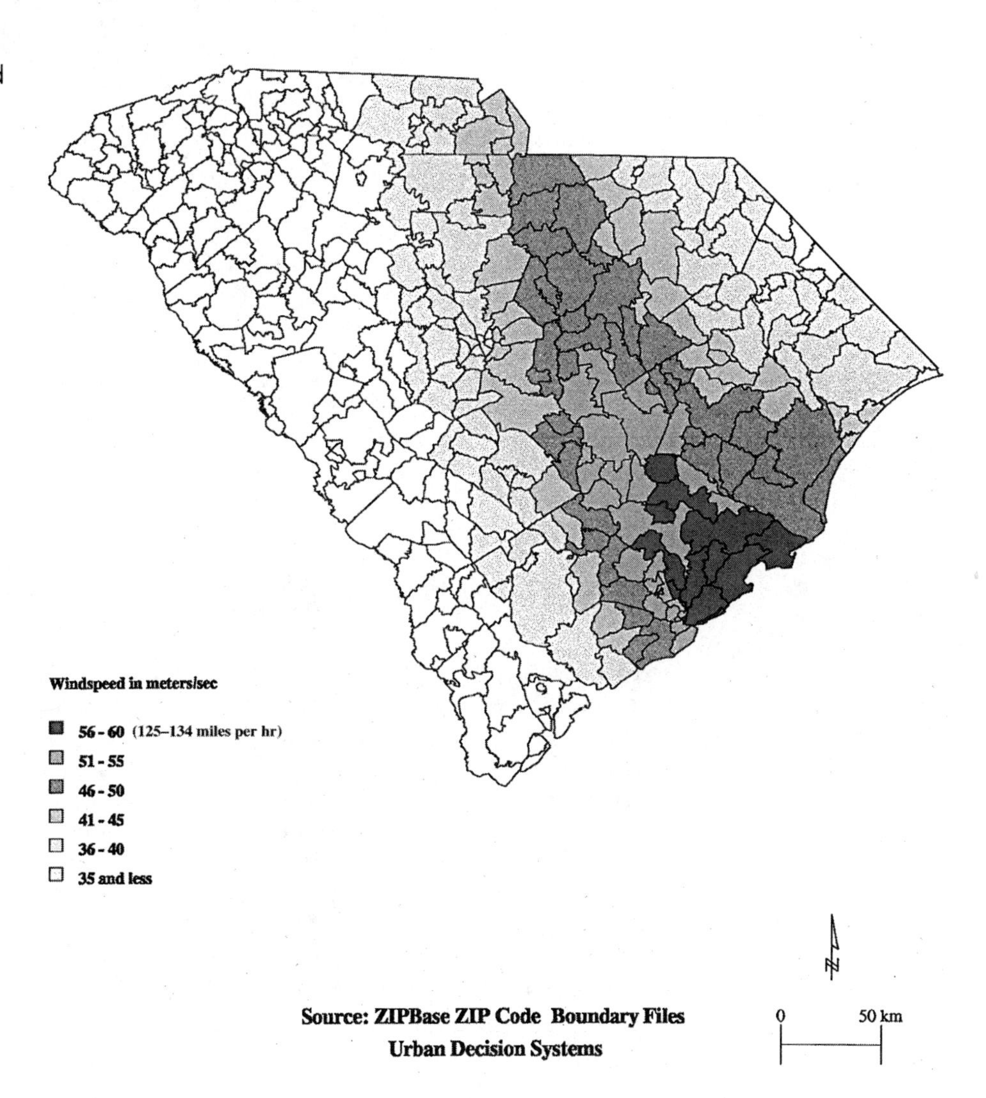

[3]D. Morris-Jones and A. Zetland, "Organizational Transitions in GIS," *Professional Surveyor*, July/Aug. 1991, vol. 11, no. 4, pp. 30–31.

throughout the state, while Figure 17-5 provides the percentages of those dwellings which were rendered initially uninhabitable by the hurricane. This information can be used to predict losses in other areas of the country in case of hurricanes.[4]

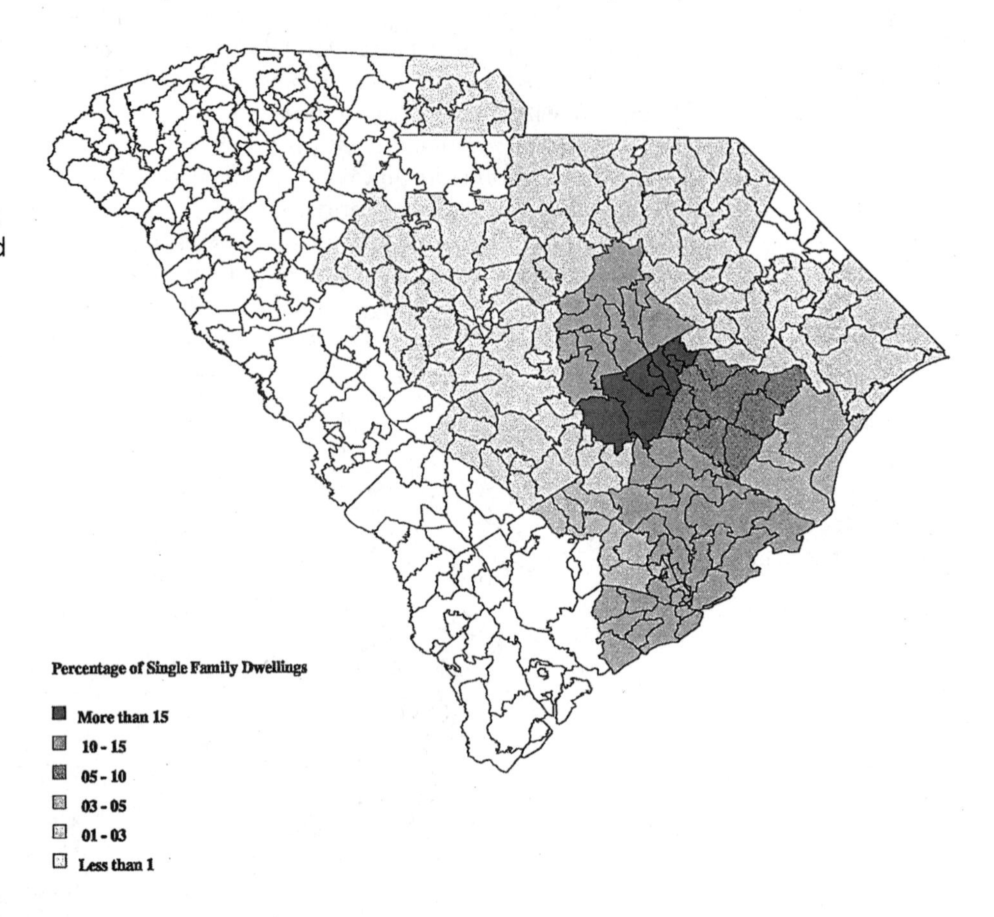

FIGURE 17-5 Percentage of single-family dwellings initially rendered uninhabitable in South Carolina by Hurricane Hugo. (The greatest damages occurred inland where there were large lakes and open agricultural areas.)

17-9 LAND INFORMATION SYSTEMS (LIS)

To some people geographic information systems (GIS) and land information systems (LIS) are thought to be identical. Even though they do have similar characteristics, they are different. A land information system should probably be thought of as a particular part or a subsidiary of a GIS. An LIS is devoted almost entirely to land data. It includes information such as size, shape, location, legal description, topography, flood plains, water resources, easements, and zoning requirements for each tract of land in the system. The legal descriptions are those given with the U.S. Public Land System, or the metes and bounds system as described in Chapter 21 of this text.

A GIS includes data for a much wider range than does an LIS. The GIS will include not only the data just mentioned for an LIS but will also include such things as soil types, depth to ground water, census data, school bus routes, tax maps, fire and police jurisdiction, and so on. With an LIS a user can determine the ownership

[4]These figures were taken from a proposed Ph.D. dissertation at Clemson University with the permission of its author V.D. Rayasam. The title of the work is "Study of Hurricane Related Insurance Losses Using Geographic Information."

of property, tax assessments, mortgages, utilities, boundaries, improvements made to the land, and other information needed for land appraisal, land acquisition, or for various uses of the land.

The United States Bureau of Land Management (BLM) and the United States Forest Service have an enormous task on their shoulders in trying to keep track of and to manage the approximately 2.3 billion acres under their jurisdiction. To attempt to handle the task, they are using an LIS called the National Integrated Land Survey (NILS).

17-10 ACCURACY IN A GIS

The surveyor thinks that the most important part of a GIS is accuracy, but there are many other persons involved with the system who don't think that way. Those other people include planners, mayors, police officers, rescue squads, firemen, city planners, and many others. Many of them couldn't care less as to whether the points are located to the nearest 0.1 ft or the nearest 50 ft. What they want to see is the overall picture, and the majority of them probably think that it would be a waste of time and money for the system to have the accuracy desired by the surveyor. (In this regard there is an old saying to the effect that "a map is a set of errors that have been agreed upon.")

The surveyor must understand what these other parties are looking for in a GIS and must learn to work with them. Perhaps this "politicking" will be the hardest thing for him or her to do in the entire development of the system, but it is essential.

To develop an ideal GIS the surveyor would like to have every last item professionally surveyed and located and then carefully described in computer form. Unfortunately, the usual GIS is not done completely in this fashion and various shortcuts are taken. As an illustration the data on existing maps (utility, zoning, etc.) for the area involved may be converted to digital values and adjusted to the coordinate system being used. Such values will often be short on accuracy.

When surveyors carefully set up a control system to 6 or 8 places and then try to fit in the various plats of land that have been previously located in that area to different degrees of accuracy, the information taken from existing maps, and so on, there is going to be problem in making things fit. As a result they are going to have to adjust or change some of the data. The different parcels of land will have to be shrunk, stretched, warped, or bent to make them fit. Such a process is hard for the surveyor because it is against all of his or her training. And yet who could do it more logically and accurately than the surveyor? Sometimes these other groups need to be reminded that putting poor data into a computer does not make it any better.

Almost all of the activities of humankind are spatial (and so are the occurrences of natural phenomena). You can see that an item of information in a GIS is meaningless unless its position is given. *Finding the position of something is a surveying specialty.*

Professional surveyors are ideally suited for obtaining much of the data used for developing geographic information systems. They have the ability and the desire to make accurate measurements and they have considerable knowledge of automated mapping procedures. In fact, it seems only logical that the various agencies preparing and using geographic information systems should not only be required to have but should also want surveyors on hand to help them develop their systems. They need people who have a passion for accuracy and *accuracy is the address of the surveying profession.* Surveyors are needed to establish control points, to make various ground and aerial surveys, to prepare maps, and to establish data in digital format for entry into a GIS.

17-11 CONTROL SURVEYING

An extremely important part of GIS is *control surveying*, which is often referred to as monumentation. A set of carefully planned and located monuments will be placed throughout the project area and measurements made with respect to them. The monuments must be located with great care because the other information to be gathered will be referred to this monument control system. If the system is established inaccurately all other information that is gathered will be at least that inaccurate or even more inaccurate.

The monuments established by the surveyor should be referenced not only to each other but also to some type of datum. The NGS has established a network of first-order control monuments throughout the country. (Usually "first-order" work is thought of as having a minimum accuracy of 1/100,000. With differential GPS measurements we are probably doing work with much better precision than this). The surveyor can obtain the positions of other points with respect to the NGS monuments. Other first-order control monuments will be carefully established. (Assuming the locations are to be established by GPS it will be necessary to locate the monuments so that the satellites may be clearly observed with no obstructions. These new monuments should be located so they are reasonably accessible and in places where they will hopefully not be disturbed by future construction work in the area.)

Not only is horizontal positioning information important to GIS work but the same is true about vertical positioning or elevations. The reference datum used will normally be the National Geodetic Vertical Datum of 1988 (NGVD88). There are approximately 600,000 vertical control stations in the United States.

The global positioning system is commonly used to establish the monumentation network for GIS. It is the backbone of GIS around the country.

17-12 LEGAL CONCERNS WITH GIS

Privacy

Various state and federal laws have been passed that enable the public to have access to information concerning the various operations of the government. However, laws have also been enacted with which our governmental bodies have attempted to prevent the public from being able to invade the privacy of individuals.

Do we want private companies to know our social security numbers, to have data on our use of credit cards, to know the magnitude of our personal debts, to have information concerning our retail purchases, etc.? Much of this type of information is contained in a GIS. Our governmental agencies are going to have to be more careful in the information they share with the public.

Liability of Companies Providing GIS Information

Companies employed to provide GIS information should be required to live up to certain levels of competence. If they don't and others suffer loss or injury as a result, they should be held responsible. If, however, those same companies can show that they have performed their work at a level at least equal to that which would have been performed by a reasonably prudent company, they can escape liability even though losses were suffered.

To avoid this gray area in the law, owners of a GIS need to use well-written contracts that clearly spell out what is expected from their contractors. Such agreements should specify what is to be done, the accuracy which is to be obtained, and the time period in which the work is to be completed.

Copyright Protection

Maps made by private entities may be copyrighted under federal law. We are prevented from making copies of such maps or of the data on those maps without permission from the owners. *Furthermore, we cannot without permission take someone else's copyrighted map, digitize it, and store it in our GIS.* We are not, however, prevented from gathering data for the same areas covered by copyrighted maps and using the information in our GIS. Furthermore, the information contained in a GIS cannot be copyrighted as it is merely a collection of data. Federal law does not permit the copyrighting of facts alone.[5]

Problems

17-1. What is an Information System as discussed in this chapter?

17-2. What is a GIS?

17-3. How did Dr. John Snow help the city of London in 1854? Explain.

17-4. Define thematic layering as it applies to GIS.

17-5. What are the objectives of a GIS?

17-6. What does the term spatial mean?

17-7. List eight different fields or disciplines which are involved in GIS activities.

17-8. What is an LIS?

[5]G. B. Korte, "Legal Aspects of GIS Data: Part II," *P.O.B. Magazine,* March 1996, vol. 21, no. 4, pp. 25–27; "Legal Aspects of GIS Data: Part III," *P.O.B. Magazine,* May 1996, vol. 21, no. 6, pp. 50–51 and 80.

CHAPTER

18 GIS, Continued

18-1 ESSENTIAL ELEMENTS OF A GIS

There are five essential elements in developing and using a geographic information system (GIS). These are: data acquisition, preprocessing, data management, manipulation and analysis, and product generation. For any given application of a geographic information system it is important to view these elements as a continuing process. Each of these elements is briefly examined in the sections that follow.

18-2 DATA ACQUISITION

The data used in a GIS is of two types—spatial or attribute. *Spatial* data describes the geographic location of various entities such as zip code areas, county boundaries, and roads in terms of latitude and longitude or other appropriate format. An *attribute* is a property or characteristic that may be ascribed to a certain thing or feature. It may be numeric (population counts, household units, and so on) or it may pertain to character (the name of a zip code, household unit, etc.) There are various ways of acquiring data as is described herein.

When a GIS project is initiated for a particular area, there will probably already be on hand or easily obtainable a great deal of information that can be used in the system. For instance, there will usually be maps and perhaps aerial photos of the area in question available from various branches of government along with much non-spatial data. Most of this data is available "off the shelf" in digital form. In fact, most users buy their data—USGS maps, census data, and TIGER files. Such data belongs to the public unless it is of importance to national security. Among the agencies which might be helpful in this regard are the U.S. Geological Survey, the National Oceanic and Atmospheric Administration, the U.S. Bureau of Census, and the Soil Conservation Service.

Should insufficient information be available from the sources just listed it will be necessary to gather more data on one's own. The first step in this process involves the location of a USGS control point (bench mark) or perhaps the use of GPS equipment to establish a location. Once this is done two GPS receivers may be used operating in the differential mode to establish other control points with sub-meter accuracy. These newly established points can then be used to extend the survey control network proceeding outward as well as in between the previously established points. Finally, other points can be located with respect to the control points by measuring angles and distances with theodolites, EDMs, total stations, etc.

Another information source for GIS systems is the so-called imagery data. Usually these are aerial photos, such as USGS's orthophotos, but they can include satellite images. The National Airphoto Program can provide photographs to various scales for much of the United States. In addition, private companies sell images.

The collection of field data for a GIS is handled with topographic mapping. As is the case with such surveys the information obtained includes the location of houses, streets, streams, lakes, power lines, sewer lines, water lines, property monuments, elevations, fire hydrants, trees, and so on. This data may be obtained with conventional surveying techniques, GPS receivers, or aerial photogrammetry. Should conventional topographic surveying performed as described in Chapter 14 of this text be used the surveyor will run traverses between the previously mentioned monuments. It will thus be necessary for the distances and directions to be balanced between the known coordinates of those monuments.

18-3 PREPROCESSING

Preprocessing involves manipulating data in several ways so that it is converted into a format that can be used by a GIS. Important preprocessing elements include data format conversion and identifying locations of the objects in original data in a systematic way. Converting the format of original data often involves the extraction of information from maps, photographs and printed records (such as demographic reports) and the recording of that information in a computer database. Some of the preprocessing items follow.

1. Matching edges of two different coverages.
2. Eliminating unnecessary lines, polygons, or points.
3. Establishing standards for maintaining spatial data in terms of projection systems, datums, etc.
4. Converting data into a format that can be used by the GIS software.
5. Establishing a consistent system for recording and specifying the location of objects in the data sets. When the task is completed, it is possible to determine the characteristics of any specified location in terms of the contents of any data layer in the system.

18-4 DATA MANAGEMENT

Data management functions govern the creation of and access to the database itself. These functions provide consistent methods of data entry, updating, deletion, and retrieval. It is obviously important to keep the data up-to-date and this task requires that data be constantly monitored and checked to see if it is obsolete. If it is obsolete, care must be taken to replace it with current data since wrong data will lead to wrong conclusions and hence poor business decisions.

Also included in data management are concerns with the issue of security. Procedures must be in place to provide different users with different kinds of access to the system and database. For example, database update may be permitted only after a control authority has verified that changes are both appropriate and correct.

18-5 MANIPULATION AND ANALYSIS

Manipulation and analysis are often the focus of attention for a user of a GIS. Many users believe incorrectly that a geographic information system consists only of this module. While this could arguably be the most important part of a GIS, it is by no means the whole thing. This part of the system contains various analytical operators, which can be used to generate sophisticated queries based on user requirements that are classified into two categories: spatial queries and aspatial queries. A discussion of these follows.

Spatial Queries. These are essentially involved with querying the geographic components of a GIS, that is, the lines, polygons, and points. Some of these queries are briefly explained below.

1. *Overlaying.* This is the process of overlaying one area of a GIS with another to retrieve information about their combined impact. A good example of overlaying is land suitability analysis. If a particular crop needs an average temperature above 75° F and less than 40 inches of rainfall. suitable pieces of land can be found that satisfy both conditions by overlaying two separate layers (average temperature and aggregate rainfall).
2. *Buffering.* This operation creates buffer zones for investigation around the geographic objects under study. An example could be the study of a potential site for construction of a nuclear plant. A buffer zone may be created in GIS to investigate the properties (number of households, types of trees, water bodies, etc.) of the surrounding area.
3. *Least cost/impedance path.* This is used to find out the shortest distance between two points based on various criteria. One good example is the routing of emergency vehicles. While routing such a vehicle it is necessary to know the shortest time required to reach the desired designation; this could be a function not only of distance but also time of day as it relates to traffic (peak traffic times could require more time), speed limits (you may have to negotiate vehicles going at lower speeds), etc. It may involve traveling longer distances on less traveled roads.

Aspatial Queries. Aspatial queries are those that involve querying attribute data concerning such items as census data, weather related data, etc. Attribute data that is related to spatial components of a GIS are queried and the resulting queries are then displayed on the spatial components, giving the user a visual representation of the query. Spatial queries are of various types such as intersection queries, union queries, and user defined queries. Given below are some examples of these types of queries.

(a) Select all zip code polygons where median income of residents exceeds $50,000 and where the average annual snowfall is less than 30 inches (intersection).

(b) Select all census tracts in South Carolina where the median price of a single family home is less than $45,000 and where the percentage of the population living under the poverty level is greater than 25% (union).

(c) Select all zip code polygons with average wind speeds greater than 12 knots (user defined).

18-6 PRODUCT GENERATION

The final element of a GIS is product generation where the results of various analyses are presented. This output could be a soft copy that is displayed on the computer screen or a hard copy that is printed on paper. Various types of products generated in a GIS are statistical reports, maps, graphs, and accompanying text. While statistical reports can be printed on white paper, maps and graphs can be printed with a regular printer or with a plotter where it can be many times larger in size than is customary with a regular computer printer.

18-7 PUTTING DATA INTO THE COMPUTER

As previously mentioned, most GIS data is purchased "off the shelf." Should new data be developed it will have to be input into the computer.

There are five general ways in which data is entered into computers. These are as follows:

1. *Keyboard entry.* With this method information is manually entered into the computer. Most attribute data is entered in this way, but a large amount of spatial data is entered in other ways.
2. *Coordinate geometry.* With coordinate geometry, usually called COGO, survey measurements are manually entered by keyboard. Then from this data the coordinates of the spatial features involved are determined by the computer.
3. *Digitizing.* This method is used to transfer the data on an existing map into the computer. The map is taped down onto a sensitized digitizing tablet. Then the map is traced over with a handheld device called a *puck* or *cursor.* The various positions are accurately measured with the cursor and the data is transferred to the computer in digital form.
4. *Scanning.* There are various types of scanners on the market (with the fax machine being the most common). Scanners use either an optical laser or some other type of electronic device to scan maps and convert their images to a raster format as described in the next section. Some computer processing is necessary to improve the quality of the work and to convert it to a vector format. The most common type of scanner used for converting maps for GIS work is the drum scanner. The maps are attached to a rotating drum and the map is scanned with very fine increments of distance and the light reflected from illuminated maps is measured. These devices are sensitive to perhaps 100 or more shades of color or brightness in the maps.
5. *Inputting existing digital files.* In both the United States and Canada map data is being prepared more frequently in digital form and as such can be input directly into computers. Not only are new maps being prepared this way but many old ones are in the process of being converted to digital form by various government agencies.

18-8 SPATIAL DATA MODELS

There are two approaches to the representation of a geographic information system: the *raster* model and the *vector* model. Part (a) of Figure 18-1 presents a real world picture of an area containing an office building and a road. In a raster model the area is subdivided into tiny cells and objects are represented in the corresponding cells. As shown in part (b) of Figure 18-1 all cells containing the letter O represent the office building that was shown in part (a) while the cells containing the letter R represent the road. These letters (O and R) have been chosen by the author in hopes that the reader will be able to clearly picture the raster situation. In all probability, numbers will be used in a real situation instead of letters.

FIGURE 18-1 Raster and vector models used to represent data.

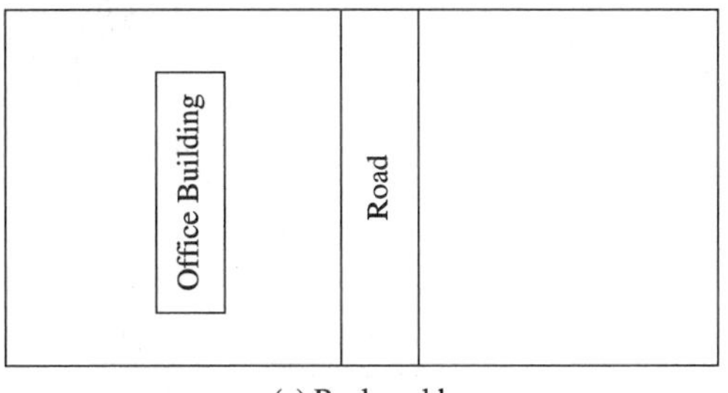

(a) Real world

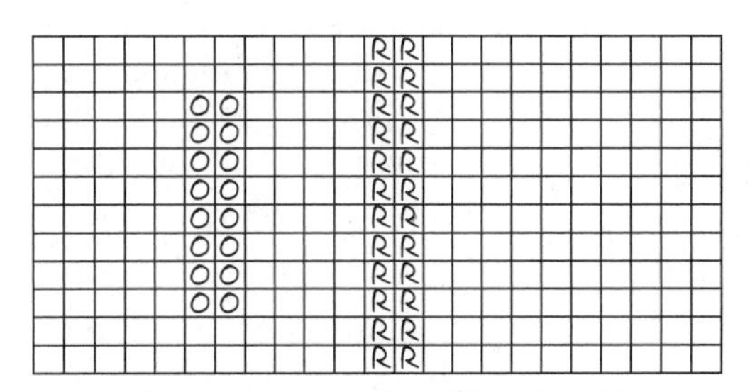

(b) Raster representation of the real world shown in part (a)

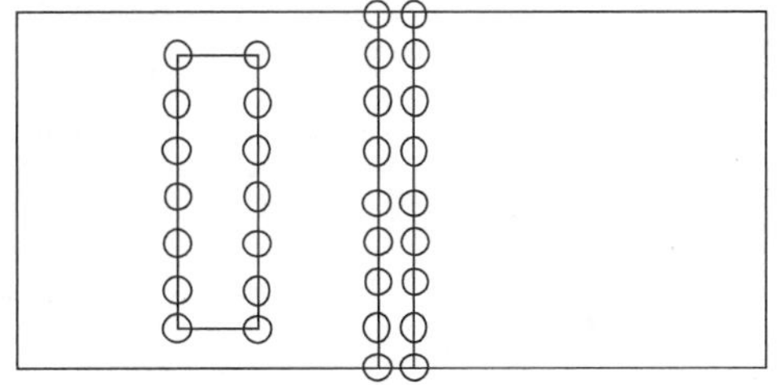

(c) Vector representation

In the vector model, objects in the real world are represented by points and lines that define their boundaries. Each of the points is located by a pair of coordinates. Part (c) of Figure 18-1 shows the representation of the same area using the vector method.

Most of the earlier attempts at GIS and thematic mapping (maps with one or more themes) used the raster model. In later years more effective vector models were developed. The following paragraphs discuss the raster and vector methods of representing data in the computer in more detail. The raster method is usually preferred when continuous data such as environmental features, air quality, natural features, and so on are being represented and when complex spatial data models are being developed.

Raster Models

Each cell in a raster file is assigned only one value. For example, vegetation and soil types for a particular area will be stored in separate files. With the raster procedure a grid of cells is formed on the map and those cells may be rectangular or square, with squares being the most common.

Each of the cells represents one map unit usually chosen to be shown on the map as one screen display unit or *pixel.* (The word *pixel* is a contraction of "picture element.") On the ground each grid cell represents a whole number increment in the coordinate system. The commonly used length of a cell is one-third to one-fourth of the length of the smallest desired map feature.[1] Raster files are often composed of quite a few million cells. The smaller the area of land in each cell the higher the resolution.

The rows of the raster are usually established parallel to the east-west direction and the columns in the north-south direction. The origin is probably the upper left-hand corner and the cells are thus numbered in the vertical direction from the top down and the horizontal position from left to right.

In Figure 18-2 (a) a map is shown of a certain land area for which the data is to be digitized for storage in the computer. In part (b) of the same figure the map is overlayed with a grid system while in part (c) a number is assigned to each cell to describe the situation therein (1 for forest land, 2 for grassland, and 3 for the lake). Finally, in part (d) of Figure 18-2 the raster information of part (c) is shown as it would appear if the computer was used to print the map.

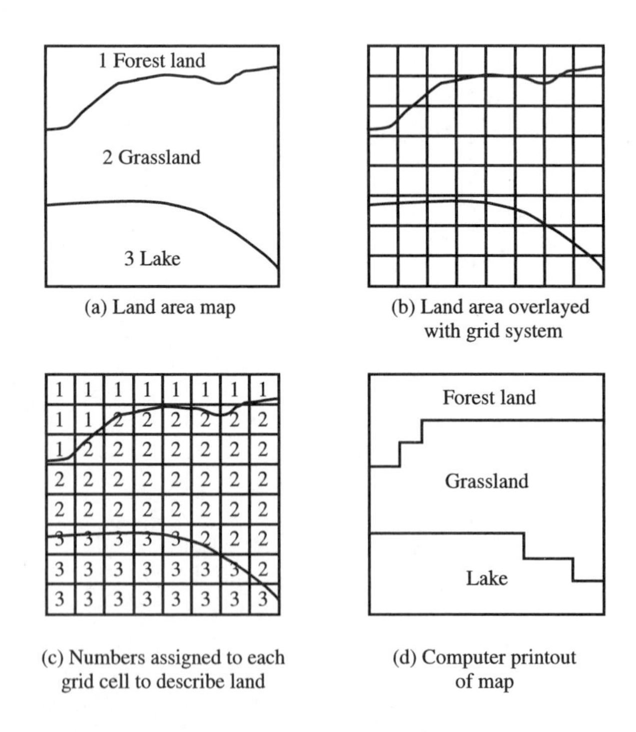

FIGURE 18-2 Example of a raster system.

[1]J. Star and J. Estes, *Geographic Information Systems* (Englewood Cliffs: N. J.: Prentice-Hall, Inc., 1990), pp. 33–42.

You can clearly see one disadvantage of the raster system in these latter parts of the figure. Each cell is given only one number and that is based on the majority-of-use rule. Along the border between the forest land and the grassland areas are quite a few cells that have some forest areas and some grass areas. If the forest covers the largest percentage of the cell it is given a 1. If the grass covers the largest part of the area it is given a 2. (It is possible to subdivide the cells and make a more accurate description, but more time and a larger disk capacity will be needed.)

The raster method has several advantages. As an illustration it is easy to understand and can be quickly retrieved from the computer. Raster systems are commonly thought to be the best for forestry, hydrology, terrain analysis, and photogrammetry.

In other words raster models are preferred for representing continuous details such as natural and environmental features.

Vector Models

With vector models geographic data is divided into points, lines, and polygons. Vector data is more satisfactory for representing features with discrete edges such as roads and power lines and for items such as rivers.

A point, which is represented by a single pair of coordinates, is used to represent an object such as a boundary or shape that is too small to be represented by a line or a polygon. Included are items such as fire hydrants, manholes, surveying monuments, houses, and advertising signs.

A line is represented by a set of ordered coordinates, which when connected will show the linear shape of an object that is too narrow to be displayed as an area. Such items as roads, streams, property boundaries, power lines, and so on fall into this class. Even curved and wiggly lines can be represented by vector models by using more points for the bends and curves and fewer for the straight line parts. The additional pairs of coordinates that are needed to represent curvilinear features are called *shape points.*

A polygon is a closed loop formed by a set of coordinates and is used to show counties, cities, lakes, golf courses, zip code areas, and areas with certain types of land covers as pine trees or grass or agricultural crops, etc. Figure 18-3 shows how points, lines, and polygons are represented by coordinates.

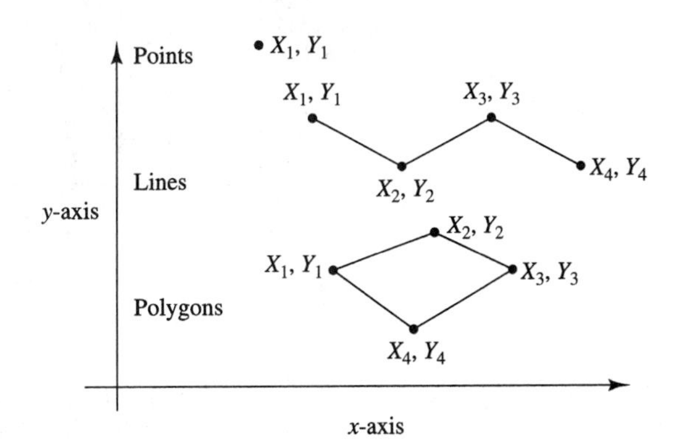

FIGURE 18-3 Points, lines, and polygons.

With a GIS the points are specified with coordinates, the lines are constructed by connecting series of points, and lines may be connected together to form polygons.

These operations are illustrated in Figure 18-3. There the reader can see how geographic features are digitally encoded using XY coordinates. The position of point A is represented by a single pair of coordinates (X_1, Y_1). To define a straight line only two pairs of coordinates are needed. These are the coordinates representing the beginning and ending points of the line. In Figure 18-3 a series of straight lines is represented by an ordered list of coordinate pairs (X_1, Y_1), (X_2, Y_2), (X_3, Y_3), (X_4, Y_4), and (X_5, Y_5). A polygon is represented by an ordered list of coordinates, which begin and end with the same pair of coordinates. In a GIS positions are usually stored in a standard geographic coordinate system like latitude and longitude.

With vectors an outline map of a fairly large area can be represented with only a few thousand points. To do the same thing with a raster model would require many more cells than the number of points required by the vector based model.

Vector models are built by creating a set of tables where points, lines, and polygons are identified spatially with cartesian coordinates. Figure 18-4 (a) shows how a line can be represented with the vector method. The same line is shown in part (b) as it would be represented with the raster method.

FIGURE 18-4
Representation of a line.

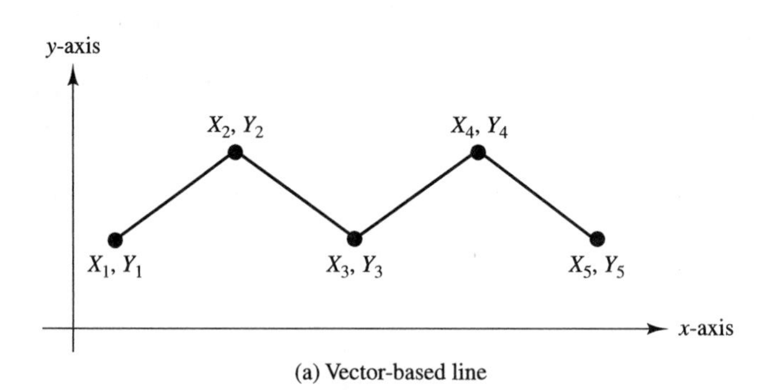

(a) Vector-based line

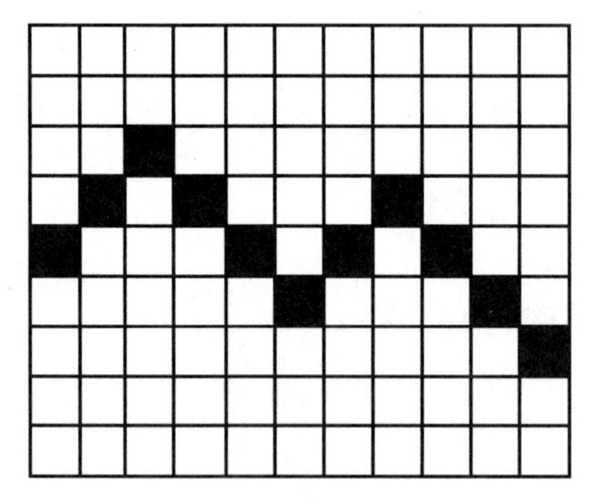

(b) Raster-based line

Maps convey information by representing geographic features with graphic map components. The three types of map components used in geographic information systems to represent various types of geographic features are points, lines, and polygons. Figure 18-5 illustrates the use of these map features to represent various geographic features. In this figure the apartments are represented by points, the roads and county boundaries by lines, and the lake with a polygon.

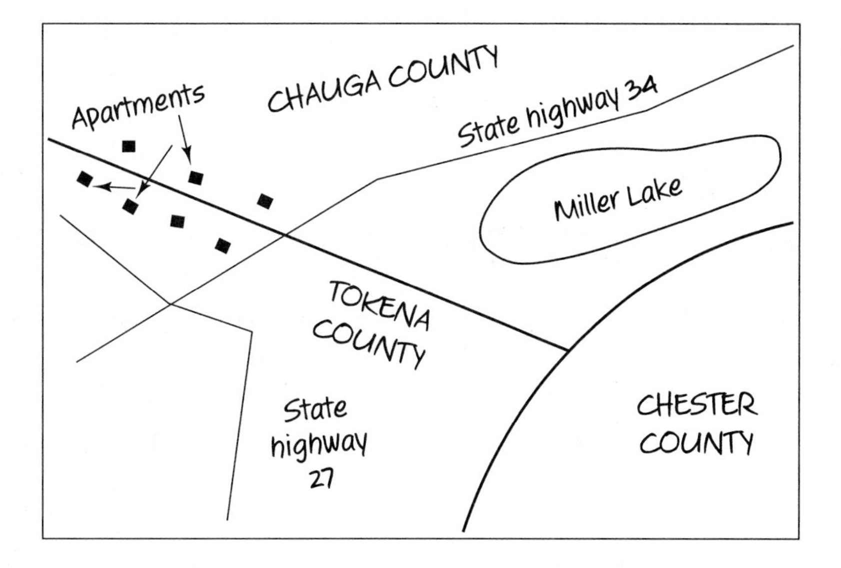

FIGURE 18-5 Representation of geographical features.

18-9 GEOGRAPHICAL DATA

We have seen that the point, the line, and the area can be used to represent topological data. (The word *topology* refers to the description of the relationship between various geographic features such as their linkage, proximity, and adjacency.) Theoretically, every geographical phenomenon can be represented by one of these items plus a label that names the item. Several illustrations follow:

1. A fire hydrant can be represented with a single pair of X, Y coordinates plus the label "fire hydrant."
2. A straight section of a railroad can be shown with a pair of coordinates at each end of the straight section plus the label "railroad."
3. A farm belonging to Mr. Smith can be represented as an area entity with the appropriate coordinates plus the label "Smith's farm."

Points. Point data applies to observations that occur only at points or at least at extremely small areas in proportion to the scale of the database. Features such as telephone booths, fire hydrants, and bus stops illustrate data that occupies a single point.

With a vector database point data can be accurately located with a pair of X, Y geographic coordinates. If a raster-based system is used, however, point data can be depicted only at the level of detail in a single cell.

Line Data. Highways, railroads, rivers, pipelines, and power lines exemplify line data. Vector systems can show this data with fine detail but a raster-based system can only depict such linear features with chains of cells. See Figure 18-4(b).

The geometry of lines is given by a series of pairs of coordinates. For a single straight line only two pairs of coordinates are needed for its description. For curved lines it is necessary to provide additional pairs of coordinates. These extra points are usually called *shape points.*

Area Data. Continuous two-dimensional items such as lakes, agricultural areas, and parking lots are depicted by area data. Such items as highways, rivers, canals, and so on may also be included in area data depending on the level of data of a GIS.

With the vector system the areas of these items can be clearly depicted, but they may not be precisely defined in a raster system. When a vector based system is used, the centroid of a particular area can be clearly shown with a pair of coordinates.

18-10 COORDINATES AND MAP PROJECTIONS

Locations in geographic information systems are generally stored with Cartesian or x and y coordinates. There is a problem, however, in using these coordinates to locate horizontal positions. The problem is that we are using a flat map to represent a part of the earth's curved surface. To remove earth's curvature and develop a flat map some things are going to have to be distorted, such as scales, areas, and directions.

Maps are flat but the surfaces they represent are curved. The transformation of three-dimensional space onto a two-dimensional map is called *projection.* Projection formulas are mathematical equations that convert the data from a geographical system (latitude and longitude) on a sphere or spheroid to a representative location on a flat surface. This process inevitably distorts at least one or more of the following properties: shape, area, distance, or direction. Since measurements of one or more of these perhaps distorted properties are often used to make decisions, anyone who uses the maps as analytical tools needs to know which types of map projections distort which properties. The various types of map projections and the properties they preserve are as follows:

1. *Conformal maps* preserve local shape. Those maps that are used for presentation to others normally use conformal projections.
2. *Equal area or equivalent maps* retain all areas at the same scale. These types of projections are normally used for thematic and distribution maps.
3. *True direction maps* are those that express certain accurate directions. Navigational maps are good examples of these types of projections.

Various methods have been developed for projecting curved areas on flat maps. One method used in the United States is the *state plane coordinate system.* The National Geodetic Survey has developed a system for each state in which plane or grid coordinates are provided at various control stations. The surveyor can merely run traverse lines (lengths and bearings) from those points to other points whose coordinates are desired. He or she will need to adjust the measured distances to sea level and apply a scale correction to them depending on the method used by the NGS in that state. The x and y components of the traverse lines (latitudes and departures) can then be computed and the plane or grid coordinates of the points in question determined. Although the work of the NGS is very complicated, the use of plane coordinates by the practicing surveyor is quite simple.

The purpose of the state plane coordinate system is to take advantage of the very precise work of the National Geodetic Survey and use it in a simple fashion to control ordinary surveying work. In other words, it is desired to use the mathematics of plane surveying, yet to take into account the earth's curvature.

The idea is to project points from the earth's spheroid to some imaginary surface that can be rolled out flat without substantially destroying its shape or size. A plane rectangular grid system is then superimposed onto that flat surface and the location of points specified with x and y components. Several of these so-called map projections have been devised through the years. These are the tangent plane projection, the Lambert projection, and the Universal Transverse Mercator (UTM) projection.[2]

For an ordinary-sized town a GIS is probably referenced to a local grid based on state plane coordinates. For larger areas and for areas that involve more than one state plane system, latitudes and longitudes should be used for consistency across the entire area.

The UTM projection is commonly used in GIS work because it is the system used on most USGS topographic maps. In this system the earth is divided into 60 rectangular zones each 6 degrees wide along the equator and running up to 84 degrees north and down to 80 degrees south. The zones are numbered 1 to 60 beginning at longitude 180 degrees west. Zones 10 through Zone 20 cover the states from the west coast to the east coast. A transverse mercator projection is centered in each of the 6 degree wide zones. The central meridian for Zone 1, which runs from 180 degrees west to 174 degrees west is at 177 degrees west. For the polar regions of the earth (above 84° north and below 80° south) the distortion of the flat projection is so great from the real earth that a different projection is used. It is the universal polar stereographic coordinate system.

Although locations in geographic information systems are generally stored in three dimensions with x, y, and z coordinates, there is another or fourth dimension frequently included. That dimension is *time*. For example, a new highway was built 6 years ago, a certain area was flooded last year and 17 years before that a dam failed and so on. Obviously such items may affect the decisions that are made.

18-11 CONCLUSION TO GIS DISCUSSION

In conclusion to this introductory GIS discussion, another example of the almost endless number of situations that may be studied with the GIS procedure is presented. The major streets and other features (such as a lake, a hospital, a golf course, etc.) of an imaginary city have been drawn by the author as they might be printed for a real city by GIS. See Figure 18-6.

The city and fire insurance companies are very much concerned with the times required for fire trucks to reach various sections of the city from a central fire station. For this figure it is assumed that fire trucks can average 30 mph within the city. It is assumed that distances are measured along the streets from the fire station for the estimated distance the trucks can travel in 5 minutes, that is, $5/60 \times 30$ miles = 2.5 miles. Then these points are connected together with a line just as equal elevation points are connected together for contour lines. This line is labeled 5 MINUTES in Figure 18-6.

In exactly the same fashion, lines are prepared for 10 minute and 15 minute travel distances and labeled 10 MINUTES and 15 MINUTES. You can see that the resulting information is not only very important to the city as they make their fire protection plans but is also very important to the insurance companies as they set their insurance rates.

In conclusion, the reader should clearly realize that all GIS applications can be handled (although impractically time wise) by draftsmen spending perhaps hundreds of hours of work for small areas. Should this data be digitally stored in the computer the same information such as that shown in Figure 18-6 could be obtained in a very few minutes.

[2]J. C. McCormac, *Surveying Fundamentals, 2nd ed.* (Englewood Cliffs, N. J.: Prentice-Hall, Inc., 1991) pp. 444–466.

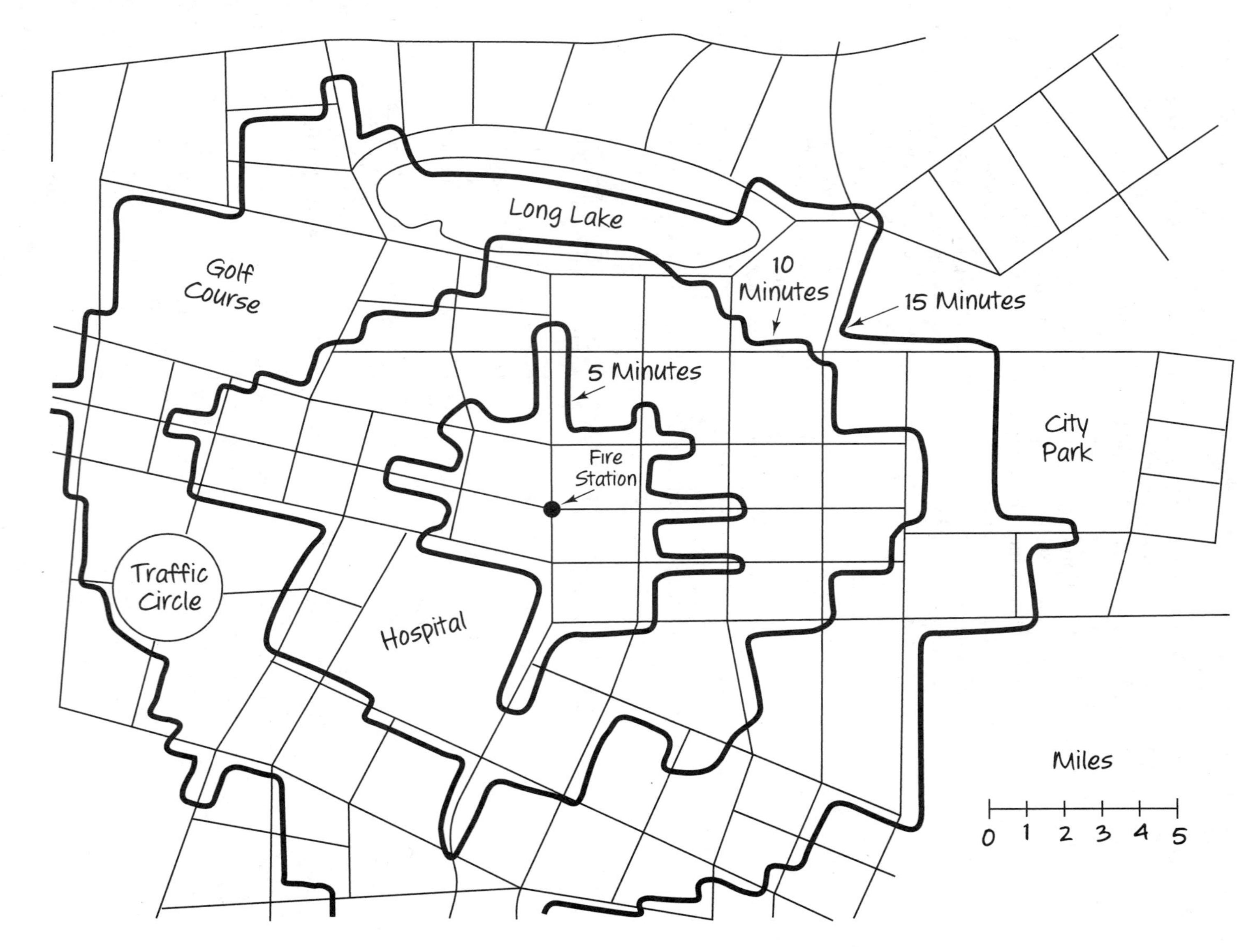

FIGURE 18-6 Estimated distances fire trucks can travel in 5 minutes, 10 minutes, and 15 minutes.

Problems

18-1. Describe how geographic features of the earth are represented in a GIS.

18-2. Distinguish between spatial data and attribute data.

18-3. With regard to geographical data, what is the meaning of the term *topology*?

18-4. What is a pixel?

18-5. For what types of data is the raster format best suited?

18-6. For what type of data is the vector format best suited?

18-7. Prepare the map shown in raster format using 10 cells vertically and 10 cells horizontally.

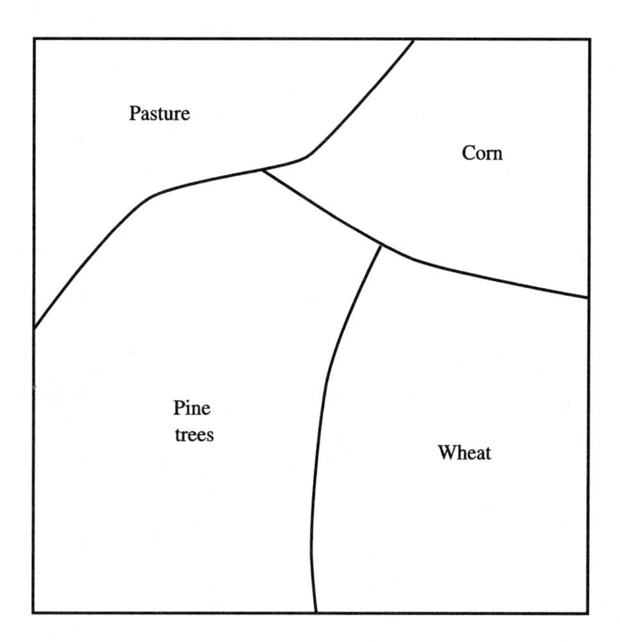

18-8. Prepare the map shown in raster format using 10 cells vertically and 10 cells horizontally.

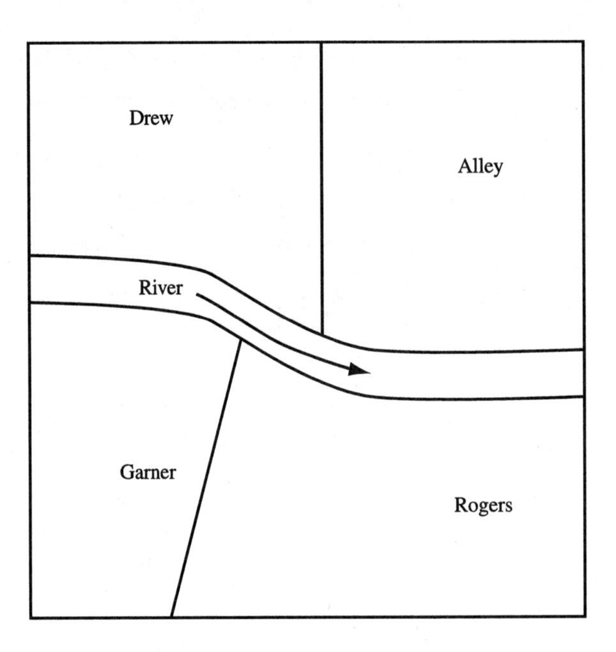

18-9. Why is the Universal Mercator System the most common type of map projection used in the United States?

CHAPTER 19 Construction Surveying

19-1 INTRODUCTION

The construction industry is the largest in the United States and surveying is an essential part of that industry. In fact, more than one-half of all surveying relates to the construction industry. To give the reader a feeling for the vast number of construction projects that need appreciable services from surveyors, the following partial list is presented:

1. Highways
2. Drainage ditches
3. Storm and sanitary sewers
4. Culverts
5. Bridges
6. Landfills
7. Pipelines
8. Railroads
9. Canals
10. Airports
11. Transmission lines
12. Buildings
13. Reservoirs
14. Water and sewage treatment plants
15. Dams
16. Mines
17. Quarries

19-2 WORK OF THE CONSTRUCTION SURVEYOR

A boundary survey and the preparation of the necessary topographic maps are the first steps in the construction process. From these maps the positions of the structures are established. When the final plans for the project are available, it is the duty of the surveyor to set the required horizontal and vertical positions for the structures. In other words, construction surveying involves transfer of the dimensions on the drawing to the ground so that the work is done in its correct position. This type of surveying is sometimes called *setting lines and grades.* The work of the surveyor for construction projects is often referred to as *layout work* and the term *layout engineer* may be used in place of the term *surveyor.*

Of necessity, construction surveying begins before actual construction and continues until the project is completed. Surveying is an essential part of the construction process and must be carried out in coordination with other operations in order to have an economical job and to prevent serious mistakes. The reader can readily understand the expense and inconvenience caused if one reinforced concrete footing for a building is placed in the wrong position. As another example, just imagine the problems involved if a few thousand feet of a sewer line are set on the wrong grade.

Construction drawings show the sizes and positions of structures that are to be erected, such as buildings, roads, parking lots, storage tanks, pipelines, and so on. The job of the construction surveyor is to locate these planned features in their desired positions on the ground. (Figure 19-1.) This is done by placing reference marks (such as construction stakes) sufficiently close to the planned work so as to permit the masons, carpenters, and other tradesmen to position the work properly using their own equipment (folding rules, mason's levels, string lines, etc.).

FIGURE 19-1 Surveying for a bridge. (Courtesy of Carl Zeiss, Inc., Thornwood, N. Y.)

Construction surveyors will find that their work is quite varied. One day they may be doing topographic work for a proposed building, while on another they may be setting stakes for pipeline excavation. It is often necessary for them to make measurements before and after some types of work (e.g., computing quantities of earthwork moved by a contractor). At other times, they will set stakes to guide the construction of foundations; they may align the columns of a steel frame building; they may check a completed structure to see that it is correctly positioned; and so on.

A construction project requires four kinds of surveys for its completion:

1. A property or boundary survey by a registered land surveyor to establish the location and dimensions of the property.
2. A survey to determine the existing conditions, such as contours, man-made and natural features, streams, sewers, power lines, roads, and nearby structures. This work may also be done by the land surveyor along with the boundary survey.
3. The construction surveys, which determine the position and elevation of the features of the construction work. These surveys include the placing of grade stakes, alignment stakes, and other layout control points. This work is often performed by the contractor.
4. Finally, there are the surveys that determine the positions of the finished structures. These are the "as-built" surveys and are used to check the contractor's work and show locations of structures and their components (water lines, sewers, etc.) that will be needed for future maintenance, changes, and new construction. These measurements should be performed by a registered surveyor.

19-3 TRADE UNIONS

In most states trade unions do not claim jurisdiction over construction surveying. Nevertheless, surveyors involved in layout work can become involved in disputes where the unions claim that surveying work should be done by union members. For instance, the unions feel that their carpenters are supposed to nail batter boards and lay out building partitions. They further claim that their ironworker foremen are supposed to check the positioning and alignment of structural steel and that their concrete finishers are to set screeds for concrete slabs. In certain areas of the country, the union operating engineers have jurisdiction over field survey work done by the contractor. If, however, the surveyor is employed as an office engineer and part of the job is to do field surveying, only the field surveying work would be under the jurisdiction of the operating engineers.[1]

19-4 PROPERTY SURVEY FROM THE CONTRACTOR'S VIEWPOINT

The contractor for a construction job should be furnished with a property survey made by a registered surveyor. In fact, the construction contract signed by the

[1]K. Royer, *Applied Field Surveying* (New York: John Wiley & Sons, Inc., 1970), p. 124.

contractor should require that the property survey be furnished by the owner. Furthermore, the contractor should be able to prove that the survey used was indeed furnished by the owner.

Even if the contractor or one of his or her employees is a registered surveyor, he or she should not make the property survey, nor should the contractor even employ a surveyor to make it. If one of the employees makes the survey, the contractor will personally be taking the responsibility of locating the structure, and if a mistake is made, the payment for the entire building project may be placed in jeopardy. Numerous lawsuits occur each year when buildings are placed in the wrong positions—frequently being located wholly or partly on other people's land. Even if the structure is only a small distance out of position, the result can be a disastrous lawsuit. This kind of mistake could possibly bankrupt a contractor.[2]

19-5 PRELIMINARY SURVEYS

To prepare the plans for a building the architect needs a map of the site so that the building will be located carefully. These maps are typically drawn to a large scale, such as 1 in. = 10 ft. or 1 in. = 20 ft. The information to be included are property lines; elevations for the preparation of contour lines; locations and sizes of existing buildings on the site or adjacent to it and materials from which they were constructed; location of any immovable objects; locations of existing streets, curbs, and sidewalks; locations of fire hydrants; sizes and locations of gas and water lines and storm and sanitary sewers, including manhole locations and *invert elevations* (these are low points on the inside circumference of the pipes); and locations of power lines, telephone lines, light poles, trees, and other items.

Before the design of the structure can begin, the above listed information needs to be furnished to the engineer-architect group. The data will ordinarily be provided on the site map; for buildings this drawing is frequently called the *building plot plan.*

The layout of the building will be based on the information presented on the plot plan and the proposed building design will be superimposed on that plan. The final drawing will show the location of the building with respect to the property lines and with respect to streets, utilities, and so on. It will probably also show the final contour lines that are to exist after the construction is complete.

The preliminary survey may include a survey of existing structures on the site and perhaps structures on adjacent property that might be affected by the new construction. Such surveys should be performed prior to the start of construction. They should include measurements of the vertical and horizontal positions of the foundations of these buildings so that it can be determined later if any lateral or vertical movement has occurred during construction. In this regard it should be noted that settlements continue for quite a few years after a building is constructed, and if the existing building or buildings on or adjacent to the site are relatively new, they may still be settling. As a result, settlement that occurs in an existing building when a new one is erected may or may not be due entirely to the new construction.

The surveys of existing buildings should also include examinations of both the exterior and interior of the buildings to record their condition. For instance, such items as the location and sizes of wall cracks should be noted.

[2]K. Royer, *Applied Field Surveying* (New York: John Wiley & Sons, Inc., 1970), pp. 110–111.

19-6 GRADE STAKES

If a project is to be constructed to certain elevations, it is necessary to set stakes to guide the contractor. Once the rough grading has been completed, it is necessary to place these stakes in order to control the final earthwork. A *grade stake* is a stake that is driven into the ground until its top is at the elevation desired for the finished job or until the elevation of the top has a definite relation to the desired elevation. Grade stakes are necessary for sewers, street pavements, railroads, buildings, and so on.

When appreciable cuts are to be made, it may not be feasible to drive stakes into the ground until their tops are at the desired final elevations because to do so they may have to be driven below the ground or buried. If appreciable fills are to be made, the tops of the grade stakes may have to be located above the ground level. Here it is the custom to drive stakes to convenient elevations above the ground and mark them with the cuts or fills necessary to obtain the desired elevations. It is very helpful to the contractor when the stakes are placed at heights such that cuts or fills are given in whole numbers of feet.

For a particular point the surveyor determines the required elevation and subtracts it from the H.I. of the level. This tells him or her what the rod reading should be when the top of the stake is at the desired elevation. Then, more or less by trial and error, the stake is driven until the required rod reading is obtained when the rod is held on top of the stake.

When earth grading operations are near their desired values, it is possible to drive stakes until their tops are at the specified final elevations. For these situations it is customary to drive the stakes to grade and color their tops with blue lumber crayons called *keel.* These stakes, known as *blue tops,* are commonly used for grading operations along the edges of highways, railroads, and so on. Some people require that grade stakes be set at road center lines instead of along shoulders.

19-7 REFERENCING POINTS FOR CONSTRUCTION

All survey stakes, even the very stoutest ones, are vulnerable to disturbance during construction. As a result, it is necessary to reference them to other points so that they may be reestablished in case they are displaced. In construction or layout surveying, the terms *stakes* and *hubs* are frequently used. A *stake* is usually thought of as an approximately $1 \times 2 \times 18$ in. (or longer) piece of wood sharpened at one end to facilitate driving in the ground. A *hub* is an approximately 2×2 in. piece of wood of variable length driven flush or almost flush with the ground, which has a tack driven into it to mark the precise position of the point. Usually, one or more stakes on which identification of the hubs is written are driven partly into the ground by the hubs. The stakes may also be flagged.

For highway construction it is common practice to reference every tenth station (1000-ft points) as well as the starting and ending points of curves and the points of intersections of the tangents to curves. For building construction it is conventional to reference building corners and even property corners if they are in danger of being obliterated.

The references used should be fairly permanent. Markers may consist of crosses or marks chiseled into concrete pavements or curbs or sidewalks, or nails and bottle caps driven into bituminous pavements. Other types are tacked wooden stakes, nails in trees, or preexisting features such as the corners of existing buildings. One disadvantage of the latter two types of points is that the tripod may not be set over them. Sometimes references may be marked on a curb, pavement, or wall with keel. In addition, directions are often written on pavements describing how to find the marks whether the marks themselves are on the pavement or wall or in some other location.

It is desirable to set a sufficient number of reference points so that if some of them are lost, the referenced point can still be reset. It is advisable that important points be recorded and described in the notes in case the markings are obliterated or the points themselves disturbed.

If the distances between survey points and their references can be kept to less than 6 ft, the points may be reestablished quickly and easily with a carpenter's rule and a plumb bob. Should reference points be set at greater distances, it will be necessary to use a tape and perhaps a total station to check and reset construction points. It is probably true that almost all construction points will be disturbed at least once and perhaps many times during the construction process. As a result, the surveyor needs to position the reference points carefully so that if they are disturbed, they can easily be reestablished. It is always desirable to put guard stakes and flags around important construction points. Despite these precautions, it is still necessary to make continual position and alignment checks. Otherwise, some of the construction work may be improperly positioned, resulting in possible removal and replacement at potentially great expense.

One very common method of referencing survey points is illustrated in Figure 19-2. In this figure, three reference points are used; if the survey point is disturbed, it can be reestablished by swinging an arc with a steel tape (held horizontally) from each of the reference points. The desirability of keeping the reference points within one tape length of the survey points is obvious (6 ft is even better, as discussed previously). To protect the hubs and make it easier to locate them, it is desirable to drive slanted stakes over their tops.

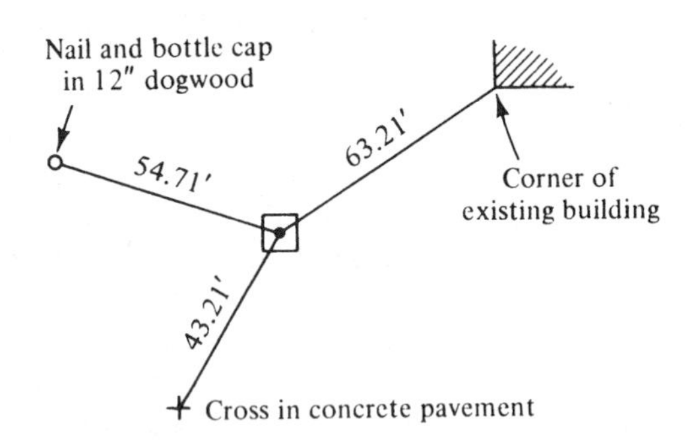

FIGURE 19-2 Three reference points.

A convenient and frequently used method for referencing building corners is shown in Figure 19-3. It is desirable, where possible, to set reference points the same distance away from the survey point being referenced so that no notes have to be used to reset it.

It is quicker to set reference points at random as shown in Figure 19-4, to make use of relatively permanent objects such as trees or buildings or pavement or poles in the vicinity. To reestablish survey points from these references, it is convenient to use three workers so that simultaneous measurements can be made from both reference points.

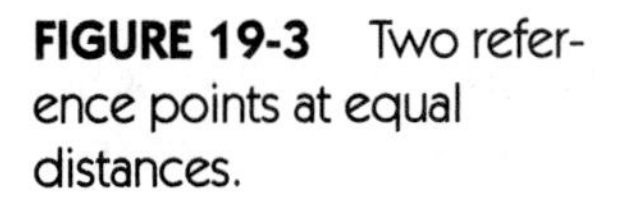
FIGURE 19-3 Two reference points at equal distances.

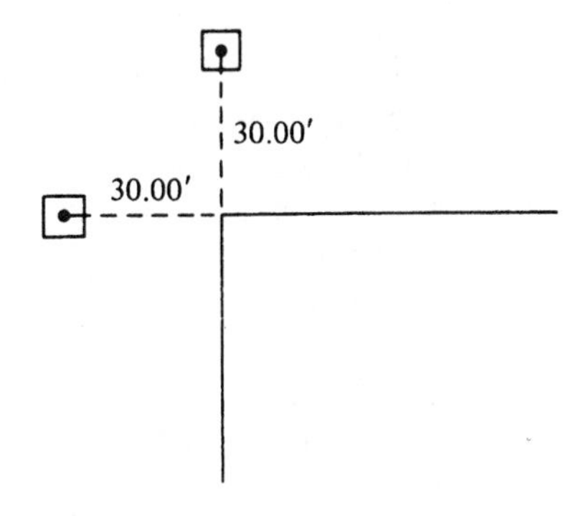

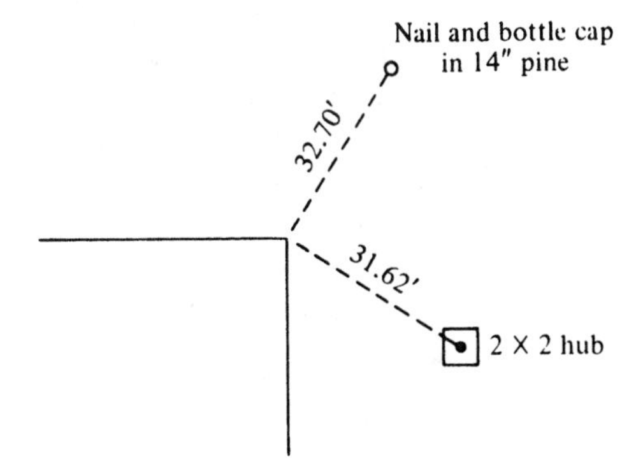

FIGURE 19-4 Setting random reference points.

19-8 BUILDING LAYOUT

The first step in laying out a building is to locate the building properly on the lot. Many cities and counties have building ordinances that provide certain minimum permissible set-back distances from the front edge of lots to the fronts of buildings. The purposes of these regulations are to promote organized growth, to improve appearances, and to enhance fire protection. The survey crew sets alignment stakes with distances and directions measured from existing buildings, or streets or curbs, so that the structure will be erected in its desired position.

After the building is located properly, it is necessary for it to be laid out with the correct dimensions. This is always important but is particularly so when some or all of the building components are prefabricated and must fit together. For most residential buildings surveying layout may consist only of surveying for footings because the contractor will be able to obtain all other measurements from the footings. Commercial construction requires detailed layout and control stakes primarily due to the length of time the project will be under construction and the size of the project.

It is obviously critical for a construction job that the various parts of the structure be placed at the desired elevations. To accomplish this goal the construction surveyor will establish one or more bench marks in the general vicinity of the project. These bench marks are placed away from the immediate vicinity of the building so they will not be destroyed by construction operations, and are used to provide vertical control for the project. Once these are set, the surveyor will establish a good many less permanent but more accessible bench marks quite close to the project (say, within 100 to 200 ft). The location of these less permanent points should be carefully selected so that turning points will ideally not be needed when elevations have to be set at the project. Such careful selection of the points may result in critical time saving, which is so important on construction projects.

For large construction projects it is desirable to establish elevations based on sea-level values. (North American Vertical Datum of 1988). This datum is particularly important for underground utilities. Should an assumed elevation datum be used, its elevations probably should be sufficiently large so that no negative elevations will occur on the project.

Another idea that may save considerable time involves establishing a permanent position for setting up the level so that the instrument can be set up in the same position every day. One way of doing this is to have actual notches for the tripod shoes cut into a concrete sidewalk or pavement or into a specially placed concrete pad. In this way the h.i. of the level will always be the same if the same instrument and tripod are used, and the rod reading for any particular desired location (say, a floor level) will be constant.

Sometimes a special concrete pier can be justified for a large job. Another case where a platform may be advisable is where it is necessary constantly to check nearby existing buildings for possible settlement during the progress of the new construction. This is particularly important during excavation, especially if there is much vibration or blasting. A set of targets can be established on nearby buildings, the level set up at the fixed point, and the telescope used to check quickly for settlements by sighting on the targets. In this regard it is necessary to use the same instrument and tripod (fixed leg, not adjustable leg) each day. It is further necessary to check the elevation of the permanent bench mark frequently against surrounding bench marks that are away from the influence of the construction activity.

Later during construction, it may be convenient to establish a bench mark at a certain point in the walls or other part of the building and use that point to measure or set other elevations in the structure. In addition, other reference points may be set within the building from which machinery or other items may be properly located once the building is enclosed. This is often done by setting brass reference points or disks in the floors or other points in the building. Once the walls are erected, the surveyor will be unable to take long sights from reference points outside the building.

19-9 BASE LINES (LAYOUT PERFORMED BY SURVEYORS)

Before the actual layout measurements can begin, it is necessary for reference lines or base lines to be carefully established. For large construction projects, the usual procedure is to set a main base line down the center line of the structure and to set stout stakes or hubs (preferably 2 × 2 in. or larger) with tacks in them at intervals not exceeding 100 ft. It is hoped that these hubs will stay in position for some time during construction. If station numbers are assigned to the points, they should preferably be large numbers so that no negative stations will be needed on the extreme ends of the system.

In addition, monuments are set along the center line at each end beyond the area of the construction work. The ends of the line are best marked with heavy cast-in-place concrete monuments with metal tablets embedded in their tops. The monuments along the ends of the line may be occupied by the surveyor and will enable him or her to check and reset, if necessary, points within the construction area.

A central base line will be disturbed so often during construction that it is common practice to set another or secondary base line parallel to the central one but offset some distance from it. In fact, sometimes two additional base lines, one in front of the building and one in the back, may be established.

Figure 19-5 shows the case where a base line is set to one side of the building. Along such a base line hubs are set where needed to enable the surveyor to set or align corners or other important features of the building. After the corners for a building are set, it is essential that their locations be carefully checked. One check that can be used for rectangular buildings is to measure their diagonals to see if they are equal (shown by dashed lines in Figure 19-5).

FIGURE 19-5 Layout of a simple building.

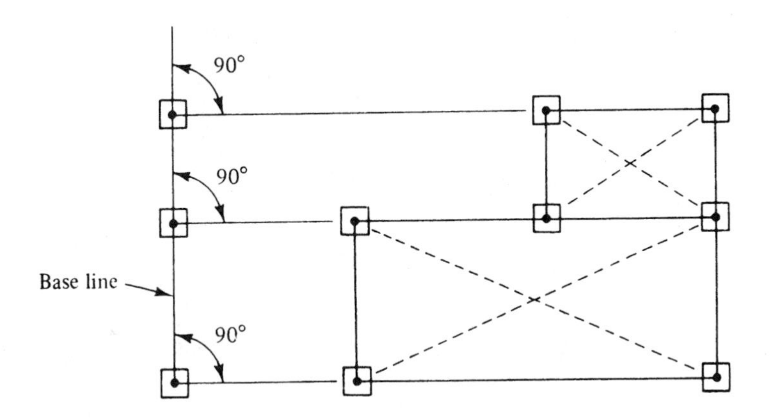

When it is constantly necessary to provide alignments for points in a building where reference markers have been disturbed (a typical situation for the construction of most buildings), the procedure described in the preceding paragraphs and illustrated in Figure 19-5 has the disadvantage that it is necessary to set up the instrument and lay off 90° angles every time a point is needed. It is preferable to use a system where the important points in the building can be checked or reestablished merely by sighting between two points or by stretching a string or wire between them. Such a system is illustrated in Figure 19-6. For this building both a base line and an auxiliary base line were established, together with the auxiliary points shown. When feasible, it is desirable to attach or paint targets on the walls of existing buildings instead of setting the auxiliary points in the ground. Such targets are much less likely to be disturbed than are stakes in the ground.[3]

[3]B. A. Barry, *Construction Measurements* (New York: John Wiley & Sons, Inc., 1973), pp. 149–155.

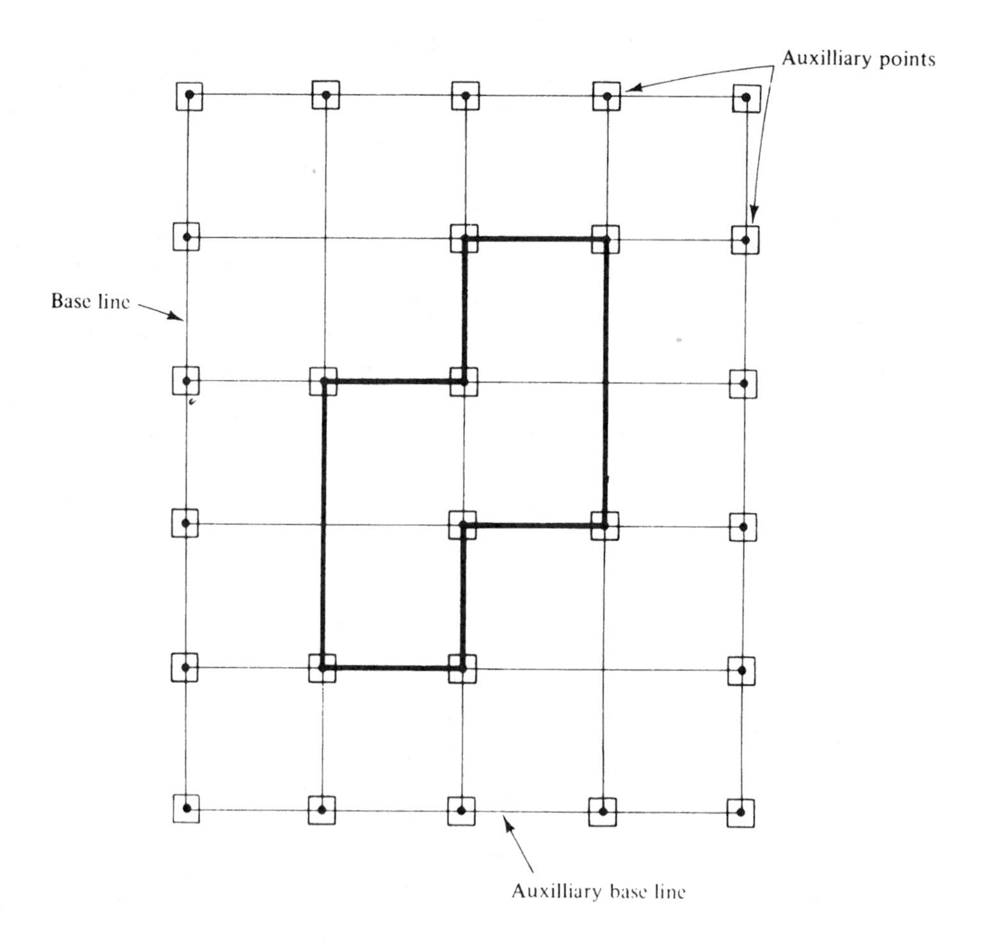

FIGURE 19-6 Layout of a building with base line, auxiliary base line, and auxiliary points.

19-10 RADIAL STAKING METHODS

Many construction plans today are prepared on the basis of radial control systems. With this type of system control points are set at important locations around the job site and used to locate the important features of the structure. These latter points are located by measuring angles and distances from the previously set control points.

The radial method is very efficient for setting large numbers of points, particularly when total stations are used. To stake out a construction project one or more control points are established on or near the job site. Then the coordinates of the various points are determined from the plans. Finally the angles and distances from the control points are computed and laid out in the field. The total station instruments can be used to make these latter calculations.

19-11 BATTER BOARDS

After hubs are set for the corners of a building, they may be secured by means of references set just outside the work area in the form of *batter boards.* If hubs were placed only at the proposed corners of a building, they would be in the way of excavation and construction operations and would not last long. For this reason, batter boards are ordinarily used. They consist of wooden frameworks that have nails driven into their tops (or have notches cut into them) from which strings or wires are placed to outline the position of the building lines and perhaps the outside of the foundation walls.

Batter boards are used not only for building corners but also for the construction of culverts and sewer lines, for bricklaying, and for many other construction jobs. Examples of batter boards are shown in Figures 19-7 and 19-8. Batter boards should be placed firmly in the ground and should be well braced. Further, they should be placed at a sufficient distance from the excavation so that they will be located in undisturbed ground, yet close enough so that strings may conveniently be stretched between them.

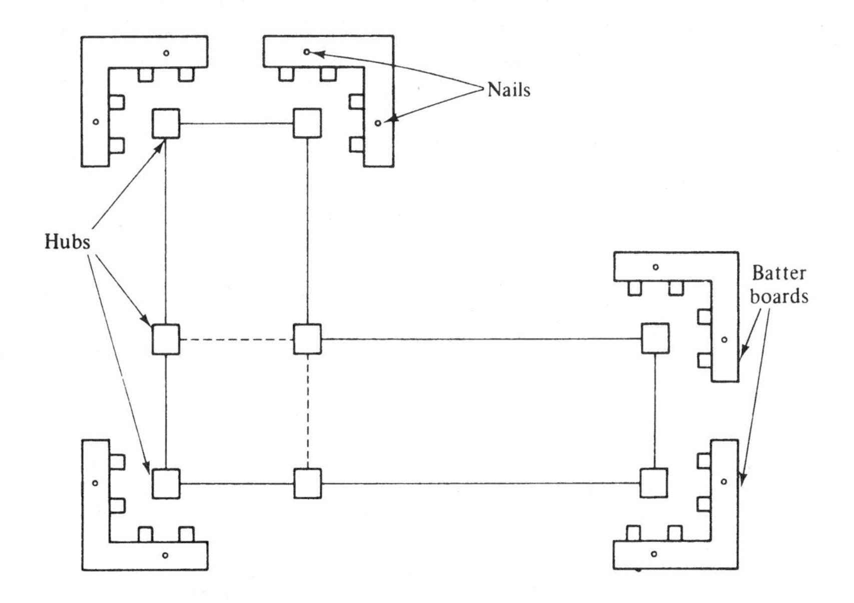

FIGURE 19-7 Batter boards.

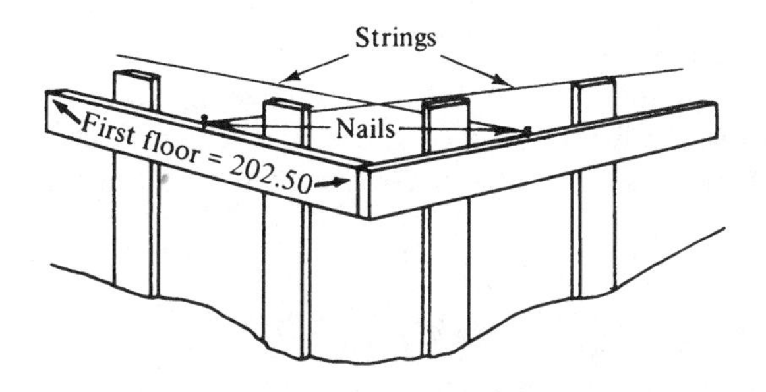

FIGURE 19-8 Batter boards showing strings and first-floor elevation.

To build the batter boards, several stout timbers are driven in the ground about 4 to 6 ft from the desired excavation. As wire lines are often used between batter boards, they have to be quite stout to withstand the pulling. The vertical posts are 2 × 4s and 1 or 2 in. boards are nailed across them. The cross boards are set between the timbers with a level rod and adjusted to the desired elevation with a level. To set the nails, which are driven vertically on the cross boards to hold the strings for alignment, the total station is used or strings are stretched across the corner points using plumb bobs as needed.

The main purpose of batter boards is to enable workers to measure values from readily accessible reference points without the necessity of having a surveyor always on the spot. Since these boards are used to provide both alignment and elevation, they must be set very carefully. It is common to set the batter boards at a prearranged height a certain number of feet above the foundation or the finished floor grade. A controlling elevation such as that of the first-floor level should be clearly marked on the batter boards, as illustrated in Figure 19-8.

Once the batter boards are set and checked, the strings or wires can be stretched between the nails (or notches) as shown in Figure 19-8. The strings can be taken down or replaced as often as desired during construction. For instance, they can be put in place and the corner of the structure reset at any time by hanging a plumb bob at the intersection of the strings. The amount of excavation needed at a particular point for the foundation can be determined by measuring down from the string. The strings can be taken down when it is necessary for workers or machines to excavate for the foundation. They can be put back up for use by the brick masons or other workers. In masonry structures a *height pole* is often used to give the elevations of various points or courses in the walls. Such a pole is shown in Figure 19-9.

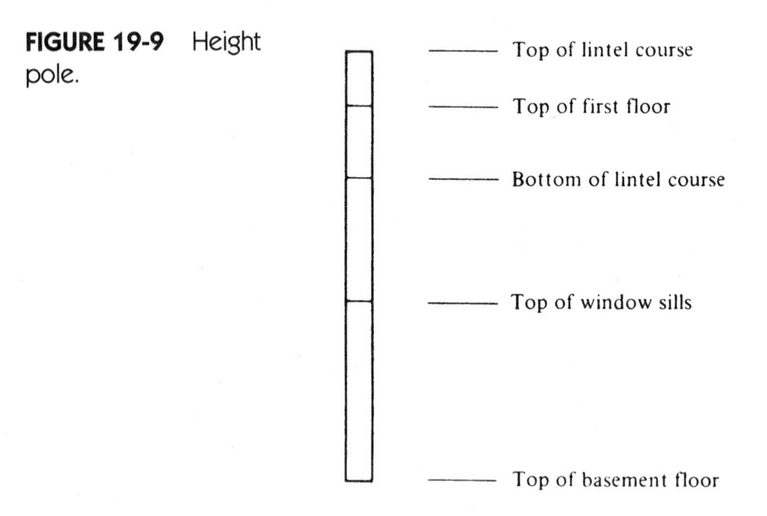

FIGURE 19-9 Height pole.

Batter boards are used quite satisfactorily for the construction of houses and small buildings. For larger structures where heavy equipment is involved, batter boards are rather vulnerable and foundation alignments and elevations are established and reestablished as necessary by using offset control points.

19-12 BUILDING LAYOUT: CONTRACTOR METHOD

For residential construction and some small commercial buildings the foundation is often subcontracted to a footing contractor. The duties of this subcontractor are to lay out, excavate, place reinforcing bars, set grade stakes, call for inspection, and place the concrete. These subcontractors are not surveyors but they use some very ancient and proven layout methods. Levels, lasers, or total stations are used only for the determination of elevations. Angles are usually not turned with total stations but are established with tapes.

A rather typical procedure followed by these subcontractors is shown in these steps:

1. The property corners are located.
2. A 1×4 stake is placed at each corner and nylon strings are stretched along the property lines.

3. The setback distances as obtained from the site and floor plans are laid off, thus establishing one of the building corners and two of its sides.
4. From the corner and sides established in step 3 the building is laid out as a large rectangular box. The 90° corners for this box are established with a tape.
5. The building corners are marked on the ground and batter boards are driven with offsets of about 6 to 9 ft.
6. Nails are driven into the batter boards and nylon strings are placed over the corners previously marked on the ground making use of plumb bobs.
7. Tapes are used to measure the exterior lines and diagonals are carefully checked. Offsets are measured from the building lines.
8. Nylon lines are carefully laid out to the outside of the foundation wall (not the outside of the footing). The earth beneath the lines is scratched with some type of rod or a reinforcing bar.
9. The batter boards are set to the desired finished floor elevations.
10. The nylon lines are then removed and the footing excavations made.
11. The placing of the reinforcing bars followed by the inspection and then the placement of the concrete finish the footing subcontractor's work.

19-13 AS-BUILT SURVEYS

As-built surveys are made after a construction project is complete, to provide the positions and dimensions of the features of the projects as they were actually constructed. These surveys not only provide a record of what was constructed but also provide a check to see if the work was done according to the design plans. The monuments set as control for the projects are checked and readjusted or replaced if necessary. It is essential to protect this control system as much as possible in case there is a future modification or expansion of the project.

From the vertical and horizontal control points a detailed map is prepared showing all changes made during construction. The usual construction project is subject to numerous changes from the original plans due to design modifications and to problems encountered in the field, such as underground pipes and conduits, unexpected foundation conditions, and other unforeseen situations.

Laser levels are particularly useful for conducting parts of as-built surveys. An example is the checking of grades for drainage. After setting up a laser level, one person can walk around with a field book and check the elevations.

The reader should note that it is necessary to survey pipelines, sewers, and other underground structures before they are backfilled so that their correct positions will be determined both horizontally and vertically. In addition, these surveys must show the locations of unexpected pipes, structures, and other features, which are encountered in the excavations. The as-built survey is a very important document and must be preserved for use in future repairs, modifications, and expansions.

Problems

19-1. What are the four kinds of surveys needed during a construction project?

19-2. Discuss the relationship of surveyors on a construction job with trade unions.

19-3. Why should a contractor for a construction job neither make nor employ someone to make a property survey for the job?

19-4. What are grade stakes?

19-5. What is the invert of pipe?

19-6. What is a blue top?

19-7. Distinguish between hubs and stakes.

19-8. What is the main purpose of batter boards?

19-9. Discuss the importance of making as-built surveys for a construction project.

CHAPTER

20 Volumes

20-1 INTRODUCTION

A tremendous volume of earthwork needs to be moved for the construction of highways, railroads, canals, foundations of large buildings, pipelines, and other projects. The surveyor is often directly involved in the determination of the amounts or quantities of this earthwork. He or she is concerned not only with quantities but also with the setting of the grade stakes needed to carry out the earthwork required to bring the ground to the desired grades and elevations.

Before a construction project involving earthwork is begun, the surveyor needs to determine the shape of the ground surface. This is necessary so that the volume of materials to be added or removed may be determined. In speaking of earthwork it is the custom to refer to excavations as *cuts* and to embankments as *fills*. The quantities of cuts and fills in the types of construction projects described herein are frequently of such magnitude as to make up appreciable percentages of the total project costs.

The principles involved in volumetric calculations are applicable not only to earthwork but also to other materials, such as to volumes of reservoirs and to stockpiles of sand, gravel, and other materials. Volumes of masonry structures may be computed directly from the dimensions on plans, but it is also quite common to check pay quantities by computing the volumes from the measurements of the completed structures made by the surveyor in the field. It should also be noted that maps produced by photogrammetric methods enable the surveyor to estimate earthwork quantities quite well. Either topographic maps prepared from photographs or stereoscopic models may be used.

20-2 SLOPES AND SLOPE STAKES

In working with cuts and fills for highway construction, the side slopes are generally based on the material involved. The slope is given as a ratio of so many units horizontally to so many units vertically. For instance, a 2-to-1, or 2:1, slope means that the bank in question goes 2 ft horizontally for each 1 ft vertically (see Figure 20-1). Perhaps a better way to indicate this, with less chance of misinterpretation, is to say 2 on 1. The $1^{1}/_{2}$ on 1 slope is probably the most commonly used, but if the material consists of solid rock, the slopes may be much steeper. For very loose material, such

FIGURE 20-1

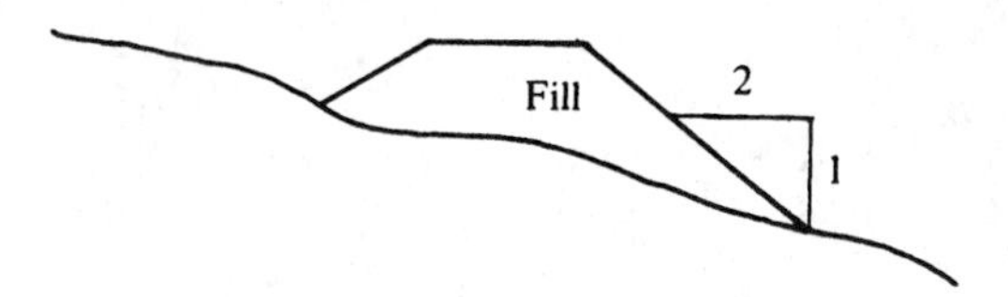

as sand, they will be much flatter (perhaps as much as 8 or 10 or even 12 on 1 in extreme cases).

After the grade of the road has been established and the material slopes decided for cuts and fills, it is necessary for the surveyor to stake out the work. The cut or fill to be made at a particular station along the center line of the road equals the difference between the ground elevation there and the final elevation from the grade line. The surveyor will need to stake the road center line and set *slope stakes* at the intersection of the natural ground line and the side slopes of the cut or fill as shown in Figure 20-2.

A few descriptive comments are made here regarding the trial-and-error

FIGURE 20-2 Slope stakes.

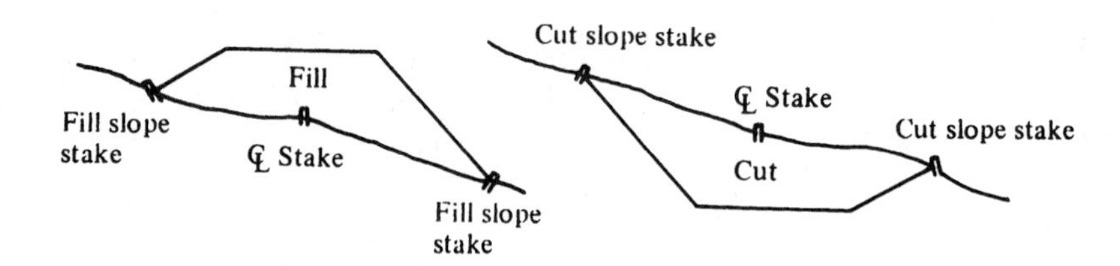

problem of setting slope stakes, because it may give the student a little trouble. If the ground surface is horizontal, the distance from the road center line to the intersection of the ground surface and the cut or fill can easily be computed, as shown in Figure 20-3.

If the ground surface is sloping, as is the case in Figure 20-4, the problem

FIGURE 20-3

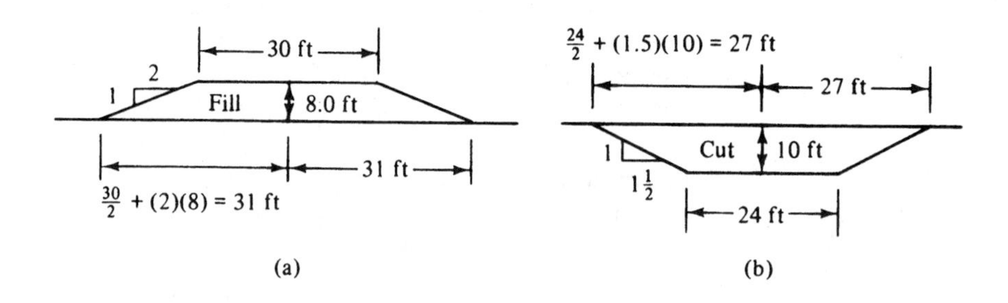

becomes a trial-and-error one because the vertical amounts of cut or fill as measured from the slope stake positions are unknown, as are the horizontal distances from the center line to the slope stakes.

One common practice is to set slope stakes with the 50-ft woven tape and a level.

FIGURE 20-4

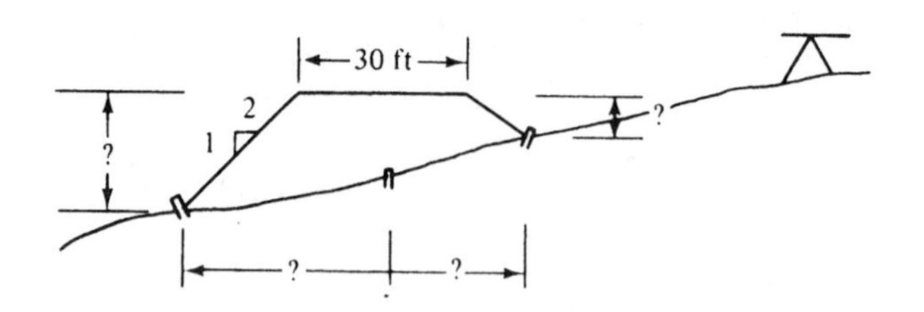

The zero end of the tape is placed at the center line. The instrumentman roughly estimates the distance to the slope stake and sends the rodman that distance from the center line. The instrumentman then takes a reading on the level rod and calculates to see if the rodman is at the desired point. For this discussion, see Figure 20-5, in which the instrumentman estimated the fill at the lower slope stake as 8.0 ft and then computed the horizontal distance from the center line as (30/2) + (2)(8) = 31 ft.

FIGURE 20-5

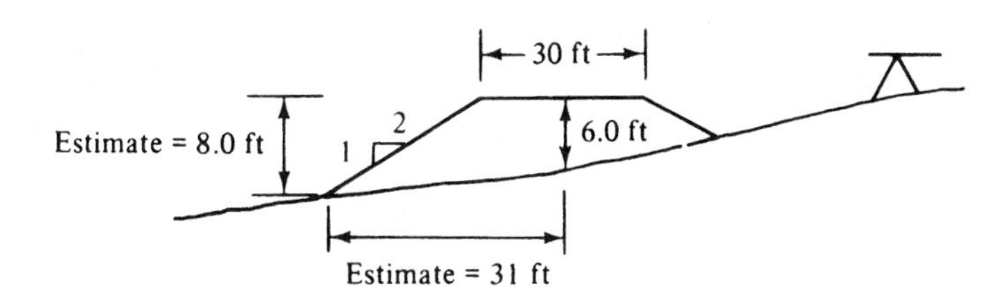

The rodman was sent out 31 ft from the center line and the level rod reading was taken. It was found that the difference in elevation from the top of the fill to the point in question was 10.0 ft. The horizontal distance from the center line for that difference in elevation should have been (30/2) + (2)(10) = 35 ft, so the rodman was not at the correct point.

Next, the rodman was sent out to the 40-ft point and the elevation difference was found to be 12.0 ft. For this point, the horizontal distance should have been (30/2) + (2)(12) = 39 ft. The rodman moved to the 39-ft point, where the elevation difference was 11.9 ft. The calculated horizontal distance equaled (30/2) + (2)(11.9) = 38.8 ft, which was just about right.

Once the trial horizontal distance is within one or two tenths of a foot of the computed horizontal distance, the slope stake is driven. An experienced person usually sets the stake within two or three trials at most, but the beginner may require a few more attempts.

Slope stakes are normally set sloping outward from the road for fills and inward for cuts. The station numbers are usually written on the outside of the stakes and the cuts or fills are given on the inside, often with the distance to the center line of the road given. Sometimes after the correct position of the slope stakes is determined, they are offset by 2 to 5 ft to try to preserve them during grading operations.

20-3 BORROW PITS

During the construction of roads, airports, dams, and other projects involving earthwork, it is often necessary to obtain or *borrow* earth from surrounding areas in order to construct embankments. These excavations are commonly referred to as *borrow pits.* The quantity of borrow material is very important because the contractor's pay is usually computed by taking the number of yards of borrow times the bid price per yard. In addition, the adjacent areas where the borrow pits are located often belong to other people, who are also paid on a quantity basis. Volumes of stockpiled materials such as sand, gravel, coal, and so on, can also be calculated using the borrow pit method.

To determine the data needed for volume calculations, elevations may be obtained at certain points before and after the earth is removed. One or more reference base lines and two or more bench marks are established at convenient and protected locations. Then some type of grid system is normally established (say, 50-ft squares) and the elevations are determined at each of the corners. This may be quickly done with a total station. When the excavation is completed, the levels are run again to the same points and from the differences between the original and final readings, the cut made at each point is obtained. Volumes may be approximately calculated by multiplying the average cut for a particular square or shape times the area of the shape.

Figure 20-6 shows a plan view of a typical borrow pit divided into convenient squares and rectangles and also into some triangles because of the irregular shape of the borrow pit. The numbers shown in the figure represent the dimensions of the figure and the cuts in feet at the corners. The value of earthwork under one of the areas can be estimated as being equal to the average of its corner cuts times the area of the shape. For instance, the estimated cut for shape *opts* is equal to

$$(40)(50)\left(\frac{2.4 + 2.8 + 2.0 + 3.0}{4}\right) = 5100 \text{ cu ft}$$

A similar calculation for shape *pqut* follows:

$$(40)(50)\left(\frac{2.8 + 2.6 + 3.0 + 2.8}{4}\right) = 5600 \text{ cu ft}$$

The total volume for the two shapes is 10,700 cu ft.

Suppose that since these two rectangles have the same horizontal dimensions (40×50 ft) we would like to combine their calculations into one set. Looking back at the preceding computations and also at Figure 20-6, we can see that to determine the sum of the corner cuts for the two shapes the values at corners *o*, *q*, *u*, and *s* appear one time, while the values at *p* and *t* (2.8 and 3.0, respectively) each appear twice. Thus, the calculations for the two figures can be handled in one step as follows:

$$(40)(50)\left(\frac{2.4 + (2)(2.8) + 2.6 + 2.0 + (2)(3.0) + 2.8}{4}\right) = 10,700 \text{ cu ft}$$

FIGURE 20-6

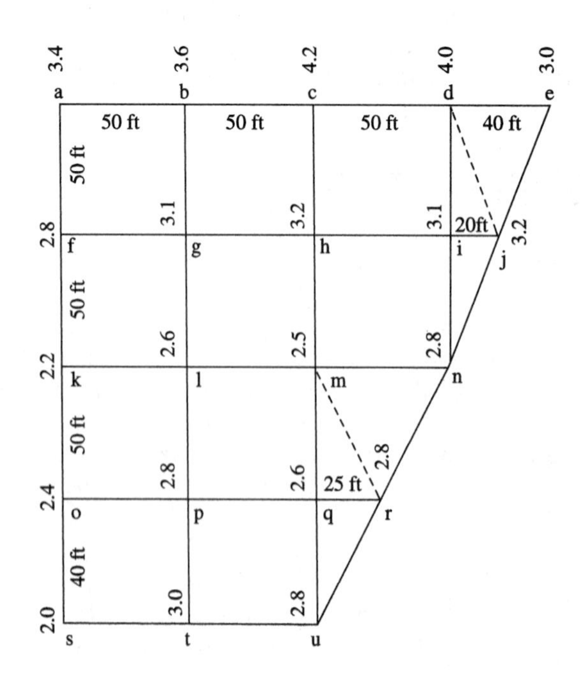

In a similar fashion all eight figures within *adnmqo* are squares (50 × 50 ft). To sum the corner cuts for all these figures the one at *a* appears in one square, the ones at *b* and *c* each appear in two squares, the ones at *g* and *h* each appear in four squares, and so on.

$$(50)(50)\left[\frac{3.4 + (2)(3.6) + (2)(4.2) + 4.0 + (2)(2.8) + (4)(3.1)}{4} \text{ etc.}\right]$$

$$= 59{,}800 \text{ cu ft}$$

The calculations for the entire borrow pit can be recorded very compactly as shown in Table 20-1. Groups of shapes that have the same horizontal dimensions are combined. Some of the computed volumes shown have been rounded off a little, as is only reasonable for earthwork calculations.

TABLE 20-1 Calculations for Borrow Pit (Figure 20-6)

Figure	Sum of corner cuts (ft)	Area of shape (sq ft)	Multiplier	Volume (cu ft)
oqsu	21.4	2000	× 1/4	10,700
adnmqo	95.7	2500	× 1/4	59,800
uqr	8.2	500	× 1/3	1,370
qrm	7.9	625	× 1/3	1,650
rmn	8.1	1250	× 1/3	3,380
nij	9.1	500	× 1/3	1,520
ijd	10.3	500	× 1/3	1,720
jed	10.2	1000	× 1/3	3,400
			Total volume =	83,540 cu ft
			=	3094 cu yd

A special comment should be made about the shapes on the side of the borrow pit, as, for example, trapezoid *deji*. The volume of this prism may be determined by taking the sum of the four corner cuts divided by 4 and multiplied by the area of the trapezoid. It will be noted that this does not give the same value as that obtained by breaking it into the two triangular figures, *ijd* and *jed*. If the ground surface across the trapezoid is not fairly constant in terms of slope (perhaps where there is a hump or dip out in the figure), a more accurate volume can be obtained by breaking the figure into triangles. For such a case the surveyor should show dashed lines on the drawing (such as *dj* and *mr* in Figure 20-6) to indicate that triangular shapes should be used.

20-4 CROSS SECTIONS

For the purpose of this discussion the construction of a highway will be considered. It is assumed that a longitudinal grade line has been selected as well as the roadway cross section. It is further assumed that cross sections of the ground surface have been taken at each station along the route by the method described in Section 8-5.

A cross section is a section normal to the center line of a proposed highway, canal, dam, or other construction project. The cross sections of the ground surface are plotted for each station and the outline of the proposed roadway as traced from a template is superimposed on the cross sections, with the elevation of the center line of the roadway obtained from the longitudinal grade line. Examples of such cross sections are shown in Figure 20-7.

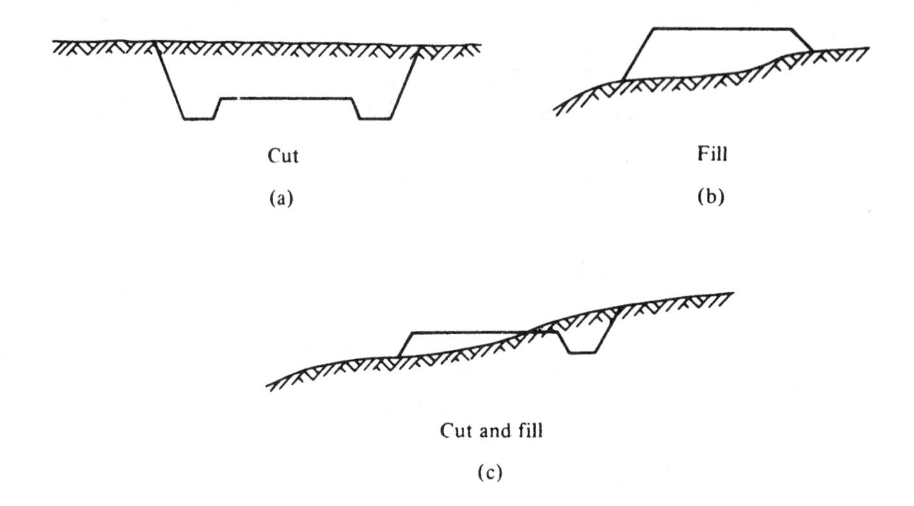

FIGURE 20-7 Typical cross sections.

20-5 AREAS OF CROSS SECTIONS

To determine earthwork quantities as described later in this chapter, it is first necessary to determine the areas of the cross sections. This may be done either by means of computations or with planimeters, as described in the following paragraphs.

Computer Programs

The calculations of cross-sectional areas and earthwork volumes are very tedious and repetitious for large projects. As such they are ideally suited for computer applications. There are available numerous "canned programs" with which the calculations can quickly be made. Most highway departments today use computers to determine their required areas and volumes for earthwork. Nevertheless, the surveyor should be familiar with the theory behind the calculations described in the remainder of this chapter.

Areas of Level Cross Sections

The area of a level cross section such as the one shown in Figure 20-8 may be calculated by multiplying the average of the top and bottom widths of the cross section by its depth, as shown in the figure. Level cross sections are quite commonly encountered for highways, railroads, ditches, and so on.

FIGURE 20-8

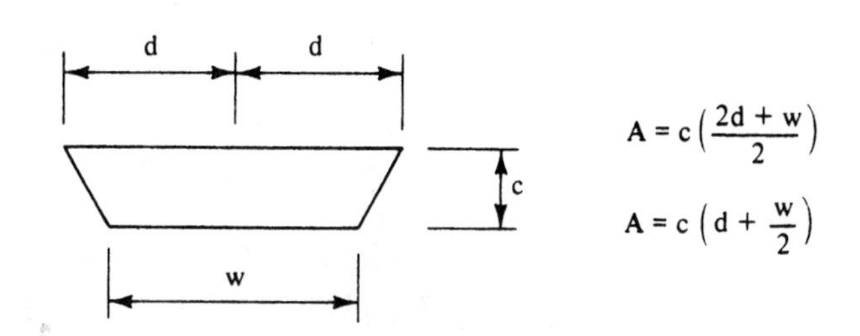

Areas of Three-Level Sections

When a three-level section is involved, such as the one shown in Figure 20-9, its area may be determined by breaking the figure down into triangles and summing up their areas as shown in the figure.

FIGURE 20-9

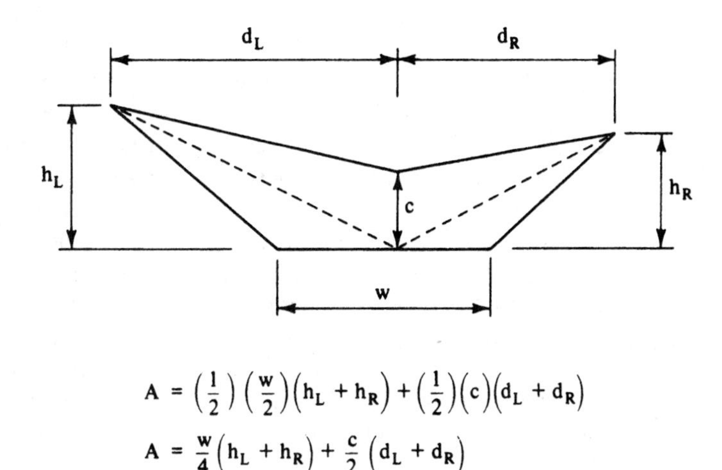

$$A = \left(\frac{1}{2}\right)\left(\frac{w}{2}\right)\left(h_L + h_R\right) + \left(\frac{1}{2}\right)(c)\left(d_L + d_R\right)$$

$$A = \frac{w}{4}\left(h_L + h_R\right) + \frac{c}{2}\left(d_L + d_R\right)$$

Areas of Five-Level or More Sections

When a section of five or more levels is encountered it is possible to compute its area by summing up the areas of the triangles and/or trapezoids that make up the figure, by using a coordinate rule, or by using a planimeter, as described in the next few paragraphs.

Measuring Areas of Irregular Cross Sections with a Polar Planimeter

An irregular cross section is defined as one where the ground surface is so irregular that neither a three-level cross section nor a five-level cross section will provide sufficient information to describe the area sufficiently. Intermediate elevations at irregular intervals are necessary between center line and slope stakes to specify the cross section adequately.

When irregular cross sections are involved it may be possible to break them down into convenient shapes such as triangles and trapezoids and compute their areas. Sometimes a form of the coordinate method may be used. These methods, however, are rather tedious to apply and it is more common to use computer programs or to plot the cross sections on cross-section paper (particularly if the sections are very irregular and also if some of the sides are curved). Once this is done the areas are determined by traversing their perimeters with polar planimeters. In almost every case the area of a cross section can be determined by planimeter with a precision as good as is justified by the precision used in making the field measurements with which the cross section was drawn. Of course, in this age of computers, use of this method has drastically declined.

20-6 COMPUTATION OF EARTHWORK VOLUMES

Earthwork volumes may be computed from the areas of cross sections by two methods, as described in this section. The distance between cross sections is dependent on the precision required for the volume calculations. Obviously, as the price per cubic yard goes up it becomes more desirable to have the cross sections closer together. For instance, for rock excavation or for underwater excavation, costs are so high that cross sections at very close intervals are required, perhaps no more than 10 ft. For ordinary road or railroad earthwork the sections are probably taken at 50- or 100-ft intervals. In addition to the cross sections taken at regular stations it is also necessary to take them at the beginning and ending points of curves, at locations where unusual changes in elevations occur, and for points where ground elevations coincide with natural grades. The latter points are called *grade points.*

The earth between two cross sections forms an approximate *prismoid.* A prismoid is a solid figure that has parallel and plane end faces (or bases) and sides that are plane surfaces. In this section two methods are presented for estimating the volume of these prismoids: the average-end-area method and a method using the prismoidal formula.

Average-End-Area Method

A very common technique used for computing earthwork volumes is the average-end-area method. In this approach the volume of earth between two cross sections is assumed to equal the average area of the two end cross sections times the distance between them (see Figure 20-10). The volume is computed as

$$V = \left(\frac{A_1 + A_2}{2}\right)\left(\frac{L}{27}\right)$$

FIGURE 20-10

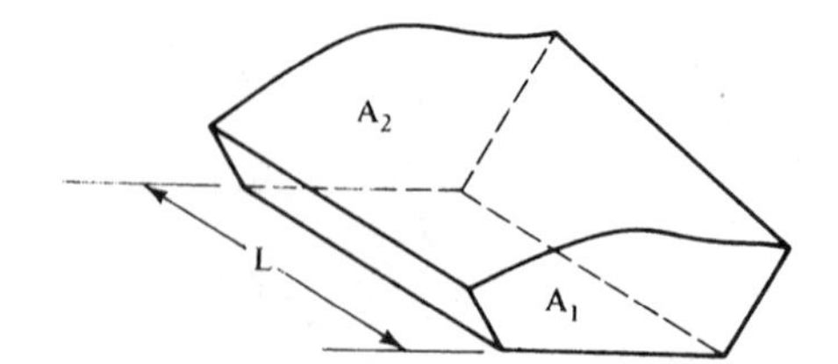

where A_1 and A_2 are the end areas in square feet, and L is the distance between the cross sections in feet. The expression has been divided by 27 to give an answer in cubic yards.

Should cross sections be taken at 100-ft stations, the expression may be reduced to the following form, the volume again being in cubic yards:

$$V = 1.85\left(A_1 + A_2\right)$$

The average-end-area method is commonly used for computing earthwork quantities because of its simplicity. It is not, however, a theoretically exact method unless the two end areas are equal, but the errors are not usually significant. Should one of the areas approach zero, as on a hillside where the cross section is running from cut to fill, the error will be rather large. For this case it may be well to calculate the volume as a pyramid, with $V = 1/3$ the area of the base times the height.

Although the average-end-area method is approximate, the precision obtained is fairly consistent with the precision obtained using the field measurements made for the cross sections. The costs per cubic yard for earthwork are usually low, and thus it is not normally justifiable economically to make refinements in earthwork calculations. Also, in most cases the method gives volumes on the high side, which is in favor of the contractor. To improve the accuracy of the average-end-area method, it is necessary to decrease the length between stations. This is particularly desirable if the ground surface is very irregular. Sometimes when a road has a very sharp curve and large cuts or fills, adjustments may be made for curvature in making volume calculations, but usually such adjustments are not considered significant.

Example 20-1 illustrates the computation of earthwork between two stations using the average-end-area method.

Example 20-1

Determine the earthwork volume between stations 100 and 101 using the average-end-area method if the road width is 30 ft. The cut side slopes are 2 horizontally to 1 vertically. Figure 20-11 shows a sketch of the cross section at station 100 + 00.

Station	Left	Center	Right
100 + 00	$\frac{C8.0}{31.0}$	$\frac{C6.0}{0.0}$	$\frac{C5.0}{25.0}$
101 + 00	$\frac{C6.0}{27.0}$	$\frac{C4.0}{0.0}$	$\frac{C2.0}{19.0}$

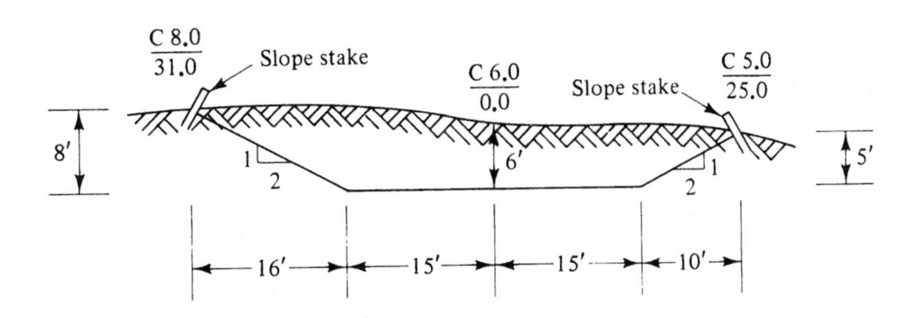

FIGURE 20-11 Cross section at station 100 + 00.

Solution

$$\text{Area at station } 100+00 = \frac{w}{4}(h_L + h_R) + \frac{c}{2}(d_L + d_R)$$
$$= \left(\frac{30}{4}\right)(8.0 + 5.0) + \left(\frac{6.0}{2}\right)(31.0 + 25.0)$$
$$= 265.5 \text{ sq ft}$$
$$\text{Area at station } 101+00 = \left(\frac{30}{4}\right)(6.0 + 2.0) + \left(\frac{4.0}{2}\right)(27.0 + 19.0)$$
$$= 152.0 \text{ sq ft}$$
$$V = 1.85(A_1 + A_2) = (1.85)(265.5 + 152.0) = \mathbf{772.4 \text{ cu yd}}$$

Volume by Prismoidal Formula

Should the ground surface be such that two adjacent end areas are quite different from each other or should a high degree of precision be desired in the calculations, as where rock quantities or volumes of concrete are being determined, the average-end-area method may not be sufficiently accurate.

Most earthwork volumes with which the surveyor deals are prismoids. The expression to follow was developed (using Simpson's one-third rule) to determine the volume of such shapes and is called the *prismoidal formula.*

$$V = \frac{L}{27}\left(\frac{A_1 + 4A_m + A_2}{6}\right)$$

In this expression A_1 and A_2 are the areas of the cross sections at the ends or bases of the prismoid, while A_m is the area of a section halfway between the two ends and L is the distance between the two end cross sections. These values are shown in Figure 20-12. *The area of the middle section A_m is determined by taking cross sections there or by averaging the dimensions of the end cross sections and using these values to calculate the area. It is not determined by averaging the end areas.*

The prismoidal formula usually yields smaller volumes than does the average-end-area method. Its use is probably justified only when cross sections are taken at

FIGURE 20-12

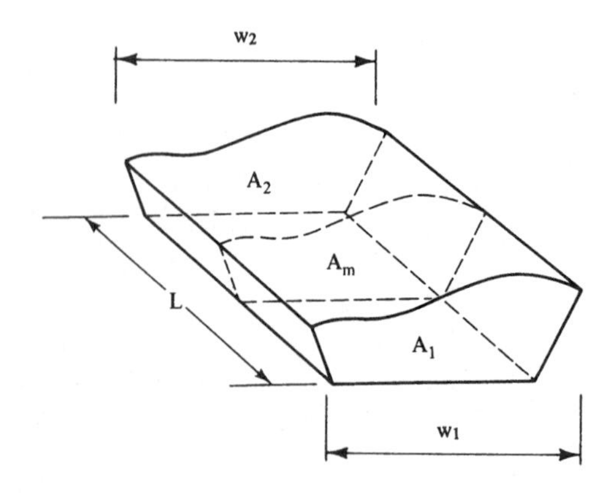

very short intervals and where the areas of successive cross sections are quite different. When earthwork is being contracted for, the method to be used for computing volumes should clearly be indicated in the contract. Should no mention be made of the method to be used, the contractor will in all probability be able to require that the owner use the average-end-area method—whatever the owner's intentions may have been.

Example 20-2 illustrates the application of the prismoidal formula.

Example 20-2

Using the prismoidal formula, compute the earthwork volume between stations 100 and 101 using the data of Example 20-1.

Solution

We determine cross-section dimensions at station 101 + 50 by averaging dimensions at stations 100 and 101:

Station	Left	Center	Right	Area by three-level section formula (sq ft)
100 + 00	$\frac{C8.0}{31.0}$	$\frac{C6.0}{0.0}$	$\frac{C5.0}{25.0}$	265.5
100 + 50	$\frac{C7.0}{29.0}$	$\frac{C5.0}{0.0}$	$\frac{C3.5}{22.0}$	206.2
101 + 00	$\frac{C6.0}{27.0}$	$\frac{C4.0}{0.0}$	$\frac{C2.0}{19.0}$	152.0

$$V = \frac{L}{27}\left(\frac{A_1 + 4A_m + A_2}{6}\right) = \left(\frac{100}{27}\right)\left(\frac{265.5 + 4 \times 206.2 + 152.0}{6}\right)$$

$$= \mathbf{766.9\ cu\ yd}$$

Although the prismoidal formula does provide better estimates of the volumes of prismoids, the average-end-area method is more commonly used because the difference between the two methods is usually quite small except where abrupt changes in cross sections occur. Furthermore, it is somewhat tedious to apply the prismoidal formula because of the extra work involved in computing the average dimensions and the area of the center cross sections. Unless rock excavation or concrete quantities are being computed, use of the prismoidal formula is not justified anyway because of the low precision of the work usually obtained in cross sectioning. Finally, if it is desired to determine the prismoidal volume, it is easier to use the average-end-area method and correct the results obtained with the *prismoidal correction formula*. This correction expression is accurate for three-level sections and is reasonably accurate for most other cross sections.

In the expression to follow for C_v, the prismoidal correction, C_1 and C_2 are the center cuts or fills at the end cross sections A_1 and A_2, while w_1 and w_2 are the distances between slope stakes (see Figure 20-12) at these sections. The correction is in cubic yards.

$$C_v = \frac{L}{12 \times 27}\left(C_1 - C_2\right)\left(w_1 - w_2\right)$$

The correction calculated is subtracted from the volume obtained by the average-end-area method unless the formula yields a minus answer, in which case the correction is added. Example 20-3 shows the application of the prismoidal correction expression to the calculations of Example 20-1, which employed the average-end-area method.

Example 20-3

Correct the average-end-area solution of Example 20-1 to a prismoidal volume using the prismoidal correction formula.

Solution

$$C_v = \frac{L}{12 \times 27}\left(C_1 - C_2\right)\left(w_1 - w_2\right)$$

$$= \frac{100}{\left(12\right)\left(27\right)}\left(6.0 - 4.0\right)\left(56.0 - 46.0\right) = 6.17 \text{ cu yd}$$

$$\text{Prismoidal volume} = 772.4 - 6.17 = 766.2 \text{ cu yd}$$

20-7 MASS DIAGRAM

For highway and railway construction it is desirable to make a cumulative plot of the earthwork quantities (designating cuts plus and fills minus) from one point to another (perhaps the beginning to the ending points). Such a plot, illustrated in Figure 20-13, is called a *mass diagram.* It is usually plotted directly below the profile of the route. The ordinate at any point on the diagram is the cumulative volume of cut and fill to that point, while the abscissa at any point is the distance in stations along the survey line from the starting point.

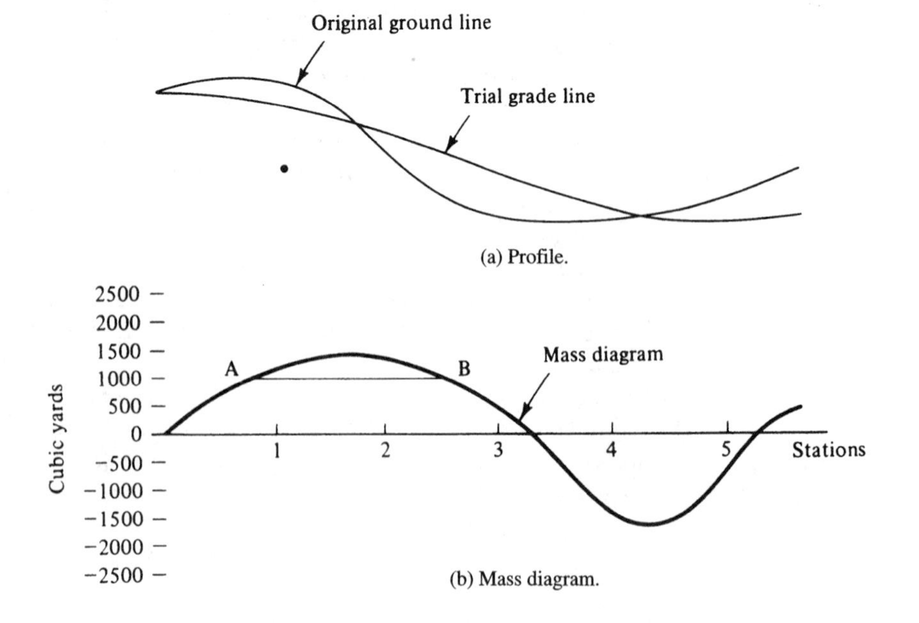

FIGURE 20-13 Mass diagram.

From a mass diagram it can readily be seen whether the cuts and fills balance throughout the job and how far the earth will have to be hauled. If there is an excess of fill needed beyond the cut quantities, it will be necessary to obtain earth from other sources, such as borrow pits. If there is more cut than fill, it may be necessary to dump the extra fill as waste, or perhaps widen the planned fills or make the valleys shallower.

When material is excavated and placed in trucks, pans or other earthmoving equipment, it will normally occupy a larger volume than it had in its original position. When solid rock is broken up, it may take up as much as two times its original volume. When the excavated material is rolled down in thin layers in an embankment at an optimum moisture content such that the greatest compaction is achieved, the resulting fill volume will be appreciably less than the cut quantities. Thus for most fills (other than those consisting of rock) it takes more cut volume than fill volume. The excess may vary from 5% to 20% depending on the character of the material involved. If a fill is made in a marshy area, the original material underneath will settle appreciably, thus requiring even more fill. As a result, it is necessary to make use of the so-called *balance factor* in figuring cut and fill quantities. A value of about 1.2 is frequently used: that is, about 1.2 yards of cut make 1 yard of fill. For rock excavation the balance factor will be less than 1.0, to allow for swell from cut to fill. This factor needs to be accounted for in constructing the mass diagram.

Referring to Figure 20-13, the existing or original ground profile is plotted and a trial grade line (shown in the figure) is drawn. In drawing the trial grade line an effort is made to balance the cut and fill quantities visually. Earthwork volumes are calculated as described in previous sections of this chapter and the mass diagram is drawn taking into account the estimated balance factor. If the cuts and fills do not balance well or if the earth has to be transported or hauled long distances, the grade line will have to be adjusted and the process repeated.

Earth grading contracts refer to a certain distance as being a *free haul* distance. It may be specified as being 500 ft, 1000 ft, 2000 ft, or some other value. If the earth is not moved more than this distance, the contractor will be paid the standard price per cubic yard agreed to in the contract. If it is moved more than this distance, the contractor will be paid extra as specified in the contract. The extra hauling is referred to as *overhaul.* The unit of measurement used for overhaul is the *station yard* where one station yard indicates the hauling of 1 cubic yard of material for one station.

From the figure it will be noted that the peaks on the diagram show where there is a change from cut to fill while the valleys show a change from fill to cut. If a horizontal line AB is drawn as shown in the figure, the cuts and fills between the two points will exactly balance.

20-8 VOLUMES FROM CONTOUR MAPS

Should an accurate contour map be available for an area being studied it can be used for computing earthwork volumes as described here. It is quite practical to estimate earthwork quantities from such a map. For this discussion, reference is made to the 5-ft contour interval map of Figure 20-14. It is assumed that the hill shown is to be graded off to elevation 515. The areas within the 525, 520, and 515 contour lines can be easily determined with a planimeter. From these values the volume of earth to be removed can be calculated from the following average-end-area expression.

$$V = (5)\left(\frac{A_{525} + A_{520}}{2}\right) + (5)\left(\frac{A_{520} + A_{515}}{2}\right) + \text{earth}$$

volume above the 525 contour line

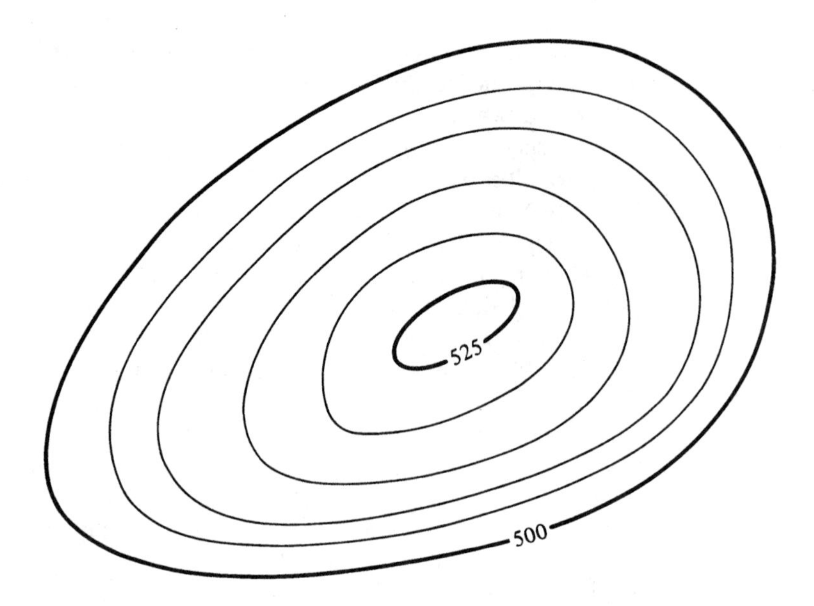

FIGURE 20-14 Contour map of hill.

20-9 VOLUME FORMULAS FOR GEOMETRIC SHAPES

In addition to earthwork quantities it is frequently necessary to compute the volumes of various construction items, such as concrete, liquids, and material stockpiles. Such items can often be broken down into standard geometric shapes such as cones, cubes, cylinders, pyramids, and spheres. Volume formulas for a few of these items are presented in Figure 20-15.

FIGURE 20-15 Geometric volume formulas.

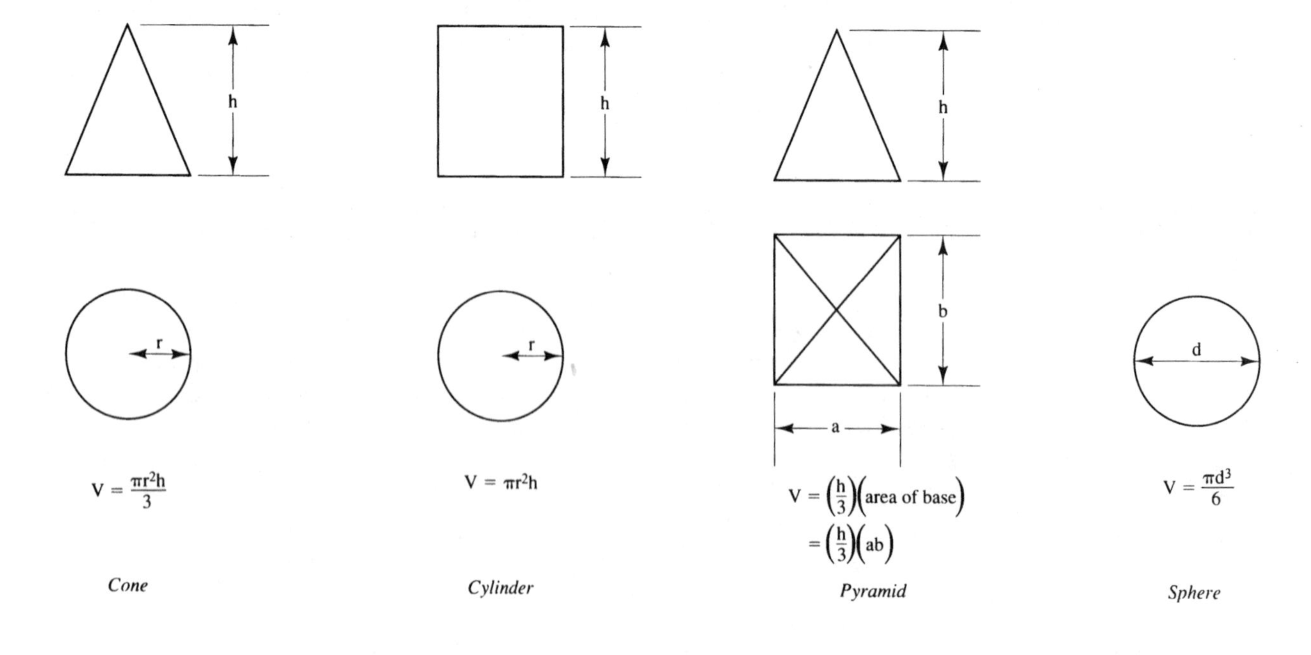

Problems

20-1. For the accompanying sketch and readings with the level placed at the center line of the road, would no. 1, no. 2, or no. 3 be at the correct location for the slope stake?

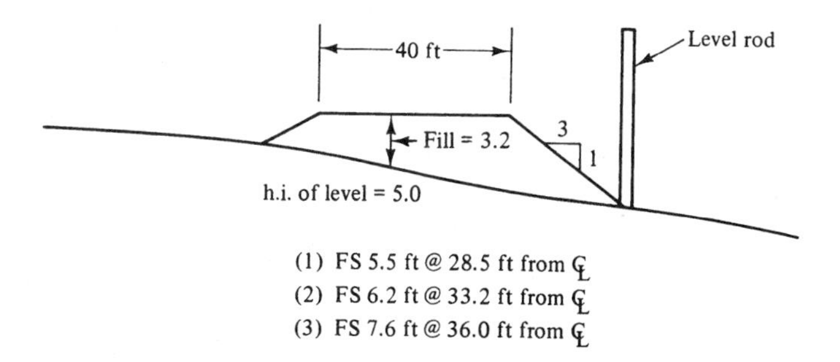

(Ans.: no. 2)

In Problems 20-2 to 20-5, find the volume in cubic yards of the excavation for the borrow pits shown. The numbers at the corners represent the cuts in feet.

20-2.

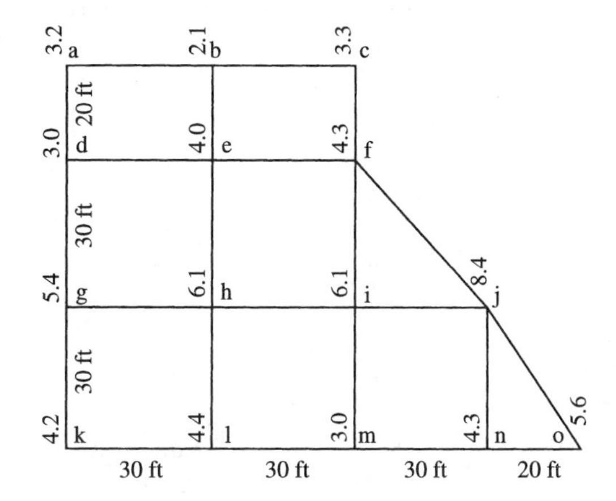

20-3.

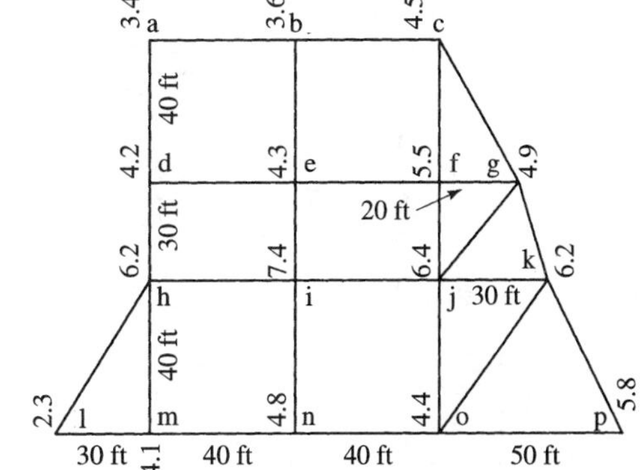

(Ans.: 2331 cu yd)

20-4.

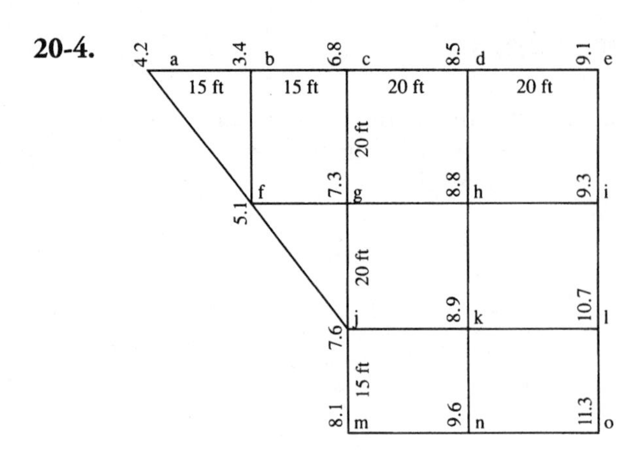

20-5.

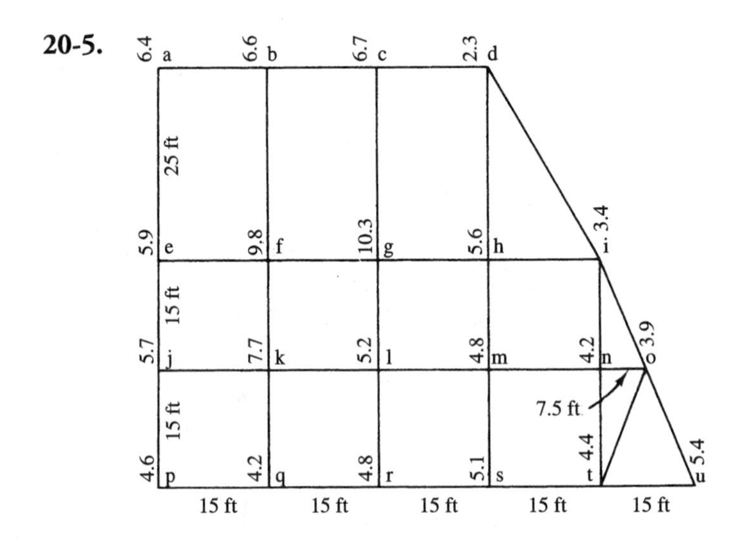

(Ans.: 756.6 cu yd)

20-6. An area has been laid out in 40-ft squares as shown in the accompanying illustration and the ground elevations are as indicated on the sketch. How many cubic yards of cut are required to grade the area level to elevation 40.0 neglecting the balance factor?

56.8 60.3
52.1 40 ft 59.6
51.1 51.0 40 ft 58.1
52.2 47.0 40 ft 59.7
60.6 49.4 40 ft 53.8
40 ft 40 ft

20-7. An area has been laid out as shown in the accompanying illustration. The elevations of the corners are indicated on the sketch. Determine how many cubic yards of cut are required to grade the area level to elevation 60.0, neglecting the balance factor.

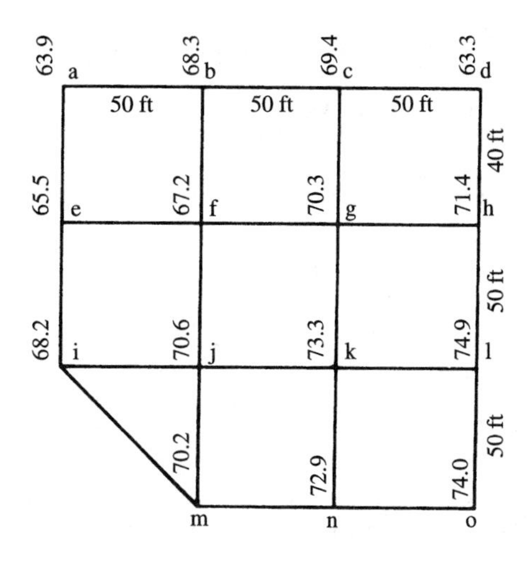

(Ans.: 7404 cu yd)

20-8. An area has been laid out as shown in the accompanying illustration and the elevations of the corners are as indicated on the sketch. How many cubic yards of cut are required to grade the area level to elevation 60.0, neglecting the balance factor?

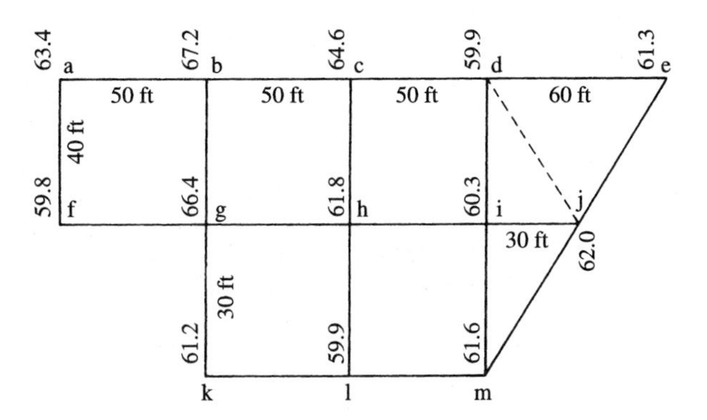

20-9. The following notes are for cross sections at stations 67 and 68. If the width of the roadbed is 30 ft, calculate the area of the two cross sections.

Station	Cross sections		
67	$\frac{C5.4}{26.6}$	$\frac{C4.2}{0.0}$	$\frac{C2.1}{23.1}$
68	$\frac{C13.6}{41.2}$	$\frac{C8.4}{0.0}$	$\frac{C2.8}{20.2}$

(Ans.: 160.62 sq ft. 380.88 sq ft)

20-10. Compute the volume of excavation between stations 67 and 68 of Problem 20-9 using:

(a) The average-end-area method.

(b) The prismoidal formula.

20-11. The cross-section areas along a proposed dike as obtained with a planimeter are as follows:

Station	End area (sq ft)
46	642
47	1484
48	762

Calculate the total volume of fill in cubic yards between stations 46 and 48 using the average-end-area method.

(Ans.: 8088 cu yd)

20-12. For a proposed highway the following planimetered areas in square feet were obtained. Using the average-end-area method, determine the total volumes of cut and fill between stations 84 and 87.

Station	Cut	Fill
84	76	
85	68	
85 + 50	0	0
86		50
87		66

20-13. It is desired to cut the hill of Figure 20-14 down to elevation 510 ft. The areas for each of the contours have been measured with a planimeter with the following results:

Contour elevation (ft)	Area by planimeter (sq ft)
525	4240
520	12,660
515	30,800
510	54,600

What will be the total excavation in cubic yards, neglecting the earthwork volume above elevation 525?

(Ans.: 13,496 cu yd)

CHAPTER

Land Surveying or Property Surveying

21-1 INTRODUCTION

Land surveying or property surveying is concerned with the location of property boundaries and the preparation of drawings (or plats) that show these boundaries. In addition, it involves the writing and interpretation of land descriptions involved in legal documents for land sales or leases. More precisely, land surveys are made for one or more of the following purposes:

1. To reestablish the boundaries of a section of land that has been previously surveyed
2. To subdivide a tract of land into smaller parts
3. To obtain the data needed for writing the legal descriptions of a piece of land

The location of property lines began before recorded history and for all of the subsequent ages it has been necessary for the surveyor to reestablish obliterated land boundaries, establish new boundaries, and prepare boundary descriptions. The need for people to locate, divide, and measure land caused the development of the land surveying profession. Today, our expanding population, the demand for second homes, and the formation of many new industries are creating a demand for more and more property surveys.

In general, land prices have been climbing rapidly and the accurate location of property lines is becoming increasingly important. A century or two ago land was so plentiful that describing a distance such as "as far as a fast horse can run in 15 minutes" might have been adequate, but it is certainly no longer true. A few decades ago the compass and chain were satisfactory for making property surveys, but this is no longer the case. Today, the transit and steel tape are still occasionally used, but the situation is rapidly changing. Recent technical advancements, such as total stations, are commonly used in the field all over the United States.

21-2 TITLE TRANSFER AND LAND RECORDS

Title to land may be transferred by deeds, wills, inheritance without a will, or by adverse possession (discussed in Section 21-8). A deed for a piece of land contains a description of the boundaries, the monuments, and important information concerning adjoining property.

For a deed to be legally effective, it must be recorded in the office of the proper public official. (Although recording is voluntary it provides public notice that someone claims an interest in the property.) Usually, this office is located in the county courthouse and is called the Registry of Land Deeds, Recorder of Deeds, or a similar name. At this office the information is copied into the records and is available for anyone to see. Legally, anyone can go to the office and see the records that show the description of anyone's land in the county and, in addition, can determine how much he or she paid for it.

These offices keep indexes so that it is reasonably easy to find the records desired. The indexes are kept chronologically and indexed in the name of both the grantor (the seller) and the grantee (the buyer). When the surveyor goes to this office, he or she is usually looking up the legal description of certain pieces of land. There the surveyor will find copies of the land description and generally copies of plats of the land which have been filed at the office.

Anyone who becomes involved in land purchases will want to be sure to obtain full or clear title to the land. Various attorneys and private title searching companies are better prepared to search these indexes for information relating to a title than is the ordinary individual. They will, for a fee, search the records and provide information regarding the title. The reader can readily understand the importance of such careful checks simply by thinking for a moment about the legal problems that could arise if a person were to build a house on a lot believing that he or she had title to the lot but actually not having it. For many important property sales the purchaser may want the seller to provide title insurance guaranteeing clear title to the property before he or she will buy. Title companies usually combine or can arrange both services (title checks and title insurance).

21-3 COMMON LAW

Most of the law pertaining to property surveying is common law or, as it is frequently called, unwritten law (i.e., it is not statute law passed by political bodies). Common law is that body of rules and principles that has been adopted by usage from time immemorial. It has been formed by transforming custom into rules of law. When disputes between two or more parties arose in early days in England, the courts decided what to do from established custom. If situations occurred that did not seem to be adequately covered by custom, the judges based their decisions on their own ideas of right and wrong.

As an illustration of the kind of problems settled by these courts, consider the case in which a person excavated for a building foundation on his land and a nearby building on another person's land caved in because of the subsidence of the ground. The injured party went to court to ask for compensation for the loss. From such cases as these the principle evolved that each landowner owes lateral support to adjoining landowners.

Beginning with the Year Book of A.D. 1272, the English courts began to keep records of their decisions and have kept them continuously since that time. Since the beginning of these records, judges have searched through them to determine if their particular case had been decided earlier.

English common law is the basis of jurisprudence in 49 of the states of the United States. Louisiana was originally French and its legal foundation is based on the Roman common law. Even though the Romans occupied most of the British

Isles for six centuries, England was the only European country to develop an independent system of common law. (English common law, however, did draw heavily from Roman common law.)

21-4 MONUMENTS

Corners are points that are established by surveys or by agreements between adjacent property owners. The usual practice is to mark these corners with a relatively permanent object called a *monument.* Monuments may be natural features such as rocks, trees, springs, and so on, or they may be artificial objects such as iron pipes driven into the ground, posts of concrete or stone, mounds of stone, wooden stakes perhaps with more permanent material buried at their bases such as charcoal or glass, or (in areas of low rainfall) mounds of earth or pits. Although wooden stakes alone would seem inadequate as monuments because of their temporary nature, some courts have held that substantial wooden stakes may be so classified.

The surveyor must be reasonable in the method of marking corners being used. He or she may describe a particular corner in a manner that is perfectly clear to everyone at the time of marking, for example, the northeast corner of Joe Smith's barn. A few decades later, however, Mr. Smith's barn may be completely obliterated and there will be no one around who can prove where it was located. There are land deeds on record that describe a corner as being that spot where so and so shot a bear on such and such a date. Needless to say, it is quite a challenge to locate that point 5 or 50 years later. Clearly, that kind of description is a good start toward future litigation.

Just about anything may be used for monuments, but long iron pipes or concrete monuments are generally more satisfactory and may even be required by law in some areas. Whatever type of monument is used, it should be described carefully in the surveyor's notes on the plats.

The life of the land surveyor and the problems of landowners (small or large) would be much simpler if they both would understand and follow this important rule: *Establish property monuments so carefully that their obliteration is unlikely, but reference them in such a manner that they may be replaced easily and economically if they are obliterated.*

Unfortunately, monuments are frequently destroyed and resurveys are required to replace them. (Deliberate damage or destruction of monuments is prohibited by law.) In those places where monuments may easily be destroyed unintentionally (e.g., near busy streets or in construction areas), it is a good practice to use *witness corners.* These might be iron pipes (or other types of monuments), placed a convenient distance back along property lines in more protected locations. These monuments are also described in the surveyor's notes and are shown on the plats. Another place where witness corners are used is where the actual corners are located in places difficult to access, for example, in streams or lakes.

If the location of a property corner can be fixed beyond any reasonable doubt, it is said to *exist.* If its position cannot be found, it is said to be *lost.* On occasions when the monument used to mark the corner cannot be found, the corner is said to be *obliterated.* This does not necessarily mean that the corner is lost, because it may very well be possible to reestablish its original position.

Corners may often seem lost when they really are not. For instance, if the corners had been marked with wooden stakes, cutting off slices of earth with a shovel might reveal change in soil color where the stake had decayed many years before. If an iron pipe or concrete monument had been removed, the void space filled in with the surrounding soil might show a slight change in coloration. Among several other methods that may be used to find old corners are the locating of old fence corners, statements by neighboring people, the use of metal detectors to locate iron pipes, and so on.

21-5 BLAZING TREES

Another aid in locating property lines and corner monuments in forest areas is blazing and hacking trees. This practice was very common with surveyors in the past, but it is not as common today because so many people object to having their trees marked. A *blaze* is a flat scar made with an ax (or machete) on a tree at about breast height. The bark and a small amount of the tree tissue are removed and the tree is marked so that it can be identified for several decades. A *hack* or *notch* is referred to as a V-shaped cut in the trunk of a tree.

One common method of marking trees is described in this section, but the reader should realize that surveyors in various parts of the country mark trees in different ways. In the author's locality it was the practice along boundary lines to put two hacks on the sides of trees nearest the boundaries. Two hacks were used to distinguish them from accidental marks from other causes. The trees hacked were those which the surveyor could reach while walking along the line. If a tree was exactly on the boundary (sometimes called a *line tree*), it was marked with a blaze with two hacks underneath. These marks were placed on both sides of the tree in line with the boundary.

It was also the custom to place three hacks on trees near property corners. These hacks were placed on the trees facing the corner and at a distance above the ground equal to the distance from the particular tree to the corner. If a tree was the corner monument, it was marked with an x and three hack marks underneath it. It is true, however, that in many cases lumber operations, forest fires, and other causes may have destroyed the marked trees.

21-6 THE LAND SURVEYOR: A SPECIALIST

Disputes over land boundaries are an everyday occurrence, as may be seen by checking court records. Although the land surveyor may on occasion be involved in original surveys or subdividing tracts of land into smaller tracts, a large proportion of his or her work deals with resurveys in which the surveyor tries to reestablish old boundaries that have previously been surveyed.

The surveyor is faced with many problems in relocating old land boundaries. Incomplete data on plats and deeds, missing monuments on the ground, conflicting claims by adjoining property owners, and inaccurate original measurements greatly magnify a surveyor's problems. These are only a few of the problems besetting the land surveyor, but they show that for the person to do the work, he or she must be something of a specialist. As a result, land surveying is a profession that is to a large extent learned only after considerable experience in a particular locality.

Land surveying may be learned by a combination of formal study and field experience and by a thorough study of the laws pertaining to the subject. For a surveyor to become familiar with the conditions in a particular area, he or she must have a great deal of experience in that locality with regard to the methods of surveying used by the previous surveyors, court interpretation of land problems, and so on. As an example, the weight given to different items varies from state to state, for instance, the relative importance of monuments when they do not agree with the recorded values.

One frequent purpose of resurveys is to attempt to settle disputes between adjacent property owners as to where the lines should actually be. *In resurveying, the land surveyor's goal is to reestablish lines and corners in their original positions on the ground, whether or not those locations are in exact agreement with the old land descriptions. The surveyor's duty is not to correct old surveys but to put the corners back in their original positions.*

Too often the inexperienced surveyor thinks that land surveying merely involves the careful measurement of angles and distances. These measurements are actually only one means of reestablishing old boundaries. This comment is not meant to imply that careful measurements are not important but rather to remind the surveyor that his or her duty is to find where the original corners and boundaries were regardless of the precision with which the original survey was conducted.

The location of corners and property lines is determined from the *intent* of the parties to the original establishment of the boundary. The law is not concerned with their secret intentions but with their intent as expressed by the action of their surveyors as evidenced in plats, deeds, existing monuments, and so on. As time goes by it becomes more and more difficult to determine the original intent of the parties.

The reader should understand that the surveyor has no legal authority to establish boundary locations. On many occasions, however, he or she is called upon to relocate lines that are so difficult to find that the results are doubtful. In such cases as these, an agreement between the parties involved is probably the most sensible and economical solution. Property owners are always free to do this as long as their agreement does not affect some other party.

In most lawsuits over land boundaries, even the winner loses unless the land is extremely valuable, because of the cost and ill-will connected with such actions. If the surveyor is able to bring about an amicable agreement between the two parties, he or she will probably have served them both from an economic viewpoint and perhaps even more in preserving friendship. An experienced land surveyor is probably in a better position to suggest a just settlement of this type of problem than is a court of law. If the surveyor is able to persuade the land owners to compromise in a boundary dispute, he or she must survey the new line and the landowners should then go through a formal and recorded acceptance of the line.[1]

21-7 MONUMENTS, BEARINGS, DISTANCES, AND AREAS

Among the factors involved in relocating boundaries are existing monuments, adjacent boundaries, bearings, distances, and areas. In considering the relative importance of these items in reestablishing property boundaries, natural monuments are given the greatest preference. Artificial monuments are given the next-highest preference. It is thought that natural objects (springs, streams, ridge lines, lakes, and beaches) offer a greater degree of permanence (they are less likely to be destroyed or relocated) than do artificial monuments such as set iron pipes, concrete markers, or

[1]A. H. Holt, "The Surveyor and His Legal Equipment," *Transactions of the ASCE,* 1934, vol. 99, pp. 1155–1169.

stones. Of course, some people will deliberately move one of the latter monuments in order to better their land position. Other people, not realizing the importance of these monuments, will pick them up and take them home for their own use. Occasionally, concrete monuments are found being used as doorsteps at nearby houses. A person who deliberately moves a property monument is breaking the law and may be fined and/or imprisoned.

The courts feel that because the original owners could *see* the monuments, they more nearly express the original intent of the parties to land agreements than do measurements of directions and distances, which are subject to so many errors and mistakes. After natural and artificial monuments come adjacent boundaries and then bearings and distances, generally in that order. It should clearly be noted that bearings and distances cannot control the reestablishment of land boundaries if the monuments that are in place actually show the original boundaries. Bearings and distances can control the outcome, however, if there have been mistakes in the placing of the monuments or if the monuments are lost.

Land area is considered the least important of the factors listed, unless the area is the very essence of the deed as, for example, when an exact amount of land is clearly conveyed. A layperson may be puzzled by the following description commonly used in deeds regarding acreage: "26 acres more or less." The purpose of the words "more or less" is to indicate that all of the land within the specified boundaries is being conveyed, even though the area stated may vary greatly from the actual area. *This does not mean that the surveyor can do a sloppy job and get by with it by using such a term as "more or less." The surveyor is responsible for the quality of his or her work, and if it is not up to the standard expected of a member of the profession, he or she is liable for any damages that result.*

21-8 MISCELLANEOUS TERMS RELATING TO LAND SURVEYING

In this section are defined several terms that frequently occur in dealing with the transfer of land title.

Adverse Possession or "Squatters' Rights"

If a person occupies and openly uses land not belonging to him or her for a specific length of time and under the conditions described by the laws of the state, he or she may acquire title to the land under the U.S. doctrine of adverse possession. The possession must be open and hostile and is usually for a period of 20 years but perhaps less under certain conditions, as for example, when the original title to the land was not clear.

Even though a surveyor is able to reestablish without question the original boundaries of a tract of land, it is possible that because of adverse possession the original lines are no longer applicable. If the owner fails to act during the designated time, he or she loses the right to act. It is generally held that a private citizen does not have the right to acquire government land by adverse possession. If the owner of the land gives permission to another person to occupy the land, that person can never acquire title to the land by the adverse possession doctrine no matter how long he or she uses the land after the permission is given.

Riparian Rights

The rights of persons who own property along bodies of water are referred to as riparian rights. Following are a few comments concerning these rights as they apply to property lines. Creeks, rivers, and lakes are very natural and convenient boundaries between various tracts of land because they are easy to describe and are easy to locate. For small streams that are not navigable, property lines usually go to the center of the stream or the "thread of the stream." The thread of the stream is generally thought of as being the center line of the main channel.

The U.S. Supreme Court has decided that it is up to the individual states to decide where private property lines run along navigable streams. Some states have decided on the center lines of streams, some on the threads of streams, and others have selected the high water marks (those marks to which the water has risen so commonly that the soil is marked with a definite character change as to vegetation, and so on).

Erosion and Avulsion

Although a stream or river is an excellent boundary, it cannot always be depended upon to stay in the same place. Its position may shift so very slowly and imperceptibly as a result of erosion, current, or the force of waves that the property owners cannot recognize the short-term change in position. When bodies of water change by this method, property lines are held to move with the change. However, if the change is very sudden and perceptible, as, for example, when a large amount of soil is suddenly moved from one landowner to another or when a stream changes its bed completely, it is called *avulsion.* When avulsion occurs, property lines are held not to change from their original positions. This description seem simple enough, but unfortunately, many years after the fact it is difficult to tell whether a change was caused slowly or suddenly. On some occasions we cannot even find the stream. Such a case might occur where beavers have built a continuous series of dams, making the entire area a swamp.

Accretion and Reliction

When water slowly and imperceptibly deposits materials on the bank of a stream or other body of water, it is called *accretion.* Should a body of water recede, as when it dries up partly or wholly, the land area adjoining it is increased and this is called *reliction.*

When land areas change by accretion or reliction the courts will attempt to establish the new boundaries of the adjoining tracts of land in an equitable manner. If water frontage is involved, it may be of the greatest importance in the decisions made. They may divide the water frontage in some proportion to the frontage of the old boundaries.

Further Comments on Riparian Rights

There are so many complicated problems that can arise regarding riparian rights that a special body of law has evolved on the subject. Some of the numerous problems that may be encountered are related to property lines, navigation, docks, water supply, fishing rights, oyster beds, artificial fills or cuts, erosion, accretion, and many others. The laws pertaining to these situations vary quite a bit from state to state, and court decisions in different states have been entirely different for similar cases.

One illustration is presented here to show how involved the legal situation can become concerning riparian rights. For this discussion reference is made to Figure 21-1, where both erosion and accretion are involved. The discussion presented here can apply to the legal situation in some states, whereas it may not apply to others.

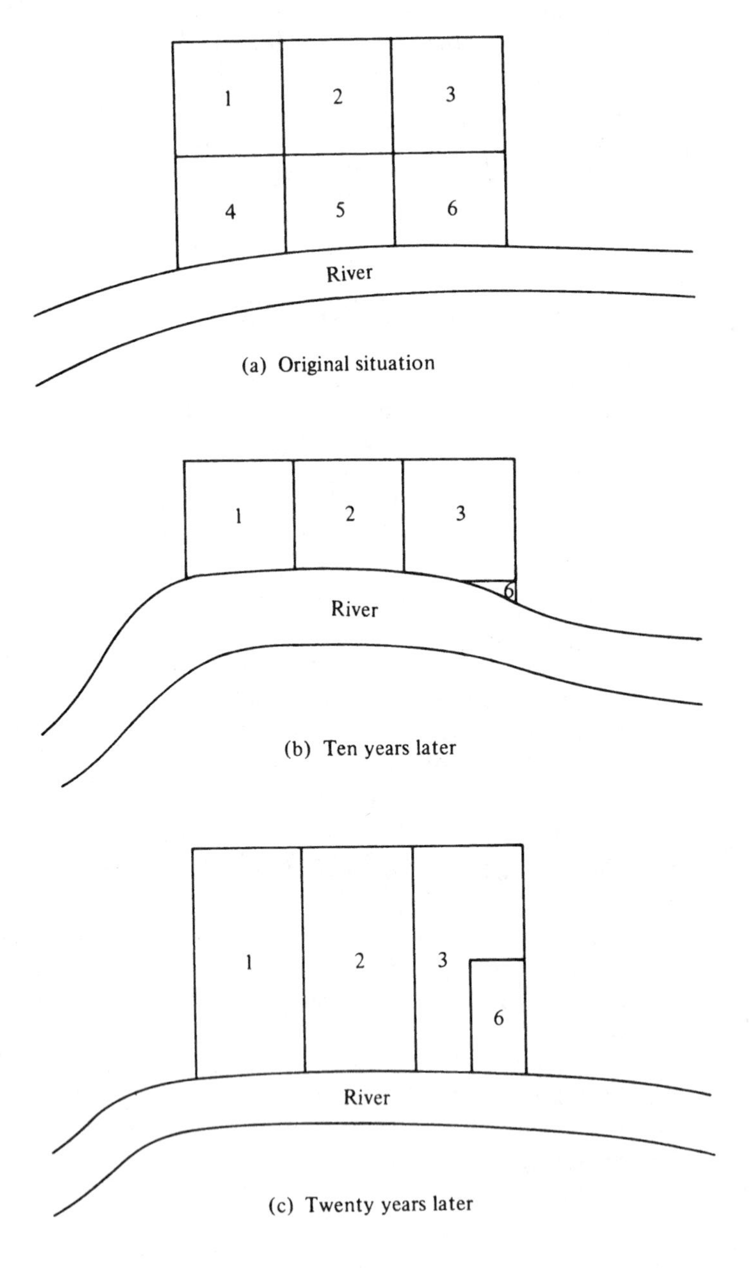

FIGURE 21-1 Possible ownership changes as water boundaries change (depending on state).

The purpose of this discussion is to show how complicated the legal problem may be. In part (a) of Figure 21-1, six lots are shown. The lots numbered 4 through 6 bordered the river and had riparian rights. During the next 10 years lots 4, 5, and most of 6 eroded away, leaving the situation shown in part (b) of the figure. At this time lots 1, 2, and the parts of 3 and 6 bordering the river have water rights. In the next 10 years it is assumed that the river gradually moved back to its original position and that the replaced land now could belong to lots 1, 2, 3, and 6, as shown in part (c) of the figure (or does it go back to 4, 5, and 6?). A similar discussion applies to beachfront property in some states.[2]

21-9 RESURVEYS

The average client of a surveyor has no idea what a time-consuming and hair-pulling job a resurvey can be. Ideally, the original surveyor painstakingly measured distances and directions and carefully set monuments and witnessed them with equal care. Sadly, the truth is that often the monuments were wooden stakes placed at points that were located by haphazardly determined distances and bearings. The result is that there is no problem in surveying so difficult and requiring so much patience, skill, and persistence as that of relocating old property lines.

If, as stated in Section 21-6, the old lines and monuments can be reestablished, they still govern the boundaries no matter how poorly the old survey was performed. For a surveyor to reestablish old property lines, he or she must follow in the footsteps of the original surveyor and these footsteps may very well be 50, 100, or more years old. To be able to do this, he or she must have a good idea of how the original survey was performed. Was a compass or transit used to measure the angles? Were distances measured on slopes or horizontally?

When the surveyor cannot definitely relocate boundaries, he has no power to establish them. If there are disputes between adjacent landowners over boundaries and if the surveyor cannot persuade the owners to compromise, the courts will have to settle the matter. In court the surveyor can only serve as an expert witness and present the evidence that he has found. Once a dispute is settled, whether by mutual agreement or by court action, the surveyor should do all he or she can to see that a precise survey is made of the settled boundaries and good monuments established.

To begin a resurvey the surveyor carefully studies available plats and deeds of the property as well as those of adjacent tracts. As a part of this study he or she typically calculates by latitudes and departures the precision obtained in the original survey of the tract in question. Surprisingly enough, the surveyor may frequently spend more than half of his or her time studying and planning, while the actual field work may not take very long at all.

Clients often do not understand the amount of research that will be needed to conduct a property survey properly. In this regard the surveyor is well advised to inform the clients of the potential time and resulting costs that may be needed for such research so that an accurate survey can be made. It may be necessary to study county records, to interview neighboring property owners, to contact other surveyors, and so on. In this regard innumerable land surveys have been made in the United States—monumented, platted, and described but never properly recorded. These surveys can provide important information as to the accurate location of land tracts, and they can be a real source of trouble if they are neglected or not discovered.

Next, the surveyor begins a careful examination of the tract of land in the field. If the original survey was done precisely and if one or more of the original corners can be found, the survey can be handled just as was the five-sided traverse used for an example in Chapters 4 and 10 in which distance and angle measurements were made.

[2]P. Kissam, *Surveying for Civil Engineers,* 2nd ed. (New York: McGraw-Hill Book Company, 1981), pp. 327–328.

The trouble, however, is that so many original surveys contain major errors and mistakes, and often several or all of the monuments are gone. One frequent cause of poor measurements was the equipment used, often the compass and link chain. If the original survey was made with a surveyor's compass, the surveyor may attempt to rerun the lines by using a compass and by making proper allowance for the estimated change in magnetic declination since the time of the original survey.

If a surveyor can definitely establish the corners at the ends of at least one line, he or she is off to a good start in the entire relocation job. He or she can measure that line and compare its distance with the original measurement and thus obtain an idea as to how the original value was obtained. Was it measured on the slope or horizontally? By calculating the proper proportions he or she can determine proportional lengths of the other sides. The surveyor can determine the astronomic bearing of the known side and from that value compute the estimated magnetic declination at the time of the original survey. With this magnetic declination he or she can compute the estimated astronomic bearings of the other sides.

With these computed distances and bearings the surveyor can start running the sides. At each estimated corner location he or she will carefully search for evidence of the original monuments. If the surveyor finds one of the old monuments, he or she sets a new one (if necessary) and carefully references it. If unable to find a monument, the surveyor sets a temporary point and moves on to the next line. If the surveyor finds monuments farther along, he or she tries to work back to those that couldn't be found by using new proportions based on the found monuments. The surveyor continues in this fashion until he or she locates all the old monuments or sets temporary monuments at all of the various corners. Several ideas will be helpful to the surveyor when looking for old corners: hacked trees, old fence lines, roadbeds, places where vegetation is different, and so on. After studying the gathered information carefully, the surveyor may very well return to the field for more measurements.

In a resurvey the surveyor may initially be able to find only one corner or perhaps, as is often the case, none at all. If only one corner is evident and the original survey was run with a compass, he or she can estimate the magnetic declination at the time of the survey, convert the bearings to estimated astronomic bearings, and try to run the lines as described in the preceding paragraph.

If no corners are evident, the surveyor will probably begin work by studying the records of the adjoining land tracts and try to run their lines in order to see if he or she can locate some of their monuments.

When the surveyor has finished working on any of the cases mentioned, he or she will give the client his or her best judgment as to the location of the lines and corners based on his or her opinion as to the true intent of the parties to the original survey.

21-10 METES AND BOUNDS

The oldest method of land surveying is the metes and bounds system. (The term *metes* means to measure or to assign a measure, and the term *bounds* refers to boundaries.) Thus a metes and bounds description gives the measurements and limits of the outside boundaries of the tract of land in question. The length and direction of each of the sides of a tract of land are determined and monuments are established at each of the property corners. Creeks, lakes, or other natural landmarks are occasionally used to define property corners and/or boundaries. Most land surveys in the original 13 states of the United States, in Kentucky and Tennessee, and in some surveys elsewhere were made by the metes and bounds method.

A *plat* is a dimension drawing that shows the data pertaining to a land survey and is primarily a legal instrument becoming a matter of public record. It provides the information necessary for finding, describing, and preparing a description of the land.

In Figure 21-2 the author has presented a portion of a plat. In addition to the information shown, there will be a title box, a scale, and a legend or any necessary special notes. Any nonstandard symbols used in the drawing and other special information will be included. The surveyor will provide a statement as to the precision of the field survey and the method used for determining the land area. Furthermore, in areas where houses or other buildings are located or are to be constructed, a note should be given as to whether or not the land is above or below the floodplain. Finally, deed book numbers and page numbers in the courthouse are often given wherein previous owners of the property are identified and previous descriptions of the property presented.

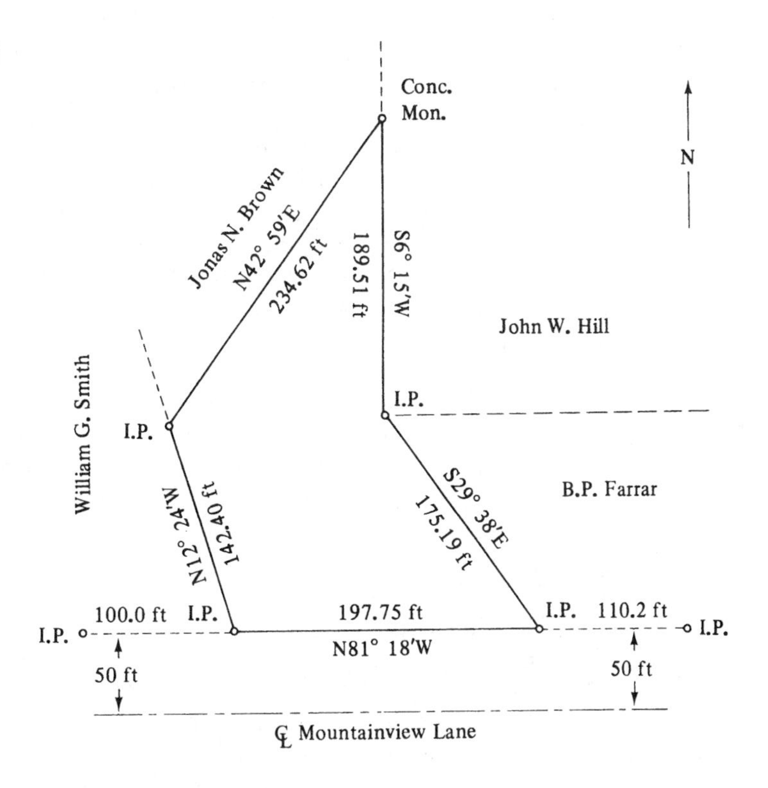

FIGURE 21-2 Portion of a plat.

Property descriptions are usually written by lawyers, real estate persons, or surveyors. The accuracy of their descriptions is of the utmost importance because one simple mistake in their writing may lead to property disputes between adjoining landowners for several generations. A person who writes the descriptions for land should try to put himself or herself in the shoes of someone else 5, 10, or 100 years later who will try to interpret the description as to ground location. Following is a typical metes and bounds deed description (see also Figure 21-2).

All of that certain piece, parcel, or tract of land situated, lying, and being in state and country aforesaid about 1/2 mile west of the town of Tamassee, and lying on the north side of Mountainview Lane. Beginning at an iron pin at the southwesterly corner of the property of the grantor located 50 ft from the center line of Mountainview Lane and running thence along the easterly line of the William G. Smith property N12°24′ W 142.40 ft to an iron pin; thence with the southerly line of the Jonas N. Brown property

N42°59′ E 234.62 ft to a concrete monument; thence ________ to the point of beginning. All bearings are referred to the astronomic meridian, the tract contains 0.83 acres more or less, and is shown on a plat dated October 16, 1981, drawn by Arthur B. McCleod, registered land surveyor #1635, state of ________.

21-11 THE U.S. PUBLIC LANDS SURVEY SYSTEM

In colonial America, the manner of obtaining land from the governments involved (Dutch, Spanish, English, French, etc.) and the methods of describing the land grants varied widely. The boundaries of these tracts of land consisted of streams, roads, fences, trees, stones, and other natural features. The tracts were generally irregular in shape except for some subdivision in towns and cities. Furthermore, no overall control method for the surveys was available. The legal descriptions of the land were often vague and the measurements involved contained many mistakes. As a result of all these factors, when boundaries were obliterated it was and is difficult, if not impossible, to restore their original positions.

Because of the problems involved in the early colonies the Continental Congress in 1785 established the U.S. Public Lands Survey System with the objective that the same mistakes would not be repeated for the remaining land that the federal government possessed. This land, called the *public domain,* is the land held in trust for the people by the federal government. About 75% of the land in the United States was once part of the public domain. Our Congress apparently thought that the sale of this land would yield sufficient funds to pay off the public debt.

The first public lands survey was begun on September 30, 1785, near East Liverpool, Ohio, under the direction of Thomas Hutchins, who was the official geographer of the United States. This point of beginning is now marked with a national historic monument. Today, approximately 30% of the country's land is still in the public domain. Of the approximately 2.3 billion of acres of land in the 50 states, over 1.35 billion have been surveyed by the public lands survey system. There are about 350 million acres in Alaska that are unsurveyed. Hawaii is not included in the public lands survey system.

The U.S. Public Lands Survey System, which is a rectangular system, has been used to subdivide the states of Alabama, Alaska, Florida, and Mississippi as well as all the states to the west and north of the Mississippi and Ohio Rivers except for Texas. Actually, Texas has a system similar to that of the U.S. Public Lands Survey System, but it was appreciably affected by the Spanish settlers before its annexation into the United States. Figure 21-3 shows the parts of the country covered by the U.S. Public Lands Survey System.

It would certainly have been ideal if all the land of the United States could have been obtained at one time and set up under one survey system. Although this was not possible, the country is indeed fortunate to have so much of its land under the public lands survey system. Although the system does have its problems, it has nevertheless been a tremendous asset to the country. The success of the system can be verified by the fact that the original procedures have not been greatly changed up to the present day.

With the public lands survey system, each parcel of land, whether it be 2 1/2, 5, 10, 40, or 160 acres, is described in such a manner that the description will not apply to any other parcel of land in the entire system. This simplicity of describing land tracts has made the system one of the most practical methods ever devised for

FIGURE 21-3 U.S. Public Lands System. (From a map by the U.S. Bureau of Land Management.)

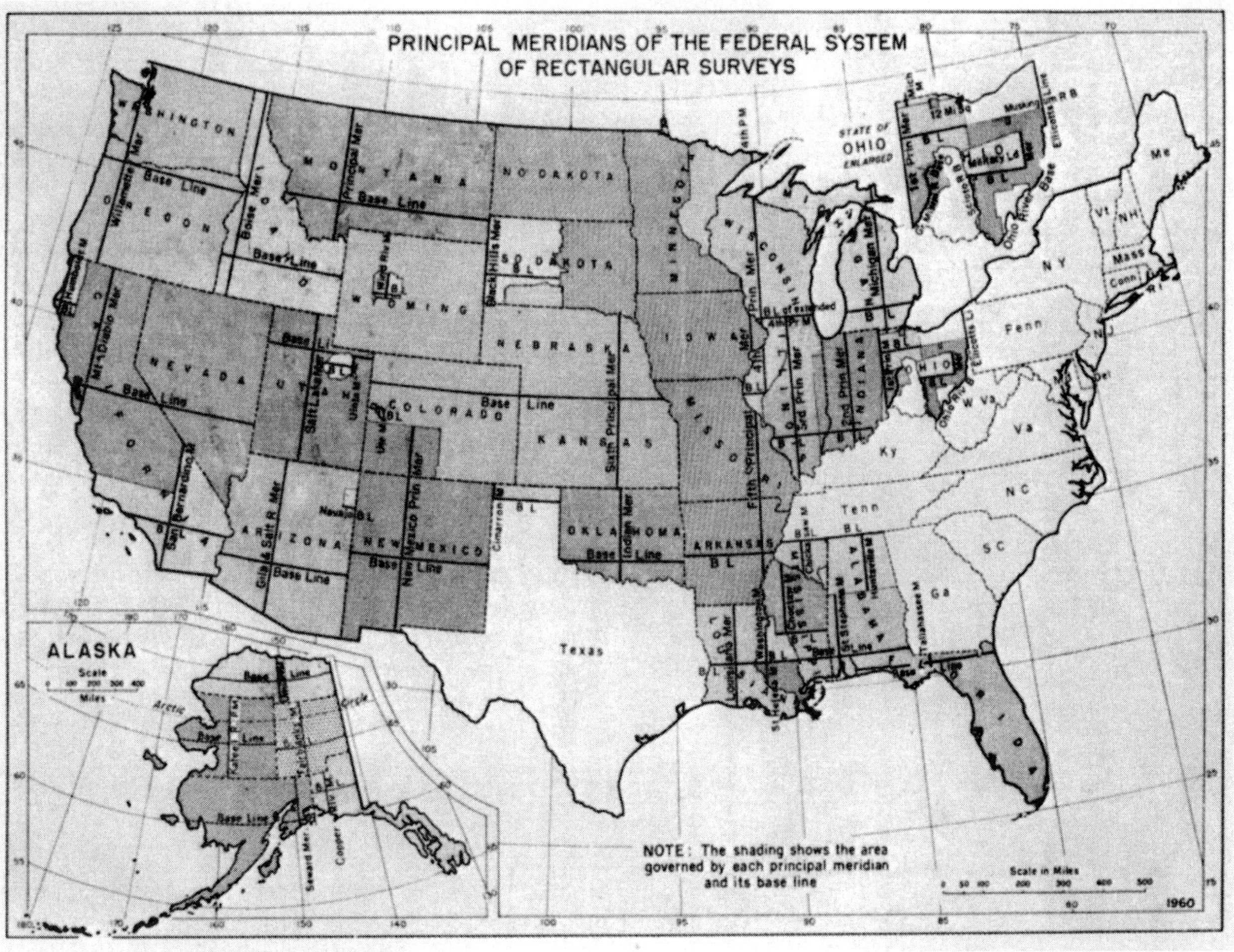

land identification and description. When the rectangular system was introduced, it was not applied to the original 13 states because of the enormous problems involved in changing the existing descriptions of the countless thousands of land tracts involved.

Most of the property surveys in the original 13 states, as well as Kentucky and Tennessee, were made by separate closed traverses, that is, by the metes and bounds system described in Section 21-10. There is no definite overall control system in those areas, and very often some boundaries, such as the edges of lakes or streams, were not even measured but were merely stated as the boundaries.

21-12 EARLY DAYS OF THE SYSTEM

The purpose of the public lands survey was to devise a rectangular system and establish it on the ground so that it would provide a permanent basis for describing land parcels. In 1796 the post of Surveyor General was established at an annual salary of $2000; the first appointment was given to General Rufus Putman, an experienced surveyor and aide to George Washington during the Revolutionary War. In 1812 the General Land Office (now the Bureau of Land Management) was established to manage, lease, and sell the vacant public lands of the United States. To identify and describe the land involved, it was of course first necessary to survey it.

At that time there was a great demand for land, yet its price was very low (about $1.25 per acre for "homesteads" of 160 acres each). For such low prices it was impossible to justify very accurate surveys. The surveying instruments available were rather crude in today's terms and the survey points were marked with wooden stakes, mounds of earth placed over buried pieces of charcoal, or pieces of stone. Many of these monuments have been destroyed through the years.

The surveys were handled on a contract basis and sometimes different rates were paid depending on the relative importance of the lines and on their characteristics (as wooded or swampy or hilly). The surveyors were paid about $2 per mile until 1796 and about $3 per mile thereafter. The amount of money surveyors could make depended entirely on how fast they could complete a survey.

Although the maximum permissible errors were specified from the early days of the system, the standards were vague and little supervision or checking was done to see if the work met the requirements until the 1880s. Another reason for large errors and mistakes in the system was the speed with which some surveys were handled in Indian territory (where often little field work was actually carried out).

Despite all of these problems a large part of the public lands survey was handled very well. Generally, inaccuracies in the old surveys cannot be corrected if there is evidence of the location of the original land corners. In other words, once corners are established and used to mark property boundaries, they cannot be changed regardless of the magnitude of the mistakes made.

21-13 OUTLINE OF THE SYSTEM

This section presents a brief summary of the procedure involved in the U.S. Public Lands Survey System, and subsequent sections present a more detailed description of the system. Beginning points called initial points were established in each area. They were selected with the intention of controlling large agricultural areas within reasonable geographical limitations. Their locations were determined on the basis of astronomical observations. There are 37 initial points, 5 of which are in Alaska.

Astronomic meridians called *principal meridians* were run through each initial point and extended as far as necessary to cover the area involved. They were identified by number or name as, say, the Sixth Principal Meridian, which is in Nebraska and Kansas, or the Williamette Meridian, which is in Washington and Oregon. Meridian divisions are shown in Figure 21-3. Each area is divided into smaller and smaller squares as follows:

1. The land is divided into *quadrangles* approximately 24 miles on each side (see Figure 21-4).
2. The quadrangles are each divided into 16 *townships,* the sides of which are approximately 6 miles (see Figure 21-5). A row of townships extending north and south is called a *tier,* and a row extending east and west is called a *range.* Figure 21-5 shows this arrangement in detail. Notice the numbering system for the crosshatched township (Tier 3 south, Range 3 west).
3. The townships are divided into 36 *sections,* each approximately 1 mile square and containing 640 acres (see Figure 21-6). The sections are numbered as shown in the figure. In addition, the government surveyors mark the quarter-section corners at 40-chain or 1/2 mile intervals and the local surveyors (state, county, or private) perform any further subdivision.

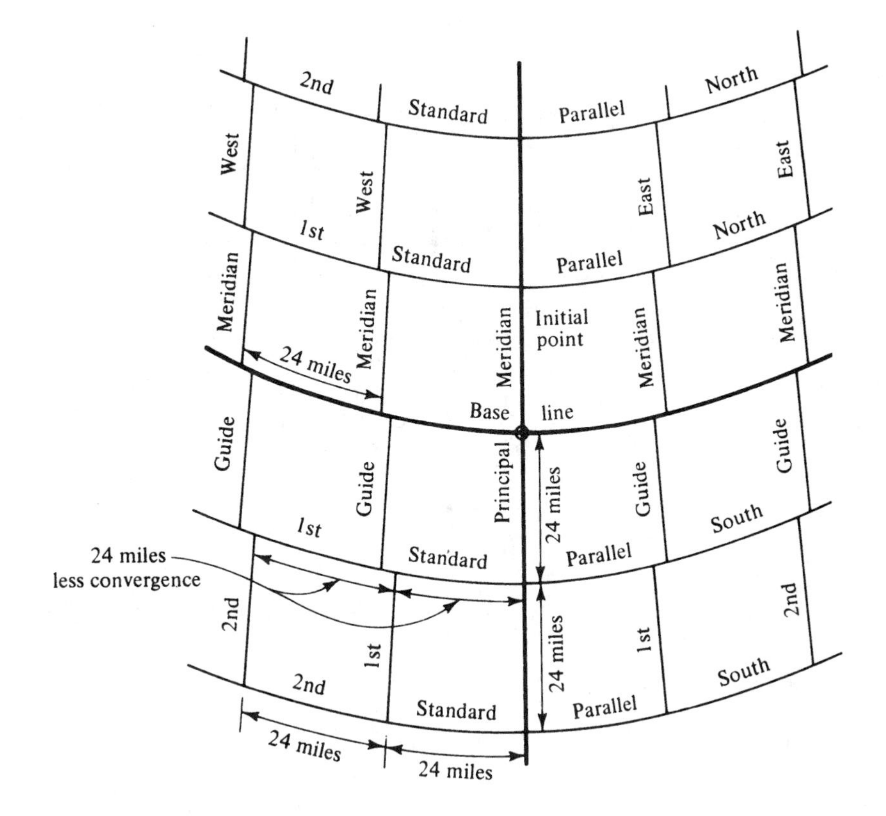

FIGURE 21-4 Subdivision of land into 24-mile quadrangles.

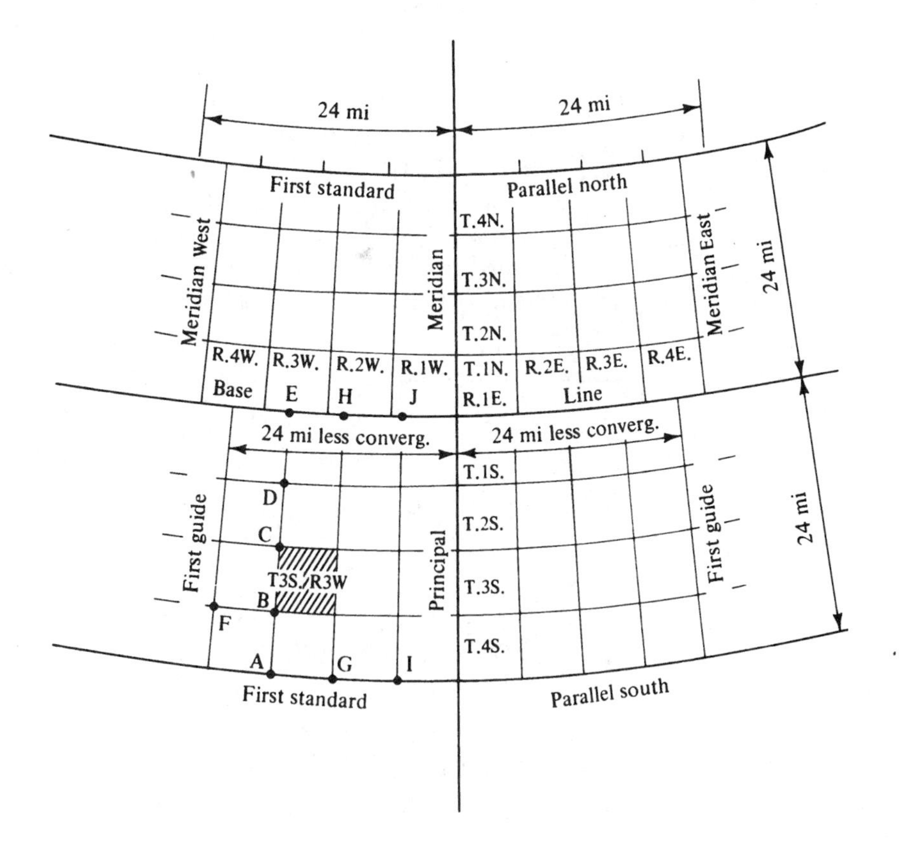

FIGURE 21-5 Division of quadrangles into townships.

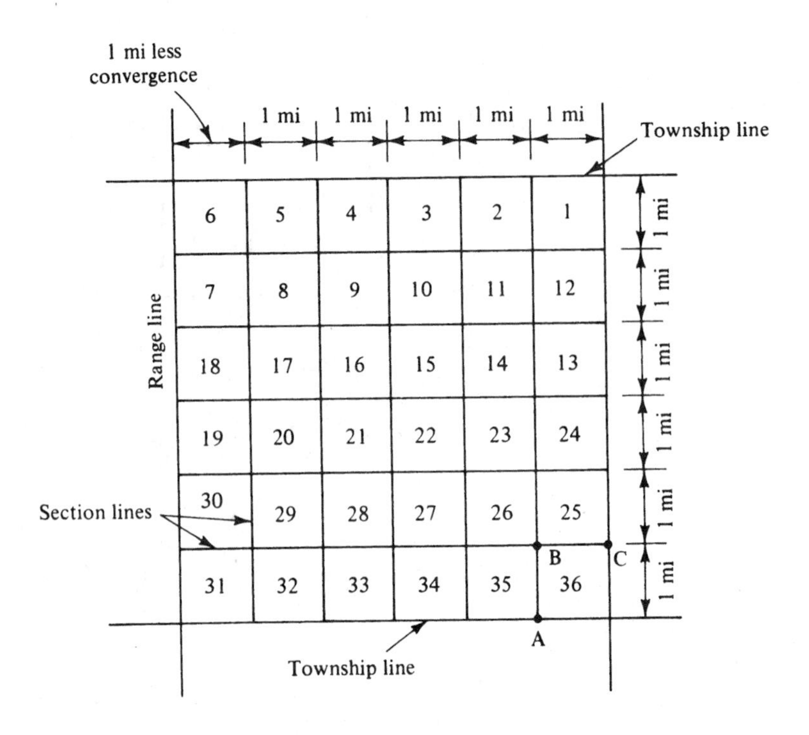

FIGURE 21-6 Subdivision of township into sections.

Since meridians converge to the north, it is impossible for all townships, sections, and so on, to be in the shape of exact squares. In the subdivision of townships it is, however, desirable to lay out as many sections as possible in 1-mile squares. To do this, convergence errors are thrown as far west as possible by running the lines parallel to the eastern boundary. In a similar manner the errors in distance measurement are thrown as far to the north as possible by locating the monuments at 40-chain (or 1/2-mile) intervals along the lines parallel to the eastern boundary so that any accumulated error will occur in the most northerly half mile.

4. The public system called for the disposal of the land in units equal to quarter-quarter sections of 40 acres each. This division is seen in Figure 21-7. It will be noted that this procedure can be continued because the quarter-quarter sections can be divided into quarters, containing 10 acres each.

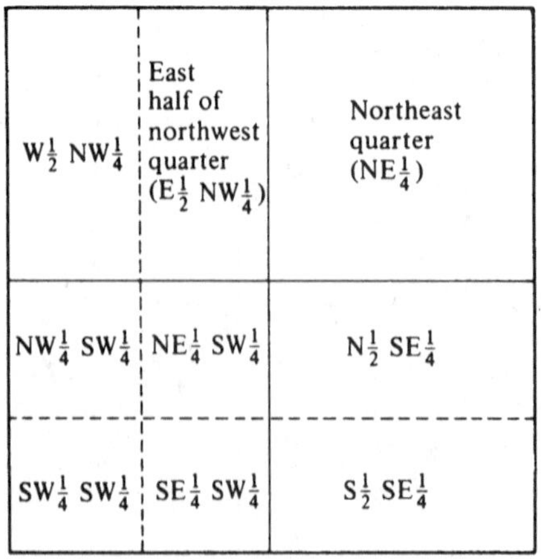

FIGURE 21-7 Subdivision of Section 22 of Figure 21-6.

W$\frac{1}{2}$ NW$\frac{1}{4}$	East half of northwest quarter (E$\frac{1}{2}$ NW$\frac{1}{4}$)	Northeast quarter (NE$\frac{1}{4}$)
NW$\frac{1}{4}$ SW$\frac{1}{4}$	NE$\frac{1}{4}$ SW$\frac{1}{4}$	N$\frac{1}{2}$ SE$\frac{1}{4}$
SW$\frac{1}{4}$ SW$\frac{1}{4}$	SE$\frac{1}{4}$ SW$\frac{1}{4}$	S$\frac{1}{2}$ SE$\frac{1}{4}$

In Figure 21-7 a particular quarter or half of a quarter section is readily identified. In the full description of one of these 40-acre pieces, the quarter of the quarter section is listed, then the quarter section, then the section number, then the township, range, and principal meridian. In this manner a tract of land could be described as the NE 1/4 SW 1/4, Section 22, T2S, R2E of the third principal meridian.

21-14 MEANDER LINES

Navigable waterways or streams of three or more chains in width, as well as lakes covering 25 acres or more (except those formed after the state in question was admitted to the union) are not part of the public domain and thus were not surveyed nor disposed of by the federal government. The individual states have sovereignty over such bodies of water.

The traverses of the margins of these bodies of water are called *meander lines.* These traverses consist of straight lines that conform as closely as possible to the mean high water marks along the banks or shore lines involved. Meander lines were not run as boundary lines and when the bed of the lake or stream changes the high water marks and the property lines change also.

21-15 WITNESS CORNERS

Should the location of a corner fall in an unmeandered lake or stream or be on a steep cliff or in a marsh or other inaccessible spot, a witness corner is established. A witness corner is usually placed on one of the regular survey lines of the property. However, if a satisfactory point for such a corner cannot be occupied within 10 chains along one of the survey lines, it is permissible to locate a witness corner in any direction within 5 chains of the corner position.

21-16 DEED DESCRIPTIONS OF LAND

To describe regular tracts of land within the U.S. Public Lands Survey System for legal purposes is quite simple. An acceptable description of a 40-acre quarter section was given in Section 21-13. When an irregular tract is involved, however, or one that is not a regular part of the public lands system, it is first necessary to tie the description carefully into the rectangular system. Then a length and bearing or metes and bounds description of each side of the tract is given. A description of such a tract might be as follows: "Beginning at a point, marked by an iron pin 300.00 ft North of the NE corner of the SE1/4 of the SW1/4 of Section 28, T3S, RIE, 3rd P.M.; thence North 998.00 ft to an iron pin; thence East 864.00 ft to an iron pin; and so on, back to the point of beginning."

Problems

21-1. Distinguish between corners and monuments.

21-2. What is an obliterated corner, and what is a lost corner?

21-3. What is a witness corner?

21-4. What is adverse possession?

21-5. What is avulsion? When it occurs, what happens to property lines?

21-6. Define the terms *accretion* and *reliction.*

21-7. Describe the metes and bounds method of surveying.

21-8. What is a plat?

21-9. Define the terms *quadrangles, townships,* and *sections* as related to the U.S. Public Lands Survey System.

21-10. In referring to townships, what are tiers and ranges?

21-11. What are the smallest land units that government surveyors mark in the U.S. Public Lands Survey System?

21-12. What is a meander line?

CHAPTER

Horizontal Curves

22-1 INTRODUCTION

The center lines of highways and railroads consist of a series of straight lines connected by curves. The curves for fast traffic are normally circular, although spiral curves may be used to provide gradual transitions to or from the circular curves. Three circular curves are shown in Figure 22-1(a). The *simple curve* consists of a single arc. The *compound curve* consists of two or more arcs with different radii. The *reverse curve* consists of two arcs that curve in different directions. The *spiral curve* (or transition curve) has a varying radius, so that the curve starts out very flat and increases in sharpness as we move into the curve. Spiral curves are illustrated in Figure 22-1(b).

FIGURE 22-1 Some types of horizontal curves.

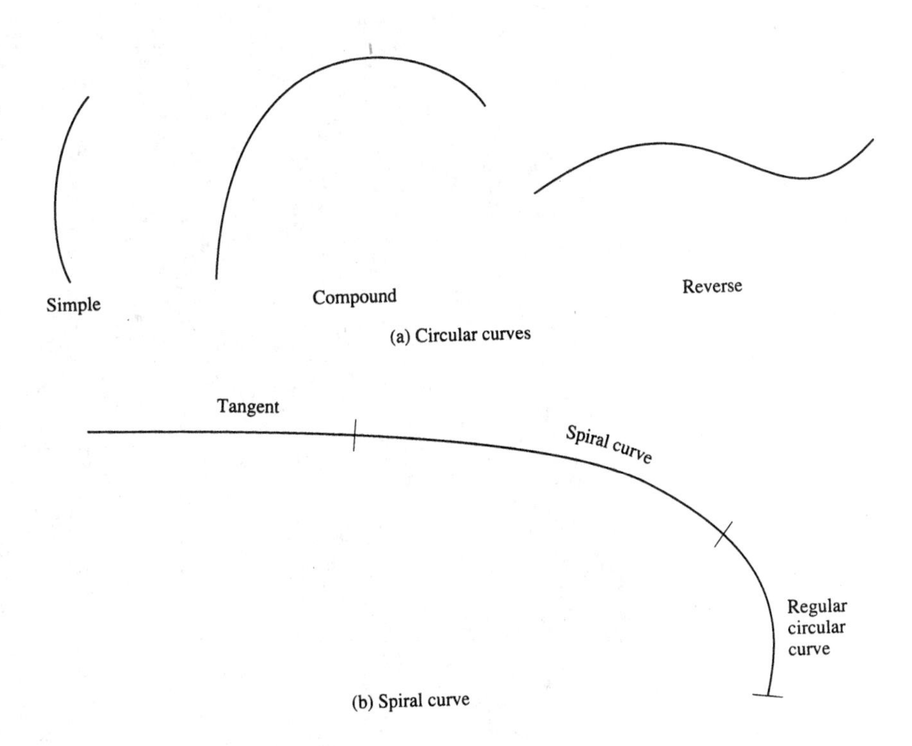

Several definitions relating to curves are presented in the next few paragraphs and are illustrated in Figure 22-2. A curve is initially laid out with two straight lines or tangents. These lines are extended until they intersect and that *point of intersection* is called the P.I. The first tangent encountered is called the *back tangent* and the second one is called the *forward tangent.*

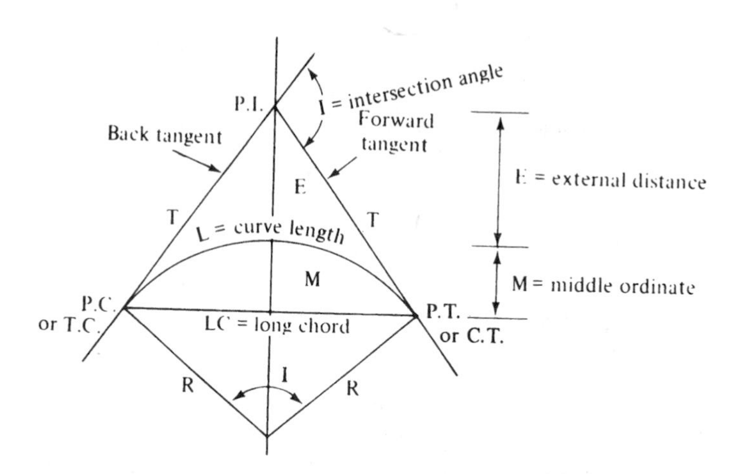

FIGURE 22-2 Curve notation.

FIGURE 22-3 Horizontal curves on I-285 in Atlanta, Georgia. (Courtesy of Georgia Department of Transportation.)

The curve is laid out so that it joins these tangents. The *points on tangents* (P.O.T.s) are the points where the curves run into the tangents. The first of these points is on the back tangent at the beginning of the curve and is called the *point of curvature* (P.C.). The second one is at the end of the curve on the forward tangent and is called the *point of tangency* (P.T.). In another notation the point of curvature may be written as T.C., indicating that the route changes from a tangent to a curve; the point of tangency may be written as C.T., indicating that the route goes from curve back to tangent.

The angle between the tangents is called the *intersection angle* and is labeled I. The *radius of the curve* is R, while T is the *tangent distance* and equal to the length of the back or forward tangents. The distance from the P.I. to the middle point of the curve is called the *external distance* and is denoted by E. Finally, the chord of the arc from the P.C. to the P.T. is called the *long chord* (L.C.) and the distance from the middle of the curve to the middle of the long chord is labeled M, the *middle ordinate*, and L is the actual curve length.

The discussion of horizontal curves presented in this chapter makes use of the foot unit. Although this is the case, it should be realized that all of the equations used herein are perfectly valid for metric distances as long as a full station is understood to be 100 m. Several horizontal curves are shown in Figure 22-3.

22-2 DEGREE OF CURVATURE AND RADIUS OF CURVATURE

The sharpness of a curve may be described in any of the following ways:

1. *Radius of curvature.* This method is often used in highway work, where the radius for the curve is frequently selected as a multiple of 100 ft. The smaller the radius, the sharper the curve. Should the degree of curvature (defined in the next two paragraphs) be specified rather than the radius of the curve, the radius can be computed. It will in all probability be an odd number of feet.
2. *Degree of curvature, chord basis.* In this method, the degree of curvature is defined as the central angle subtended by a chord of 100 ft, as illustrated in Figure 22-4. The radius of such a curve may be computed with the following equation where D is the intersection angle in units of degrees:

$$R = \frac{50}{\sin \frac{1}{2} D}$$

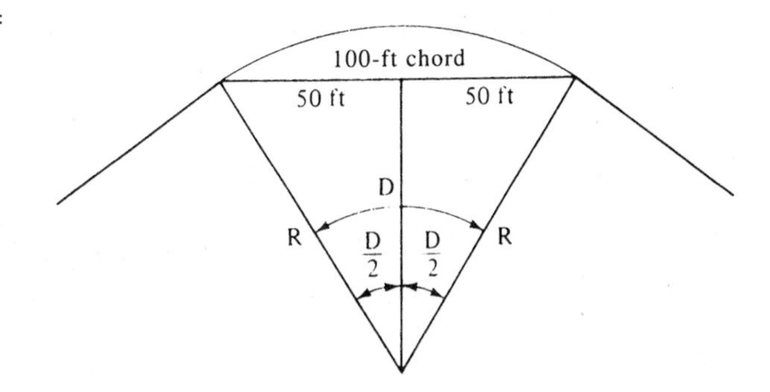

FIGURE 22-4 Degree of curvature, chord basis.

3. *Degree of curvature, arc basis.* As shown in Figure 22-5, the degree of curvature on the arc basis is the central angle of a circle that will subtend an arc of 100 ft. It will be noted that a sharp curve has a large degree of curvature and a flat curve a small degree of curvature. For a particular curve with a degree of curvature D (degrees), the radius of the curve, R, can be computed as follows:

$$\text{circumference of circle} = \left(\frac{360^\circ}{D}\right)(100) = 2\pi R$$

$$R = \left(\frac{360^\circ}{D}\right)\left(\frac{100}{2\pi}\right) = \frac{5729.58}{D}$$

FIGURE 22-5 Degree of curvature, arc basis.

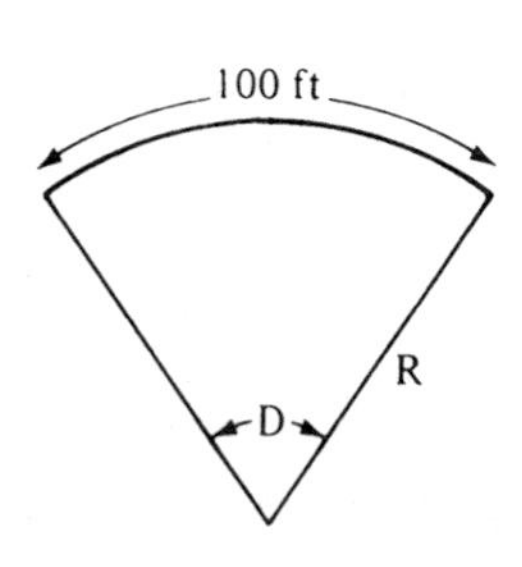

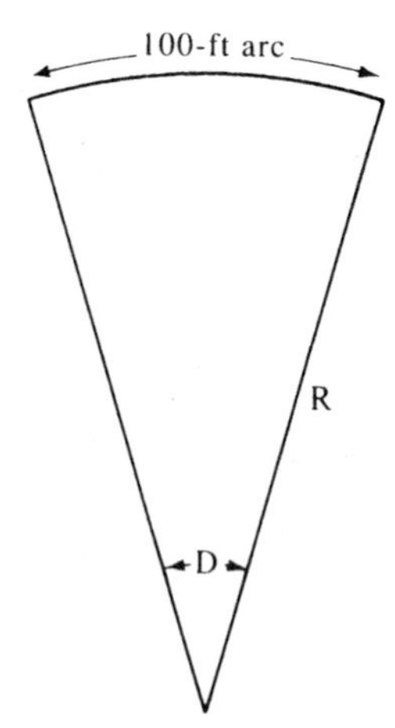

The arc basis is used for the calculations presented in this chapter. Actually, both the chord and arc methods are used extensively in the United States. The method selected often depends on the experience of the surveyor involved. For long gradual curves, which are common in railroad practice, the chord basis (where the lengths along the arcs are considered to be the same as along the chords) is normally used. For highway curves and curved property boundaries, the arc basis is more common.

It will be noted that the difference between the chord and arc bases is normally not large. For instance, for a 1° curve, R on the arc basis is 5729.58 ft and on the chord basis is 5729.65 ft. Corresponding values for a 4° curve are 1432.39 ft and 1432.69 ft, respectively. *From a practical standpoint a horizontal curve defined by the arc basis would appear to be the same as a horizontal curve defined by the chord basis. However, the parameters for the two curves would be significantly different when computed to the nearest hundredth of a foot.*

22-3 CURVE EQUATIONS

The formulas needed for circular computation are presented in this section, with reference to Figure 22-2. The radius of curvature was given previously as

$$R = \frac{5729.58}{D}$$

The tangent distance T is the distance from the P.I. to the P.C. or P.T. and can be computed from

$$T = R \tan \frac{1}{2} I$$

The length of the long chord is

$$\text{L.C.} = 2R \sin \frac{1}{2} I$$

The external distance E is

$$E = R\left(\sec \frac{I}{2} - 1\right) = R \text{ external secant } \frac{I}{2}$$

where

$$\text{external secant} = 1 - \text{secant}$$

A more convenient form with today's calculators is

$$E = R\left[\frac{1}{\cos \frac{I}{2}} - 1\right]$$

The middle ordinate M equals

$$M = R - R\cos \frac{1}{2} I = R \text{ versine } \frac{I}{2}$$

where

$$\text{versine} = 1 - \cos$$

or, more conveniently,

$$M = R\left(1 - \cos \frac{I}{2}\right)$$

The length of the curve is

$$L = \frac{100I}{D}$$

For the chord definition of D,

$$L = \frac{RI\pi}{180}$$

where I is measured in degrees.

For surveyors who frequently work with curves, the use of versines and external secants simplifies the calculations. Various books provide tables of their values.[1] Use of these terms is becoming obsolete, however, because the equations using the secant and cosine values can be so easily solved with modern handheld calculators (or with computer programs) without reference to tables.

[1]For example, T. F. Hickerson, *Route Location and Design,* 5th ed. (New York: McGraw-Hill Book Company, 1967), pp. 559–588.

Curves are staked out using straight-line chord distances. If the degree of curvature for a particular curve is 3° or less, the curve can be staked out using 100-ft chords as shown in Figure 22-6, yet keep the arc length and chord length values sufficiently close to each other so as to be within the precision of tape measurements. As the usual curve will be of an odd length (i.e., not a whole number of hundreds of feet), there will certainly be subchord lengths at the ends of the curve as indicated in the figure. For curves from 3 to 7°, it is necessary to go to 50-ft chords and for those from 7 to 14°, to 25-ft chords to maintain satisfactory precision.

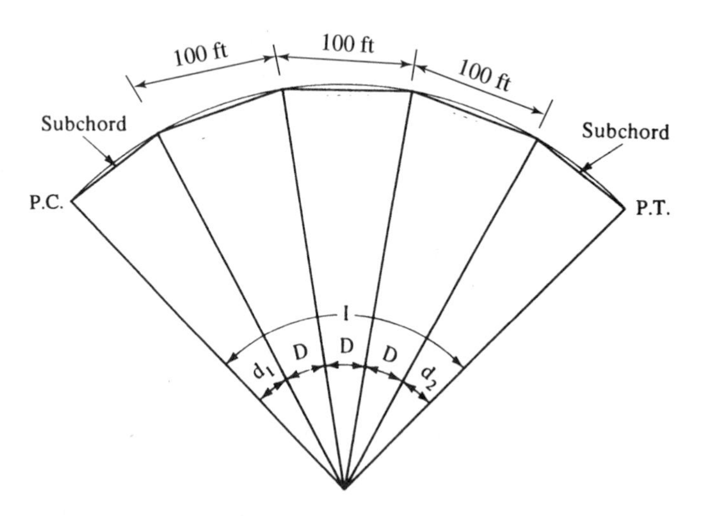

FIGURE 22-6 Length of curve, chord basis.

For flat curves, say 2 or 3°, the chord distances are almost the same to two places beyond the decimal, whereas for sharper curves the difference is more pronounced. For a 100-ft long 2° curve, the chord is 99.995 ft, and for a 100-ft 3° curve, it is 99.989 ft. For a 6° curve, the chord is 99.954 ft, and for a 10° curve, it is 99.873 ft.

A better procedure to follow than the one described above is to stake out the actual chord lengths as 99.995 ft for a 2° curve, etc.

22-4 DEFLECTION ANGLES

The angle between the back tangent and a line drawn from the P.C. to a particular point on a curve is called the *deflection angle* to that point. Circular curves are laid out almost entirely by using these angles. From the geometry of a circle, the angle between a tangent to a circular curve and a chord drawn from that point of tangency to some other point on the curve equals one-half of the angle subtended by that chord. Thus for a 100-ft chord, the deflection angle is $D/2$, and for a 50-ft chord is $50/100 \times D/2$. These values are illustrated in Figure 22-7, where D is 3°. It will be noted that the deflection angle from the P.C. to each succeeding 100-ft station can be calculated by adding $D/2$ to the last deflection angle, or to each succeeding 50-ft station by adding $D/4$ to the last deflection angle.

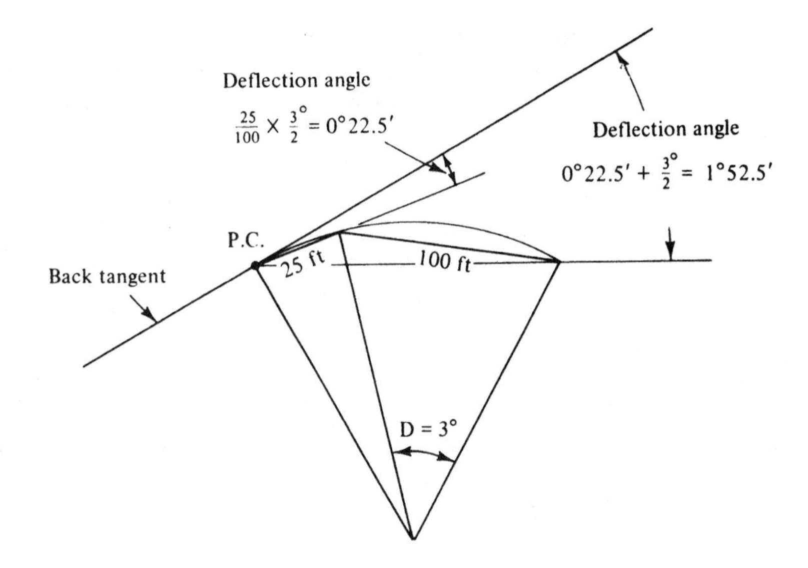

FIGURE 22-7 Staking out a horizontal curve.

22-5 SELECTION AND STAKING OUT OF CURVES

Before a horizontal curve can be selected, it is necessary to extend the tangents until they intersect, measure the intersection at angle *I*, and select *D*, the degree of curvature. For any pair of intersecting tangents, an infinite number of curves can be selected, but the field conditions will narrow down the choices considerably. For high-speed highways the degree of curvature is generally kept below certain maximums, while for twisting mountain roads the lengths of tangents may be severely limited by the topography. Should a road be running along the bank of a river, the external distance *E* may be restricted.

Actually, the curve to be used can be selected by assuming a value for *D, R, T, E,* or *I*. This permits the calculation of the other four items since they are dependent on each other. Usually, the radius of the curve or the degree of curvature is the value assumed. The necessary computations can be made as illustrated in Example 22-1. In addition to the determination of *R, T,* L.C., *E, M,* and *L*, these computations include the calculations of the deflection angles and the setting up of the field-book notes for staking out the curve.

The positions of the P.C. and P.T. are determined by measuring the calculated tangent distance *T* from the P.I. down both tangents. The reader should note that the station of the P.C. equals the station of the P.I. minus the distance *T*, and that *the station of the P.T. equals the P.C. plus the curve length L (not the P.I. plus T).* This information can be expressed as follows:

$$\text{P.C.} = \text{P.I.} - T$$
$$\text{P.T.} = \text{P.C.} + L$$

The instrument is set up at the P.C. or P.T. and the curve staked out. As the work proceeds, the instrument can be set up at intermediate points as well. The latter procedure will be described after Example 22-1.

Example 22-1 illustrates the calculations and field-book notes needed for the staking out of a horizontal circular curve. It will be noted that these notes are arranged and the stations numbered from the bottom of the page upward. This practice, which is common for route surveys such as this one, enables the surveyor to look forward along the route and follow his or her notes while going in the same direction. In the same manner, as the surveyor looks at the sketch page and forward along the route, items to the right of the project center line on the ground are to the right on the sketch page, and vice versa.

As regards stationing the reader will remember that if we say *T* is 308.82 ft that's the same as 3 + 0.0882 stations (written as 3 + 8.82) or if *L* is 982.36 ft that's 9 + 0.8236 stations (written as 9 + 82.36).

Example 22-1

For a horizontal circular curve, the P.I. is at station 64 + 32.2, *I* is 24°20′, and a *D* of 4°00′ has been selected. Compute the necessary data and set up the field notes for 50-ft stations.

Solution

The values of *R, T,* L.C., *E, M,* and *L* are computed by the formulas presented previously, although tables are given in many books for simplifying the calculations.[2] For horizontal curves it is common to carry calculations to the nearest ±0.01 ft.

$$R = \frac{5729.58}{4} = 1432.39\text{ ft}$$

$$T = (1432.39)(0.21560) = 308.82\text{ ft}$$

$$\text{L.C.} = (2)(1432.39)(0.21076) = 603.78\text{ ft}$$

$$E = 1432.39\left[\frac{1}{\cos 12^\circ 10'} - 1\right] = 32.91\text{ ft}$$

$$M = 1432.39 - (1432.39)(\cos 12^\circ 10') = 32.17\text{ ft}$$

$$L = \frac{(100)(24.333333)}{4} = 608.33\text{ ft}$$

From these data, the stations of the P.C. and P.T. can be calculated as follows:

$$\begin{aligned} \text{P.I.} &= 64 + 32.2 \\ -T &= -(3 + 08.8) \\ \hline \text{P.C.} &= 61 + 23.4 \\ +L &= 6 + 08.3 \\ \hline \text{P.T.} &= 67 + 31.7 \end{aligned}$$

As the degree of curvature is between 3° and 7°, the curve will be staked out with 50-ft chords. The distance from the P.C. (61 + 23.4) to the first 50-ft station

[2]T. F. Hickerson, *Route Location and Design,* 5th ed. (New York: McGraw-Hill Book Company, 1967), pp. 396–458.

(61 + 50) is 26.6 ft, and the deflection angle to be used for that point is (26.6/100) × (*D*/2) = (26.6/100) × (4°/2) = 0°31′55″. For each of the subsequent 50-ft stations from 61 + 50 to 67 + 00, the deflection angles will increase by *D*/4 or 1°00′00″. Finally, for the P.T. at station 67 + 31.7, the deflection angle will be 11°32′ + (31.7/100)(4°/2) = 12°10′00″. This value equals 1/2 *I*, as it should.

The field notes are set up as shown in Figure 22-8.

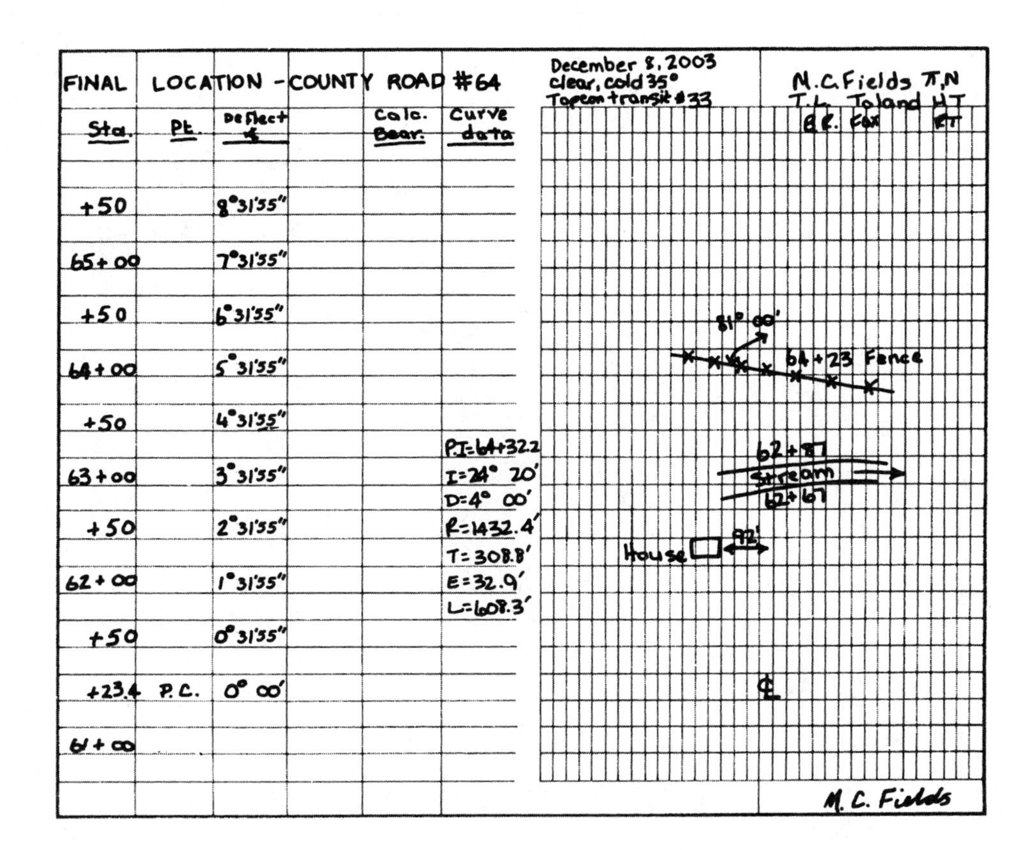

FIGURE 22-8 Field notes for a horizontal curve.

22-6 COMPUTER EXAMPLE

To apply SURVEY to a horizontal curve problem, the user need only move the cursor to Program 5 (Horizontal Curves), press the ENTER key, and supply the data requested. That data will include the station of the P.I., the intersection angle *I*, and either the degree of curvature *D* *or* the radius of curvature *R*.

EXAMPLE 22-2

Repeat Example 22-1 using the SURVEY CD-ROM.

Solution

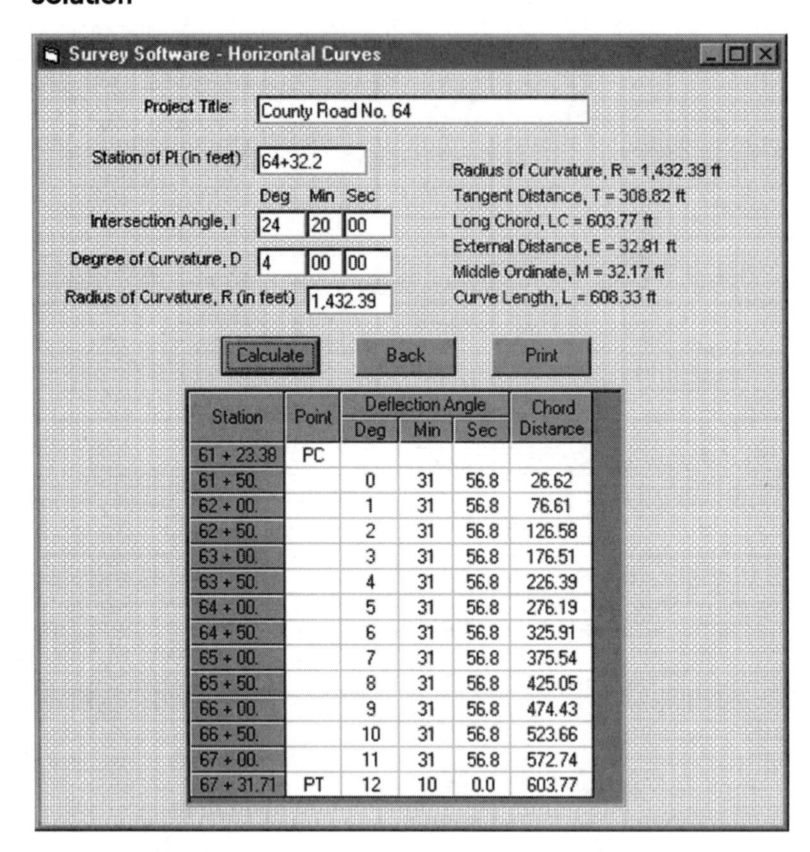

Station	Point	Deflection Angle			Chord Distance
		Deg	Min	Sec	
61 + 23.38	PC				
61 + 50.		0	31	56.8	26.62
62 + 00.		1	31	56.8	76.61
62 + 50.		2	31	56.8	126.58
63 + 00.		3	31	56.8	176.51
63 + 50.		4	31	56.8	226.39
64 + 00.		5	31	56.8	276.19
64 + 50.		6	31	56.8	325.91
65 + 00.		7	31	56.8	375.54
65 + 50.		8	31	56.8	425.05
66 + 00.		9	31	56.8	474.43
66 + 50.		10	31	56.8	523.66
67 + 00.		11	31	56.8	572.74
67 + 31.71	PT	12	10	0.0	603.77

FIGURE 22-9 Printout from SURVEY.

22-7 FIELD PROCEDURE FOR STAKING OUT CURVES

The tangent distances are measured from the P.I. down both tangents to locate the P.C. and P.T. For this discussion the instrument is assumed to be set up at the P.C. and the curve and numbers of Example 22-1 used. It should be noticed, however, that if the entire curve can be seen from the P.T., it is possible to avoid one instrument setup by setting up there, because after the curve is completed, the survey can proceed down the forward tangent from the same setup.

The instrument is set up at the P.C. with the scale reading zero while sighting on the P.I. The first deflection angle (0°31′55″) is turned and the chord distance of 26.6 ft is taped to locate station 61 + 50. The instrument is turned to the next deflection angle (1°31′55″) and 50 ft is taped beyond station 61 + 50 to stake station 62 + 00. *Once again a better and more accurate procedure is to compute the chord lengths for these 50-ft arcs and use those values in the field.* This procedure is continued until the P.T. is reached. These points should check very closely. For long curves it is considered better to run in the first half of the curve from the P.C. and the second half back from the P.T. so that small errors that occur can be adjusted at the middle of the curve, where a little variation is not as important as it would be near the points of tangency to the curve. Taping is quite practical for the short distances involved here but total stations are being used more and more frequently.

Very often it may not be possible to set all the points on the curve from one position. The instrument may have to be moved up to one of the intermediate stations on the curve and the work continued. To do this, the instrument is set up on the intermediate station and is backsighted on the P.C. with the telescope inverted and the scale reading 0°00′. The telescope is reinverted, the deflection angle that would have been used for that next station turned, and the process continued.

For an illustration Example 22-1 is considered. There it is assumed that the instrument is moved up to station 64 + 00 and backsighted on the P.C. with the telescope inverted. The telescope is reinverted and turned to an angle of 6°31′55″, the distance of 50 ft taped, and station 64 + 50 set.

Another horizontal curve situation is considered in Example 22-3. A surveyor is assumed to be working along the boundary of a piece of private property that adjoins a curved section of a state highway. He would like to determine the information necessary to describe correctly the curved boundary of the land.

EXAMPLE 22-3

As shown in Figure 22-10, a surveyor has determined that the intersection angle for a particular highway curve is 28°00′. In addition, he or she has measured the external distance *E* from the P.I. to the highway center line for this curve and found it to be 73.6 ft.

(a) Determine the values of *D, R,* and *T* for the highway center line.
(b) If the highway right-of-way is 50 ft from the highway center line as shown in the figure, determine for the property on the inside of the curve the values of *R* and *D.*

Solution

(a)

$$R = \frac{E}{\left[\dfrac{1}{\cos\dfrac{I}{2}} - 1\right]} = \frac{73.6}{0.03061} = \mathbf{2404.4\ ft}$$

$$D = \frac{5729.58}{2404.4} = 2.383^\circ = \mathbf{2^\circ 23'}$$

$$T = (2404.4)(0.24933) = \mathbf{599.5\ ft}$$

(b)

$$\text{R of property line or right-of-way line} = 2404.4 - 50 = \mathbf{2354.4\ ft}$$

$$\text{D of property line curve} = \frac{5729.58}{2354.4} = 2.434^\circ = \mathbf{2^\circ 26'}$$

The surveyor who is staking out new lots or resurveying old ones along a highway will often find that parts of the lot boundaries adjacent to the right-of-way (ROW) of the highway will be curved. His or her work should reflect these curved property lines.

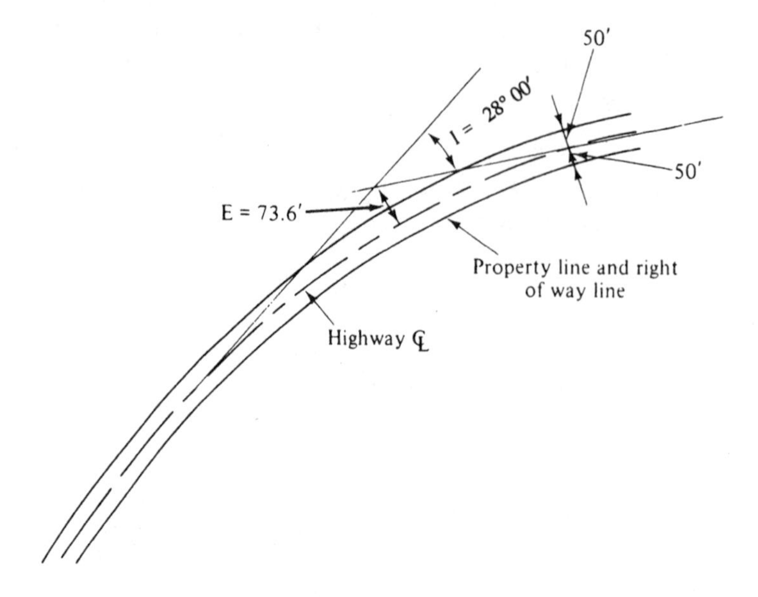

FIGURE 22-10 Computing horizontal curve data for highway center line and right-of-way.

From the equations given earlier in this chapter it is evident that only two of the elements (R, T, L.C., E, M, and L) of a curve are necessary to define the curve. It is often desirable, however, for the surveyor to provide three or even more of these elements. If more than two elements are provided, they must be compatible with each other. Among the curve elements most commonly provided are the radius, intersection angle, curve length, and sometimes the length and bearing of the chord.

Example 22-4 illustrates the information necessary to stake out 100-ft lots along a right-of-way curve for a highway.

Example 22-4

It is assumed that the curve considered in Example 22-1 represents the center line of a highway with a 50-ft right-of-way on each side. It is desired to lay off 100-ft lots on the inside of the curved right-of-way as shown in Figure 22-11. Compute the necessary data for laying off the lots assuming that the last corner on the ROW back tangent is 53.7 from the P.C.

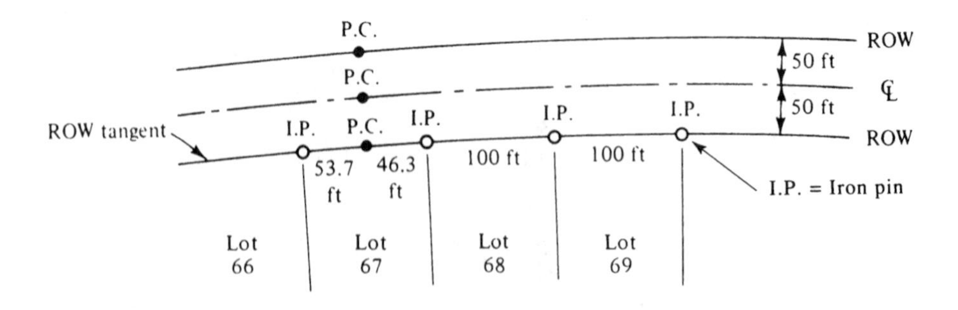

FIGURE 22-11 Laying off lots along a horizontal curve.

Solution

The radius of the ROW curve is 50 ft less than the radius of the center-line curve. Therefore,

$$R = 1432.39 - 50.00 = 1382.39 \text{ ft}$$

$$D = \frac{5729.58}{R} = \frac{5729.58}{1382.39} = 4.1446914^\circ = 4^\circ 08' 41''$$

$$L = \frac{100I}{D} = \frac{(100)(24.333333)}{4.1446914} = 587.10 \text{ ft}$$

Setting up the instrument at the P.C. on the ROW curve the deflection angle to the first lot corner on the curve (a distance equal to 100.0 – 53.7 = 46.3 ft) is

$$\left(\frac{46.3}{100}\right)\left(\frac{4.1446914}{2}\right) = 0.95949606^\circ = 0^\circ 57' 34''$$

From the same position of the instrument, the deflection angle to the next lot corner is

$$\left(\frac{146.3}{100}\right)\left(\frac{4.1446914}{2}\right) = 3.0318418^\circ = 3^\circ 01' 55''$$

or it can be calculated as

$$0^\circ 57' 34'' + \frac{D}{2} = 0.95946606^\circ + \frac{4.1446914}{2} = 3.0318418^\circ = 3^\circ 01' 55''$$

The deflection angle for the P.T. can be determined by taking the initial deflection angle (0°57.6′) and adding $D/2$ to it for each 100 ft plus the angle value for the partial lot width before the P.T.; or it equals

$$\left(\frac{587.10}{100}\right)\left(\frac{4.1446914}{2}\right) = 12.166742^\circ = 12^\circ 10' 00''$$

which is half of I of 24°20′.

Note: Since the degree of curvature is between 3° and 7°, the curve should be laid out with 50-ft chords with the iron pin corner monuments placed at the 100-ft points.

22-8 CIRCULAR CURVES USING THE SI SYSTEM

It appears that those countries who have adopted the SI system of measurements for highway work are today using 1000 m as a full station (that is 1 + 000 m). They further seem to be using the radii of curves for their calculations rather than using the degrees of curvature. These practices are followed for the example problem presented in this section.

The formulas used for circular curves are the same as those used with U.S. customary units as previously described in this chapter. It should be noted, however, that if the degree of curvature is used it will have to be redefined as, say, the central angle of a circle which subtends an arc of 100 m (rather than 100 ft as previously used).

Using the SI system the computation of the deflection angles for laying out the curves is based on $I/2$. For instance, the deflection angle for a chord length of 20 m from the PC will be $\left(\frac{20}{L}\right)\left(\frac{I}{2}\right)$. Example 22-5 which follows illustrates circular curve calculations in SI units.

Example 22-5

Two highway tangents intersect with a right intersection angle $I = 12°30'00''$ at station 0 + 152.204 m. If a radius of 300 m is to be used for the circular curve, prepare the field book notes to the nearest full minute as needed to lay out the curve with stakes at 20 m intervals.

Solution

$$T = R \tan\frac{I}{2} = (300)(\tan 6°15') = 32.855\text{m}$$

$$L = \frac{RI\pi}{180} = \frac{(300)(12.50)(\pi)}{180} = 65.450\text{m}$$

$$\text{Station of P.I.} = 0 + 152.204$$

$$-T = -(0 + 32.855)$$

$$\text{Station of P.C.} = 0 + 119.349$$

$$+L = +(0 + 65.450)$$

$$\text{Station of P.T} = 0 + 184.799\text{m}$$

The field notes are set up as shown in Table 22-1 with stakes to be placed at 22-m intervals.

TABLE 22-1 Circular Curve Notes Using SI Units

Station	Chord distance from P.C.	Point	Deflection Angle
0 + 184.799 m	65.450 m	P.T.	6°15′00″
0 + 180 m	60.651 m		5°47′30″
0 + 160 m	40.651 m		3°52′55″
0 + 140 m	20.651 m		1°58′19″
0 + 120 m	0.651		00°03′44″
0 + 119.349 m	0	P.C.	0°00′00″

22-9 HORIZONTAL CURVES PASSING THROUGH CERTAIN POINTS

Sometimes it is necessary to lay out a horizontal curve that passes through a certain point. For instance, it may be desired to establish a curve that passes no closer than a certain distance to some feature such as a building or stream. The distance from the P.I. to the point in question and the angle from one of the tangents can be measured. Such a situation is considered in Example 22-6 and illustrated in Figure 22-12 where it is desired to run the curve through point A.

To solve the problem, the distances x and y shown in Figure 22-12 can be computed by trigonometry. With reference to the figure it can be seen that the radius of the curve R can be determined by considering the right triangle ABC. For this triangle the following quadratic equation applies:

$$R^2 = (R - y)^2 + (T - x)^2$$

The value of T can be expressed in terms of R as $R \tan \frac{1}{2} I$ and substituted into the preceding equation leaving only R as an unknown. Then R can be determined with the quadratic equation, by trial and error with a calculator, or by completing the squares. If the quadratic equation is used, the solution will yield two answers, but one of them will be seen to be unreasonable. Hickerson[3] provides a solution to this type of problem which does not involve the use of a quadratic equation.

FIGURE 22-12

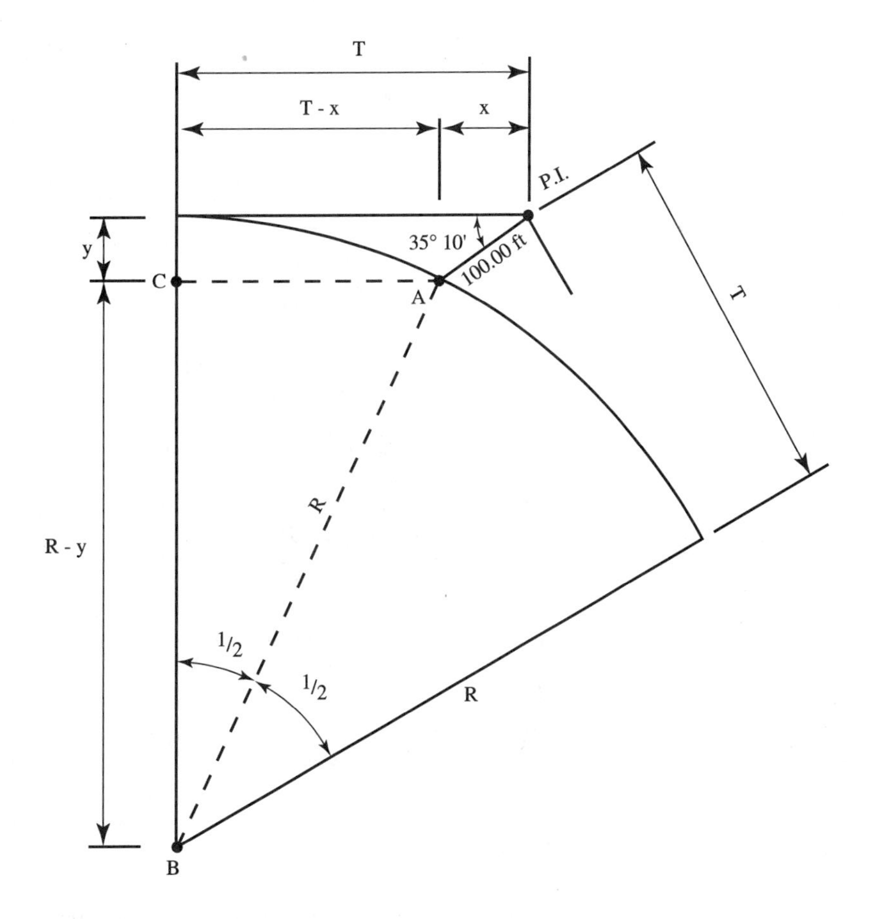

[3]T. F. Hickerson, *Route Location and Design,* 5th ed. (New York: McGraw-Hill Book Company, 1967), pp. 90–91.

Example 22-6

A horizontal curve is to be run through point *A* as shown in Figure 22-12. From the P.I. the distance to point *A* is 100.00 ft and the angle from the back tangent to a line from the P.I. to point *A* is 35°10′. If *I* is 65°00′, determine the values of *R, D, T,* and *L.*

Solution

$$y = (100.00)(\sin 35°10') = 57.596\ \text{ft}$$

$$x = (100.00)(\cos 35°10') = 81.748\ \text{ft}$$

$$T = (R)\left(\tan \frac{65°}{2}\right) = 0.63707R$$

$$R^2 = (R-y)^2 + (T-x)^2$$

$$= (-57.596)^2 + (0.63707R - 81.748)^2$$

Squaring the terms and simplifying yields

$$R^2 - 540.46R + 24.639137 = 0$$

Using the quadratic equation with $a = 1.00$, $b = -540.46$, and $c = 24.639137$:

$$R^2 = \frac{540.56 \pm \sqrt{(-540.46)^2 - (4)(1.00)(24.639137)}}{(2)(1.00)}$$

$$= 490.20\ \text{ft or } 50.26\ \text{ft (not feasible)}$$

$$D = \frac{5729.58}{490.20} = 11.688°$$

$$T = (490.20)(\tan)\left(\frac{65°00'}{2}\right) = 312.29\ \text{ft}$$

$$L = \frac{(100)(65.00)}{11.688} = 556.13\ \text{ft}$$

22-10 SPIRAL CURVES

A spiral curve is used to provide a gradual transition from a straight line or tangent to the full curvature of a circular curve. It starts out very flat with a radius of infinity and increases in sharpness as the circular curve is approached. When the circular curve is reached the spiral curve will have a degree of curvature equal to that of the circular curve. Although spirals are not used universally for highway curves, they are used extensively by railroads. There are other transition curves that can be used such as the compound ones described in Section 22-1.

When spirals are not provided on a highway and a circular curve is reached, the driver of a vehicle will have to quickly turn the steering wheel to the full curve. When this is done there will be a sudden lateral thrust or centrifugal force applied to the vehicle and the result may be that the vehicle will deviate or veer somewhat in its path. Some railroad accidents in the 19th century were attributed to the omission of spirals by their designers. If a spiral is properly designed it will provide an easy path for vehicles such that the centrifugal forces will increase and decrease gradually as the vehicles enter and exit a circular curve.

For many high-speed highways with very flat or low degree curves spirals are not used because their designers feel that the changes in direction are so small as to be negligible. It is to be realized that superelevations may be introduced on the tangent portions before the circular curves are reached and this practice will help vehicles ease into the circular curves.

When transition curves are not used and where lane width on the highway is available, drivers will often create their own spirals by "shortcutting the curves." That is, they may gradually start turning their vehicles while they are still on the straight part of the road, thus cutting the corners a little and working their way gradually into the full circular curve. On concrete highways the oil stains going around curves will often show that many drivers have followed this practice.

In Figure 22-13 equal spiral curves are shown that join a circular curve to the main tangents. In this figure the following abbreviations are used:

T.S. = the beginning point of the first spiral (tangent to spiral point)

L_S = the length of the spiral curve

S.C = the ending point of the first spiral or the beginning point of the circular curve (spiral to circular curve point)

C.S. = ending point of circular curve or starting point of second spiral (curve to spiral point)

S.T. = ending point of the second spiral (spiral to tangent point)

T_S = tangent distance from the T.S. or the S.T. to the P.I.

R = the radius of the circular curve

The spiral curve shown on the left side of Figure 22-13 is redrawn in Figure 22-14 to a much larger scale. The symbols needed for calculating the spiral's properties and for laying it out in the field are shown in the figure.

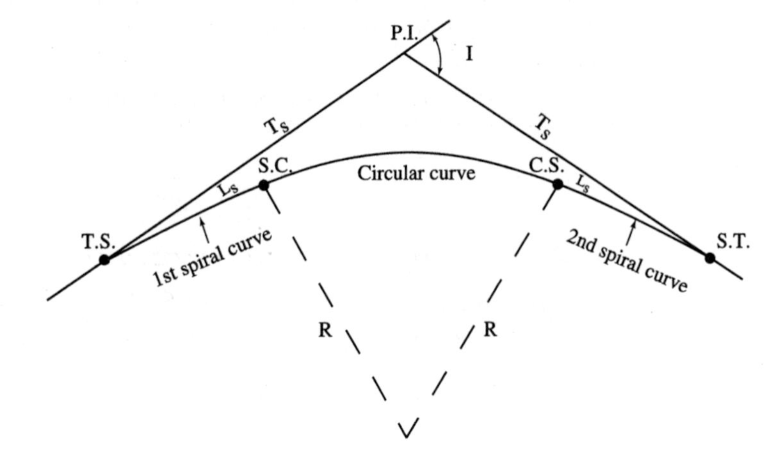

FIGURE 22-13 Spiral curves.

The equation for computing the radius of the circular curve R was previously given in Section 22-3.

$$R = \frac{5729.58}{D}$$

The additional abbreviations shown in Figure 22-14 are defined below and the necessary equations for their calculations are provided. Although the expressions for X_S and Y_S are somewhat abbreviated from their theoretical values, they are satisfactory for practical purposes.

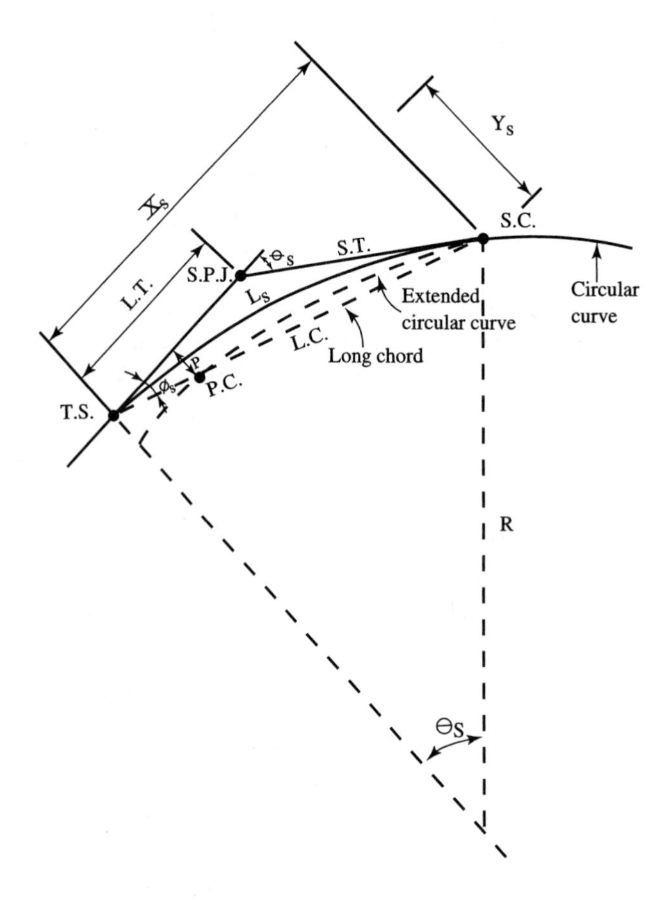

FIGURE 22-14 First spiral curve from Figure 22-13.

If a tangent (S.T.) to the curve at the S.C. is run back until it intersects a tangent (L.T.) run forward from the T.S. the intersection point is called the S.P.I. (the spiral point of intersection).

The intersection angle in degrees between the tangents just mentioned is called θ_S (the spiral angle). It equals

$$\frac{L_s D_c}{200}$$

where D_c is the degrees of curvature of the circular curve.

X_s = the distance measured from the T.S. along the long tangent (and its extension) to a point where a perpendicular line to the tangent hits the S.C. =

$$L_s \left(1 - \frac{\theta_s^{\,2}}{10} + \frac{\theta_s^{\,4}}{216}\right)$$

where θ is in radians.

$$\text{One radian} = \frac{180^\circ}{\pi}$$

Y_s = the distance measured perpendicular from the X_s coordinate to the S.C.

$$= L_s\left(\frac{\theta_s}{3} - \frac{\theta_s^{\,3}}{42}\right)$$

L.C. = The long chord from the T.S. to the S.C. = $\sqrt{X_s^{\,2} + Y_s^{\,2}}$

S.T. = the short tangent or the distance from the S.P.I. to the S.C.

$$= \frac{Y_s}{\sin\theta_s} \text{ with } \theta_s \text{ in degrees}$$

L.T. = the long tangent from T.S. to S.P.I. = X_s − S.T. $\cos\theta_s$ with θ_s in degrees

$\varnothing_s$ = the deflection angle from the T.S. to the S.C. = $\frac{\theta_s}{3} - C_s$

(The term C_s is a correction factor which is negligible when the degree of curvature of the matching circular curve is less than 15°)

Should a circular curve be extended back from its S.C. it will fall inside the spiral. At the point where the circular curve becomes parallel to the spiral (called the P.C.) the curves will be a distance p apart as shown in Figure 22-14. This distance, which is perpendicular to the back tangent, is called the *throw* and it may be calculated with the following equation:

$$p = Y_s - R(1 - \cos\theta_s)$$

With the value of p available it is possible to compute T_S, the length of the tangent from the *T.S.* to the *P.I.*

$$T_s = X_s - R\sin\theta_s + (R + p)\tan\frac{I}{2}$$

To lay out a spiral curve in the field the following steps are taken:

1. A length (L_S) is selected for the spiral giving consideration to the traffic design, speed, the number of lanes, the degree of curvature of the circular curve, and the length needed for superelevation. An empirical equation is provided by the AASHTO for computing the minimum length of spirals.[4]
2. The values for R, θ_S, X_S, Y_S, L.C., S.T., L.T., ϕ_S, p, and T_S are computed with the equations just given.

[4]A Policy on Geometric Design of Highways and Streets (Washington, D.C.: AASHTO, 2001)

3. The chord lengths to be used in staking out the curve are assumed and the deflection angle (ϕ) to be used at each point is determined from the expression to follow:

$$\phi = \left(\frac{L}{L_s}\right)^2 \phi_s$$

4. The curve will be staked out in a manner almost identical to that used for circular curves, the difference being that the deflection angles will change as each chord length is laid out. In the field the computed tangent distance (T_S) is laid off from the P.I. back to the point where the spiral starts. With reference to Figure 22-14 the distances X_S and Y_S can be laid out to establish the position of the S.C. Then with the instrument at the T.S. the spiral is laid out from the main tangent using the chords and computed deflection angles until the beginning point of the circular curve (S.C.) is reached. If it is necessary to move up on the spiral because of slight distance problems, it will be handled in the same manner as that used for circular curves. The position of the S.C. determined with the chords and deflection angles should agree reasonably well with the position obtained with the X_S and Y_S values. If it does not, the work will have to be repeated.

Next the instrument is moved up to the S.C. and the circular curve is laid out to its end at the C.S. Finally the distance T.S. is measured from the P.I. to the S.T. and the second spiral is laid out from the S.T. back to the C.S. The points obviously should check.

Example 22-7 which follows presents the calculations needed for laying out a spiral curve in the field.

Example 22-7

It is desired to stake out a 300 ft spiral curve for transition into a 4°00′00″ circular curve (arc basis). The P.I. is at station 70 + 00 and the intersection angle I has been measured and found to be 50°00′00″. Compute the deflection angles and chords for staking out the curve at 50 ft stations for transition into the circular curve.

Solution

$$R = \frac{5729.58}{D} = \frac{5729.58}{D} = 1432.39\text{ ft}$$

$$\theta_s = \frac{L_s D_c}{200} = \frac{(300)(4°00')}{200} = 6°00' = 0.10472\text{ radians}$$

$$X_s = L_s\left(1 - \frac{\theta_s^{\,2}}{10} + \frac{\theta_s^{\,4}}{216}\right) = 300\left(1 - \frac{0.10472^2}{10} + \frac{0.10472^4}{216}\right) = 299.67\text{ ft}$$

$$Y_s = L_s\left(\frac{\theta_s}{3} - \frac{\theta_s^{\,3}}{42}\right) = 300\left(\frac{0.10472}{3} - \frac{0.10472^3}{42}\right) = 10.46\text{ ft}$$

$$\text{L.C.} = \sqrt{X_s^2 + Y_s^2} = \sqrt{(299.67)^2 + (10.46)^2} = 299.85\text{ ft}$$

$$\text{S.T.} = \frac{Y_s}{\sin\theta_s} = \frac{10.46}{\sin 6°00'} = 100.07\text{ ft}$$

$$\text{L.T.} = X_s - \text{S.T.}\cos\theta_s = 299.67 - (100.07)(\cos 6°00') = 200.15\text{ ft}$$

$$ø_s = \frac{\theta_s}{3} = \frac{6°00'}{3} = 2°00'00''$$

$$p = Y_s - R(1 - \cos\theta_s) = 10.46 - 1432.39(1 - 0.994521895) = 2.61\text{ ft}$$

$$T_s = X_s - R\sin\theta_s + (R + p)\tan\frac{I}{2}$$

$$= 299.67 - (1432.39)(\sin 6° + (1432.39 + 2.61))\left(\tan\frac{50°00'00''}{2}\right)$$

$$= 819.09\text{ ft}$$

Computing the stations of the T.S. and T.C. from this data:

$$\begin{aligned} \text{P.I.} &= 70+00 \\ -T_s &= -\ (8+19.09) \\ \text{T.S.} &= 61+80.91 \\ +L_s &= 3+00 \\ \text{S.C.} &= 64+80.91 \end{aligned}$$

The field notes are set up as shown in Table 22-2.

TABLE 22-2 Field Notes for a Spiral Curve

Station	PT	Total chord distance L from	Deflection angle $ø = \left(\frac{L}{L_s}\right)^2 ø_s$
64 + 80.91	S.C.	300.00	2°00′00″
64 + 50		269.09	1°36′33″
64 + 00		219.09	1°04′00″
63 + 50		169.09	0°38′07″
63 + 00		119.09	0°18′55″
62 + 50		69.09	0°06′22″
62 + 00		19.09	0°00′29″
61 + 80.91	T.S.	0.00	0°00′00″

Problems

All the problems listed here are to be solved on the arc basis.

22-1. For a particular horizontal curve the degree of curvature is 3°30′. Compute its radius of curvature. (Ans.: 1637.02 ft)

22-2. Repeat Problem 22-1 if the degree of curvature is 7°30′.

In Problems 22-3 to 22-5, a series of horizontal curves are to be selected for the data given. Determine the stations for the P.C.s and P.T.s for each of these curves.

	P.I.	Intersection angle, I	Degree of curvature, D	
22-3.	10 + 16.56	9°20′	4°00′	(Ans.: 8 + 99.64, 11 + 32.97)
22-4.	32 + 54.92	18°30′	4°30′	
22-5.	87 + 09.20	14°24′	3°20′	(Ans.: 84 + 92.06, 89 + 24.06)

22-6. If I is 30°0′30″ and the maximum value of E is 85.0 ft, determine the degree of curvature D to the nearest full minute that will provide this E.

22-7. Two highway tangents intersect with a right intersection angle of 32°20′ at station 62 + 46.40. If a 3°00′ horizontal circular curve is to be used to connect the tangents, compute R, T, L, and E for the curve.
(Ans.: 1909.86 ft, 553.66 ft, 1077.78 ft, 78.63 ft)

22-8. A horizontal circular curve having a radius of 650.00 ft is to connect two highway tangents. If the chord length L.C. is 800.00 ft, compute E, L, M, and the intersection angle I.

In Problems 22-9 to 22-11, prepare the field book notes to the nearest full minute for laying out the curves with stakes at full stations for the data given.

22-9. P.I. at 56 + 48.60, I = 18°00′, and D = 4°00′.
(Ans.: R = 1432.39 ft, deflection angle at 58 + 00 = 7°33′55″)

22-10. P.I. at 53 + 61.80, I = 26°20′, and D = 5°00′.

22-11. P.I. at 117 + 16.60, I = 16°16′, and D = 4°30′.
(Ans.: L = 361.48 ft, deflection angle at 117 + 00 = 3°43′14″)

For Problems 22-12 to 22-16, repeat the problems given using the SURVEY program provided with this book.

22-12. Problem 22-3.

22-13. Problem 22-4. (Ans.: 30 + 72.96, 34 + 36.88)

22-14. Problem 22-5.

22-15. Problem 22-7. (Ans.: R = 1909.86 ft, E = 78.63 ft)

22-16. Problem 22-10.

For Problems 22-17 and 22-18 prepare the field book notes to the nearest full minute for laying out curves with stakes at 20 m intervals for the data given.

22-17. P.I. @ 1 + 320.26 m, I = 16°30′, R = 400 m.
(Ans.: L = 115.192 m, deflection angle @ 1 + 320 =4°08′)

22-18. P.I. @ 2 + 644.53 m, I = 14°40′, R = 350 m.

22-19. It is necessary to have a horizontal curve pass through a specified point. The point was located by distance and angle from the P.I. of the curve (150.00 ft and 32°00′ to the left of the back tangent with the instrument located at the P.I.). If the intersection angle between the back and forward tangents is 50°00′ to the right, determine the curve radius required (Ans.: R = 1189.75 ft)

For Problems 22-20 to 22-22 compute the deflection angles and chords for staking out spiral curves for transition into the given circular curves.

	For circular curves			L_s for spiral	
	D	P.I.	I	curve	
22-20.	3°00′	122 + 00	40°00′	250 ft	
22-21.	3°30′	62 + 50	46°30′	300 ft	(Ans.: deflection angle @ 55 + 50 = 0°27.8′)
22-22.	5°00′	37 + 11.2	52°00′	350 ft	

CHAPTER 23 Vertical Curves

23-1 INTRODUCTION

The curves used in a vertical plane to provide a smooth transition between the grade lines of highways and railroads are called *vertical curves.* These curves are parabolic rather than circular. Figure 23-1 shows the nomenclature used for vertical curves. When moving along the road, the first of the grade lines that will be encountered is called the *back tangent.* The other one is called the *forward tangent.* To distinguish the points of tangency and their intersection from the similar terms used for horizontal curves, the letter *V* (for vertical) is added to their abbreviations. For instance, the point of intersection for the tangents is called the P.V.I. (point of vertical intersection) and the points of tangency are referred to as the P.V.C. (point of vertical curvature at start of curve) and P.V.T. (point of vertical tangency at end of curve). The *tangent offsets* are distances measured from the tangents in a vertical direction to the curve.

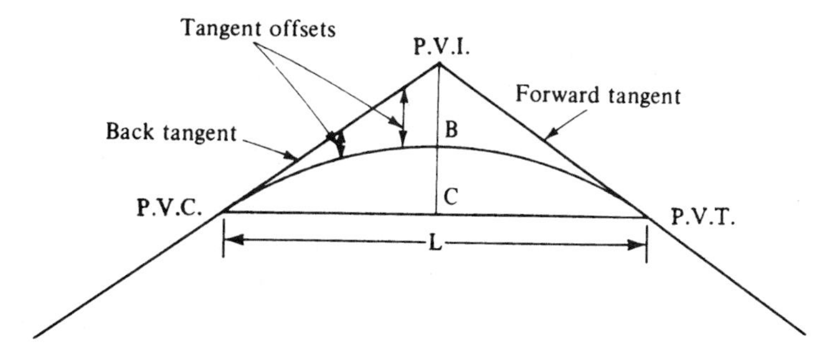

FIGURE 23-1 Vertical curve nomenclature.

The center-line stakes for the project are set, the profile level made, and the profile drawn with an exaggerated vertical scale as described in Chapter 8. Then trial tangents or grade lines are selected with the idea of establishing a design with optimum cut and fill and hauling costs, reasonable percent grades, and with satisfactory sight distances at the top of hills. Adjustments to the curve are made by changing the curve length *L*.

Several different methods are available for making the necessary calculations. One very satisfactory method involves the use of tangent offsets from the grade lines and it is the method that is first described. An alternative method involving the equation of a parabola is presented in Section 23-6. The parabola has three mathematical properties that make its application very convenient for vertical curves. These are as follows:

1. The curve elevation at its midpoint will be halfway from the elevation at the P.V.I. to the elevation at the midpoint of a straight line from the P.V.C. to the P.V.T.
2. The tangent offsets vary as the square of the distance from the points of tangency.
3. For points spaced at equal horizontal distances, the second differences are equal. This is a useful property for checking vertical curve calculations as is illustrated in Example 23-1. The differences between the elevations at equally spaced stations are called the *first differences.* The differences between the first differences are called the *second differences.*

23-2 VERTICAL CURVE CALCULATIONS

The station of the vertex or P.V.I. can be determined from the grade lines. Then the length of the curve is selected—usually as some whole number of hundreds of feet or stations. The length of a vertical curve is defined as the horizontal distance from the P.V.C. to the P.V.T. This length is normally controlled by the specification being used. For instance, the length required for a highway vertical curve must be sufficient to provide a sufficiently flat curve over the top of a hill so that certain minimum sight distances of oncoming vehicles or objects in the road are provided. The AASHTO (American Association of State Highway and Transportation Officials) provides this information depending on the design speed of the highway.[1] This topic is continued in Section 23-3.

Example 23-1 illustrates the calculations required for determining the elevations along a vertical curve. For this example the given grade lines intersect at station 65 + 00 at an elevation of 264.20 ft, and the curve is assumed to be 800 ft long. From this information the stations of the P.V.C. and P.V.T. can be determined as follows:

$$\text{P.V.C.} = \text{P.V.I.} - \frac{L}{2}$$

$$\text{P.V.T.} = \text{P.V.I.} + \frac{L}{2} \text{ or P.V.C.} + L$$

Then the elevations of the intermediate stations along the grade lines or tangents are calculated. The elevation at the midpoint of the long chord point, C in Figures 23-1 and 23-3, is equal to the average elevation of the P.V.C. and the P.V.T.

Next, the elevation of the curve midpoint (point *B* in Figures 23-1 and 23-3) is determined by averaging the elevations of point *C* and the P.V.I. Finally, the tangent offset distances from the grade lines are calculated, thus giving the elevations of the intermediate stations on the curve.

Example 23-1

Figure 23-2 shows the known data for a vertical curve. Assume a curve length of 800 ft and compute the elevation on the curve at each full station by using the tangent offset method.

[1] *A Policy on Geometric Design of Highways and Streets* (Washington, D.C.: AASHTO, 2001), pp. 128–131, 269–282, 682.

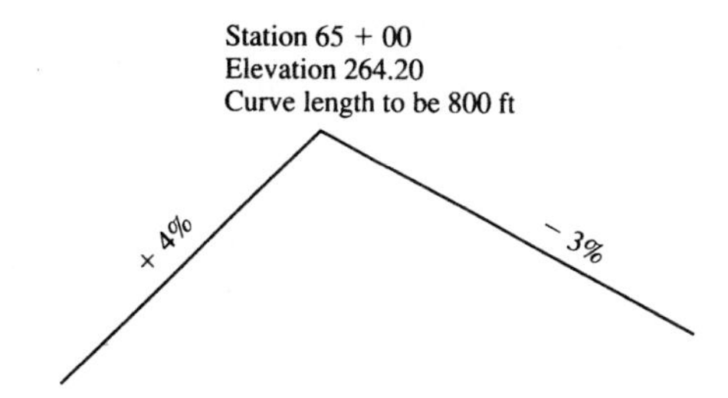

FIGURE 23-2 Data needed for a vertical curve.

Solution

From the length of the curve the stations of the P.V.C. and P.V.T. are determined (61 + 00 and 69 + 00, respectively) and from the tangent grades the station elevations along the tangents or grade lines are obtained as shown in Figure 23-3. Then the elevation of point *B*, the midpoint of the curve, is calculated as follows:

$$\text{Elevation of P.V.I.} = 264.20\text{ ft}$$

$$\text{Elevation of } C = \frac{248.20 + 252.20}{2} = 250.20\text{ ft}$$

$$\text{Elevation of } B = \frac{264.20 + 250.20}{2} = 257.20\text{ ft}$$

The tangent offset values are calculated as follows:

$$\text{Tangent offset at station } 62 + 00 = \frac{(100)^2}{(400)^2}(7.00) = 0.44\text{ ft}$$

$$\text{Tangent offset at station } 63 + 00 = \frac{(200)^2}{(400)^2}(7.00) = 1.75\text{ ft}$$

The elevations of the stations are computed by subtracting the tangent offsets from the grade elevations. These values are shown in Figure 23-3.

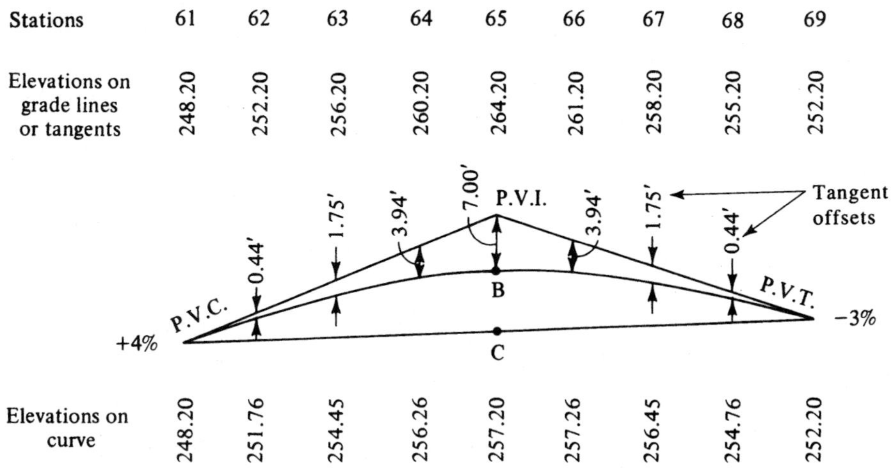

Stations	61	62	63	64	65	66	67	68	69
Elevations on grade lines or tangents	248.20	252.20	256.20	260.20	264.20	261.20	258.20	255.20	252.20
Tangent offsets		0.44′	1.75′	3.94′	7.00′	3.94′	1.75′	0.44′	
Elevations on curve	248.20	251.76	254.45	256.26	257.20	257.26	256.45	254.76	252.20

P.V.C. +4% P.V.I. B C P.V.T. −3% Tangent offsets

FIGURE 23-3 Vertical curve elevations for a crest curve.

In Table 23-1 all of the preceding information is recorded, and in addition, the first and second differences are computed as a math check.

TABLE 23-1 Vertical Curve Calculations Using Tangent Offsets

Station	Point	Elevation on grade lines	Tangent offset	Curve elevation	First difference	Second difference
61	P.V.C.	248.20	0.00	248.20		
					+3.56	
62		252.20	0.44	251.76		0.87
					+2.69	
63		256.20	1.75	254.45		0.88
					+1.81	
64		260.20	3.94	256.26		0.87
					+0.94	
65	P.V.I.	264.20	7.00	257.20		0.88
					+0.06	
66		261.20	3.94	257.26		0.87
					–0.81	
67		258.20	1.75	256.45		0.88
					–1.69	
68		255.20	0.44	254.76		0.87
					–2.56	
69	P.V.T	252.20	0.00	252.20		

Example 23-2 provides another vertical curve example. The calculations are made exactly as they were for the curve in Example 23-1. For all vertical curve problems it is necessary for the person making the calculations to be very careful to use the correct signs for determining the elevations from the tangents. He or she must be equally careful with the signs of the offset distances from these tangents, which are used to determine the curve elevations. In this particular case the offsets are measured up from the tangent or grade lines. Space will not be taken here to show the second difference check, which is nevertheless always desirable.

Example 23-2

A +3% grade line intersects a +5% grade line at station 62 + 00, where the elevation is 862.30 ft as shown in Figure 23-4. Determine the elevations at full 100-ft stations along the curve if a 600-ft long curve is to be used.

Solution

The elevations along the grade lines are computed and shown in the figure.

$$\text{Elevation of point } C = \frac{853.30 + 877.30}{2} = 865.30 \text{ ft}$$

$$\text{Elevation at midpoint of curve} = \frac{862.30 + 865.30}{2} = 863.80 \text{ ft}$$

The tangent offsets are computed and subtracted from the grade-line elevations, giving the final curve elevations as shown in the figure.

FIGURE 23-4 Elevations for a vertical curve where both tangents have a plus grade.

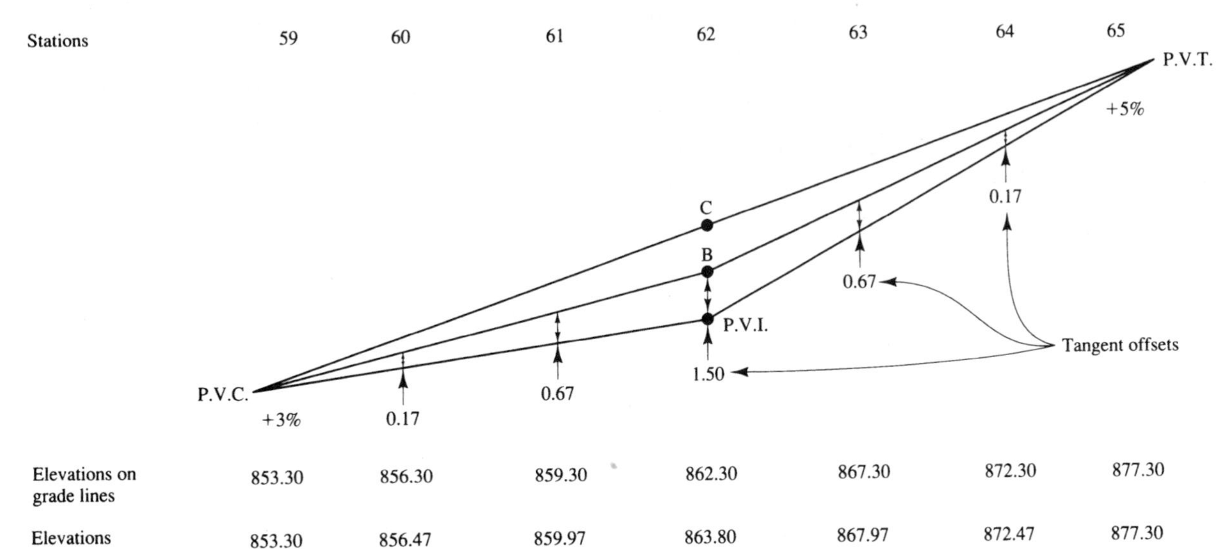

23-3 MISCELLANEOUS ITEMS RELATING TO VERTICAL CURVES

Elevation of Intermediate Points on Curves

It is often necessary to compute the elevation of points on vertical curves at closer intervals than full stations, for example, at 50 ft or 25 ft, as well as the values for some occasional intermediate point. For such cases as these, the tangent offset method will work as well as it did at full stations, although the numbers are not quite so convenient. As an illustration, the elevation of station 63 + 75.2 for the curve of Example 23-1 equals

$$248.20 + (.04)(275.2) - \frac{(275.2)^2}{(400)^2}(7.00) = 255.90 \text{ ft}$$

Highest or Lowest Point on Curve

When the two grade lines for a vertical curve have opposite algebraic signs, there will be either a low point or a high point between the P.V.C. and the P.V.T. This point will probably not fall on a full station, and yet its position will often be of major significance, as, for example, where drainage facilities are being considered.

For a particular vertical curve the total change in grade (A) between the tangents is determined from the following equation in which g_1 and g_2 are the percent grades of the back and forward tangents, respectively:

$$A = g_2 - g_1$$

The rate of change of grade r is determined by the following equation, in which L is the curve length in stations:

$$r = \frac{A}{L}$$

For this discussion it is assumed that a vehicle moves on to a curve that has an initial grade g_1. As the vehicle moves along the curve it will rotate vertically until finally at the far end it will be sloped at g_2. During the vehicle's rotation it will eventually become level. If it rotates at a rate of r percent per station, the number of stations to the level point (which is the high or low point) is calculated by dividing r into g_1:

$$X = \frac{-g_1}{r}$$

For the curve of Example 23-1, the station and elevation of the high point are determined as follows:

$$r = \frac{g_2 - g_1}{L} = \frac{-3 - 4}{8} = -\frac{7}{8}$$

$$X = -\frac{g_1}{r} = -\frac{4}{-7/8} = +4.57 \text{ stations} = (61 + 00) + (4 + 57) = 65 + 57$$

$$\text{Elevation at station } 65 + 57 = 264.20 - (0.03)(57) - \frac{(3.43)^2}{(4)^2}(7.00) = 257.34 \text{ ft}$$

Sight Distance and Curve Length

As mentioned previously, it is necessary for vertical curves to be so constructed that drivers will have certain minimum sight distances for seeing cars or other objects on the road. The driver should be able to see an object of a given height at no less than the estimated distance that he or she would travel while reacting to put his or her foot on the brake pedal plus the distance required for the car to stop.

The AASHTO Specification goes into great detail in describing minimum lengths needed for sag and crest vertical curves and minimum sight distances. The length of vertical curves is generally predicated on minimum sight distances measured from an assumed eye height of 3.50 ft above the road looking at an object with a 6 in. height in the road.

At least four different criteria are used in establishing the lengths of sag vertical curves. These are (1) headlight sight distance, (2) rider comfort, (3) drainage control, and (4) general appearance. The lengths selected for crest vertical curves are based upon safety, comfort, and appearance with some other special considerations. The reader can refer to the AASHTO for this detailed information.

23-4 UNEQUAL-TANGENT VERTICAL CURVES

Almost all vertical curves have equal-length tangents. All those considered so far in this chapter fall into this class. Sometimes, however, it is desirable to use unequal-tangent curves to make them fit unusual topographical situations. An unequal-tangent vertical curve consists of two equal tangent vertical curves each with a different r value (rate of change of grade). The end of the first curve (its P.V.T.) coincides with the start of the second curve (its P.V.C.). This point is referred to as the *compound vertical curvature point* or C.V.C. This point is shown in Figure 23-5.

The C.V.C. is located by drawing a straight line from the midpoint of the back tangent to the midpoint of the forward tangent (represented by the dashed line *DE* in Figure 23-5). The C.V.C. is located on this dashed line directly above or below the P. V. I. The two equal-tangent curves will be selected so as to be tangent to line *DE* at the C.V.C. As a result, there will be a smooth transition from the first curve to the second curve.

The elevation of the C.V.C. can be determined by proportions or it can be obtained from the grade of line *DE* as follows, noting the elevation of the P.V.I. and the percentages of the grade lines given in Figure 23-5. Here the back tangent is 600 ft long, while the forward tangent is 400 ft long.

$$\text{Elevation of } D = 642.40 - (0.04)(300) = 630.40\text{ ft}$$

$$\text{Elevation of } E = 642.40 - (0.02)(200) = 638.40\text{ ft}$$

$$\text{Grade of line } DE = \frac{638.40 - 630.40}{500} = +0.016 = +1.60\%$$

$$\text{Elevation of C.V.C.} = 630.40 + (0.016)(300) = 635.20\text{ ft}$$

Now it is possible to proceed with the calculations for the two curves noting that for the first curve the length is 600 ft, $g_1 = +4\%$ and $g_2 = +1.6\%$. For the second curve of 400-ft length, $g_1 = +1.60\%$ and $g_2 = -2\%$. In Example 23-3 that follows, elevations are determined along this unequal-tangent curve.

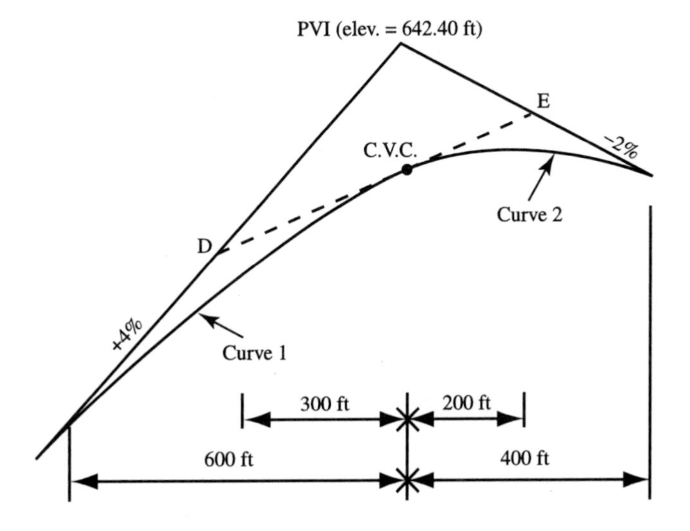

FIGURE 23-5 Compound vertical curvature (C.V.C.) point for an unequal-tangent curve.

Example 23-3

A +4% grade line intersects a –2% grade line at station 68 + 00, where the elevation is 642.40 ft, as shown in Figure 23-5. Determine elevations of 100-ft stations for this 1000-ft unequal-tangent curve.

Solution

The elevation of the C.V.C. and the percent grade of line *DE* determined before this example are actually the first part of this solution. The elevations along the grade lines are computed and shown in Table 23-2. In computing these elevations, tangent elevations from stations 62 to 65 are changing at +4%, while from stations 65 to 70 they are changing at +1.60%, and from stations 70 to 72 the change is –2%.

TABLE 23–2

	Station	Point	Elevation on grade lines	Tangent offset	Curve elevation	First difference	Second difference
Curve 1	62	P.V.C.$_1$	618.40	0.00	618.40	+3.80	
	63		622.40	0.20	622.20	+3.40	0.40
	64		626.40	0.80	625.60	+3.00	0.40
	65	P.V.I.$_1$(*D*)	630.40	1.80	628.60	+2.60	0.40
	66		632.00	0.80	631.20	+2.20	0.40
	67		633.60	0.20	633.40		
Curve 2	68	P.V.T.$_1$ and P.V.C.$_2$ (C.V.C.)	635.20	0.00	635.20	+1.15	
	69		636.80	0.45	636.35	+0.25	0.90
	70	P.V.I.$_2$(*E*)	638.40	1.80	636.60	–0.65	0.90
	71		636.40	0.45	635.95	–1.55	0.90
	72	P.V.T.$_2$	634.40	0.00	634.40		

Curve 1:

$$\text{Elevation of P.V.I.} = \text{elevation of D} = 630.40 \text{ ft}$$

Elevation of midpoint of straight line from station 62 to C.V.C.

$$= \frac{618.40 + 635.20}{2} = 626.80 \text{ ft}$$

$$\text{Elevation of midpoint of curve} = \frac{630.40 + 626.80}{2} = 628.60 \text{ ft}$$

Difference in elevation from P.V.I. to midpoint of curve

$$= 630.40 - 628.60 = 1.80 \text{ ft}$$

$$\text{Tangent offset at station 63} = \frac{(1)^2}{(3)^2}(1.80) = 0.20 \text{ ft}$$

Curve 2:

$$\text{Elevation of P.V.I.} = \text{elevation of E} = 638.40 \text{ ft}$$

Elevation of midpoint of straight line from station 68 to station 72

$$= \frac{635.20 + 634.40}{2} = 634.80 \text{ ft}$$

$$\text{Elevation of midpoint of curve} = \frac{638.40 + 634.80}{2} = 636.60 \text{ ft}$$

Difference in elevation from P.V.I. to midpoint of curve

$$= 638.40 - 636.60 = 1.80 \text{ ft}$$

$$\text{Tangent offset at station } 69 = \frac{(1)^2}{(2)^2}(1.80) = 0.45 \text{ ft}$$

23-5 VERTICAL CURVE PASSING THROUGH A CERTAIN POINT

A problem commonly faced in working with vertical curves is that of passing a curve through a definite point. For instance, as shown in part (a) of Figure 23-6, it may be desired to have a vertical curve pass a certain distance over the top of a culvert. The problem is to determine the correct length of a curve that will pass through the point in question. In a similar fashion a vertical curve on a bridge may need to pass a certain distance for clearance over an existing roadway, railroad, or navigable stream as shown in part (b) of the figure. To solve the problem, the tangent offset equation is written from both sides of the curve using the length L as one unknown and the tangent offset at the P.V.I. as the other. The two equations may be solved simultaneously for the unknown values, as illustrated in Example 23-4.

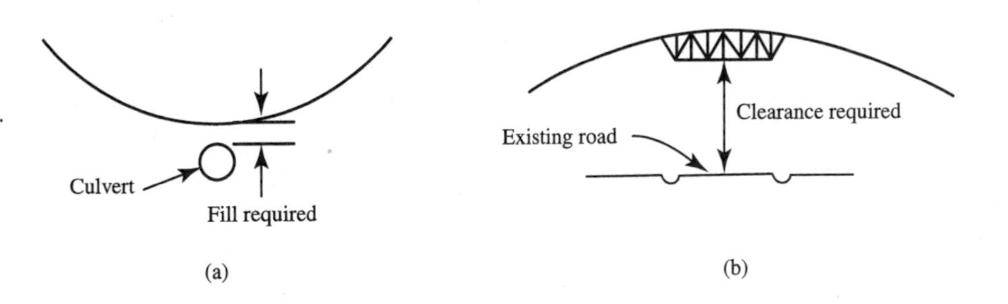

FIGURE 23-6 Vertical curves passing through points of given elevations.

Example 23-4

A −6% grade line intersects a +5% grade line at station 11 + 00 (elevation 43.00 ft). The top of a 4 ft culvert oriented at 90° to the road center line at station 10 + 00 is to be located at elevation 53.00 ft. Determine the vertical curve length in whole stations such that there will be from 1.0 to 3.0 ft of cover material over the top of the culvert.

Solution

Assume the material covering the top of the culvert is 2.0 ft thick and determine the theoretical curve length required. Thus the desired elevation on the curve at that station (10 + 00) is 53.00 + 2.00 = 55.00 ft.

With reference to Figure 23-7, equations for the offsets at station 10 + 00 are written with respect to both tangents and labeled y_{10}.

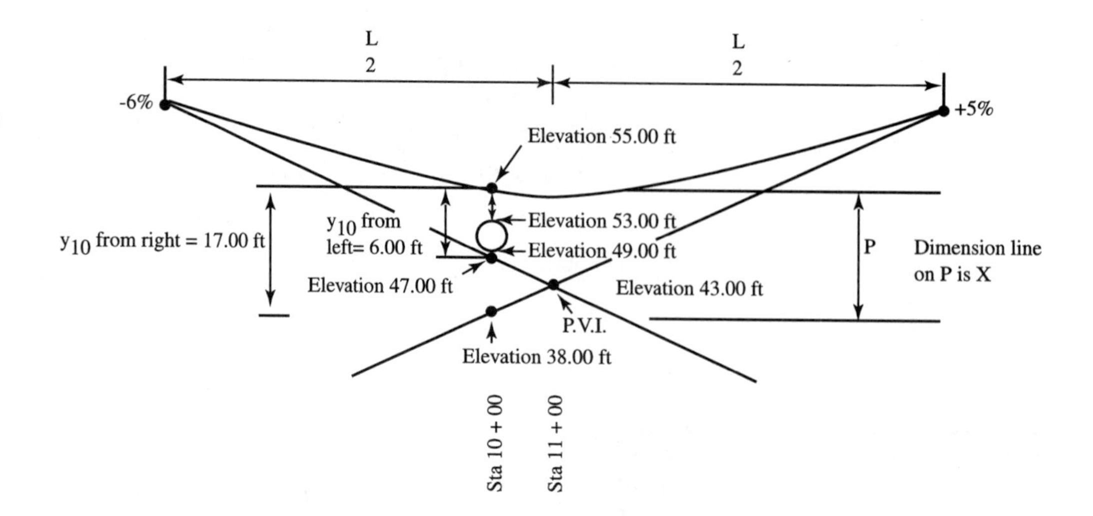

FIGURE 23-7 Vertical curve designed to provide 2 feet of fill above an existing culvert.

1. The vertical distance from the back tangent to the desired curve elevation = 55.00 – 49.00 = 6.00 ft = y_{10}.
2. The distance from the forward tangent (extended to station 10 + 00) to the desired curve elevation = 55.00 – 38.00 = 17.00 ft = y_{10}.

In the following two offset equations, P is the tangent offset at the P.V.I. and L the desired curve length in stations.

For the back tangent:

$$\frac{6.00}{P} = \frac{\left(\frac{L}{2} - 1.00\right)^2}{\left(\frac{L}{2}\right)^2} \qquad P = \frac{6.00\left(\frac{L}{2}\right)^2}{\left(\frac{L}{2} - 1.00\right)^2} \tag{1}$$

For the forward tangent:

$$\frac{17.00}{P} = \frac{\left(\frac{L}{2} + 1.00\right)^2}{\left(\frac{L}{2}\right)^2} \qquad P = \frac{17.00\left(\frac{L}{2}\right)^2}{\left(\frac{L}{2} + 1.00\right)^2} \tag{2}$$

Equating the values for P [Eqs. (1) and (2)] and solving for L yields

$$L = 4.83 \text{ stations}$$

Use $L = 5$ stations and calculate elevations along the curve, making sure that the culvert cover is between 1 and 3 ft.

23-6 PARABOLIC EQUATION

The elevations of points along a vertical curve can be determined using the equation of a parabola instead of the tangent offset method. To use the parabolic equation that follows, a coordinate system is set up with x = the horizontal distance from the P.V.C. to a particular point, y = the elevation of the point in question in feet, and r = the rate of grade change = $(g_2 - g_1)/L$:

$$y = \frac{1}{2}rx^2 + g_1x + \text{elevation of P.V.C.}$$

In Example 23-5 the elevations along the curve of Example 23-1 are recalculated using the preceding equation.

Example 23-5

Repeat Example 23-1 using the parabolic equation.

Solution

$$\text{From Example } 23-1, g_1 = +4\%, g_2 = -3\%, \text{and } L = 8 \text{ stations}$$
$$\text{Elevation of P.V.C.} = 264.20 - (0.04)(400) = 248.20 \text{ ft}$$

Being very careful with signs, the value of r is determined.

$$r = \frac{-3 - (+4)}{8} = -\frac{7}{8}$$

The elevations at full stations are computed and shown in Table 23-3.

TABLE 23-3 Vertical Curve Calculations with Parabolic Equation

Station	x (stations)	$1/2\ rx^2$	g_1x	$y = 1/2\ rx^2 + g_1x +$ elevation P.V.C.
61	0	0.000	0.0000	248.20
62	1	−0.4375	4.0000	251.76
63	2	−1.7500	8.0000	254.45
64	3	−3.9375	12.0000	256.26
65	4	−7.0000	16.0000	257.20
66	5	−10.9370	20.0000	257.26
67	6	−15.7500	24.0000	256.45
68	7	−21.4380	28.0000	254.76
69	8	−28.0000	32.0000	252.20

23-7 COMPUTER EXAMPLE

Example 23-6

Repeat Example 23-2 using the SURVEY program on the enclosed CD-ROM.

Solution

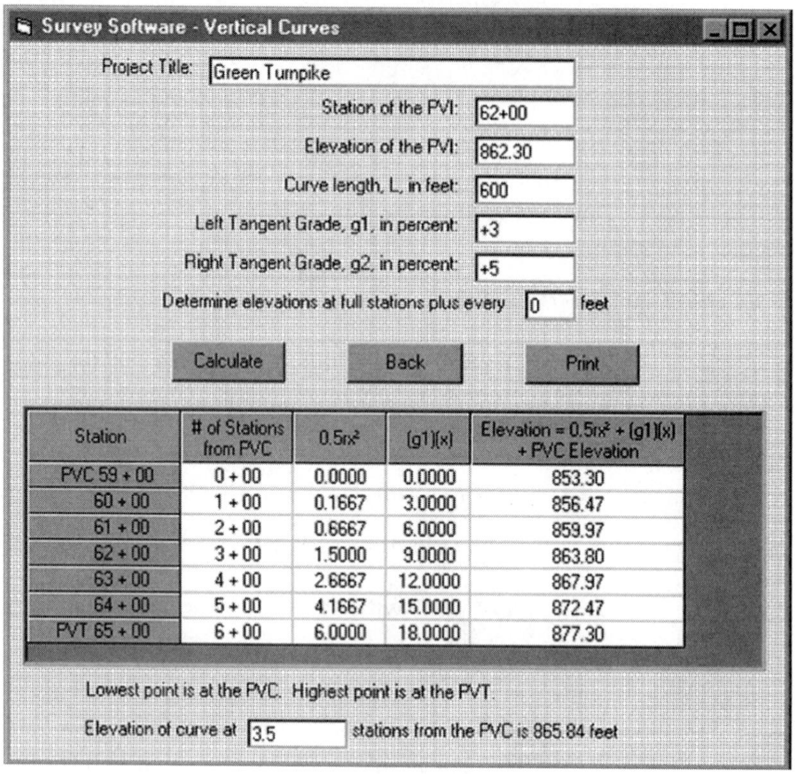

FIGURE 23-8 Computer printout.

Survey Software - Vertical Curves

Project Title: Green Turnpike

Station of the PVI: 62+00

Elevation of the PVI: 862.30

Curve length, L, in feet: 600

Left Tangent Grade, g1, in percent: +3

Right Tangent Grade, g2, in percent: +5

Determine elevations at full stations plus every 0 feet

Calculate Back Print

Station	# of Stations from PVC	0.5rx²	(g1)(x)	Elevation = 0.5rx² + (g1)(x) + PVC Elevation
PVC 59 + 00	0 + 00	0.0000	0.0000	853.30
60 + 00	1 + 00	0.1667	3.0000	856.47
61 + 00	2 + 00	0.6667	6.0000	859.97
62 + 00	3 + 00	1.5000	9.0000	863.80
63 + 00	4 + 00	2.6667	12.0000	867.97
64 + 00	5 + 00	4.1667	15.0000	872.47
PVT 65 + 00	6 + 00	6.0000	18.0000	877.30

Lowest point is at the PVC. Highest point is at the PVT.

Elevation of curve at 3.5 stations from the PVC is 865.84 feet

23-8 CROWNS

In establishing the elevations for the construction of highways and streets, two factors in addition to the vertical curve values can affect the elevations at particular points. These factors, which are discussed in this section and the next, are *crowns* and *superelevations.*

To provide adequate drainage for the pavement surface, it is necessary to raise the center of the pavement with respect to the edges. Crowns or cross slopes usually run from about 0.015 to 0.020ft/ft for heavy-duty pavements. For lower-duty pavements the slopes may be slightly higher.

The crown can be formed by two planes, but more commonly it has a parabolic or circular shape. Since the width of the pavement is large compared to the height of the crown, there is almost no difference between a circle and a parabola. The parabolic vertical curve formula can be used to calculate the amount of crown at a particular point with respect to the pavement edge, as shown in Figure 23-9.

23-9 SUPERELEVATION

For horizontal curves, highway and street pavements are built with a transverse slope called *superelevation* that is usually given as a percent. A fairly common value

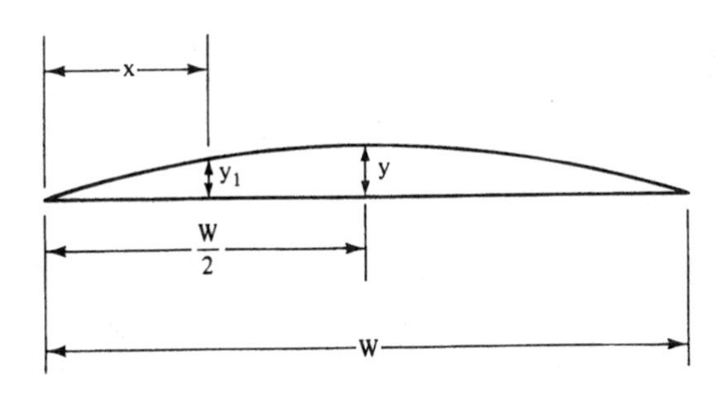

FIGURE 23-9 Pavement cross section showing pavement crown.

used is 8%. This means that a 24-ft-wide pavement would have its outside edge = (0.08)(24) = 1.92 ft higher than its inside edge. A superelevated curve is often said to be *banked.*

The amount of superelevation used is controlled by the design speed of the highway and the side friction factor between tires and pavement. The maximum amount that can be used practically depends on the climate and the classification of the area as urban or rural. The maximum amount usually encountered is about 10%. Sometimes, however, it will be a little higher—perhaps up to 12% for highways not subject to appreciable snow and ice and also for infrequently traveled gravel roads to help with cross drainage. These values are normally suited for the southern states. Situations where combinations of design speeds and degrees of curvature require greater superelevations than the values given here should be avoided.

In parts of the country where snow and ice conditions occur frequently over a period of several months (usually, the northern states), maximum rates of 7% to 8% are used to prevent vehicles from slipping sideways across the pavement when they are stopped or moving slowly. For urban streets, where there may be considerable traffic and where speeds often have to be reduced, maximum superelevations of 4% to 6% are common. Sometimes where there are long radius or flat horizontal curves with considerable turning and crossing of traffic, superelevation may be omitted.

Some highways have both fast and slow traffic, and varying weather situations occur during the different seasons. These facts make it impossible to select superelevations that are appropriate for all traffic situations and weather conditions.

There is a transitional area between a normal crown cross section and full superelevation where traffic is on a tangent section of the highway approaching a horizontal curve. On this straight section we will start gradually working into the superelevation and work our way up to the maximum superelevation value, as shown in Figure 23-10. In a similar fashion we will work our way from full superelevation to the normal cross section on the other end. Notice in this figure how the pavement cross section is revolved around the pavement center line. (Sometimes for divided highways with narrow medians the pavement may be revolved around one of its edges.)

The transition from the normal crown cross section on a tangent to the fully superelevated cross section is referred to as *runoff.* The purpose of runoff is to spread the transition between the two different pavement cross sections over a sufficient distance so as to provide a safe design. Runoff lengths probably range up to as much as 500 or 600 ft. Minimum runoff values vary from about 100 to 250 ft, regardless of superelevation values or pavement widths. These minimum values are used to improve the appearance of the highway and to smooth the edge profile of the pavement.

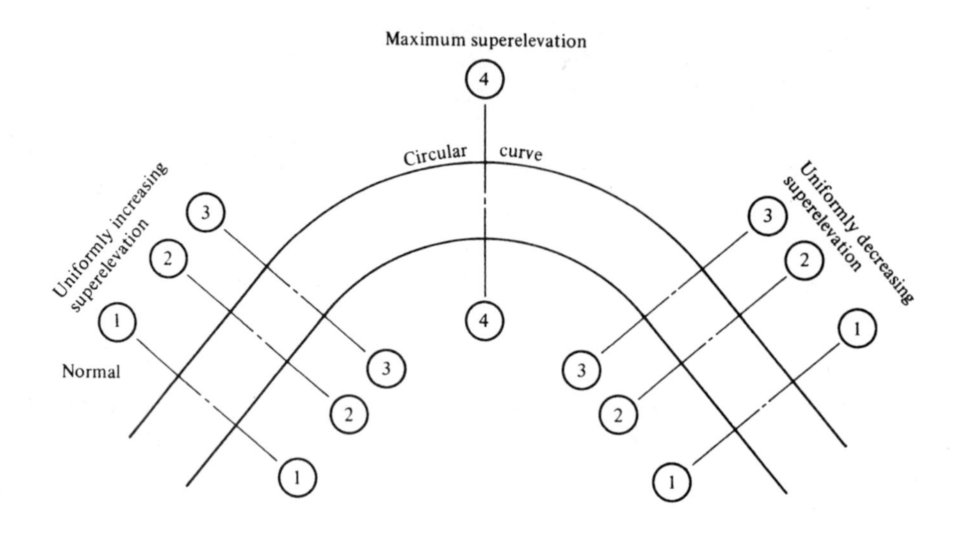

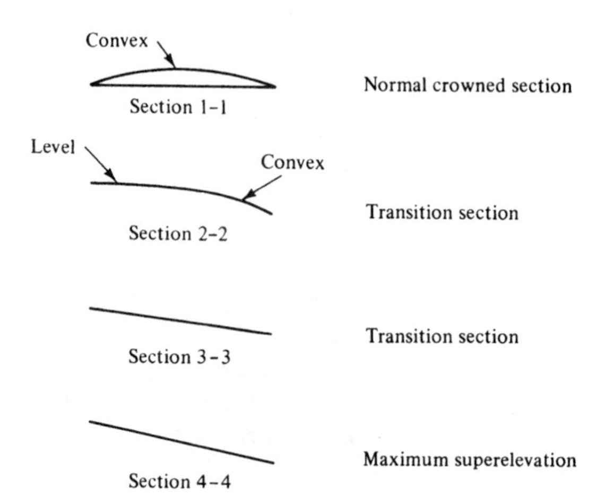

FIGURE 23-10 Superelevation for a highway curve.

Problems

23-1. A 1000-ft vertical curve is to be used for joining a +4.0% grade line and a –3.0% grade line. The vertex or intersection of the grade lines is at station 62 + 00 and the elevation there is 462.75 ft. Compute the elevations at 100-ft stations throughout the curve.

(Ans.: Curve elevation at stations 58, 60, and 64 = 446.40 ft, 451.60 ft, 453.60 ft)

23-2. Repeat Problem 23-1 if the curve is 800 ft long and the first grade is +3.6% instead of +4.0%.

23-3. A +4.0% grade meets a –2.2% grade at station 46 + 50, where the elevation is 469.88 ft. Compute the elevation of all full stations if the curve is to be 800 ft long.

(Ans.: Curve elevation at stations 44, 46 + 50, and 49 = 459.00 ft, 463.68 ft, and 463.50 ft)

23-4. Repeat Problem 23-3 if the first grade is +4.0% and the second +1.8%.

23-5. A –3.2% grade line meets a +4.10% grade line at station 103 + 00. Elevation = 1424.44 ft. Determine the elevation of all full and half stations if a 650-ft curve is to be used. (Ans.: Curve elevations at stations 102 and 104 + 50 = 1430.48 ft and 1432.31 ft)

23-6. Compute the location and elevation of the high point of the curve of Problem 23-1.

23-7. At what stations is the elevation of the curve of Problem 23-1 equal to 446.80 ft? (Ans.: 58 + 12)

23-8. A grade of +4.0% passes station 36 + 00 at an elevation of 622.80 ft. A grade of –5.0% passes station 52 + 00 at elevation 623.40 ft. Compute the station and elevation of the vertex or P.V.I. of these grades.

In Problems 23-9 to 23-14, compute the elevations at 100-ft stations throughout the given curves.

	Initial grade (%)	Final grade (%)	Station of P.V.I.	Elevation of P.V.I. (ft)	Length of curve (ft)
23-9.	–3	+5	62 + 00	600.60	800
23-10.	–6	+2	103 + 00	354.42	1000
23-11.	–4	–2	31 + 00	1208.80	1200
23-12.	–3	–5	43 + 00	632.64	600
23-13.	+2	+4	91 + 00	1442.84	600
23-14.	+4	+3	52 + 50	282.60	800

(Ans.: 23-9. Curve elevations at stations 59 and 64 = 610.10 ft and 612.60 ft)
(Ans.: 23-11. Curve elevations at stations 27 and 36 = 1225.13 ft and 1199.88 ft)
(Ans.: 23-13. Curve elevations at stations 90 and 92 = 1441.51 ft and 1447.51 ft)

23-15. A +3% grade intersects a –2% grade at station 0 + 550 m at elevation 322.200 m. Compute the elevation at 50 m stations for a 600-m curve. (Ans.: Curve elevations at stations 0 + 450 and 0 + 700 = 317.53 m and 318.26 m)

23-16. A +4% grade intersects a +2% grade at station 1 + 320 m at elevation 264.44 m. Determine the elevations at 20-m stations for a 160-m long curve.

23-17. A –6% slope meets a +2% grade at elevation 436.66 ft at station 61 + 00. Assuming a 1200-ft vertical curve, determine the elevation on the curve at each 100-ft station and find the station and elevation at the low point on the curve. (Ans.: Low point at 64 + 00, elevation 445.66 ft)

23-18. A –2.4% grade intersects a +3.2% grade at station 103 + 00 at an elevation of 153.20 ft. If a 1200-ft long vertical curve is to be used to connect the tangent grades, determine the elevation at stations 99 + 00 and 104 + 00.

23-19. A +4% grade (600 ft long) intersects a –2% grade line (400 ft long) at station 52 + 00, where the elevation is 462.90 ft. Determine elevations at 100-ft stations for this 1000-ft unequal tangent curve.
(Ans.: Elevation at station 50 = 451.70 ft, at station 53 = 456.85 ft)

23-20. A –4% grade line (400 ft length) intersects a –2% grade line (200 ft length) at station 63 + 00, where the elevation is 949.50 ft. Determine elevations at 100-ft stations for this 600-ft unequal-tangent curve.

23-21. A –4% downgrade intersects a +3% upgrade at elevation 120.00 ft at station 67 + 00. It is desired to pass a vertical curve through a point of elevation 128.00 ft at station 68 + 00. Determine the curve length required.
(Ans.: $L = 928.3$ ft)

23-22. A –4% grade intersects a +3% grade at station 62 + 00 (elevation = 642.70 ft). Determine the elevations at 100-ft stations for an 800-ft long curve using the parabolic equation.

23-23. Repeat Problem 23-11 using the parabolic equation.
(Ans.: Elevation at station 32 + 00 = 1208.88 ft)

For Problems 23-24 to 23-26, repeat the problems given using the SURVEY program provided with this book.

23-24. Problem 23-2

23-25. Problem 23-3 (Ans.: Elevation at station 49 + 00 = 463.51 ft)

23-26. Problem 23-13

CHAPTER 24

Surveying—The Profession

24-1 SURVEYING LICENSES

The purpose of building codes, medical codes, surveying registration requirements, and so on, is to protect the public from unqualified and/or unscrupulous people in these fields. Perhaps the first regulations of this type were contained in the code of laws of Hammurabi, who was king of Babylon in the eighteenth century B.C. His code covered many different subjects (including prohibition), but only his building code is mentioned here. It was famous for its "eye for an eye and tooth for a tooth" section. In general, the code said that if a builder constructed a house that collapsed and killed the owner, the builder would be killed. If the collapse caused the death of the owner's son, the builder's son would be killed, and so on. Those laws were presumably very effective in making builders put forth their best efforts.

Through the centuries since the time of Hammurabi many codes and laws have been established to guide the various professions, but it was not until 1883 that registration requirements reached the United States. At that time registration of dentists began and as the years went by doctors, lawyers, pharmacists, and others were required to obtain licenses before they could practice their professions.

Licensing of surveyors is not new. George Washington and Abraham Lincoln held surveyor's licenses, but licensing as it is known today for engineers and surveyors was begun in Wyoming in 1907. Today, the law in all 50 states, as well as the District of Columbia, Canada, Puerto Rico, and Guam, requires that a person must meet certain licensing requirements before practicing land surveying. For the purposes of such licensing, land surveying is generally said to include the determination of areas of tracts of land, the surveying needed for preparing descriptions of land for deed conveyance, the surveying necessary for establishing or reestablishing land boundaries, and the preparation of plats for land tracts and subdivisions. *A license is usually not required for construction surveys and for route surveys (roads, railroads, pipelines, etc.) unless property corners are set.*

24-2 REGISTRATION REQUIREMENTS

To obtain a land surveyor's license, it is necessary for a person to meet the requirements of the applicable state board of examiners. Registration for engineers and registration for land surveyors are handled in most states by a single board, but fourteen states have a separate land surveying board. Although these requirements vary considerably from state to state, they in general require (1) graduation from a school or college approved by the state board, including the completion of several approved courses in surveying, (2) two or more years of surveying experience of a character suitable to the board, and (3) successful completion of written examinations under the supervision of the state board. If the applicant is unable to meet the formal education requirements listed in (1), he or she will probably be required to obtain additional experience (perhaps as much as four or more years) before being permitted to take the written exam. The reader must realize, however, that most states are gradually increasing their formal education requirements. As a result, the door to registration through experience and exams alone is nearly closed.

In 1973 the National Council of Examiners for Engineering and Surveying (NCEES) began offering semiannual national surveying exams. Today they provide an 8-hour Land Surveyor-in-Training exam (LSIT). After a person passes this exam and works under a practicing Registered Land Surveyor for a period of years specified in each state, they must take additional exams to become licensed. These exams are a 6-hour Principles and Practice of Land Surveying exam (PLS) offered by the NCEES and an approximately 2-hour exam offered by the individual states. This last exam is given because the laws pertaining to land description are different from state to state.

Almost every state uses the national exams, with the result that it is easier for a person registered in one state to become registered in another. Eventually, it is hoped that a surveyor who passes the national exam in one state will be permitted to obtain with little or no delay and few additional requirements a license in any other state. Such reciprocity is a reality for professional engineering registration as a result of the national exams given by the NCEES in which all states participate.

A large and increasing number of states now require that registered land surveyors obtain a certain number of continuing education units (CEUs) each year to maintain their licenses. These units may be obtained in various ways, including the following:

1. Successful completion of applicable continuing education courses.
2. Attendance at workshops or professional or technical seminars.
3. Completion of college courses related to surveying.
4. Teaching in 1–3 above.
5. Publishing papers, articles, or books related to surveying.

24-3 PENALTIES FOR PRACTICING SURVEYING WITHOUT A LICENSE

The various states have laws prohibiting a person from practicing land surveying without a license, from using a false or forged license or registration number for the same, or from using an expired license. These laws usually call for fines or imprisonment, or both, for such violations. Of course, the ordinary person can measure land as long as he or she neither represents himself or herself as a registered surveyor nor uses or permits the measurements made to be used for deed conveyance or other formal legal purposes.

A surveyor is expected to be competent in his or her work. The person cannot be perfect, of course, but if grossly negligent, incompetent, or guilty of misconduct, that person may lose his or her license. Furthermore, the state board may revoke a land surveyor's license if it discovers that there was any fraud or deceit involved in obtaining the license.

24-4 REASONS FOR BECOMING REGISTERED

There are several reasons why a person interested in surveying should work toward becoming licensed as soon as possible. These include the following:

1. The desire for obtaining a license encourages a person to study and improve his or her technical ability and aids in his or her professional development.
2. A surveyor may be offered a job that requires registration.
3. Registration raises the status of the profession as a whole.
4. If a surveyor is registered, he or she may be able to perform part-time professional surveying or even enter into full-time practice in his or her own business.
5. Registration gives a person status as a professional in his or her community.

24-5 A PROFESSION

The term *profession* has many definitions. In a narrow sense the professions are limited to doctors, lawyers, engineers, and clergymen, but in a broad sense they include almost any occupation. *In general, a professional is a person who has acquired some special knowledge that he or she uses to instruct, help, advise, or guide others.*

The primary objective of a profession is service to humanity without regard to financial reward. Not only does this mean that a professional has a responsibility to place his or her job above the amount of money he or she is to receive, but it also means that the person has a responsibility to give voluntary service to the community when needed without any financial compensation. The surveyor should clearly understand that a professional who performs free services (or acts as a gratuitous agent) is legally required to exercise the same high level of care, skill, and judgment that the person would if he or she were being paid.

A true profession is said to have four basic elements: *organization, education, experience,* and *exclusion.* Organization means membership and participation in a professional organization. For surveyors, this organization might be the American Congress on Surveying and Mapping (ASCM) and/or the state's surveying society. The young surveyor might feel that the dues for these organizations are too high, but membership in them is a first step toward obtaining the recognition and status of a true professional. As the old saying goes, "You get out of an organization what you put into it," and such participation will lead to that most important feeling of belonging.

Generally, education means the completion of as many surveying courses as possible and obtaining formal school degrees, but it can be, and often is, *self-education.* In addition, education must be continuing through self-study, short courses, and perhaps more formal study. Experience is obtained over the years and is a gradual transformation obtained by undertaking specific tasks.

A real profession should require the strict exclusion of those who are unfit or unworthy. Exclusion may be accomplished by the state surveying licensing requirements and by a code of ethics or code of professional conduct. Undoubtedly, this has been a weak point in the surveying profession throughout most of the country—the machinery for expelling registered land surveyors for incompetence or unethical behavior is not often applied.

24-6 CODE OF ETHICS

The most important possession a person owes himself or herself is a spotless reputation. No amount of money, fame, or knowledge is an adequate substitute. A person who can be depended on and who will work hard and learn as much about a profession as he or she can is going to be successful.

The reader is well aware that a person may live strictly within the confines of the law and yet be an undesirable citizen. Sadly, there are always people who will operate up to the very limits of and through any loopholes they can find in the law. But to a true professional, a legal right is not a right unless it is also morally right.

Ethics may be defined as the duties that a professional owes to the public and to fellow professionals. Such a code is a helpful guide to the members of a profession to help them know what standards they should live up to and what they can expect from their fellows.

The earliest known code of ethics is the Hippocratic oath of the medical profession, which is attributed to Hippocrates, a Greek physician (460–377 B.C.). The present form of the Hippocratic oath dates from approximately A.D. 300, but a detailed code of ethics for the medical profession did not appear in the United States until 1912. The code of ethics for the American Society of Civil Engineers was adopted in 1914.

A code of ethics is not intended to be a lengthy detailed statement of "thou shall nots," but rather a few general statements of noble motives expressing concern for the welfare of others and the standing of the profession as a whole. The classic comparison of law and ethics is made by relating them to medicine and hygiene. The object of medicine is to cure diseases, and the object of law is to cure or repress evil. The object of hygiene is to prevent illness, and the object of ethics is to prevent evil by raising the moral plane on which people deal with each other.

Another classic comparison of law and ethics is frequently made with relation to the Old and New Testaments. Essentially, the Old Testament is a list of "thou shall nots," such as "thou shall not kill, thou shall not steal," and so on; the New Testament is based on a concern for the welfare of our fellows and a desire to do good because it is a privilege and not a necessity. Many groups of land surveyors, primarily state societies, have published codes of ethics. In general, these codes are similar to each other and to the codes of the various engineering societies.

Perhaps the purpose and heart of a code of ethics may be summarized in a few sentences: The surveyor must faithfully and impartially perform work with fidelity to clients, employers, and the public. (For instance, when locating a property corner, its position will be the same regardless of which property owner is paying the surveyor.) The person will be seriously concerned with the standing of the profession in the public's eye and will not only strive to live and work according to a high standard of behavior, but, in addition, will avoid association with persons or enterprises of questionable character. In other words, the person will be concerned not only with wrongdoing but with the very appearance of wrongdoing. Further, he or she will be actively interested in the welfare of the public and will always be ready to apply his or her knowledge for the benefit of humankind.

When a Roman soldier encountered a Jew in biblical times, he could require the Jew to carry his pack for a mile. Jesus said (Matthew 5:41) that he should carry it for 2 miles (thus the origin of the expression "go the extra mile"). This idea is the theme of a code of ethics.

Instead of reproducing one of the codes of ethics, the author is devoting the remainder of this section to the following general statements that summarize what a code of ethics means to him:

1. The surveyor must not place monetary values above other values. Although it may seem difficult to apply this to specific cases, it simply means that the surveyor should never recommend to a client a course of action that is based on the amount of money that the surveyor will thereby receive.
2. In the course of work the surveyor may very well acquire knowledge that could be detrimental to the client if it were revealed to others. The surveyor's responsibility to the client goes beyond the immediate job and he or she must not reveal private information concerning the client's business without the client's permission.
3. In concern for the reputation of the profession, the surveyor must refrain from speaking badly of other surveyors, or he or she will be lowering the profession in the eyes of the public. This does not mean that there is not a time and place for an honest appraisal of other surveyors. A surveyor is far better able to judge the work of a fellow surveyor than is anyone else, whether lawyer, judge, or layperson.
4. In further concern for the standing of the profession, the surveyor must not become professionally associated with surveyors who do not conform to the standards of ethical practice discussed in this section. Furthermore, the surveyor will not become involved in any partnership, corporation, or other business group that is a cloak for unethical behavior. The surveyor must accept full responsibility for his or her work.
5. The surveyor will not be too proud to admit that he or she needs outside advice in order to solve a particular problem.
6. The surveyor will admit and accept his or her own mistakes.
7. When on a salaried job the surveyor will not do outside work to the detriment of his or her regular job. Further, the surveyor will not use such a job to compete unfairly with surveyors in private practice.
8. The surveyor will not agree to perform free surveying (except for community service) because he or she would be taking work away from the profession.
9. The surveyor will not advertise in a self-laudatory or blatant manner or in any other way that might be detrimental to the dignity of the profession.
10. The surveyor has a duty to increase the effectiveness of the profession by cooperating in the exchange of information and experience with other surveyors and students, by contributing to the work of surveying societies, and by doing all he or she can to further the public's knowledge of surveying.
11. The surveyor will encourage employees to further their education, to attend and participate in professional meetings, and to become registered. The surveyor will do all he or she can to provide opportunities for the professional development advancement of surveyors under his or her supervision.
12. The surveyor will not review the work of another surveyor without the knowledge or consent of that surveyor or unless the work has been terminated and the other surveyor has been paid for same.
13. The surveyor will always be greatly concerned with the safety and welfare of the public and his or her employees.
14. The surveyor will conform with the registration laws of the state.

Many people say that codes of ethics have not accomplished their purpose. The author likes to think they have helped because they give surveyors a goal to strive for and because they make them look toward a higher moral plane. The late Winston Churchill said that the flame of religious ethics is our highest guide and that to guard it and cherish it is our first interest in the world.

24-7 TO BE CLASSED AS A PROFESSIONAL

For surveying to be truly classed as a profession, the average standing of the group as a whole must be raised in the eyes of the public. This can only be done by having the title bestowed upon the group by the public because the public recognizes the group's high level of technical and ethical performance.

Although surveyors and engineers are highly respected by the community, they have not fully arrived as a profession in everyone's eyes. As a matter of fact, only doctors, lawyers, and clergymen have achieved this status. These three groups do not have to call themselves professional doctors or professional lawyers or professional clergymen because everyone recognizes them as professionals. But surveyors and engineers have persuaded government bodies to legislate titles for them as "professional surveyors" and professional engineers."

No amount of self-proclamation or legislative action can achieve a true title as a profession. It can only be obtained by the actions of the group over a long period of time.

24-8 CONCLUSION

This book is an introduction to surveying and those planning to follow the subject as a career must continue their studies. Only a small percentage (if any at all) of those entering the surveying field have a completely adequate background for the work that they will face. The answer for most lies in many hours of *self-study.*

Problems

24-1. List four reasons for becoming a registered surveyor.

24-2. What is the primary objective of a profession?

24-3. What are the four basic elements of a profession?

24-4. What is the purpose of a code of ethics?

24-5. Why are codes of ethics often compared to hygiene or the New Testament of the Bible?

24-6. Let's assume that when you graduate from school you receive an offer to work for ACX, Inc., and you agree to take the job. Two days later you receive an offer of employment you much prefer to take with Surveying Unlimited. Discuss the steps you can ethically take (if any) in an attempt to obtain the second job.

24-7. Suppose that you have a job with a surveying company during regular working hours. Another surveying company asks you to work for them at times other than your regular employment. Discuss the conditions under which you can ethically do this extra work.

24-8. You are asked to look over the work of a fellow surveyor. What procedure should you follow?

24-9. Suppose that you are working for a state highway department measuring quantities of earthwork, paving, and so on, for payment to a contractor and the contractor gives you a Christmas present of several thousand dollars. What should you do?

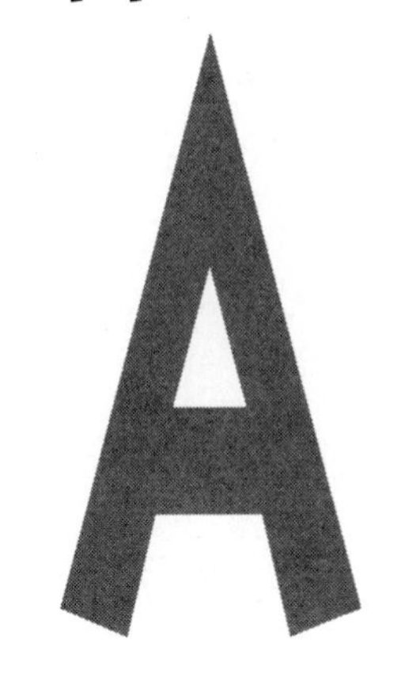

Appendix A

Some Useful Addresses

American Congress of Surveying and Mapping
5410 Grosvenor Lane
Bethesda MD 20814-2122
http://www.survmap.com

American Society of Civil Engineers
Surveying and Mapping Division
1801 Alexander Bell Drive
Reston VA 20191
http://www.asce.org

American Society for Photogrammetry and Remote Sensing
5410 Grosvenor Lane
Suite 210
Bethesda MD 20814-2122
http://www.asprs.org

Earth Science Information Center
509 National Center
Reston VA 22092
http://www.usgs.gov

GPS World
Advanstar Communications Inc.
131 West First Street
Duluth MN 55806-2065
http://www.editorial-gpsworld.com

National Council of Examiners for Engineering and Surveying
280 Seneca Creek Road
Clemson SC 29633-1686
http://www.ncess.org

National Geodetic Survey Division
National Geodetic Information Branch N/CG17
1315 East-West Highway, Room 9218
Silver Spring MD 20910-3282
http://www.mgs.nooa.gov

USGS Geography
Eastern Region Mapping
National Center
12201 Sunrise Valley Drive
Reston VA 22092
http://www.erg.usgs.gov

POB Publishing Company
PO Box 7069
Troy MI 48007
http://www.pobonline.com

Professional Surveyor
Suite 501
2300 Ninth Street
South Arlington VA 22204-2320
http://www.profsurv.com

USGS Geography
Rocky Mountain Mapping Center
Box 25046
MS 968
Denver Federal Center
Denver CO 80225-0046
http://www.rmncweb.cr.usgs.gov

Appendix B

Baccalaureate Degree Programs in Surveying

As of 1996 there were 13 ABET-accredited four-year surveying programs in the United States.[1] They are as follows:

- Alaska, University of Anchorage - Ph. (907) 786-6430; Fax (907) 786-6429
- California State Polytechnic University, Pomona - Ph. (909) 869-2645; Fax (909) 869-4370
- California State University, Fresno - Ph. (209) 278-4827; Fax (209) 278-2889
- East Tennessee State University, Johnson City - Ph. (423) 439-5497; Fax (423) 439-6200
- Ferris State University, Big Rapids, MI - Ph. (616) 592-2360; Fax (616) 592-2931
- Florida, University of, Gainesville - Ph. (904) 392-9492; Fax (904) 392-4957
- Michigan Technological University, Houghton - Ph. (906) 487-2259; Fax (906) 487-2583
- New Jersey Institute of Technology, Newark - Ph. (201) 596-5808; Fax (201) 242-1827
- New Mexico State University, Las Cruces - Ph. (505) 646-5375; Fax (505) 646-3549
- New York, Alfred State College, S.U.N.Y., Alfred - Ph. (800) 4 ALFRED or (607) 587-4698; Fax (607) 587-4620
- Oregon Institute of Technology, Klamath Falls - Ph. (800) 422-2017; Fax (541) 885-1687
- Purdue University, West Lafayette, IN - Ph. (317) 494-2166; Fax (317) 496-1105
- Wisconsin, University of, Madison - Ph. (608) 262-3542; Fax (608) 262-5199

Many other schools offer surveying programs (called *geomatics* in some institutions), including schools of civil engineering, civil technology, forestry, and two-year technical schools.

[1] J.K. Crossfield, "The Future of Surveying Education" *P.O.B. Magazine*, May 1996, vol. 21, no. 6, pp. 30–44.

Appendix

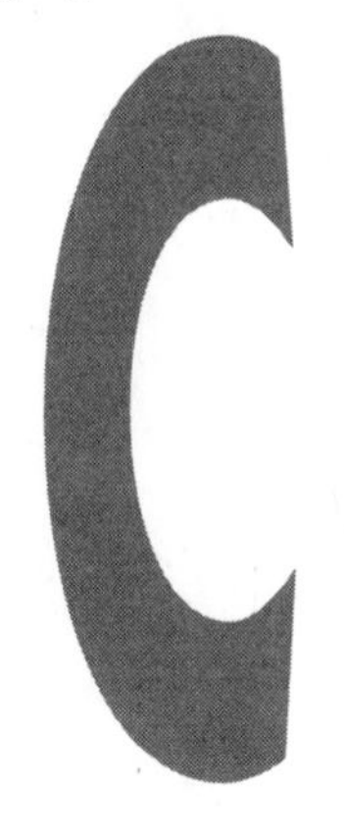

Appendix C Some Useful Formulas

TABLE C-1 Trigonometric Formulas for the Solution of Right Triangles

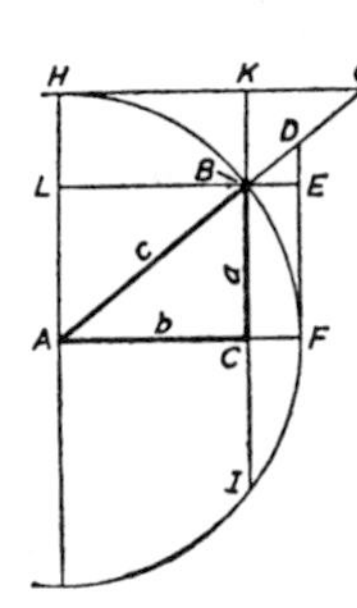

Let A = angle BAC = arc BF, and let radius $AF = AB = AH = 1$. Then,

$\sin A = BC$ $\quad \csc A = AG$

$\cos A = AC$ $\quad \sec A = AD$

$\tan A = DF$ $\quad \cot A = HG$

$\text{vers } A = CF = BE$ $\quad \text{covers } A = BK = LH$

$\text{exsec } A = BD$ $\quad \text{coexsec } A = BG$

$\text{chord } A = BF$ $\quad \text{chord } 2A = BI = 2BC$

In the right angled triangle ABC, let $AB = c$, $BC = a$, $CA = b$. Then,

1. $\sin A = \dfrac{a}{c}$
2. $\cos A = \dfrac{b}{c}$
3. $\tan A = \dfrac{a}{b}$
4. $\cot A = \dfrac{b}{a}$
5. $\sec A = \dfrac{c}{b}$
6. $\csc A = \dfrac{c}{a}$
7. $\text{vers } A = 1 - \cos A = \dfrac{c-b}{c} = \text{covers } B$
8. $\text{exsec } A = \sec A - 1 = \dfrac{c-b}{b} = \text{coexsec } B$
9. $\text{covers } A = \dfrac{c-a}{c} = \text{vers } B$
10. $\text{coexsec } A = \dfrac{c-a}{a} = \text{exsec } B$
11. $a = c \sin A = b \tan A$
12. $b = c \cos A = a \cot A$
13. $c = \dfrac{a}{\sin A} = \dfrac{b}{\cos A}$
14. $a = c \cos B = b \cot B$
15. $b = c \sin B = a \tan B$
16. $c = \dfrac{a}{\cos B} = \dfrac{b}{\sin B}$
17. $a = \sqrt{c^2 - b^2} = \sqrt{(c-b)(c+b)}$
18. $b = \sqrt{c^2 - a^2} = \sqrt{(c-a)(c+a)}$
19. $c = \sqrt{a^2 + b^2}$
20. $C = 90° = A + B$
21. $\text{Area} = \frac{1}{2}ab$

TABLE C-2 Trigonometric Formulas for the Solution of Oblique Triangles

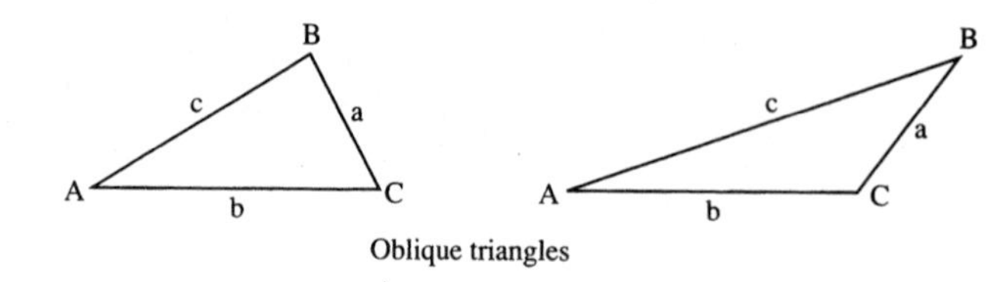

Oblique triangles

Given	Sought	Formula
A, B, a	C, b, c	$C = 180° - (A + B)$ $b = \dfrac{a}{\sin A} \times \sin B$ $c = \dfrac{a}{\sin A} \times \sin(A + B) = \dfrac{a}{\sin A} \times \sin C$
	Area	$\text{Area} = \frac{1}{2}ab \sin C = \dfrac{a^2 \sin B \sin C}{2 \sin A}$
A, a, b	B, C, c	$\sin B = \dfrac{\sin A}{a} \times b$ $C = 180° - (A + B)$ $c = \dfrac{a}{\sin A} \times \sin C$
	Area	$\text{Area} = \frac{1}{2}ab \sin C$
C, a, b	c	$c = \sqrt{a^2 + b^2 - 2ab \cos C}$
	$\frac{1}{2}(A + B)$	$\frac{1}{2}(A + B) = 90° - \frac{1}{2}C$
	$\frac{1}{2}(A - B)$	$\tan \frac{1}{2}(A - B) = \dfrac{a - b}{a + b} \times \tan \frac{1}{2}(A + B)$
	A, B	$A = \frac{1}{2}(A + B) + \frac{1}{2}(A - B)$ $B = \frac{1}{2}(A + B) - \frac{1}{2}(A - B)$
	c	$c = (a + b) \times \dfrac{\cos \frac{1}{2}(A + B)}{\cos \frac{1}{2}(A - B)} = (a - b) \times \dfrac{\sin \frac{1}{2}(A + B)}{\sin \frac{1}{2}(A - B)}$
	Area	$\text{Area} = \frac{1}{2}ab \sin C$
a, b, c	A	Let $s = \dfrac{a + b + c}{2}$ $\sin \frac{1}{2}A = \sqrt{\dfrac{(s - b)(s - c)}{bc}}$ $\cos \frac{1}{2}A = \sqrt{\dfrac{s(s - a)}{bc}}$ $\tan \frac{1}{2}A = \sqrt{\dfrac{(s - b)(s - c)}{s(s - a)}}$ $\sin A = \dfrac{2\sqrt{s(s - a)(s - b)(s - c)}}{bc}$ $\cos A = \dfrac{b^2 + c^2 - a^2}{2bc}$
	Area	$\text{Area} = \sqrt{s(s - a)(s - b)(s - c)}$

Appendix D

Transits and Theodolites

D-1 TRANSITS

The surveyor's transit (Figure D-1) is a versatile instrument that may be used to measure vertical and horizontal angles, prolong straight lines, perform leveling of a precision almost equivalent to that obtained with levels, determine magnetic bearings, and measure distances by stadia. This universal instrument was long used for land surveying, mapping, construction surveys, and astronomical observations. For decades it was recognized as the American surveyor's most important instrument.

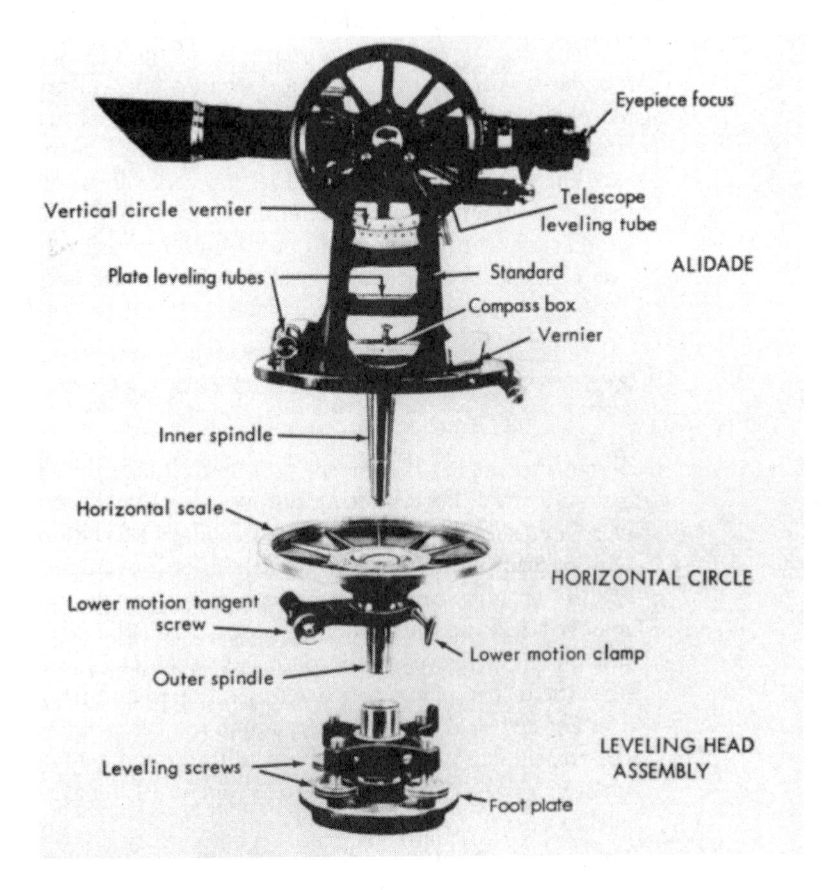

FIGURE D-1 Transit showing three fundamental parts. (Courtesy of Keuffel & Esser—a Kratos Company.)

The transit was typically used to measure angles to the nearest 1′ of arc, but better instruments were available (at somewhat higher costs) with which angles could be read to the nearest 30″, 20″, 15″ or even 10″. Theodolites with which angles could be read to the nearest 1″ are common but some theodolites were manufactured with which angles could be read to the nearest 0.2″. Sections D-1 to D-6 are devoted to transits and Sections D-7 to D-9 are devoted to theodolites.

The various parts of a transit and their use may best be learned by handling and working with them. For this reason, only a brief introduction to these parts is presented here. The transit consists of three fundamental parts: the alidade, the horizontal circle, and the leveling head assembly. See Figure D-1 for a picture of a transit and these parts.

The *alidade,* or upper plate, is the top part of the instrument and includes a circular cover plate that is rigidly connected to a solid conical shaft called the *inner spindle.* It also includes the telescope (with a magnification power of approximately 18 to 28 diameters), the telescope leveling tube, the vertical circle (which is mounted on the telescope for measuring vertical angles), two plate leveling tubes (one parallel to the telescope and one perpendicular to it), two pairs of verniers for reading the horizontal circle, and probably a compass. The telescope rotates up or down on a pair of side shafts called *trunnions.* This word was originally used to refer to the similar-side shafts on cannons.

The *horizontal circle* or lower plate is the scale with which horizontal angles are measured. It is usually graduated in degrees and halves of degrees and numbered every 10° clockwise and counterclockwise. The underside of the horizontal circle is attached to a hollow, vertical, tapered spindle called the *outer spindle* into which the inner spindle fits. In addition, there is a clamp (called the *lower motion clamp*) that is used to permit or prevent rotation of the outer spindle.

The *leveling head assembly* consists of the four leveling or foot screws and the footplate. The leveling screws are set into cups to prevent them from scoring the footplate. Included in the assembly is a device that permits the transit to be moved laterally from 1/4 to 3/8 in. without movement of the tripod.

The tripods used with transits are generally about 4 1/2 or 5 ft in height, but there are available types of extension legs with which this height can be increased another 4 or 5 ft. These might be useful on occasions where sights need to be taken over obstacles such as bushes, walls, and mounds.

D-2 SETTING UP THE TRANSIT

To measure angles, the transit is set up over a definite point such as an iron pin or a tack in a stake. For centering purposes, a plumb bob is suspended from a hook or chain beneath the instrument (see Figure D-2). Many later transits are equipped with optical plummets. Nevertheless, it is convenient to use plumb bobs to center transits roughly before the optical plummets are used. The plumb bob is then removed and the fine centering is done with the plummet. After it is centered the instrument is releveled, the plummet recentered, and the procedure is continued until the instrument is concurrently centered and leveled.

The transit is leveled in very much the same way as the four-screw level. Because the transit has two plate bubble tubes at right angles to each other, the instrumentman has only to turn the telescope until the axis of each of these tubes is parallel to a line through opposite leveling screws. The surveyor may then proceed with the leveling operation without having to turn the telescope as he or she works from one pair of leveling screws to the other.

FIGURE D-2 Transit set on a steep slope. Notice the manner in which the tripod is set for stability with one leg uphill and two legs downhill.

When both bubble tubes are properly centered, the telescope may be turned through 180° to see if the bubbles remain centered. If the bubbles move, the instrument is out of adjustment, but if the movement is slight (for example, less than one division), the instrumentman is wasting time in trying to recenter the bubble every time the telescope is moved.

To center the transit over a particular point with a plumb bob, the instrumentman places the tripod so that the plumb bob hangs approximately over the point. In doing this he or she tries to keep the leveling head approximately level. The instrumentman roughly levels the instrument to see where the plumb bob will be. If it is an appreciable distance off center, the instrumentman picks up the tripod without moving the relative position of the legs and moves the instrument in the desired direction. The instrumentman repeats the rough leveling process, after which the instrument may have to be moved again. When the instrumentman gets the plumb bob fairly close to center, he or she may move it closer and closer by pressing one or two tripod legs more firmly into the ground. (Some surveyors use transits that have adjustable leg tripods that permit the lengthening or shortening of the legs as required to center the plumb bob.)

Finally, when the plumb bob is within approximately 1/4 to 3/8 in. of its proper position, two adjacent leveling screws are loosened. This loosens all four of the leveling screws and the head of the instrument may be pushed or slid over the required distance. When the plumb bob is centered, the instrumentman retightens the screws, relevels the instrument, and checks to see if the plumb bob remains centered. If it does not, the instrumentman may have to repeat this last part of the procedure.

D-3 READING TRANSIT VERNIERS

In Section 7-5 the reading of level rod verniers was discussed. In this section the subject is continued but in relation to angle measurements. The verniers used for reading horizontal and vertical angles are identical in principle with target rod verniers. In other words, a person who can use the level rod vernier can use a transit vernier just as well.

The graduated scale on the lower plate is provided with two sets of numbers, one increasing in a clockwise direction from 0 to 360° and the other increasing from 0 to 360° in a counterclockwise direction. For reading these scales, two pairs of verniers (*A* and *B*) are provided. (Verniers that can be read in either direction are called *double-direct verniers.*) These are located on opposite sides of the upper plate, the *A* vernier being located below the eyepiece for convenience. There are two *A* verniers and two *B* verniers because one vernier is needed at each location for the clockwise numbers and one for the counterclockwise numbers. These verniers are located in windows in the cover plate covered with glass.

For most angle measurements, the *A* vernier is the only one used, but for more precise work, both *A* and *B* verniers are used and the mean of the readings calculated. The purpose of taking the two sets of readings is to reduce instrumental errors caused by imperfections in the scale and vernier.

For the average transit scale, the smallest divisions are $^1/_2°$ and the verniers are constructed so that 30 divisions on the vernier cover 29 divisions on the main scale. As a result, the smallest subdivision that can be read is

$$D = \frac{s}{n} = \frac{\frac{1°}{2}}{30} = \frac{30'}{30} = 1'$$

FIGURE D-3 Styles of graduations. (Courtesy of Keuffel & Esser, a Kratos Company.)

STYLES OF GRADUATIONS

The circles and verniers of transits are graduated in various ways. The usual styles of graduation and the method of numbering the horizontal circle and vernier are shown below.

GRADUATED 30 MINUTES READING TO ONE MINUTE
DOUBLE DIRECT VERNIER

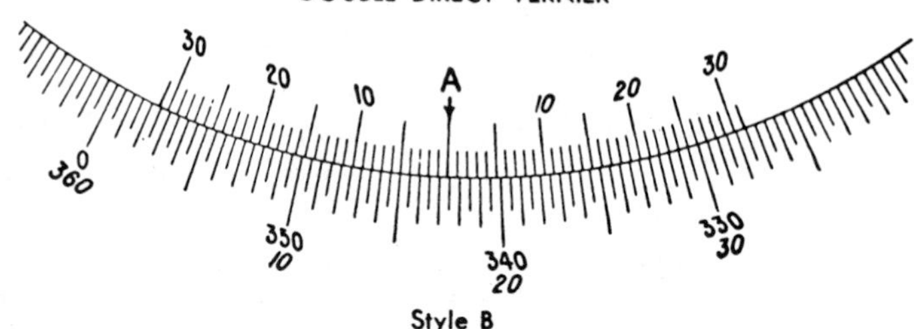

Style B

Style B represents the usual graduation of the horizontal circle of a transit with its vernier, as furnished on K & E Transits Nos. P5081C and P5136. This is an ordinary double direct vernier, reading from the center to either extreme division (30). The circle is graduated to half degrees, and the vernier (from 0 to 30) comprises 30 divisions; therefore, the value of one division on the vernier is 30 minutes ÷ 30 = 1 minute.

The figure reads 17° + 25′ = 17° 25′ from left to right and 342° 30′ + 05′ = 342° 35′ from right to left.

GRADUATED 20 MINUTES READING TO 30 SECONDS
DOUBLE DIRECT VERNIER

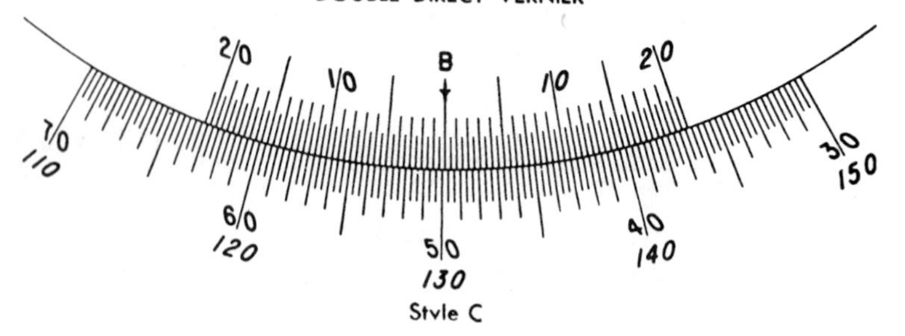

Style C

Pictures of two transit verniers, together with example readings, are illustrated in Figure D-3. When the telescope has been turned in a clockwise direction, the numbers increasing in that direction will be read on the horizontal scale. Similarly, the verniers whose numbers are increasing in the same direction (clockwise) will be used. The letter *A* at the 0° mark signifies which of the verniers is being used.

To reduce the possibility of mistakes, it is wise to make a rough estimate by eye of the fractional part of a circle division involved to check against the vernier reading. This practice will greatly reduce the number of mistakes occurring in angle measurement. Some transits are equipped with attached magnifying glasses to aid the instrumentman in reading the verniers. If this is not the case, a hand magnifying glass or reading glass is used. A surveyor using a magnifying glass can estimate readings to the nearest $^1/_2$ of the vernier's least division.

D-4 MEASUREMENT OF HORIZONTAL ANGLES WITH A TRANSIT

For this discussion it is assumed that a transit is located at point *A*, as shown in Figure D-4, and it is desired to measure the angle between the lines *AB* and *AC*. Both the upper motion and the lower motion clamps are released and the horizontal scales are adjusted by turning the instrument on its spindle with the hands until the *A* vernier is near zero. The clamps are then tightened and the upper motion tangent screw is turned until the vernier reading is zero. The lower motion is released and the telescope is sighted close to a range pole or other object at point *B*, after which the lower motion is clamped and the line of sight is pointed precisely on point *B* with the lower motion tangent screw. The reading at the *A* vernier should still be zero.

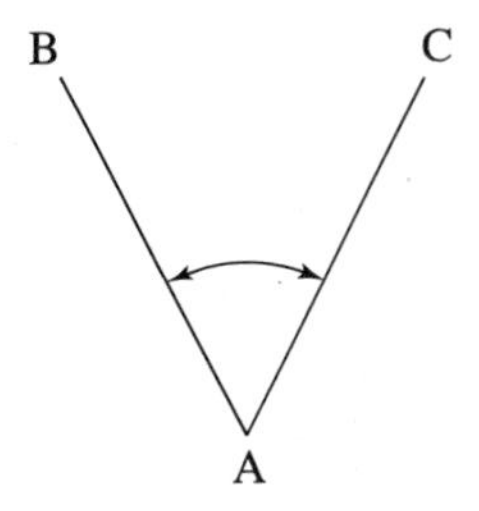

FIGURE D-4 Measuring horizontal angle.

To turn the angle, the upper motion clamp is released and the telescope pointed toward *C*. The fine adjustment of the line of sight to *C* is made by clamping the upper motion and turning the upper motion tangent screw, after which the angle is read at the *A* vernier, as described previously.

D-5 MEASURING ANGLES BY REPETITION WITH A TRANSIT

A method that nearly always eliminates mistakes in angle measurement is that of *measuring angles by repetition*. After an angle is measured, the lower motion clamp and the lower motion tangent screw are used to sight the telescope on the initial point again. This means that the reading at the vernier should be the same as when the angle was initially measured.

With the aid of the upper motion clamp and tangent screw, the telescope is sighted on the second point and the reading is taken with the vernier. Obviously, the measured value should equal twice the value obtained initially for the angle. If not, a mistake has been made. The instrumentman should do his or her best to forget the value recorded for the single angle and not compare the values until the double angle has been read and recorded. Otherwise, the instrumentman will have lost much of the value of this check.

For more precise surveying, horizontal angles may be repeated six or eight times. The value of the angle is added each time to the previous value on the vernier. The resulting value is divided by the number of measurements in order to obtain the angle. Actually, an angle can be measured with a 1′ transit to approximately the nearest ±30″. If an angle is turned six times and if the final total is read, it should contain roughly the same error. Dividing it by 6 should then give a reading within ±05″. Actually, repeating an angle more than six or eight times does not appreciably improve the precision of its measurement because of accidental errors in centering and pointing and instrumental errors of the transit.

When sights are relatively short, 300 ft or less, there is actually little advantage in repetition because the errors made in pointing the telescope and in setting up over the points take away all or almost all of the increased precision supposedly obtained by repetition. Similarly, the precision of a 1″ theodolite is probably wasted if the sight distances are short.

D-6 MEASURING VERTICAL ANGLES WITH A TRANSIT

Vertical angles may be measured with the transit in much the same way that horizontal angles are measured. The horizontal plate levels are carefully leveled, the horizontal cross hair is sighted on the point to which the vertical angle is to be measured, and the vertical scale and its vernier are used to read the angle.

As described in Chapter 14, the usual procedure involves (1) measuring the height or the h.i. of the transit telescope above the ground by holding a level rod on the ground by the instrument, (2) sighting the telescope on the level rod at the other point with the center horizontal cross hair adjusted on the rod to the h.i. of the instrument, and (3) reading the vertical angle. This procedure is illustrated in Figure D-5.

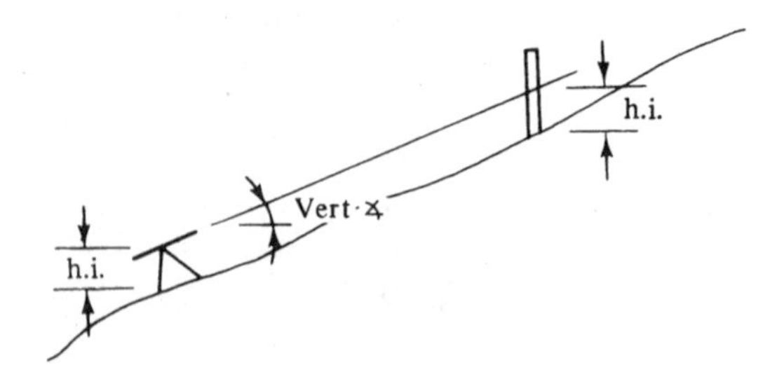

FIGURE D-5 Measuring a vertical angle.

The transit is not constructed to permit the measurement of vertical angles by repetition. Vertical angles are referred to as either plus or minus (the sign must be carefully recorded), depending on whether the telescope is sighted above or below the horizontal.

A transit has a bubble tube attached to its telescope. The bubble tube is as sensitive as the plate bubble tube. When the telescope bubble is centered, the vernier on the vertical scale should read 0°00′. Actually, it is easy for these scales to get out of adjustment. The discrepancy is called the *index error* of the vertical circle.

For an instrument that has a full vertical circle, the index error problem is easily handled by the so-called *double-centering* procedure. The angle is measured once with the telescope in its normal position and once with the telescope inverted. The average of these two readings is taken. For transits that have vertical arcs instead of full vertical scales, it is necessary to correct the readings by an amount equal to the index error. Great care must be taken to use the correct sign for such corrections.

D-7 THEODOLITES

Theodolites are manufactured to accomplish the same purposes as transits, that is, to determine horizontal and vertical angles and prolong straight lines. They do, however, enable users to make single observations of angles as precisely or more precisely and in less time than those that can be made by several repetitions with a transit.

Theodolites have horizontal and vertical circles for angle measurements, as do transits, but the circles are made of glass instead of metal. Light passes through the glass circles, and with the aid of glass prisms the readings from the circles are reflected into the eyepiece. The values are greatly magnified, enabling the user to make readings without eyestrain. All the readings are taken from the eyepiece end of the telescope so that the user does not have to keep moving around the instrument. The tribrach, or the base of the theodolite, has three leveling screws and a bull's-eye level. In addition, there is a level on the alidade.

Theodolites today usually have short telescopes that can be transited. These telescopes are internally focusing and enable the user to have clear sharp views for both long and short sight distances. Several types of theodolites have been used through the years, as listed:

1. The *scale-reading theodolite* has vertical and horizontal circles that are read directly with an optical microscope.
2. Angle-measuring instruments are often referred to as repeating instruments or direction instruments. The transits discussed in the last few sections of this chapter are repeating instruments. As with transits the user of a *repeating optical theodolite* can measure angles as many times as desired by adding them successively on the instrument circle.
3. The *directional optical theodolite* is a nonrepeating instrument that does not have a lower motion. The horizontal circle remains fixed during a series of readings. The telescope is sighted on each of the points in question and directions to those points are read. Horizontal angles are determined by calculating the differences between the directions.
4. During the past few decades advances in electronics have led to modern *electronic* transits and *theodolites* (Figure D-6). As these instruments provide a visual display of horizontal and vertical angles, there is no need for microscopes. They were extremely popular with their users. With electronic theodolites the measurements are displayed digitally and may be recorded in field books; they can be used with electronic data recorders to store the information for later use.

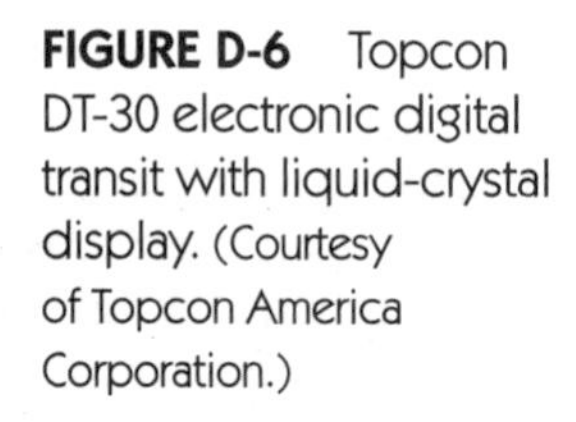

FIGURE D-6 Topcon DT-30 electronic digital transit with liquid-crystal display. (Courtesy of Topcon America Corporation.)

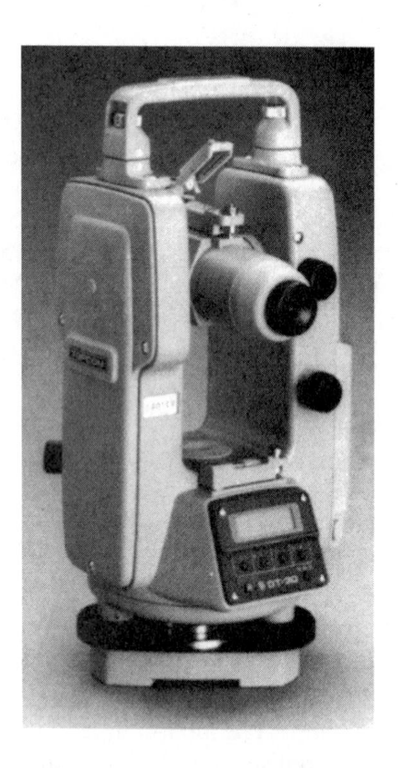

The circle readings can be set to zero by touching a button. For some instruments the horizontal circle can be set to any desired value (0° or otherwise). Then the angle is measured by sighting to the next point and its value is automatically displayed by the instrument. The angles can be repeated any desired number of times with the instrument in the normal or reversed positions and the averages taken. Some of these instruments can be read to about ±0.2″, but when atmospheric and pointing errors are considered the readings are good to about ±0.4″ or 0.5″.

A large percentage of these instruments have vertical axis compensators (i.e., devices that cause vertical circles to be vertical even if the theodolites are not quite level). Only a small percentage of them have horizontal axis compensators.

D-8 SETTING UP THE THEODOLITE

The purpose of this section is to provide a general discussion applicable to the setting up of all theodolites. More detailed information for a specific type of theodolite can be found in its operation manual. Before the theodolite is unpacked, the tripod should be carefully set over the point to be occupied with the plate approximately level. The theodolite is carefully removed from its carrying case by picking it up using its standards, or the handles if provided. The instrument is set up on the tripod, and the centering screw, which is located under the tripod head, is screwed up into the base of the instrument—otherwise, the instrument will fall off the tripod when picked up. The instrument can be shifted laterally when the centering screw is loose. It is centered over the tripod and the screw is tightened.

The tripod is located over the point to be occupied as closely as possible, the same as described earlier for the transit. Once the instrument is approximately centered over the point with the tripod legs, it can be centered more exactly with a plumb bob or preferably with an optical plummet. If the plumb bob is not over the point or if the cross hairs of the optical plummet are not centered on the point, the centering screw can be loosened and the instrument shifted over the point.

Next the theodolite is approximately leveled with the rather insensitive bull's-eye or circular level and then it is very carefully leveled with the plate level. The alidade is turned so that the level vial is parallel to two of the three leveling screws and the bubble is brought to the center with these two screws. The instrument is rotated 90° and the bubble is centered using the third screw. Then the instrument is rotated back 90° to its position parallel to the two leveling screws to see if the bubble is still centered. After a few trials the bubble is centered and the alidade is rotated through 180°. If the bubble remains centered, the instrument is level. If not, it is out of adjustment.

D-9 MEASURING ZENITH ANGLES WITH A THEODOLITE

As described previously, a vertical angle is the plus or minus angle from a horizontal plane to the point in question. Sometimes a plus angle is referred to as an *elevation angle* and a negative angle is called a *depression angle.* A *zenith angle* is the angle from a vertical line to the point in question. These angles are illustrated in Figure D-7. Engineer's transits are usually constructed to measure vertical angles, while theodolites are usually constructed to read zenith angles. As Section D-6 was devoted to the measurement of vertical angles with a transit, this section is devoted to the measurement of zenith angles. It should be noted that theodolites have a 90° reading when the telescope is horizontal. Some of them have the 0° reading mark at the zenith (high point), whereas others have it at the nadir (low point).

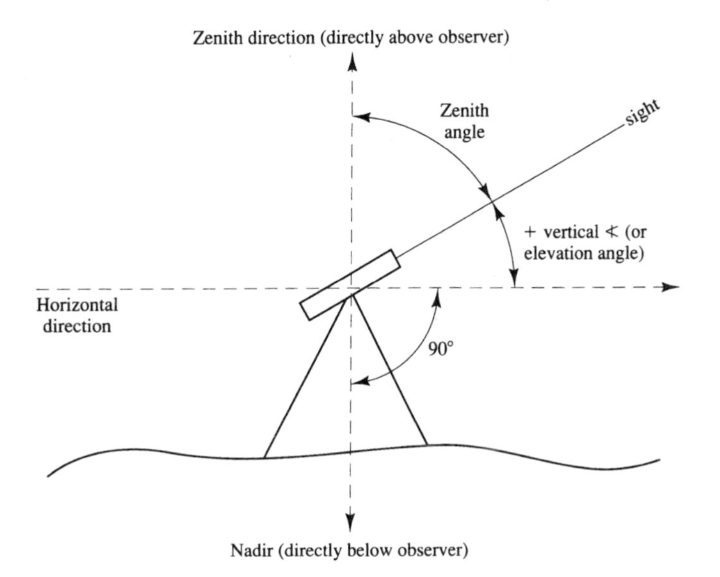

FIGURE D-7 Measuring a zenith angle.

Newer theodolites have an *automatic compensator* (controlled by gravity), which will cause the vertical scale to be in its proper position when the instrument is correctly leveled. Some theodolites have a spirit level attached to their vertical circles for the purpose of getting the circle in its proper position when the instrument is leveled.

The instrument is leveled and sighted on the point in question using the horizontal cross hair with the fine setting made with the vertical tangent screw. If the theodolite does not have an automatic compensator, the spirit level is centered and the zenith angle is read. The telescope is rotated 180° about its vertical axis, inverted, and the procedure repeated. The average of the two angles is computed and the index error is theoretically removed. For greater accuracy several direct readings (telescope in normal position) and several reverse readings (telescope inverted) may be taken and averaged.

Glossary

AASHTO American Association of State Highway and Transportation Officials.

Accuracy Degree of perfection obtained in measurements. It denotes how close a given measurement is to the true value.

ACSM American Congress of Surveying and Mapping.

Alidade The upper part of a total station that includes the telescope, scales, and other components involved in measuring angles and distances.

Apparent Accuracy *See* Precision.

As-built surveys Surveys made after a construction project is complete to provide the positions and features of the project as they were actually constructed.

Atmospheric refraction The bending down of light rays as they pass through air strata of different densities (thus bending down our telescope lines of sight).

Automatic level A surveyor's level where the line of sight is automatically kept in a horizontal position once the instrument is approximately leveled.

Avulsion The sudden removal of a portion of land belonging to one person and its addition to the land of another by, for example, the sudden change in position of a stream or river.

Azimuth Clockwise angle from the north or south end of a reference direction (or meridian) to a line in question.

Batter Boards Wooden frameworks from which strings or wires are placed to outline positions.

Bearing Smallest angle that a line makes with the north or south end of a reference direction (or meridian).

Bench mark Relatively permanent point of known elevation.

Borrow pits Sources of earth from areas near our projects–the earth to be used for the construction of fills or embankments.

Cartography The art or science of making maps.

Centroid The center of a mass of an object or area having density.

Collimation The lining up or adjusting the line of sight of an instrument.

Contour line Line of equal elevation.

Cut Excavation.

Departure of a line Projection of the line on the east-west reference direction. It equals its length times the sine of its bearing angle.

Differential GPS A GPS survey where one receiver is located at a known position while the other receivers are moved to other points whose positions are to be determined.

Differential leveling The measurement of vertical distances with respect to a horizontal line for the purpose of determining elevations.

Double centering Measurements of a direction with the telescope in its normal or upright position and then with the telescope inverted or upside down.

Dumpy level Older level that has a dumpy appearance (i.e. a short telescope at least as compared to even older levels with long telescopes). In a dumpy level the telescope, vertical supports, horizontal bar, and vertical spindle are all made in one casting, or those parts are fastened rigidly together.

Dynamic taping Distances measured on a slope with a tape, vertical angles measured, and horizontal distances computed.

EDM Electronic distance measuring instrument.

Elevation Vertical distance of a particular point above or below a reference level surface (normally sea level).

Ellipsoid A figure formed when an ellipse is rotated about its minor axis. The ellipse referred to in this text is one that approximates the shape and size of the earth. The surface of this ellipsoid is said to be equipotential in that the potential due to gravity is equal at all points on its surface.

Ephemeris An astronomical almanac that provides the positions of various bodies in space such as the sun, moon, stars, and satellites.

Errors Difference between a measured quantity and its true value caused by the imperfection of the surveyor's senses, by the imperfection of the equipment, or by weather effects.

Ethics Duties that a professional owes to the public and to fellow professionals.

Fill Embankment.

Forced Centering The interchanging of theodolites, prisms, EDMs, and targets into tribrachs, which are left in position over stations.

Geoid A hypothetical figure that represents the spheroidal shape of the earth but whose surface is mean sea level (MSL).

Geodetic survey Survey adjusted for the curved shape of the earth's surface.

Geodesy The science of studying the form and dimensions of the earth.

Geomatics Term applying to the work of persons involved in surveying, mapping, GIS, and remote sensing. It includes the measurement, representation, and display of information concerning the natural and man-made features of the earth's surface.

GIS Geographic information system: a system for storing, recalling, and displaying detailed information regarding the natural and man-made features of a certain piece of land.

GPS Global positioning system: artificial satellites and various ground equipment used to convert radio signals from satellites into three-dimensional positions on the earth's surface.

Grade Number of feet of vertical change in elevation of a line per 100 feet of horizontal distance (expressed as a percentage).

Grade stake Stake driven into the ground until its top is at the elevation desired for that point for a particular construction project (or until its top is at a certain vertical distance from the elevation desired).

Gunter's chain A chain with 100 links totaling 66 ft in length, which was formerly used for measuring distance.

Hachures Short lines of varying widths used on a drawing to indicate topography (the wider lines are made, the steeper are the slopes).

Hectare An area equal to 10,000 square meters or 2.47104 acres.

Invar tapes Alloy steel tapes composed of 35% nickel and 65% steel. These tapes have very small changes in length due to temperature variations.

Invert of pipe The inside bottom of a pipe.

Isogonic line Line of equal magnetic declination (the difference between astronomic north and magnetic north is the same at every point on the line).

Keel Colored lumber crayon often used in surveying for marking on pavements and other surfaces.

Laser Device with which low-intensity light waves are generated and amplified into a very narrow and intense beam. (The word *laser* is an acronym for "light amplification by stimulated emission of radiation.")

Latitude (geographic) Angular distances (0 to 90°) that a particular point is above or below the equator.

Latitude of a line Projection of the line on the north-south reference direction. It equals its length times the cosine of its bearing angle.

Leveling Determination of elevations.

Longitude Distance (0 to 180°) that a particular point is east or west of the meridian through Greenwich, England.

Magnetic declination Angle between astronomic north and magnetic north at a particular location.

Mask angle A vertical or cutoff angle (specified as about 10 to 20 degrees by the observer) below which the observation of satellites in GPS work is deliberately blocked because of large atmospheric errors that would occur with such observations.

Mass diagram Cumulative plot of the earthwork quantities (designating cuts plus and fills minus) from one point to another on a route project such as a highway or canal.

Meridian Reference direction (can be astronomic, magnetic, or assumed).

Meridian distance Distance (parallel to the east-west direction) from the midpoint of a line to the reference meridian.

Meter Since 1960, equal to 3.280840 ft. Before 1960, was 39.37 in or 3.280833 ft in the United States.

Metes and bounds A description of land in which the lengths and directions of the land boundaries are given.

Mistake Difference between a measured quantity and its true value caused by the carelessness of the surveyor.

Optical plummet Special telescopic device with which the surveyor can sight vertically from the center of an instrument to the ground below.

Parallax Apparent movement of telescope cross hairs on a sighted object due to improper focusing.

Photogrammetry Science of making measurements from photographs (usually, aerial).

Planimeter An instrument used to measure the areas of figures on drawings by tracing the boundaries of the figures (the planimeter mechanically integrates the area).

Plat A dimension drawing which shows in plan view the data pertaining to a land survey.

Plumb bob Pear-shaped or globular weight suspended on a string or wire used to establish a vertical line. The bottom of the bob is pointed.

Plunge To invert the telescope or to turn it so that it is upside down.

Precision Degree of refinement with which a measurement is taken. It is the closeness of one measurement to another of the same quantity. Also called *apparent accuracy.*

Prismoid Solid figure with parallel end faces joined by plane or continuously warped surfaces.

Profile Graphical intersection of a vertical plane along a certain route with the earth's surface.

Radar Any of several systems using transmitted and reflected radio waves for locating objects. (The word *radar* was taken from the first letters of the words "radio detecting and ranging.)

Radian Measure of angles frequently used for calculation purposes; $360°/2\pi$.

Radiation Location of a series of points by measurements of distances and directions from one convenient point.

Remote sensing A broad area in which images produced by electronic sensing devices or aerial or space photographs are studied.

Riparian rights "Water rights" or the legal rights along waterways (lakes, rivers, etc.) belonging to the persons who own land thereon.

SI New metric system or international system of units (from the French term Le Systeme International d'Unities).

Significant figures Number of digits in a measured value that can be stated with reasonable certainty.

Slope stakes Stakes set at the intersection of the natural ground line and the side slopes of a cut or fill for a grading operation (stakes may be offset).

Spatial Refers to particular positions on the earth's surface.

Spiral curve A curve which provides a gradual transition from a straight line or tangent to the full curvature of a circular curve.

Stadia Method used for quickly measuring distances that makes use of a special configuration of telescope cross hairs (see Section 3-4).

Standardization Comparison of equipment against a standard (e.g., comparing the length of a 100-ft tape against a standard 100-ft distance).

Stations 100-ft distances along the center lines of a road, railroad, or other strip conveyance. Stations are numbered from some arbitrary starting point.

Superelevation Transverse slope of streets and highways on horizontal curves.

SURVEY Name of the computer program enclosed with this book.

Surveying Science of determining the dimensions and three-dimensional characteristics of the earth's surface by the measurement of distances, directions, and elevations.

Tachymetry or tacheometry Swift measurements (see Section 3-4).

Theodolite In this book, an angle-measuring instrument with three leveling screws and horizontal and vertical glass circles that may be read directly or with an optical micrometer. Also, those same instruments, that provide digital displays of angle readings.

Topography Shape or configuration or relief or roughness or three-dimensional quality of earth's surface.

Total station Instrument that combines a theodolite and EDM (thus having both distance and angle-measuring capabilities). Also called a tacheometer or tachymeter.

Transit In this book, an American-style angle-measuring instrument with four leveling screws and silvered horizontal and vertical angle scales.

Traversing Process of measuring the length and direction of the sides of a traverse (open or closed).

Tribach Base of a theodolite. It has a feature that enables the surveyor to interchange the instrument with electronic distance-measuring instruments, targets, subtense bars, and so on.

Troposphere The part of the atmosphere of the earth that extends for about 80 km from the earth's surface. Its outer part is referred to as the stratosphere.

U.S. survey foot Definition of the foot (1 meter = 3.280833 ft).

Vernier Scale device used for making readings on a divided scale closer than the smallest divisions on the scale.

Zenith angle An angle measured vertically down from a vertical or plumb line.

Index

What more fitting way can be found to start the index of a surveying textbook than with the most important word in surveying? That word is accuracy.